Groundwater Contamination and Analysis at Hazardous Waste Sites

Environmental Science and Pollution Control Series

Additional Volumes in Preparation

Groundwater Contamination and Analysis at Hazardous Waste Sites

edited by

Suzanne Lesage

National Water Research Institute
Environment Canada
Burlington, Ontario, Canada

Richard E. Jackson

Intera Inc.
Austin, Texas

Marcel Dekker, Inc.   New York • Basel • Hong Kong

Library of Congress Cataloging-in-Publication Data

Groundwater contamination and analysis at hazardous waste sites /
   edited by Suzanne Lesage, Richard E. Jackson.
        p.   cm. -- (Environmental science and pollution control)
   Includes bibliographical references and index.
   ISBN 0-8247-8720-X
   1. Hazardous wastes--Environmental aspects.   2. Water,
Underground--Pollution.   3. Water, Underground--Quality-
-Measurement.   4. Hazardous waste sites--Case studies.   I. Lesage,
Suzanne   .   II. Jackson, R. E. (Richard E.)   III. Series.
   TD427.H3G76   1992
   628.4'2--dc20   92-20758
                                                                             CIP

This book is printed on acid-free paper.

Marcel Dekker, Inc.
270 Madison Avenue, New York, New York 10016

Current printing (last digit):
10 9 8 7 6 5 4 3 2 1

PRINTED IN THE UNITED STATES OF AMERICA

Preface

Groundwater contamination at hazardous waste sites has proved to be a challenge to hydrogeologists and chemists because of the complexity and diversity of wastes and of the sites where these wastes are deposited. Our intention was to gather in one text information regarding the investigation of the fate of toxic chemicals emanating from hazardous waste sites, thereby contaminating the groundwater.

The book is divided into four main sections: analytical methodologies, monitoring strategies, site investigations, and geochemical investigations. Its main emphasis is on the chemicals found in hazardous wastes and the potential for groundwater contamination. Discussions include the behavior of toxic chemicals, the methods used for analysis, and their validity, and the interpretation of the data. Authors from diverse backgrounds were asked to describe hazardous waste sites investigations from their own perspective: this is not proceedings from a conference. Some are chemists and have substantial experience in analyzing samples of contaminated groundwater; others are hydrogeologists who have specialized in contaminant hydrogeology and have vast experience in investigating chemical plumes. The contributors include authors in the United States, Canada, Germany, and Australia, to reflect the different approaches used in different parts of the world. The result is a book we hope will appeal to hydrogeologists and chemists alike. Some chapters are relatively basic and will be appreciated by the reader who is totally new to the subject. Other chapters are much more advanced and describe the state of the art in hydrogeological investigations. We include topical information such as the dissolution of DNAPLs while also covering the basics of environmental analytical chemistry and the standards methods used at RCRA and CERCLA sites in the United States. There are no fewer than 12 case studies discussed either as whole chapters or as illustrations of the topics described. Information is presented on virtually every known type of organic contaminant plume. Factual data from large surveys will be useful to regulators in assessing the extent of contamination in the major industrialized countries. Different site investigations and monitoring strategies are also discussed.

While the text was still in preparation, several requests were received for information contained in many of the chapters. We are pleased to finally publish the completed text. We thank all the contributors, who by showing their immediate enthusiasm have given us the encouragement to get started. We are grateful for their fine contributions.

Suzanne Lesage
Richard E. Jackson

Contents

Contents

IV. Geochemical Investigations

Introduction

Suzanne Lesage

National Water Research Institute, Environment Canada,
Burlington, Ontario, Canada

Richard E. Jackson

Intera Inc., Austin, Texas

This book is divided into four parts: analytical methodologies, monitoring strategies, site investigations, and geochemical investigations.

ANALYTICAL METHODOLOGIES

The first part contains chapters describing mostly chemical analytical techniques used for hazardous waste site investigation. Because they are the primary problem in the majority of sites, organic contaminants are the focus of these chapters.

The first chapter is an overview of organic analytical techniques most commonly utilized by commercial laboratories to analyze groundwater samples. A brief technical discussion of the analytical methods is provided to familiarize the reader with the technical jargon of the analytical laboratory. The basic concept of quality assurance/quality control is introduced with typical examples of what constitutes acceptable quality control. This is not meant to supersede legislated quality control programs but is meant as a guideline of principles to be generally followed. A case study is described in which the whole problem of groundwater contamination turned out to be a quality control problem compounded by a failure to select enough analytes to properly assess the problem. The chapter as a whole is intended as an introduction for students and practitioners of hydrogeology and environmental science and engineering who may have had little exposure to organic analytical chemistry. A glossary of terms is included.

Readers who are already familiar with this material may want to skip directly to Chapter 2, where K. C. Swallow discusses the importance of not restricting the analyses to priority pollutants in hazardous waste site investigations. She reiterates some of the basic principles outlined in the first chapter, in particular, that a priority pollutant GC/MS scan may be of little value in these investigations. A more global approach to the problem is warranted where GC/MS is one of the tools. Indeed, the spectra obtained with the mass spectrometers can be used as a fingerprint, even when the exact identity of the compound cannot be ascertained.

Jerry Parr, Gary Ward, and Gary Walters of Enseco introduce the reader to the U.S. Environmental Protection Agency's SW-846 methods of analysis for groundwater samples taken at hazardous waste sites under RCRA control. The list of target chemicals on the Appendix IX list has been shortened since the Appendix VIII list. Many compounds were withdrawn because the analytical methodology provided was simply not adequate. There are still some problems with the methods, but the SW-846 manual is evolving as new methods are being developed. These methods are, however, a good starting point for any investigator.

The fourth chapter, by Larry Barber, is a detailed account of the analytical methodologies used by the U.S. Geological Survey in its in-depth study of the contaminant plume at its Cape Cod, Massachusetts, site. This case history takes the reader from the general problems discussed by Parr et al. in their survey of analytical methods to suit many different types of sites into the application of analytical chemistry to one particular site. In this chapter, not only are analytical methodologies described in detail, but the results are applied to the site. The need to employ a wide arsenal of analytical techniques is well exemplified. The extent of the plume was characterized by using indicator parameters such as dissolved organic carbon, specific conductance, chloride, and boron. Samples were analyzed by closed-loop stripping, purge-and-trap, liquid–liquid extraction, and solid-phase extraction followed by GC/MS analysis. Anionic surfactants were determined as methylene-blue-active substances. Surfactant-derived compounds were further characterized by extraction followed by derivatization and analysis by GC/MS and carbon-13 nuclear magnetic resonance.

In Chapter 5, a specific technique is studied in detail, again for a field application. Hughes and colleagues discuss the use of the Photovac gas chromatograph to study a vapor-phase plume of trichloroethylene. They not only describe the performance of the instrument, but also give readers practical tips as to where they may encounter more problems that lead to variability in the data. This chapter is very important because on-site techniques are much less expensive than regular laboratory analysis and provide instantaneous information, allowing the investigators to alter their sampling as they go. Also, soil-gas investigation is being used to measure shallow groundwater plumes and for this purpose is a very cost-effective alternative.

MONITORING STRATEGIES

These chapters focus on monitoring strategies—not on the details of how the analyses should be performed, but rather, on which parameter should be investigated and for what purpose.

In Chapter 6, using PAHs as an example of a mixture of target contaminant, Terri Bulman discusses the analytical requirements for three types of investigations: qualitative surveys, regulatory compliance, and research programs. Polynuclear aromatic hydrocarbon contamination is shared by many different types of industries that are based on fossil fuels. For example, oil and gas extraction, refining, and distribution and the production and end use of coal, coal tar, and creosote all share the potential for contamination by PAHs. Bulman also stresses that although PAHs may be the compound of concern, it is not wise to restrict the analyses to those parameters. She also introduces the concept of the use of statistics in site investigations, a concept described in more detail in Chapter 8. Data interpretation techniques such as multivariate plots are also introduced.

Chapter 7, by Russell Plumb, emphasizes the importance of volatile organic compounds as the primary target group of compounds at waste sites. Using data from 500 waste disposal sites, the author discusses the statistical validity of his premise. This leads him to recommend a sampling strategy based on using volatiles as the first parameter indicative of groundwater contamination instead of the more comprehensive Appendix IX list. This is a very cost-effective approach that should seriously be considered by regulatory authorities as well as by the individual investigator.

In Chapter 8, Robert Gibbons invokes statistical methods for groundwater detection monitoring at waste disposal facilities. The purpose of this approach is to minimize the number of samples for which chemical analysis must be performed while maximizing the chances of finding groundwater contamination if it is occurring. If a site were operated in isolation from any other potential source of contamination, in a perfectly homogeneous hydrologic medium with a well-defined gradient, such an approach would not be necessary. However, since actual sites are never so simple, a statistical approach to what constitutes background and what constitutes contamination is very much warranted. This approach also leads to monetary savings, since sampling and analyzing samples at random can generate astronomical costs.

Chapter 9, by Helmūt Kerndorff and his colleagues in collaboration with Russell Plumb, compares the geochemistry of over 200 German sites with some 500 American sites. As in the United States, German sites are frequently contaminated with tetrachloroethylene and trichloroethylene. Important inorganic fingerprints of contamination in Germany are arsenic, cadmium, nickel, boron, chromium, and sodium and ammonium salts. The authors introduce the concept of a contamination factor (KF): the ratio of contaminant concentration in a downgradient sample to that of an upgradient sample. Contamination is deemed to have occurred when the ratio exceeds 1. They combine this number with the percentage of sampling events for which a specific contaminant exceeds this ratio (e.g., 200 of 250 samples). A very large KF combined with a high frequency of occurrence is an indication of a severe problem. This system is used to identify high-priority chemicals. Also, multivariate plots are used to find clusters of contaminants characteristic of different types of waste.

In Chapter 10, Schleyer and his colleagues describe a hierarchical protocol to be applied to groundwater investigations. In this tiered approach, the investigator is invited to conduct a preliminary reconnaissance in which the analysis is limited to a few target parameters: boron, sulfate, AOX (total organic halogens), and a GC/FID fingerprint. This will identify locations that require a more comprehensive analysis. The second step includes a GC/ MS analysis of a list of priority substances. Only in a few cases will a complete investigation using a combination of the most sophisticated analytical techniques be warranted. This is an entirely logical approach that is based on experience gathered from several hundred sites.

SITE INVESTIGATIONS

This group of chapters on the investigation of hazardous waste sites consists of contributions from workers in the United States, Canada, and Australia. Given the multidisciplinary nature of investigations of groundwater contamination, it is not surprising that they represent backgrounds as diverse as chemistry, hydrogeology, and chemical engineering.

Chapter 11, by Don Goerlitz, might be considered a retrospective review. He describes five case histories spanning some 30 years of activities by the USGS's Organics in Water project, including studies of pesticide, munitions, and wood preservative contamination of groundwaters. The earlier studies, from the 1960s and 1970s, involve the use of the recently developed electron-capture detector with gas chromatography. By the mid-1970s, Goerlitz was using HPLC in the field to measure pentachlorophenol, polyaromatic hydrocarbons, and trinitrotoluene residues in groundwater. The more recent studies have benefited from the development of GC/MS and HPLC/MS instruments with their immense capabilities for identification and quantitation.

These studies by the USGS provided some very important early information about the migration and fate of contaminants in the subsurface. The studies of wood-preserving chemicals at Visalia, California, and St. Louis Park, Minnesota, in the early 1980s demonstrated that in situ anaerobic biodegradation was an important process resulting in the transformation of phenols to methane. Furthermore, these studies demonstrated that dense nonaqueous-phase liquids (DNAPLs) such as creosote could travel considerable distances in the subsurface both vertically and laterally—for example, some 250 m in St. Louis Park.

The second chapter in this group is from the Commonwealth Scientific and Industrial Research Organization's (CSIRO) groundwater group in Perth, Western Australia. Although a relative newcomer to the world of contaminant hydrogeology, CSIRO played a pioneering role in the development of pH electrometry and atomic absorption spectrometry in the 1950s and 1960s. Chris Barber and his colleagues present their results from a detailed study of the leachate plume from the Morley landfill near Perth. Of particular interest is the description of the integrated sampling and analytical methods and their observation that microbial utilization of such volatile organic chemicals as PCE and TCE occurred only in the presence of nonvolatile organics, that is, they were cometabolized.

In Chapter 13, Hooshang Pakdel and his colleagues present a detailed analysis of a DNAPL recovered from an unlined lagoon near Montreal that was filled with 40,000 m^3 of liquid industrial wastes during the years 1968–1972. Because the lagoon was situated in an excavation within a sand and gravel deposit, an immense plume of organic chemicals was found in groundwaters migrating away from the site. It is of great interest to note that while the DNAPL contained 4% 1,2-dichloroethane (i.e., 40 million μg/L), the groundwaters in nearby monitoring wells displayed dissolved concentrations of less than 1% of the aqueous solubility of 1,2-dichloroethane.

GEOCHEMICAL INVESTIGATIONS

The final chapters are concerned with certain specific geochemical processes observed in groundwaters at hazardous waste sites. Chapter 14 addresses the primary inorganic processes of importance—acid–base, redox, and isotopic equilibria. Chapter 15 concerns the dissolution of DNAPLs in the subsurface and the resulting groundwater contamination—a phenomenon identified in Chapters 11 and 13 and now widely perceived to be the cause of the persistence of organic contamination at sites undergoing pump-and-treat remediation. The two following chapters that consider the fate and mobility of two typical DNAPLs: polychlorinated biphenyls (PCBs) and a halogenated solvent known as CFC-113.

The biogeochemistry of organically contaminated groundwaters is the subject of the contribution from Mary Jo Baedecker and Isabelle Cozzarelli, who point out that the presence of oxidizable organic compounds, such as creosote or landfill leachate, in groundwaters tends to induce microbially mediated inorganic (i.e., biogeochemical) reactions that affect the redox and acid–base states of the groundwater. Because these are the master variables (pH and Eh) controlling the solubility and speciation of inorganic solutes and the transformation of organic solutes, Baedecker and Cozzarelli's study provides an exemplary case history of the biogeochemistry of the ''unstable constituents'' of groundwater. Furthermore, they show that the redox reactions also affect the fractionation of carbon isotopes as carbon is transformed from aqueous carbon dioxide to methane.

It is of interest to note that Baedecker and Cozzarelli report significant quantities of aqueous sulfide in the groundwaters at Pensacola, Florida (see their Figs. 7 and 8). In the very few sites where sulfide and heavy metals have both been measured in contaminated groundwaters (e.g., the Borden, Chalk River, and Gloucester sites in Canada), metal sulfide precipitation has been identified as the principal chemical process limiting the solubility of heavy metals. Thus it is reasonable to conclude that the unimportance of toxic heavy metals with insoluble sulfide phases, such as Pb, Cu, Ag, Zn, Cd, and Hg, at hazardous waste sites in North America is due to the presence of significant amounts of aqueous sulfide. The inorganic species that is of greatest toxicological significance at North American hazardous waste sites is arsenic, which tends to form an extremely soluble oxyanion as well as a sparingly soluble sulfide mineral at very low Eh.

Chapter 15 formally addresses the critical issue of the dissolution of DNAPLs such as halogenated solvents and PCBs. Rick Johnson discusses dissolution in aquifer sediments in which the dense organic solvents are present both as ganglia, or disconnected liquid drops held in place by capillary forces, and as thin pools perched on aquitards within contaminated aquifers. Whereas concentrations of dissolved DNAPL constituents observed in monitoring wells are usually much less than the constituents' aqueous solubility, Johnson points out that the investigator should not draw the conclusion that DNAPLs are absent from the subsurface. Owing to a combination of mass transfer limitations in the dissolution process, the dispersion of dissolved constituents in the space between the DNAPL source and the monitoring well, and dilution within the monitoring well caused by mixing along the screened interval, ''concentrations far below the solubility limits are most common.'' Therefore, it is necessary to conclude with Johnson that ''it is more often correct to assume that immiscible-phase product is present in the subsurface whenever aqueous solvent plumes are observed.''

The final two chapters have much in common and serve to introduce the issues of DNAPL dissolution using Raoult's law, biotransformation of the dissolved DNAPL constituents, and their subsequent sorption to aquifer materials. Both chapters also use Canadian case histories to demonstrate these principles. The first, by Stan Feenstra, assesses the transport and transformation of polychlorinated biphenyls at a site in southern Ontario. Feenstra also accounts for the role of matrix diffusion in retarding the transport of PCB solutes through fractured rock. The second, by the editors of this volume and Mark Priddle, addresses the dissolution, migration, and fate of a chlorofluorocarbon solvent, CFC-113, or Freon-113. Like other two-carbon halogenated solvents—for example, PCE, TCE, and TCA—the biotransformation products of CFC-113 are often more toxic than the parent compounds.

I
ANALYTICAL METHODOLOGIES

1

Practical Organic Analytical Chemistry for Hazardous Waste Site Investigations

Suzanne Lesage

National Water Research Institute, Environment Canada, Burlington, Ontario, Canada

Richard E. Jackson

Intera Inc., Austin, Texas

This chapter describes briefly the analytical techniques used by chemists in analyzing hazardous waste samples. The emphasis is placed on organic compounds because in the vast majority of cases they pose the main threat to aquifers. The main techniques described are **GC/MS**, GC, and **HPLC** with a mention of more novel hyphenated techniques. Because most hydrogeologists and environmental engineers use the services of contract laboratories for their analytical needs, the relationship with laboratories and the need for strict **quality control** is also emphasized. Finally, a case study is described where the perceived problem rested entirely on analytical chemistry for its resolution.

Terms set in boldface type are defined in the Glossary at the end of the chapter.

I. INTRODUCTION

Many treatises have been written on the topic of aqueous analytical chemistry, but most of them are far too comprehensive for the user of analytical results, or are silent on how to interpret analytical data produced by others. Unfortunately, most laboratory clients—hydrogeologists and environmental engineers—do not understand the jargon of analytical chemistry.

Most introductory courses are long on nomenclature and short on the practical information needed by groundwater contamination investigators. This chapter approaches nomenclature in a different way: just know enough to distinguish between compounds and to recognize synonyms. We hope it will also furnish insight into the behavior of a component in the subsurface of a waste site. Finally, it provides advice on field methods of sampling, preservation, and analysis.

The analytical chemical techniques are not described for the operator of analytical instruments, but more from the viewpoint of the user of the results. Therefore emphasis is placed on the output and on understanding the different levels of confidence that can be attached to laboratory reports. The largest problems usually arise not from what is written down, but from what is not. Assumptions are often the source of misunderstandings; the quality of the client's interaction with laboratory personnel is often key to the success of a groundwater investigation.

An example of a field investigation is described to illustrate the importance of adequate analysis in hazardous waste investigations.

II. LABORATORY SELECTION

Most hydrogeologists have to deal with a contract laboratory when investigating a site. How can one choose a laboratory? This section describes a few indicators that may be used in the selection of a laboratory that will answer specific needs of the hazardous waste site investigator.

A. Interaction with Laboratory Personnel

Although suitable analytical instrumentation and adequate facilities are obviously essential components of a successful laboratory, the personnel will often make the difference between an average facility and a superior facility. It is important to be able to discuss the sampling program with the managers, identify specific needs, and be satisfied that the laboratory personnel will be attentive to them. It should be possible to meet the personnel or at least be informed of which individuals will be involved and what their qualifications (both academic and experiential) are.

The laboratory should have examples of their reports, be willing to do a **blind sample**, and share their **round-robin** results. A protocol should be established ahead of time to deal with unexpected problems, such as loss of samples or nonattainment of quality control criteria, which are possibly due to **matrix effects** but may also be due to the laboratory. Will a second analysis be performed at no extra cost to the client? A comprehensive discussion with several laboratory managers prior to awarding the contract is therefore the first step in any site investigation. It is important to remember that groundwater sampling is expensive and that there often are time constraints that preclude resampling.

B. How Long Will It Take for the Samples To Be Processed?

The best laboratories are often the busiest. It is therefore important to discuss with the laboratory managers how your samples fit into their schedule and what priorities they will receive. Will the laboratory managers take a contract with penalties for late reports, or do they charge a premium for short turnaround times? If there is no provision for timeliness, you must decide whether fast service is important enough for you to look elsewhere.

Many analytical procedures prescribe a maximum allowable storage time for the samples. It is important to be aware of such requirements and to make sure the laboratory respects them. Because the holding time for volatile organic materials is shorter than for semivolatile extracts, the results for **volatiles** should be available more rapidly than for extractable semivolatiles. The laboratory may choose to send you all the results in a final report. If that isn't suitable to you, it should be possible to receive results as they are being produced.

C. Number of Significant Figures Reported

The number of significant figures reported should reflect the **precision** of the analysis. The basic premise of the degree of precision of an analytical procedure is that the results are only as precise as the least precise measurement during that procedure. For most organic analyses, that measurement is either the volume of the sample, measured in the highly imprecise graduated cylinder, the final volume of solvent for the extract, or the amount injected into the analytical instrument—for example, 1 μL in a 10-μL syringe. If the measurement is carried out manually, the best operators claim 5% error; on the average, 10–15% is probably closer to reality. Modern autosamplers can do better, but the flaw is then usually in the measurement of the volume of the final extract, 1 mL by pipette or syringe.

Therefore, the next time you see on a lab report benzene = 5.245 μg/L, use this as a mental flag, and raise the question of the number of significant figures with the chemist. Inexperienced analysts have a tendency to report to the client all the digits printed out by integrators, without considering the meaning of these numbers. If an inordinate number of significant figures are being reported, the instrument output probably received very little review.

D. Professional References

To find a suitable laboratory, as for any other professional service, ask other users for their opinion. It is also very acceptable to ask the laboratory to provide a list of their satisfied clients. This doesn't constitute complete assurance, because laboratory performance can fluctuate. Summer is, of course, a critical period; the sample load increases, regular staff take their annual leave, and summer students learn how to do their first extraction. This is why **QA/QC** has to be an ongoing, integrated process for any investigation. For certain critical investigations, it is possible to request that the same group handle all your samples. A well-run analytical laboratory should be able to integrate new personnel without a decline in performance. The user can ensure him/herself of this continued quality by the application of a rigorous **quality assurance**/quality control program as discussed in Section VI.

E. Suggested Reading

In his book *Quality Assurance of Chemical Measurements*, J.K. Taylor [1] devotes an entire chapter to the issue of laboratory selection.

III. SELECTION OF ANALYTES

The selection of **analytes** in a hazardous waste investigation can be very difficult. At some sites, little or no information is available as to the quantities and types of wastes involved; however, as Plumb shows in Chapter 7 of this volume, volatile organic chemicals are frequently present at hazardous waste sites and it is relatively inexpensive to measure them. If information is available, it often has to be translated into single components. For example, a "total coal tar" analysis is viewed as impossible. However, most laboratories are able to analyze for the components of coal tar, such as benzene, toluene, the xylenes, and the polynuclear aromatic hydrocarbons (**PAHs**). This is why, in this section, nomenclature is discussed before analytes.

A. Nomenclature

Most hydrogeologists have a relatively good background in inorganic chemistry and know which metals and ions are expected to be found in the subsurface. Also, the nomenclature of inorganic compounds is relatively simple because the use of trivial names was discontinued a long time ago; for example, hydrochloric acid is now seldom referred to as muriatic acid. Therefore, only the topic of organic chemical nomenclature is discussed here.

Organic compounds are compounds based on carbon. Additional elements that may be present in naturally occurring organic compounds, in order of abundance, are hydrogen, oxygen, nitrogen, and sulfur. There are a multitude of naturally occurring organic compounds that contain some or all of these elements in combination, and an equally astounding number of synthetic ones. In synthesis, halogens (F, Cl, Br, I) are most commonly added.

For the hydrogeologist, knowing the name of the chemical is not as important as being able to predict its behavior in the subsurface. Therefore, the first thing to do when confronted with a new name is to find out the corresponding structure, because it contains information that will allow the prediction of the chemical's mobility and solubility. Where can this information be found? It can be requested from the chemist who reported the data. Alternatively, it can be found in a number of handbooks such as the *CRC Handbook* [2], *Groundwater Chemicals Desk Reference* [3], the *Merck Index* [4], or even the Aldrich Chemical Company catalog [5]. The use of **CAS numbers** is a good way to ensure that the same compound is referred to.

Compounds that have only C and H atoms are called hydrocarbons. Of all the organic compounds, they are the least soluble in water. They can be saturated or unsaturated, aliphatic or aromatic. Aliphatic compounds have chains of carbon atoms that may be branched or even cyclic. *Saturated* means that carbon is bonded to four different atoms. *Unsaturated* means that two adjoining atoms share more than one pair of electrons and form a so-called double bond. If several unsaturated carbons are linked in a cyclic structure in which the electrons may be shared over several carbons, the compound is *aromatic*. The term *aromatic* comes from the fact that certain hydrocarbons do have a pleasant odor; for example, benzene, toluene, the xylenes, and naphthalene fall into this category. All these compounds are cyclic, and this special configuration of electrons confers more stability. The word *aromatic* should thus be associated with stability. Examples are shown in Fig. 1.

As other atoms such as oxygen, nitrogen, and sulfur are added to the hydrocarbons, their solubility in water increases. Indeed, to be solubilized, molecules have to form weak hydrogen bonds with water. The more similar a compound is to water, the greater is its the solubility. For example, ethanol CH_3CH_2OH) is more soluble than ethane (CH_3CH_3), and in turn ethylene glycol (or 1,2-dihydroxyethane; CH_2OHCH_2OH) is more soluble than ethanol.

Often, students claim that they do not understand organic chemical nomenclature. One problem is the coexistence of several nomenclature systems. This most often arises in connection with biological molecules, whose structures are often too complex to describe simply in chemical terms. It is not hard to imagine what **IUPAC** nomenclature would produce for a protein with a molecular weight of 2000.

As it is easier to learn chemical names and structures if they can be associated with a problem, the wisest learning technique is to look up names and structures as you meet them. Grouping them according to structure is also

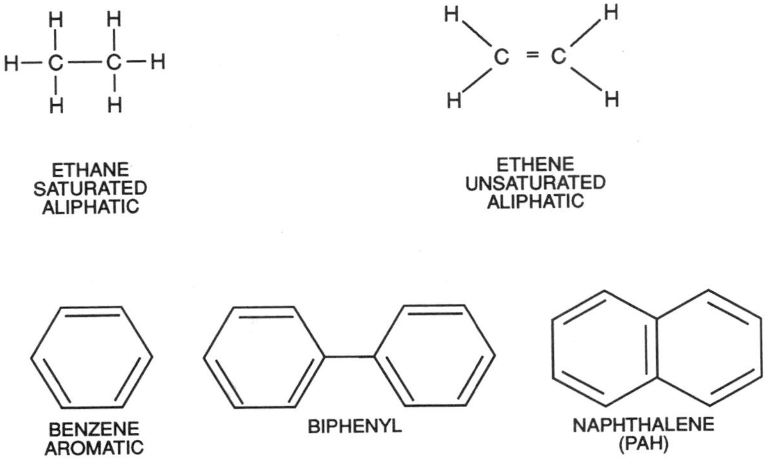

Figure 1 Examples of hydrocarbons.

helpful. For instance, polychlorinated biphenyls are a group of biphenyls (Fig. 1) that differ only in the number of hydrogens that have been substituted by chlorine atoms and thus have similar properties. Polynuclear aromatic hydrocarbons (PAHs) are aromatic compounds that simply differ by the number of fused aromatic rings (rings that share two or more carbons).

B. Target Compound Analysis Vs. Complete Analysis

At the first investigation, the more that can be determined, the better. However, analytical chemistry is expensive, and a rational approach is to be recommended. It has been recognized that the analysis for volatiles can be a very relevant cost-effective technique for assessing groundwater contamination problems [6] (see also Chapter 7, this volume). The reason is very simple: Small molecules are more soluble than their higher molecular weight analog (i.e., methanol is more soluble than octanol), hence they are more likely to have migrated away from the source. Fully automated purge-and-trap systems are found in the vast majority of North American environmental laboratories; alternatively, headspace analysis can be done rapidly. Portable field gas chromatographs (see Chapter 5) are becoming commonplace in hazardous waste site investigations.

Water-miscible compounds, such as acetone, ethanol, acetic acid, 1,4-dioxane, and aniline, are a particular problem. Because of their high solubilities, they are very mobile in the subsurface and are the most likely to move off site, but at the same time the analytical methodologies to address them are often inadequate [7].

Volatiles should therefore be screened in the first analysis, but a more complete analysis should be done closer to the source. Quantitative analysis of U.S. EPA Appendix IX compounds (see Chapter 3, this volume) should be mandatory for selected samples, but it should be remembered that no single list is comprehensive and that most analytical methods are selective. Also, a good proportion of the chemicals found in a typical hazardous site are not on any priority list. There are major difficulties inherent to their measurement, as discussed by Swallow (Chapter 2, this volume). Metal ions should not be neglected, nor should basic field measurements such as pH, Eh, and dissolved oxygen. The problem of selection of parameters is addressed in Part II of this book.

IV. GC/MS ANALYSIS

At the vast majority of hazardous wastes sites, organic chemical analysis is conducted by gas chromatography coupled with mass spectrometry (GC/MS). This is primarily because, although it is relatively expensive, it is the most cost-effective analysis if the amount of information obtained per analytical dollar spent is considered. This section covers the basic principle of GC/MS analysis and its use as both a qualitative and quantitative analytical instrument. Other mass spectrometric techniques are described briefly. The emphasis is placed on the quadrupole instruments because they are currently the most popular in environmental analysis.

A. Trace Organic Analysis

The principle behind most organic analyses is the same and is illustrated in Fig. 2. Compounds dissolved in water are extracted (by a gas for volatiles or a solvent for semivolatiles) and separated by chromatography. Their presence is detected by a detector, the output of which is proportional to the total amount of each component. Quantitation of components is done by integrating the area under the chromatographic peak and comparing it to that of a standard. The specificity of the analysis depends on the type of detector used. It may be relatively nonselective such as the flame ionization detector (FID), where essentially anything that will burn is detected, or fairly specific, such as the electron capture detector (ECD), which will almost only detect halogenated compounds. A mass spectrometer can be used as both a nonselective and a specific detector.

SAMPLE   GAS CHROMATOGRAPHY   MASS SPECTROMETRY

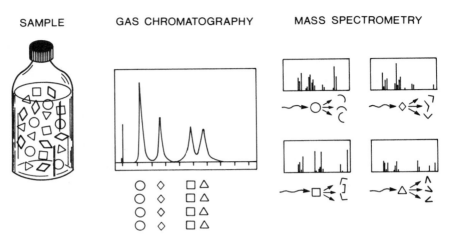

Figure 2   Components of a GC/MS system. (Reprinted from Swallow et al. [8].)

B. Basic Principle

A mass spectrometer is an instrument that differentiates compounds according to their mass. It is composed of three main parts: the ionization chamber, a mass filter (magnetic field or electronic mass filter), and a signal amplifier (electron multiplier). The compound enters the ionization chamber and is bombarded by a current of electrons at 70 eV in a vacuum. This causes the molecule to lose an electron and acquire a positive charge (z). The charged molecule is then destabilized and tends to break down into several smaller fragments along its weakest bonds. These ionic fragments will be attracted to the detector that is of the opposite charge. In a quadrupole instrument the ionized particles travel through a set of four rods (two negative and two positive), to which dc voltages are applied, which act as an electrostatic filter. The ratio mass/z transmitted is proportional to the amplitude of the applied radio frequency, which is scanned with time to allow a selected range of ionized molecules to reach the electron multiplier. For a more detailed discussion, readers are referred to a publication by Haas and Norwood [9]. The mass/z ratio is representative of the molecular mass of the compounds and is thus usually referred to in atomic mass units (**amu** or daltons).

C. Data Acquisition

Gas chromatography-mass spectrometric analyses can be done in two main modes, full **scan** or selected ion monitoring (**SIM**). In the first mode, a range of masses (typically 45–450 amu) are acquired at the rate of one scan per second; that is, every second a full mass spectrum is obtained and stored in the computer for later retrieval (Fig. 3). The word scan refers to scanning of the radio-frequency voltages as noted above, in a range proportional to the selected masses. In the selected ion mode, only a group of ions, which are typical fragments of the analytes of concern, are acquired (Fig. 4). The chromatogram will reflect the intensity of only these ions. Because usually a maximum of 10-15 ions are monitored simultaneously, the scan rate—the number of scans per second—can be increased, resulting in an increase in sensitivity. Because only a few masses are acquired, it is not possible to identify unknowns in the sample. Also, in heavily contaminated samples, interferences can cause problems. If a large quantity of analytes are requested, the gains in sensitivity are relatively small, and thus it is preferable to acquire a full scan.

Even when a full spectrum is acquired, quantitative analysis is done on extracted ions (Fig. 5). One ion is selected for each compound for quantitation. Each ion is, in essence, a specific detector. As in any chromatographic analysis, the area under the curve of a specific time window is integrated, then the area of the unknown is compared to that of the standard. In addition, the area of one or two other ions characteristic of the analyte are also integrated and their ratio compared to that of the primary ion; these are termed *qualifying ions* because they allow for qualitative identification of a compound. Under the same operating conditions, the spectrum obtained from a given compound is always the same; hence the ratios of the qualifying ions to the primary ion are constant and, in addition to the retention time, are the criteria that are used to ensure correct identification of the analytes (Fig. 6).

The strength of GC/MS over conventional GC detection is therefore two-fold. Quantitation using a selected mass, which is characteristic of the target

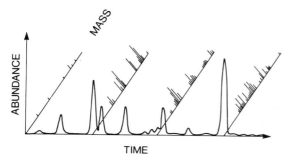

Figure 3   Total ion chromatogram.

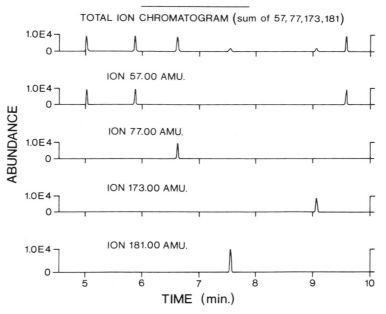

- ONLY SELECTED IONS ARE ACQUIRED
- MAXIMIZES SENSITIVITY AND SELECTIVITY
- DOES NOT ALLOW THE IDENTIFICATION OF UNKNOWNS

Figure 4   Selected ion monitoring: each channel is a specific detector.

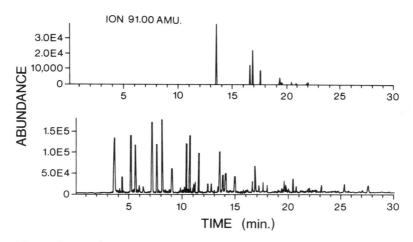

Figure 5   Extracted ion chromatogram: a range of ions (e.g. 45 to 450 a.m.u.) is acquired, but selected ions are quantitated.

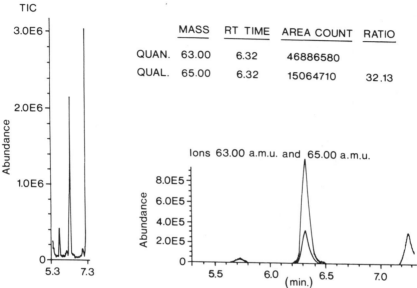

Figure 6 Unknown plus standard quantitation and qualifying ion.

analyte, ensures selectivity and reduces the problem of coeluting interferences. For example, if two peaks were to elute closely, but one is from benzene (mass 78) and the other one from carbon tetrachloride (mass 117), an unspecific detector would see the summed response from the two compounds in the overlapping region. With the mass spectrometer a different chromatogram can be drawn for the two masses in which interference from the coeluting compounds is totally eliminated (Fig. 4). The acquisition of a complete mass spectrum allows for the unambiguous identification of the correct analyte peak and for the tentative identification of peaks arising from nontarget analytes.

D. Library Searches

The identification of the components in a mixture is done by comparing the spectrum of the unknown to that of a library of spectra stored in the GC/MS computer. There are two main methods of doing this comparison: forward search and reverse search (Table 1).

During a *forward search*, the spectrum of the unknown is compared to spectra contained in a library, either a commercially available library such

Table 1 Forward Search Vs. Reserve Search

Forward search
    The spectrum of an unknown is recorded.
    It is compared to a library of spectra.
    A tentative identification is made.
    This is the only option available when no standards are available.
Reverse search
    A standard is analyzed.
    A retention time window and a spectrum are obtained.
    Masses characteristic of the compound will be searched in the same retention window.
    This is the technique used for quantitative target compound analysis.
      It is possible only if a standard is available.
      It allows for a much lower detection limit even in complex samples.

as the Wiley library of spectra, which now contains the spectrum of over 100,000 compounds, or a user-created library, which, although usually more modest in size, can be tailored to specific types of compounds and be more efficient. For instance, a pharmaceutical company may wish to use a library containing drugs only. In groundwater analysis, the broader library is the most useful because the contaminants may originate from very diverse sources such as agricultural runoff, industrial spills, and landfill leachates. The output from the search consists of a list of best matches ranked in order of best fit. A perfect match would carry a fit index of 100; this is seldom observed, although matches of 90–95% are not uncommon. When the match is of lesser quality, the chemist must interpret the differences and make a tentative identification. Often the spectrum of the unknown is not present in the library and the library search will give at best an indication of the chemical class of the compound. The mass spectrometer is not very effective at distinguishing between different isomers, because they usually have very similar spectra. Thus, if a series of dimethyl naphthalenes are present in a sample from a wood preservative site, the isomers can be identified only by comparing the spectra with standards.

In a *reverse search*, a group of spectra contained in a user-created library are compared to all the spectra found in a sample. This type of routine is used in target compound analysis. The advantage of reverse search is the ability to identify much smaller quantities of a compound than is possible in a forward search because, even if the peak is of very low intensity or is buried in a group of poorly resolved peaks, it will still be found. This is the type of search that is used in priority pollutant analysis by most laboratories. The danger of restricting oneself to this type of analysis is to miss some nontarget compounds that may be very important. Therefore, whenever a GC/MS scan is requested from a laboratory, it is essential to clarify whether a forward search will also be carried out to identify "what else may be there." Because forward searches are much more time-consuming, the cost is likely to be higher, but it is the only way to ensure that important constituents are not overlooked. The importance of nonpriority pollutant analysis is also discussed by Swallow in Chapter 2.

E. Problem Samples

Analytical methods are validated for a given analyte or group of analytes in a specific **matrix** (e.g., water, soil, sludge) and for a determined concentration range. When the concentration in the samples falls outside of this range, it becomes necessary to either increase the sample size if the analyte is too dilute, or decrease the sample size when the sample is too heavily contaminated. Either way, this affects the precision and accuracy of the method and should be taken into consideration when evaluating the data.

At many hazardous waste sites, the concentration of one or more analytes often exceeds the working range of methods designed for relatively clean water and soils, and therefore the samples must be diluted. The major effect is, of course, to correspondingly reduce the detection limit of all the analytes present in the samples. Minor constituents will therefore be lost. It is sometimes possible to analyze the sample at two different dilutions, but this will be effective only if the high-concentration contaminants elute relatively far from the analytes present in lower concentrations (i.e., are well separated on the chromatograms). The potential for contamination of the analytical instrument makes this approach impractical. For semivolatile compounds, a liquid chromatographic cleanup can be done on the extract. This fractionates the sample according to chemical classes and is useful only in target compound analysis. For volatiles, no such scheme is possible, because most of the analytes would be lost during processing. It is thus customary to simply dilute the sample with "organic-free" water. Unfortunately, it is virtually impossible to obtain water that is totally free of organic compounds. Also, the polymers used in the traps of the purge-and-trap system are also organic and can bleed small quantities of compounds such as benzene and toluene.

This problem can be corrected by analyzing appropriate **blanks**, but, as there is always a slight variation between analytical runs, there is often a residual amount carried over, usually close to the detection limit. When a sample has been diluted 100-fold, a residual of 0.3 μg/L of benzene will become a reported concentration of 30 μg/L, which may seem significant. In evaluating data arising from diluted samples, it is important to remember to readjust the detection limit and the associated analytical error accordingly. In this case, the detection limit would be 10 μg/L (not 0.1 μg/L). To reflect the actual **accuracy** of the data, it would be best reported as 0.03 mg/L with a detection limit of 0.01 mg/L. Unfortunately, not all analytical laboratories follow this policy, and the client has to exercise his/her own judgment in evaluating the data.

V. OTHER ORGANIC ANALYTICAL METHODS

As discussed above, GC/MS analysis has a lot to offer because it can provide qualitative and quantitative analysis as well as confirmation of the identity of compounds. It does, however, have limitations. Only compounds that are sufficiently volatile or that can be made volatile through chemical derivatization are amenable to GC/MS analysis. For semivolatile compounds, an extraction is usually done with dichloromethane prior to introducing the sample into the instrument; thus, only extractable compounds are suitable for GC/MS analysis. In environmental samples, this can be as little as 10–15% of the total dissolved organic carbon.

Some samples are sufficiently unstable to warrant immediate on-site analysis. At the moment there are very few portable GC/MS systems that can be brought to the field. Some mobile laboratories have been equipped with mass spectrometers (the Ontario Ministry of the Environment has small vans equipped with **TAGA** systems, a product of Sciex Instruments, Mississauga, Ontario), but in general they are restricted to air sampling where no sample preparation is necessary. Portable gas chromatographs (see Chapter 5) are invaluable tools for field monitoring. Analysis either on-site or close to the site has definite advantages in hazardous waste site investigations, and at the moment this need is best filled with gas chromatography (GC) and high-pressure liquid chromatography (HPLC) (see Chapter 11).

A. Gas Chromatography

Gas chromatographs with specific detectors can outperform GC/MS systems in terms of specificity and detection limit. The electron capture detector was for a long time the sole instrument used by pesticide chemists and is still routinely utilized in analyzing for chlorinated pesticides and PCBs. One of its advantages is its specificity for certain groups of compounds, mostly halogenated hydrocarbons and nitro-substituted compounds. It is also at least 100 times as sensitive as a mass spectrometer, even in the selected ion (SIM) mode. Furthermore, it is much less expensive and thus is a very cost-effective means of analysis. Other commonly used detectors include the flame ionization detector (FID), a good multipurpose detector often used in hydrocarbon analysis; the thermal conductivity (TC) detector, used in gas analysis at the percentage level; the photoionization detector (PID), used mostly for aromatic hydrocarbons but also in portable gas chromatographs such as the Photovac (Thornhill, Ontario); and the nitrogen/phosphorus detector, also called the alkali flame detector, which is used for pesticide compounds containing nitrogen and phosphorus.

In the gas chromatograph, the separation is done in a column (a tube of glass or metal) filled with a packing material coated with an adsorbent. The separation is based not only on the volatility of the compounds but also on their relative affinity for the packing. Since the early 1980s these columns are being gradually replaced with capillary columns, which provide a much better separation. The most common columns are hollow fused silica tubing coated with a polymeric material. They are typically 30 m in length and have diameters ranging from 0.25 to 0.53 μm.

B. High-Pressure Liquid Chromatography (HPLC)

High-pressure liquid chromatography (HPLC) is the analytical tool of choice for most thermally labile (unstable) compounds. In HPLC, the carrier gas of the gas chromatograph is replaced by a solvent mixture, usually containing a large proportion of water. It is thus quite logical that when the sample matrix is water, HPLC is the instrument of choice. Several detectors have been developed for the liquid chromatographs, the most popular being the ultraviolet, the fluorescence, and, to a lesser extent, the electroconductivity detectors. Ion chromatographs are in essence liquid chromatographs with conductivity detectors. There are two main reasons why HPLC is not used for all analyses: (1) The efficiency of the separation is not as good as what can be achieved with a capillary gas chromatograph and (2) except for the fluorescence detector, the detection systems are not as sensitive, but then, not all molecules fluoresce or can be derivatized to fluorescent species. The derivatization reaction can be carried out postcolumn, that is, following sample elution from the chromatography column. A good example of this in groundwater contamination is the analysis for the pesticide aldicarb and its two toxic metabolites, aldicarb sulfone and aldicarb sulfoxide [10]. The three compounds are separated on the HPLC column and then hydrolyzed and derivatized to the same fluorescent species.

Diode array detectors are now often used in HPLC analysis instead of ultraviolet detectors (UV). They are scanning UV detectors and can provide conformatory data for HPLC analysis in the same way the mass spectrometer does for the gas chromatograph. A UV spectrum is not nearly as useful as a mass spectrum in the identification of organic compounds because it is not as detailed, but there are situations where it is the only possible choice. An example of this is the case of a hazardous waste site where the presence of both phenol and aniline was suspected [11]. These two compounds elute very closely in a gas chromatograph. Phenol has a molecular weight (M) of 94, and aniline, one of 93. However, because carbon has both ^{12}C and ^{13}C isotopes, there is always a significant $M + 1$ peak in a mass spectrum of organic compounds. Therefore aniline also had a peak at 94. The fragment ions are also similar, which means that mass spectrometry is useless in telling them apart. Aniline also interferes with the total phenols measurement by the 4-aminoantipyrene colorimetric test. The two compounds also elute closely in HPLC, but their UV spectra are sufficiently different that it is possible to distinguish them by diode array detection (Fig. 7).

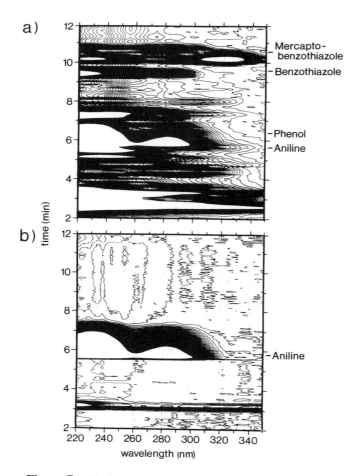

Figure 7 Diode array spectra: Top view, contour. (a) Contaminated groundwater sample: phenol is seen as a peak (lightly shaded area) at 6.8 min and 260 nm. (b) Aniline standard: elutes at 6 minutes.

C. Other Hyphenated Methods

There are also liquid chromatography-mass spectrometry (LC/MS) instruments on the market, but they are not nearly as widespread as GC/MS systems, partly because they are more recent. Indeed, interfacing a liquid chromatograph with a mass spectrometer is a significantly more difficult task than coupling a gas chromatograph to a mass spectrometer. The spectrometer needs to operate in a vacuum, whereas the LC effluent is a solvent, mostly water, mixed with buffer salts. Many ingenious interfaces have been devised, but there is invariably a trade-off in terms of sensitivity [12].

There are several other hyphenated instrumental methods of analysis now available on the market. Most of these are not accepted in regulatory methods because they are either too scarce or still considered too experimental. One of these promising tools is **GC/FT/IR** [13,14]. The GC/FT/IR instrument, which gathers infrared spectra instead of mass spectra, is an excellent complementary tool to GC/MS. It addresses the same range of compounds, but the spectrum obtained gives information on the functional groups of the molecule. For instance, with IR spectrometry it is very easy to distinguish alcohols from ketones and esters. It is also possible to distinguish between **isomers,** for example, between *ortho-, meta-,* and *para*-xylenes.

GC/MS/MS systems are used to distinguish between very similar compounds by allowing the analyst to get the mass spectrum of one spectral peak of the first mass spectrum. These instruments contain three quadrupoles in series, operating as magnetic/electrostatic/magnetic fields (B/E/B configuration). The molecule in the ion source fragments along its most vulnerable bonds first, and these are separated by the first quadrupole. The second quadrupole acts as a second ionization chamber and further fragments the molecule. These fragments, also called daughter ions, are then separated by the third quadrupole and detected as in the simple mass spectrometer. This is useful in the case where similar molecules have initial fragments of the same mass/charge ratio. However, when these initial fragments are isolated and bombarded a second time, the secondary fragments will differentiate them. The first mass spectrometer can be tuned to allow compounds of only a certain molecular weight to reach the second mass spectrometer. In effect, it acts as an electronic cleanup system.

VI. QUALITY ASSURANCE/QUALITY CONTROL

The term quality assurance/quality control (QA/QC) finds its way into most analytical contracts, yet a little probing showed that it means different things to different people. The often heard statement "We sent duplicate samples to another lab and their results are different than yours. Why? " summarizes the frequent failure of what is assumed to be QA/QC by many inexperienced investigators. At that point, there is very little that can be done to salvage the data. The best approach is to design the QA/QC program before the first sample is taken.

A. Definition

The importance of a quality assurance/quality control program is being increasingly recognized, primarily through the efforts of the U.S. Environmental Protection Agency (EPA) [15], which laid out precise guidelines for the implementation of QA/QC for the validation of analytical data. Unfortunately, these have not made their way into the curriculum of most university analytical chemistry courses. Thus, even the meaning of the words quality assurance or quality control is subject to different interpretation. Regardless of how one wants to label them, QA/QC programs need to be operated at two levels. First, within the laboratory, to ensure that good laboratory practice is consistently followed and that the results achieved are of consistent and measurable quality: quality control. This means that, for example, the chemist reporting the results needs to ensure that a concentration of 5.2 μg/L of benzene is accurate (i.e., it wasn't in fact 2 or 10) and that the degree of precision of this measurement (i.e., 5.2 ± 0.2 or 5 ± 1 μg/L) and the detection limit for the sample analyzed can be given to the client. Second, external control by the client has to be implemented to ensure **comparability** of data obtained from different sources and over different time periods. This is called quality assurance and is the responsibility of the site investigator. The sections that follow describe how this can be achieved.

B. Precision and Accuracy

Precision is the degree to which data generated from replicate or repetitive measurements differ from one another [16]. It is obtained by the repetitive analysis of **standards,** to determine the method precision, and of samples, to assess the degree of variation between **replicates.** Any reputable commercial laboratory will provide data on the precision of their analytical methods, but it is the responsibility of the client to request the appropriate number of replicates.

Accuracy is the degree of agreement of a measured value with the true or expected value of the quantity of concern [1]. Accuracy is more difficult to assess than precision but can be evaluated by analyzing certified reference materials and participating in interlaboratory comparisons, also known as round-robins. Certified reference materials, which are generally available from government agencies, are not available for many analytes and matrices, and interlaboratory studies are infrequent. It is possible to get a measure of accuracy by having another laboratory send some **spiked samples** mixed in with the batch of samples submitted.

So, what are the practical tools to ensure that the quality of data obtained meets expectations? Split samples? As stated above, the two main factors that need to be assessed are precision and accuracy; split samples will do neither. If the two data sets agree, it may be fortuitous. If they do not agree, it will not be possible to measure whether the difference falls within normal analytical error or to know which value is the accurate one. The only answer is to prepare a QA/QC program that suits the size and schedule of the sampling program.

C. QA/QC Programs

For large field projects, the regulatory agency will often dictate the QA/QC program. However, for smaller investigations or in countries other than the United States, this task is typically the responsibility of the investigator. An example of an adequate QA/QC program is outlined in Table 1. The total number of QA/QC samples needs to be approximately 10–15% of the total sample load, relatively more if the batch size is small. This includes blanks, replicates, and spiked samples. Blanks should be reported, and it is important to find out whether or not the data were corrected for blanks. Replicates are important because they will be the only measure of sample variability. For this reason, it is preferable to do one sample in triplicate rather than two samples in duplicate. The cost will be the same, but statistical evaluation cannot be done on duplicates. In the example given in Table 2, only three samples were quality assurance samples: the **field blank** and two of the three replicates. This should be sufficient in a batch of 12 samples. The laboratory inserted **reagent blanks** after the standards and after a series of samples to find out about possible carryover. Carryover can occur in certain procedures during which the chemical to be analyzed is adsorbed and then desorbed. If the desorption is not complete, the rest of the compound will be desorbed in the next sample. If this happens to a mixture of standards, it would cause the analyte to be reported in a sample analyzed right after it. The standards were measured at two different concentration levels to allow the construction of a standard response curve (the blank is used for the third data point). This does not always happen in organic analysis because if 50 or more analytes are measured simultaneously, the calculations may be somewhat cumbersome. In that case, a single response factor is measured (area count per unit concentration), and the response of the detector is assumed to be linear.

Spiked samples are used to measure matrix effects and possible bias on the data. This is most important in hazardous waste samples because many

Table 2 Example of a Laboratory Load with Very Good Quality Control

Run no.	Analysis
1	Standard, concentration level 1
2	Standard, concentration level 2
3	Blank (reagent grade water)
4	Sample 1
5	Sample 2
6	Sample 3
7	Sample 4
8	Sample 5 replicate
9	Sample 5 replicate
10	Sample 5 replicate
11	Blank (field or **trip**)
12	Standard, concentration level 1
13	Standard, concentration level 2
14	Sample 6
15	Sample 7
16	Sample 8
17	Sample 9
18	Sample 10
19	Sample 11
20	Sample 12
21	Blank (reagent)
22	Standard, concentration level 1
23	Standard, concentration level 2

Note: **Surrogates** are added to all samples.

of the contaminants may be present in high enough concentrations to act as cosolvents and change the composition of the solvent used to extract the analyte from the sample. Indeed, an analytical method developed for water may not be suitable for a concentrated landfill leachate. Unfortunately, unless the expected concentration of the analyte is known ahead of time, it is very difficult for the laboratory to spike the sample at a realistic level.

The matrix effect is measured by comparing the percentage recovery of the spiked compound in the sample with the amount found in distilled water. The percentage recovery of the analyte in the sample is calculated as follows:

$$\text{Recovery (\%)} = \frac{\text{weight}_{spsamp} - \text{weight}_{ussamp}}{\text{weight spiked}} \times 100$$

where the subscripts spsamp and ussamp denote spiked sample and unspiked sample, respectively.

In highly contaminated samples for which the analytical error is large, it is not unusual to find more of the analyte in the unspiked sample than in the spiked sample and hence get a negative recovery. For example,

$$\text{Recovery} = \frac{624 \pm 230 \ \mu g \ - \ 750 \pm 250 \ \mu g}{100 \ \mu g} \times 100$$

The large standard deviation will often be due to the fact that the sample had to be diluted to stay within the capacity of the chromatographic column and hence is 23×10 because of the dilution factor. A standard deviation of 30% is not at all rare for organic analysis. Had the sample not had to be diluted, 100 μg would have been a very realistic spiking range. Unfortunately, only after the sample has been analyzed once is it possible to gauge the appropriate dilution range.

Because the same amount of information can be gathered with one analysis instead of two, and because they are not likely to be found in the samples, spiked samples have been largely replaced by surrogates.

D. Surrogates and Internal Standards

A surrogate is a compound that is an **analog** of the analytes and is added to the sample prior to the extraction step. It can be a deuterated analog (one in which deuterium has replaced hydrogen), most commonly used in GC/MS, or a chemical analog that would not be expected in the samples. It is carried through the entire analytical procedure, and its recovery is reported as a measure of matrix effect and analytical error. The ultimate would be to have a deuterated analog for each analyte (isotope dilution) as described in EPA methods 1624 and 1625 [14], but cost precludes its widespread use.

Surrogates are not to be confused with internal standards. Internal standards are also usually deuterated analogs, but they are added in the final analytical step and are used for quantitative analysis, where their purpose is to account for instrument variability from one sample to the next. The area of the peak of the internal standard in the sample run is compared to the area of the internal standard in the standardization run, and all the concentrations are corrected for any discrepancy.

$$\text{Concentration of } x = \frac{\text{area } x_{\text{samp}}}{\text{area } x_{\text{std}}} \times \frac{\text{area I.S.}_{\text{std}}}{\text{area I.S.}_{\text{samp}}} \times \text{conc } x_{\text{std}}$$

where x denotes the analyte (or parameter), I.S. is internal standard, and the subscripts samp and std denote sample and standardization run, respectively.

It is possible to use more than one internal standard in an analysis to account for the possible behavior of either early- or late-eluting compounds or of acidic or neutral compounds. During purge-and-trap analysis for volatiles, surrogates and internal standards are added together because there is no other preliminary extraction step. Then the only difference between the two is that internal standards are used in the calculations whereas surrogate concentrations are simply reported.

In a typical laboratory report, the percent recovery for surrogates is reported for each sample. Obviously, 100% is the target, but acceptable recovery ranges are much wider, as shown in Table 3.

VII. SAMPLING AND FIELD METHODS

The sampling of contaminated groundwater requires a careful choice of monitoring instruments, the principal criterion of choice being that the individual hydrostratigraphic units within a groundwater flow system must be sampled individually. Wells that penetrate more than a single unit provide little useful information [18,19]. This is because the sample mixing and dilution that occurs in a fully penetrating well imply a greater hydrodynamic dispersion than actually takes place in the aquifer itself. It is also because this integrated (and diluted) sample may indicate contaminant concentrations within the acceptable limits of the guidelines, although these limits may in fact be exceeded within a particular hydrostratigraphic unit.

Figure 8 shows two commonly used devices for sampling groundwater quality: the cluster-type multilevel sampler and the 2-in. (5-cm) i.d. piezometer.

Table 3 Surrogate Spike Recovery Limits for Water and Sediment Samples for Neutral, Acidic, and Volatile Compounds

Surrogate	Water (%)	Sediment (%)
Nitrobenzene-d5	35–114	23–120
2-Fluorobiphenyl	43–116	30–115
p-Terphenyl-d14	33–141	18–137
Phenol-d6	10–94	24–113
2-Fluorophenol	21–100	25–121
2,4,6-Tribromophenol	10–123	19–122
4-Bromofluorobenzene	86–115	74–121
Dibromofluoromethane	86–118	80–120
Toluene-d8	88–110	81–117

Source: U.S. EPA [17].

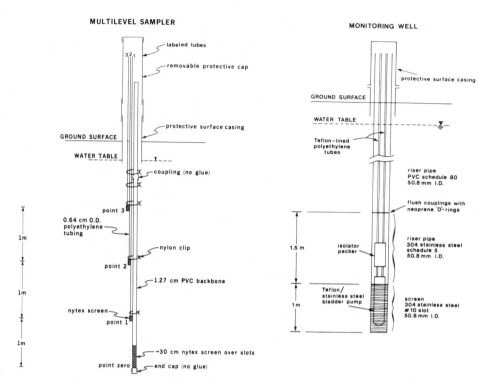

Figure 8 Typical monitoring wells.

Both provide the capability of sampling discrete zones of potentially contaminated groundwater. The first is used to map the outlines of contaminant plumes in three-dimensional details [20]. The second is generally used for monitoring groundwater quality where it is necessary to detect groundwater contamination or to establish that this quality is in compliance with regulated or guideline values. Consequently, the screen material is made of an essentially inert material (e.g., stainless steel). Cowgill [21] presents evidence that the well-casing materials and sampling devices, except those made of PTFE (Teflon), should be steam-cleaned prior to use or installation, respectively.

Samples for detection or compliance monitoring are best collected by using dedicated submersible pumps with PTFE bladders operated by compressed air or nitrogen that does not come into contact with the groundwater sample [22]. The pumps (see Fig. 8) are located at the depth of the well screen and can be isolated from the stagnant water in the well bore by inflating a

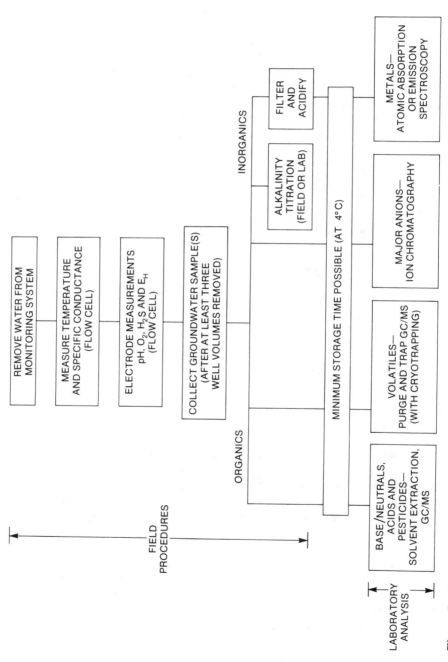

Figure 9 Flowchart for sampling and analysis of groundwater.

packer system, located immediately above the pump. Generally, two or three well-screen volumes of stagnant water are pumped before sampling begins [23,24].

Figure 9 shows a typical sequence of operations conducted in the field to collect samples and the subsequent distribution of aliquots for analytical purposes. Samples are collected in precleaned (and baked) amber glass bottles, with no headspace for volatile organic samples, at a pump delivery rate of 100 mL/min or less. The bottles are allowed to overflow by at least 1.5 volume and are then rapidly capped and stored at about 4°C until analyzed [25]. Sample bottles should not be rinsed out with the sample because a film of organic compounds from any nonaqueous-phase liquids present may adhere to the glass and artificially increase the concentration in the sample.

Measurements of certain unstable constituents—for example, pH, redox potential (Eh), temperature (T), and specific electrical conductance (SEC)—must be conducted in the field. Although referred to as "well purging parameters" [25], in that they are monitored during purging operations and stable readings indicate the appropriate time for sampling, these parameters are critical in the quantitative assessment of water quality [26]. Baedecker and Cozzarelli discuss the collection of groundwaters for dissolved gas analysis, in particular O_2, H_2S, CH_4, NH_3, and CO_2 (i.e., alkalinity), which are critical to understanding the nature of redox processes and metal-ion speciation within the subsurface, in Chapter 14 of this volume.

Preservation techniques for organic constituents of groundwater differ from those for inorganics. In particular, it is inadvisable to subject samples collected for volatile organic analysis to vacuum filtration because of the potential losses by volatilization. Samples for the analysis of volatile aromatic hydrocarbons (e.g., benzene, toluene, ethylbenzene, and xylenes—BTEX) should be preserved with HCl to prevent biodegradation [15]. Inorganic cations are usually filtered and acidified to pH <2 to prevent precipitation or sorption to the container walls, whereas samples for inorganic anions are usually only filtered and refrigerated. Kent and Payne [27] show the importance of the filtration of suspended solids in inorganic analysis and present evidence that frozen icepacks do not adequately chill water samples to 4°C unless the samples are prechilled with wet ice.

Apart from *precision* and *accuracy* (see Section VI.B), groundwater samples must also be *representative* of the hydrostratigraphic formation from which they were collected, *complete* in the sense that the groundwater flow system at the site in question has been fully examined, and, finally, *comparable* with other data collected from the same site. These five attributes define the quality of environmental data and are known as the PARCC attributes [28]. **Representativeness, completeness,** and **comparability** are more the responsibility of the hydrogeologist than of the analytical chemist, who is responsible for proper precision and accuracy. Good examples of hydro-geochemical studies of hazardous waste sites that observe the PARCC requirements are those of Barcelona et al. [29], Jackson and Patterson [26], and several of the chapters in this book (e.g., Chapters 4 and 11–13). The reader may find more information on groundwater sampling in Reference 25.

VIII. CASE STUDY: NEWCASTLE, NEW BRUNSWICK, CANADA

A. The Problem

This example is taken from the report of a review team composed of hydro-geologists and chemists from WMS Associates, Intera Technologies, and the National Water Research Institute of Canada [30] and was chosen for several reasons. It illustrates clearly the need for strict quality control, it is a good example of how to choose the correct target analytes, and it demonstrates the necessity for hydrogeologists to establish good communication with the analytical laboratory and to include the chemist's viewpoint in the overall assessment of the problem.

The drinking water supply of the town of Newcastle (population 7000) is derived from pumping wells that draw water from a buried-channel sand and gravel aquifer underlying the town. A wood preservation plant that used **PCP** and creosote for 50–60 years is situated on the outskirts of the town. This plant has been regarded as a potential threat to the municipal water supply, and the plant site is scheduled for cleanup. In the interim, the provincial government has been monitoring the water supply wells for the presence of PAHs, compounds known to be present in creosote and feared because of their toxicity.

In 1988, low levels of PAHs (< 1 μg/L total) were reported to be present in certain wells at certain times. During 1988 and 1989, PAHs were found intermittently during detection monitoring, and both wells were closed in early 1989 because water from them exceded the 10 ng/L (10 ppt) World Health Organization (WHO) drinking water limit for benzo[α]pyrene. The source of PAH contamination had not been definitively identified, although both the provincial government and the community suspected the wood preservation plant.

Examination of the chemical analysis data revealed a number of QA/QC and other problems:

There was no proper protocol for the sampling and analysis of compounds at or near detection limit or the WHO guideline value of 10 ppt.

No field blanks had been taken to assess the possibility of other sources such as diesel fumes in the pump house or cigarette smoke, and no replicate sample had been analyzed to measure the precision of the data.

The occurrence of the PAHs was sporadic, with different compounds being identified during different sampling rounds, and the more soluble, more mobile compounds present in creosote were not detected together with the PAHs.

The chemist who had reported the data suggesting low-level PAH contamination of the wells had expressed some concern over the validity of his data at the low level of detection requested by the regulators and admitted having had very little experience doing this type of analysis.

The water supply wells used for detection monitoring did not appear to be suitable as monitoring wells with which to establish the level of contamination of the aquifer.

B. The Plan

A complete study was therefore commissioned to determine whether the aquifer was indeed contaminated with PAHs and, if so, to identify the source of the contamination. A survey of the area was conducted to determine all the possible sources of PAHs. Apart from the plant mentioned above, there were several possibilities: a dump site, a small lumber company using wood preservatives, and several potentially leaking fuel storage tanks.

A new set of monitoring wells were drilled with a cable-tool rig (i.e., one needing no compressor or other component producing PAHs), and a rigorous cleaning protocol was followed for the drilling equipment to prevent contamination with any oil or grease. The steel casing was washed with degreasing solvent, detergent, a hot water rinse, acetone, and finally a distilled water rinse prior to use. Samples of the rinse waters were collected and analyzed. Each monitoring well was provided with a dedicated submersible pump to prevent cross-contamination between wells.

Three rounds of sampling were done. The QA/QC program included equipment rinse samples, drill water samples, trip blanks, field blanks, and replicates. The samples were also sent to two laboratories: the one that had published the initial results and a commercial laboratory that was deemed reliable by the review team.

The list of selected analytes was expanded to include, in addition to the PAHs, the volatile organic compounds benzene, toluene, and xylene as well as EPA priority pollutant phenols in order to assess the source of the contaminants. As explained above, the more mobile compounds are the most soluble ones. It was known from studies done by Goerlitz at the USGS (see Chapter 11) that water contaminated by wood preservatives would indeed contain PAHs and that naphthalene would be the predominant species; phenols, mostly methylphenol and dimethylphenol, would also likely be found. From other work [31], it was known that the volatile compounds benzene, toluene, and xylenes should be also present because they are components of creosote and are even more soluble than naphthalene. Nitrogen-containing heterocyclic compounds should also have been used, but it was impossible to find a commercial laboratory that was immediately able to carry out this analysis quantitatively. Because the city water supply was shut down, bottled water was being used at great expense and inconvenience, and the replacement aquifer supply was still under development, it was important to carry out this project as diligently as possible.

C. The Results

Equipment rinse samples were generally found to contain low levels of phenols and occasionally some PAHs. Drill water samples showed trace levels of phenol (<0.2 μg/L) and cresol, but no PAHs. Distilled water exposed to diesel engine exhaust fumes resulted in low levels of fluorene and phenanthrene (0.17 and 0.29 μg/L, respectively). No PAH was detected in the water exposed to gasoline engine exhaust or to cigarette smoke.

Polynuclear aromatic hydrocarbons were found in two of the three sampling rounds, but the results of the two laboratories did not corroborate each other. One of the laboratories found phenanthrene, fluorene, and fluoranthene, whereas the other found only naphthalene. As with all split samples, this did not tell the reviewing team anything other than that the two labs did not agree. Both laboratories were requested to participate in a round-robin specifically designed to test for the analysis of low levels of PAHs in water, the results of which could be used to assess their capabilities and weaknesses. The next step was to question both laboratories about their procedures to try to explain these discrepancies.

Laboratory 1 did admit to having blank problems at the low level. A close look at their raw data and their procedure showed that although naphthalene was listed on their report sheet, their procedure did not measure it. For them, somehow, "not detected" and "not reported" were equivalent. Their internal quality control was insufficient; they did not add any surrogates to their samples, and the precision of their results could therefore not be assessed. Laboratory 2 used naphthalene-d8 as a surrogate and was reporting constant recoveries for all their samples. They also had a measurable naphthalene concentration in their laboratory blanks. The naphthalene concentration reported for the samples were three times as high. When questioned as to whether they regarded this as significant, they replied affirmatively. Upon further prodding by the review team, they explained that somehow they had not been told in advance that they were expected to analyze for PAHs and phenols and they had to carry out this analysis on 300 mL instead of the usual liter. In the review team's opinion, the naphthalene found in the blank was most likely to be from an impure deuterated surrogate standard. Because the sample size was one-third of normal, this number was multiplied by the dilution factor, 3, which accounted for the apparently significant naphthalene concentration. The dilution factor should have been noted on the laboratory report, but it wasn't.

Was this sufficient evidence to say that the earlier reports were false positives? Probably; but, in addition, no phenols or volatiles were detected in any of the samples. Without the presence of such mobile contaminants in the monitoring well network, the review team suspected that the reported occurrences of PAH "hits" were simply false positives.

D. Conclusion

This very expensive exercise, which cost several million dollars, could have been avoided if the initial sampling had been conducted with proper quality control. The choice of target parameters that are complementary is very important, as no single piece of evidence, even if it is acquired by GC/MS, is sufficient to state a conclusion with certainty. Maintaining a very close dialogue with the analyst and the sampling team is very important for the hydrogeologist, in order for her or him to know whether the samples were effectively collected, shipped, and analyzed.

GLOSSARY

Terms listed in the glossary are set in **boldface** type at first use in text.

Accuracy The amount of bias that a sample may be exposed to during sampling and laboratory analysis [27]; the degree of agreement of a measured value with the true value or expected value of the quantity of concern [1].

amu Atomic mass unit. (1 amu = 1 dalton (Da) = 1/12 mass of one ^{12}C atom.) The mass of a compound calculated from its formula using the lowest mass isotopes expressed in daltons is not to be confused with its molecular weight, which takes isotopic abundances into account.

Analog A compound that differs from another by only a few atoms and thus usually has very similar properties. For example, bromodichlorobenzene is an analog of trichlorobenzene; d-10 anthracene is an analog of anthracene.

Analyte The compound or parameter to be analyzed.

Blanks Samples of the matrix only, used to evaluate possible sources of contamination during sampling and analysis.

Blind sample A sample spiked with a known amount of a standard solution and sent to the laboratory within a batch of samples. Its use helps to detect any bias or systematic error of the laboratory. However, blind samples are usually very easily spotted by the laboratory because of their unusual mixtures. The same results can be obtained by sending a set of standards.

CAS number   In an ultimate attempt to solve the nomenclature problem, the Chemical Abstract Service has come up with a numbering system based on structure. It is an excellent idea to double check the CAS number to ensure that the same chemical is being referred to. One drawback is that very few people can construct a structure from the number. However, structures are easily retrieved from a computer.

Comparability   The ability to fairly compare sample test results taken from the same facility at different times [28].

Completeness   The number of samples that must be taken and analyzed before a confident judgment can be made that the groundwater conditions at a facility have been adequately assessed [28]; a measure of the amount of data obtained from a measurement process compared to the amount that was expected to be obtained under the conditions of measurement [1].

Field blank   A sample of purified water that is transferred to a sample bottle at the same time as the samples are collected, the purpose of which is to ensure that air contaminants are not introduced into the samples at the site. Fuel fumes from pumps and gasoline generators are a common source of problems.

GC/FT/IR   Gas chromatography coupled to Fourier-transformed infrared spectrometry. Spectra obtained in the Fourier domain can be easily accumulated; hence this is a method of enhancing the signal from the infrared spectrometer. When coupled to a gas chromatograph, the system can be used in much the same fashion as GC/MS is, with comparison of unknowns with a standardized library of spectra. Infrared spectra allow for the identification of types of functional groups in a molecule (alcohols, amines, ketones, etc.) as well as distinguishing between positional isomers such as *o*- and *m*-xylene).

GC/MS   Gas chromatography coupled to mass spectrometry. The gas chromatograph separates mixtures in the gas phase into their components. The mass spectrometer detects them and allows for their quantitation and identification.

HPLC   High-pressure liquid chromatography. The HPLC instrument separates mixtures of compounds dissolved in liquids. It is used mostly for aqueous samples and for thermally unstable compounds.

Isomers   Compounds that have the same molecular formula (and molecular weight, of course) but in which the functional groups or atoms are arranged differently, giving them different structures.

IUPAC   International Union of Pure and Applied Chemistry. This organization has proposed a standardized system of nomenclature.

Matrix   The material in which the analyte is dissolved. Water, soil, effluents, landfill leachates, and sludges are typical matrices in environmental analysis.

Matrix effect   Interference in the analysis due to interaction between the analyte and the matrix.

PAH   Polynuclear aromatic hydrocarbon—one of a group of compounds commonly found in fossil fuels; some have been found to be carcinogenic. The smallest of the group is naphthalene, used in crystal form as mothballs. Benzo[a]pyrene is usually the regulated one, with a drinking water limit of 10 ng/L (World Health Organization). Also referred to as polynuclear aromatics (PNAs).

PCP   Pentachlorophenol; compound used as a pesticide, mostly as a wood preservative.

Precision   The average amount of variability experienced in collecting and analyzing a sample, expressed as the relative standard deviation [28]; the degree to which data generated from replicate or repetitive measurements differ from one another [16].

QA/QC   See *Quality assurance* and *Quality control*.

Quality assurance   A system of activities whose purpose is to provide the producer or user of a product or a service the assurance that it meets defined standards of quality with a stated level of confidence [1].

Quality control   The overall system of activities whose purpose is to control the quality of a product or a service so that it meets the needs of the users. The aim is to provide quality that is satisfactory, adequate, dependable, and economical [1].

Reagent blank   An analysis carried out using all the reagents but no sample. It is used to ascertain the purity of the analytical reagents used in a specific method.

Replicate   In an experiment, a repeat using the same variables. In analysis, a repeat of the same sample using the same analytical conditions. "Duplicate" means two replicates.

Representativeness   A subjective assessment of whether the sample truly reflects the groundwater in a particular hydrogeologic unit at a particular location [28].

Round-robin   Interlaboratory study in which a group of laboratories all analyze a subsample from a large homogeneous sample in order to compare their methodology and their accuracy.

Scan   In mass spectrometry, a given range of masses (e.g., 50–450 amu or daltons) acquired in a specified time (e.g., 1 sec). In other spectroscopies, the acquisition of a range of wavelengths; for example, an ultraviolet scan refers to the acquisition of an adsorbance spectrum from 200 to 650 nm.

SIM   In mass spectrometry, selected ion monitoring. In contrast to the scan mode, only a few masses (e.g., 78,151,153,173) are acquired. This takes only a fraction of a second and allows several acquisitions to be performed in 1 sec, enhancing the sensitivity for these ions. Not to be confused with SIMS.

SIMS   Secondary ion mass spectrometry, a technique in which nonvolatile compounds are ionized on a metal surface [9].

Spiked sample   A sample to which a known amount of analyte is added to measure the matrix effects on the analytical methodology. A large sample is divided in two; half is spiked, the other half is left intact, and both samples are analyzed in parallel.

Standard   A solution of known composition made from the purest available chemical and used to calibrate an instrument for quantitative analysis.

Surrogate   A compound that is representative yet different (often deuterated or fluorinated) from the target analytes and is added to the sample prior to extraction. Its recovery is indicative of a matrix effect.

TAGA   Target atmospheric gas analyzer, a mass spectrometer with the source at ambient pressure, which allows instant on-site monitoring of atmospheric gases.

Trip blank   A sample of laboratory-purified water that travels unopened from and back to the laboratory along with the sample bottles, to find out about possible contamination in transit. See also *field blank.*

Volatiles   A group of compounds that can be analyzed by purging them out of water with a stream of inert gas at room temperature. Their boiling point is below 150°C.

REFERENCES

1. Taylor, J.K. *Quality Assurance of Chemical Measurements*, Lewis, Chelsea, MI, 1987.
2. *CRC Handbook of Chemistry and Physics*. Published yearly by The Chemical Rubber Co., Cleveland, OH.
3. Montgomery, J.H., and Welkom, L.M. *Groundwater Chemicals Desk Reference*, Lewis, Chelsea, MI, 1990.
4. Budavari, S. (ed.) *The Merck Index*, 11th ed. Merck, Rathway, NJ, 1989.
5. Aldrich Chemical Company Inc. *Catalog Handbook of Fine Chemicals*, Milwaukee, WI, 1988/89.
6. Koehn, J.W., and Stanko, J.H., Jr. Groundwater monitoring. *Environ. Sci. Technol., 22*:1262–1264, 1988.
7. Priddle, M.W., Lesage, S., and Jackson, R.E. Analysis of oxygenated solvents in groundwater by dynamic thermal stripping-GC-MSD. *Int. J. Environ. Anal. Chem.*, 1992, in press.

8. Swallow, K.C., Shiffrin, N.S., and Doherty, P.J. Hazardous organic compound analysis. *Environ. Sci. Technol., 22*:136–142, 1988.

9. Hass, J.R., and Norwood, D.L. Organic mass spectrometry. In *Water Analysis*, Vol. 3 (R.A. Minear and L.H. Keith, eds.), Academic, Orlando, FL, 1984, pp. 253–316.

10. Jackson, R.E., Mutch, J.P., and Priddle, M.W. Persistence of aldicarb residues in the sandstone aquifer of Prince Edward Island, Canada. *J. Contaminant Hydrol., 6*:21, 1990.

11. Lesage, S., Ritch, J.K., and Treciokas, E.J. Characterization of groundwater contaminants at Elmira, Ontario, by thermal desorption, solvent extraction GC-MS and HPLC. *Water Pollution Res. J. Canada, 25*:275–292, 1990.

12. Covey, T.R., Lee, E.D., Bruins, A.P., and Henion, J.D. Liquid chromatography/mass spectrometry. *Anal. Chem., 58*:1451–1461, 1986.

13. Gwka, D.F., and Hiatt, M. Analysis of hazardous wastes and environmental extracts by GC/FTIR and GC/MS. *Anal. Chem., 56*:1102, 1984.

14. Shafer, K.H., Hayes, T.L., Brasch, J.W., and Jakobsen, R.J. Analysis of hazardous wastes by fused-silica capillary GC/FTIR and GC/MS. *Anal. Chem., 56*:237, 1984.

15. U.S. Environmental Protection Agency. *Federal Register* Part VIII. 40 CFR Part 136, Guidelines establishing test procedures for the analysis of pollutants under the Clean Water Act; final rule and interim final rule and proposed rule, Vol. 49, No. 209, Oct. 26, 1984.

16. American Chemical Society Committee Report. Principles of environmental analysis. *Anal.Chem., 55*:2210–2218, 1983.

17. U.S. Environmental Protection Agency. *Test Methods for Evaluating Solid Wastes Physical/Chemical Methods (SW-846)*, 3rd ed., 1987.

18. Grisak, G.E., Jackson, R.E., and Pickens, J.F. Monitoring ground-water quality: the technical difficulties. In *Establishment of Water Quality Monitoring Programs* (L.G. Everett and K.D. Schmidt, eds.), AWRA, Minneapolis, MN, 1978, pp. 210–232.

19. Reilly, T.E., Franke, O.L., and Bennett, G.D. Bias in ground-water samples caused by wellbore flow. *J. Hydraul. Eng., 115*:270–276, 1989.

20. Lesage, S., Jackson, R.E., Priddle, M.W., and Riemann, P.G. Occurrence and fate of organic solvent residues in anoxic groundwater at the Gloucester landfill, Canada. *Environ. Sci. Technol., 24*:559–566, 1990.

21. Cowgill, U.M. Sampling waters. In *Principles of Environmental Sampling* (L.H. Keith, ed.), American Chemical Society, Washington, DC, 1988, pp. 171–189.

22. Barcelona, M.J., Helfrich, J.A., Garske, E.E., and Gibb, J.P. A laboratory evaluation of ground water sampling mechanisms. *Ground Water Monit. Rev., 4*:32–41, 1984.

23. Barcelona, M.J., and Helfrich, J.A. Well construction and purging effects on ground-water samples. *Environ. Sci. Technol., 20*:1179–1184, 1986.

24. Robin, M.J.L., and Gillham, R.W. Field evaluation of well purging procedures. *Ground Water Monit. Rev., 7*:85–90, 1987.

25. Barcelona, M.J., Gibb, J.P., Helfrich, J.A., and Garske, E.E. *Practical Guide to Ground-Water Sampling*, Illinois State Water Survey, Champaign, IL, 1985.

26. Jackson, R.E., and Patterson, R.J. Interpretation of pH and Eh trends in a fluvial sand aquifer system. *Water Resources Res., 18*:1255, 1982.
27. Kent, R.T., and Payne, K.E. Sampling groundwater monitoring wells. In *Principles of Environmental Sampling* (L.H. Keith, ed.), American Chemical Society, Washington, DC, 1988, pp. 321-246.
28. GAO. *Ground-Water Conditions at Many Land Disposal Facilities Remain Uncertain*, U.S. General Accounting Office, Washington, DC, 1988.
29. Barcelona, M.J., Holm, T.R., Schock, M.R., and George, G.K. *Water Resources Res., 25*(5):991-1003, 1989.
30. WMS Associates Ltd., Intera Technologies, and National Water Research Institute. Investigation of suspected PAH contamination of groundwater, Newcastle, New Brunswick. Report to the New Brunswick Ministry of the Environment, WMS file 2988, Fredericton, New Brunswick, Canada, 1989.
31. Lesage, S., and Lapcevic, P.A. Differentiation of the origins of BTX in ground water using multivariate plots. *Ground Water Monit. Rev., Spring*:102-106, 1990.

2
Nonpriority Pollutant Analysis and Interpretation

Kathleen C. Swallow

Merrimack College, North Andover, Massachusetts

I. INTRODUCTION

Characterization of the nature and extent of the groundwater contamination at a hazardous waste site is a difficult task requiring the qualitative and quantitative analysis of complex mixtures of compounds in dilute aqueous solutions. The preliminary chemical analysis of water samples from a hazardous waste site often involves application of broad-spectrum gas chromatography/mass spectroscopy (GC/MS) methods such as those developed and mandated by the U.S. EPA for priority pollutants [1]. These broad-spectrum methods were developed to "encompass a wide range of target and nontarget analytes having a broad range of chemical structures [and] allow the identification and measurement of hundreds of compounds in several types of water samples with a few standardized extraction, concentration, and GS/MC procedures" [2].

Although GC/MS screening methods can provide excellent qualitative and quantitative data on the priority pollutants, these compounds are infrequently detected in landfill leachate. A review of the groundwater monitoring data from 156 Resource Conservation and Recovery Act (RCRA) sites and 178 Comprehensive Environmental Response, Compensation, and

Liability Act (CERCLA) sites [3] showed that of the compounds included in the broad-spectrum GC/MS methods, only compounds in the volatiles fraction were routinely detected, with frequencies of 12% in the CERCLA site data and 4% in the RCRA site data. The acid-extractables were detected with a frequency of 2% at CERCLA sites and 4% at RCRA sites, whereas pesticides and base/neutral compounds were detected with a frequency of 1% at both types of sites.

The data from a broad-spectrum screening program cannot completely or even adequately describe the nature and extent of the contamination at a hazardous waste site. Most of the compounds present in contaminated groundwater are not the priority pollutants for which these analytical procedures were developed. An abundance of water-soluble compounds containing polar functional groups, especially phenols and organic acids, is characteristic of leachate from both sanitary and hazardous waste landfills. Only a few of the original 229 priority pollutants are in the acid-extractable fraction that contains the phenols and organic acids. Yet in a comparative study of eight landfill leachates it was found that nonpriority pollutant organic acids made up approximately 90% of the total organics [4].

Nonpriority pollutants are detected in broad-spectrum screening analyses, but their identities are unknown. Tentative identifications of nonpriority pollutants can be made by computerized matching of their mass spectra to library spectra. The tentative identifications are often inaccurate, inconsistent, and misleading, with the result that most of the nonpriority pollutant compounds detected in landfill leachate are misidentified or unidentified [5]. In one set of landfill leachate analyses, for example, the 1,1,3-trichlorobutadiene present in the contaminated groundwater was detected in many of the wells and gave a clear, recognizable mass spectrum. The compound was correctly identified by the computer in some of the wells, but in other wells it was identified as 2,3-dichloro-1,4-dioxane, dichlorophenylborane, or bromobenzene.

These misidentifications are in part due to the inherent problems with computerized library matching. Successful library matching depends upon the close similarity between the mass spectrum generated for a compound in the sample and the spectrum for the pure compound stored in the computer's library. Relatively small differences between the mass spectrum of the pure compound and the spectrum of the compound in the sample matrix can lead to misidentification of the compounds. Such differences result from background interference, poor chromatographic resolution of closely eluting compounds, or instrument variability. The libraries stored in the computer also contain a limited number of spectra, and therefore not all of the possible compounds in a leachate sample can be accurately identified by computer matching.

Failure to identify nonpriority pollutant compounds may also be a consequence of the application of GC/MS analysis to samples for which this analytical technique has serious drawbacks. At a southern California dump site, the California Department of Health Services found that 95% of the organic compounds found in the groundwater were not amenable to GC/MS analysis. Using liquid chromatography/mass spectroscopy (LC/MS), they determined that a nonpriority pollutant, p-chlorobenzenesulfonic acid, which is used in DDT production, was the principal contaminant [6].

Water-soluble compounds often have relatively low volatilities, and gas chromatography, which depends on compounds being transported in the gas phase, is not the best analytical method available for them. Even volatile water-soluble compounds, however, contain polar functional groups that cause problems in broad-spectrum GC methods designed to detect and measure a wide range of compounds in a single analysis. The recovery of polar compounds in the extraction step of these methods is poor, often less than 20%. Polar compounds also produce broad, tailing gas chromatographic peaks that give poor quantitative results and are difficult to identify by mass spectrometry.

Although polar compounds can be successfully analyzed using gas chromatography, special techniques and equipment are required for optimum results. Chemical derivatization of sample components has been used to eliminate much of the peak tailing caused by strong interactions between polar functional groups and the chromatographic stationary phase. Column packings and coatings specifically formulated to minimize interactions between either acids or bases and the column can also minimize peak tailing without derivatization. The broad-spectrum GC/MS screening methods are not optimized for polar compounds, which comprise the bulk of the contaminants in landfill leachate.

The results from initial broad-spectrum screening analyses should be used to develop analytical programs for ongoing monitoring that are optimized for the major components found in the contaminated groundwater. Analytical techniques other than GC/MS can often provide better data faster and at a considerably lower cost. Too often, however, ongoing monitoring programs rely on periodic repetition of the broad-spectrum screening methods. The practice is costly and ineffective [7]. Moreover, the data acquired for site assessment, risk assessment, assignment of liability, and decisions on cleanup strategies are limited and often woefully inadequate.

II. PRIORITY POLLUTANT VERSUS NONPRIORITY POLLUTANT ANALYSIS

The priority pollutants are compounds that may pose a threat to human health and the environment because they are toxic. The ubiquitous and abundant nonpriority pollutant organic acids and phenols found in landfill leachate are probably not as toxic. Because a few toxic priority pollutants may pose a greater risk than a stew of relatively harmless acids and phenols, more effort is put into the qualitative and quantitative analysis of the priority pollutants. Although this approach is efficient for regulatory purposes, it may be misleading in assessing the risk to human health. Toxicity testing is expensive and time-consuming, and only relatively few of the compounds to which human beings may be exposed have been tested. Nonpriority pollutants may also be toxic, and they are frequently present in higher concentrations than the priority pollutants. Unless they are properly identified, the extent of risk is unknown.

Adequate analysis of the nonpriority pollutants in contaminated groundwater is essential for the determination of the extent of the contamination and the best approach to remediation and ongoing monitoring programs at a hazardous waste site. The processes that produce landfill leachate favor production of organic acids and phenols, many of which are daughter compounds of the original wastes deposited in the landfill.

A. The Nature of Landfill Leachate

The process of contaminant plume formation and migration at a hazardous waste site explains the observation that 90% of the organic compounds detected in landfill leachates are organic acids and phenols. The contaminant plume from a landfill emanates in two stages [8]. When groundwater first comes into contact with the buried waste, water-soluble compounds are rapidly leached out and dissolved in the water. Because polar organic compounds such as acids and phenols are more water-soluble than nonpolar compounds such as hydrocarbons, they are leached out more rapidly and efficiently and comprise the bulk of the organic load of the initial plume.

The contaminant plume extends outward from the point of deposition as the groundwater flows, carrying the dissolved compounds with it. The initial surge of water-soluble components in the plume moves away from the site fairly quickly, whereas the hydrophobic compounds move away more slowly. This occurs because as the groundwater moves away from the site, the compounds dissolved in it partition between the moving water and the stationary soil components in processes analogous to chromatography. Nonpolar, hydrophobic compounds in the plume have a high affinity for the naturally occurring soil organic carbon fraction and tend to be strongly sorbed to the stationary phase, whereas polar, hydrophilic compounds tend to remain dissolved in the aqueous phase. Because sorption by the soil retards the rate of migration of compounds away from the site, the nonpolar compounds move away more slowly than the polar, water-soluble compounds. The plume is not a homogeneous solution of compounds leached from the site but contains relatively higher concentrations of water-soluble compounds such as acids and phenols as it moves further from the site.

In the second stage of leachate formation, the water-soluble compounds in the plume are continually replenished by chemical hydrolysis and biodegradation. In chemical hydrolysis, molecules react with water and acquire polar functional groups that increase their water solubility. In the process of biodegradation, soil microorganisms convert nonpolar, hydrophobic organic compounds into water-soluble compounds as a first step in the mineralization process.

For example, both aerobic and anaerobic organisms can degrade toluene, a nonpolar, relatively water-insoluble compound. The aerobic processes proceed through production of dihydroxylated intermediates such as catechol.

Toluene   Catechol

Anaerobic processes appear to proceed either through oxidation of the methyl group to produce benzyl alcohol,

Toluene   Benzyl Alcohol

or through oxidation of the ring to produce *ortho-* or *para*-cresol.

Toluene   o-Cresol   p-Cresol

Further reactions produce a diverse mixture of polar compounds including benzaldehyde, benzoic acid, 2-hydroxybenzoic acid, phenol, and aliphatic acids until the organic compounds are mineralized or completely degraded to carbon dioxide [9]. The production of water-soluble compounds from hydrophobic compounds continually renews the leachate plume and allows the partially degraded, polar, water-soluble contaminants to move away from the site as they dissolve in the groundwater.

B. The Use of Nonpriority Pollutants in a Landfill Study

The value and importance of good nonpriority pollutant analysis is illustrated by a complex site study involving adjacent landfills. The discovery of anthropogenic organic compounds in the groundwater in a neighborhood located downgradient from a permitted RCRA hazardous waste landfill led to allegations that the facility was leaking. An investigation of the allegation was complicated by the presence of an old, unlined hazardous waste dump immediately upgradient of the RCRA landfill. The old landfill was known to have produced a leachate plume in the aquifer. The source of the contaminants in the groundwater could have been either the old landfill, the RCRA landfill, or both or neither.

The RCRA landfill had been monitored regularly, and data were available for several years. The landfill was separated into a number of cells containing different types of hazardous waste. Within each cell samples had been collected from standpipes connected to the leachate collection system. Samples had also been collected from monitoring wells outside the landfill cells, both upgradient and downgradient. The samples had been analyzed for the full spectrum of priority pollutants. Organic unknowns present in significant concentrations had been tentatively identified by computer matching of their mass spectra to library spectra. Although some priority pollutants had been detected in the samples, most of the compounds present were tentatively identified as nonpriority pollutants.

The old landfill had also been monitored. There was no leachate collection system, so all samples were from monitoring wells, either within the landfill area or downgradient from it. These samples had also been analyzed for priority pollutants, and the unknowns had been tentatively identified.

To differentiate the plume emanating from the old landfill from any possible leakage from the RCRA facility, it was necessary to find separate suites of indicator compounds unique to each landfill. Then the compounds found in the wells in the neighborhood downgradient from the two sites could be compared to the indicator lists to determine which landfill was the potential source of the contamination.

The priority pollutants detected at each site were compared to develop a list of compounds unique to each. With very few exceptions, identical suites of priority pollutants, mostly common industrial solvents found in the volatiles fraction, were found in samples from both landfills. Using only the priority pollutants, it was impossible to distinguish the plume from the old landfill from any potential leakage from the RCRA facility. It was necessary to include some nonpriority pollutants, the tentatively identified compounds, to develop suites of indicator compounds unique to each of the landfills.

The gas chromatograms from the semivolatile fraction of samples from six monitoring wells located downgradient from both landfills were visually compared. All of the samples had strikingly similar elution patterns, shown in Fig. 1, suggesting that the samples contained similar suites of compounds. Between 15 and 20 peaks were found that appeared at the same retention times and the same positions relative to the other peaks in most of the chromatograms.

The only priority pollutant detected in these samples was phenol, which was found in a mixture with other polar compounds in the large tailing peak at the beginning of each chromatogram. Except for the phenol detected in the initial peak, all of the compounds in these samples were nonpriority pollutants.

The nine numbered peaks in Fig. 1 are all nonpriority pollutants that produced virtually identical mass spectra in every sample. The computer-generated tentative identifications of these compounds are given in Table 1. Except for peaks 3 and 7, each compound was identified as two or three different compounds in different samples, despite the fact that their mass spectra looked identical.

Table 2 gives the tentative identifications for the numbered compounds in duplicate samples from well E. Even in duplicate samples from the same well, whose mass chromatograms had the same peaks, different tentative identifications of the compounds were made by computerized library matching.

The reason for this inconsistency is illustrated in Fig. 2. The first mass spectrum is that of the compound producing peak 1 in the mass chromatograms. The other three spectra are computer-generated library matches for this compound. Although each matched spectrum has striking similarities to the unknown spectrum, none is identical. Small differences in the sample spectrum can cause the computer to choose a different compound as the best match in different samples. In fact, none of these tentative identifications appears to be correct. The library probably did not contain a known spectrum for this compound, and its identity remains unknown. The same is true for the other peaks; this led to the overall conclusion that none of the tentative identifications of any of the unknown compounds was likely to be correct.

Even without knowing the identities of the nonpriority pollutants, however, it was possible to recognize their chromatographic peaks and their mass spectra, which were highly reproducible in samples taken from the different

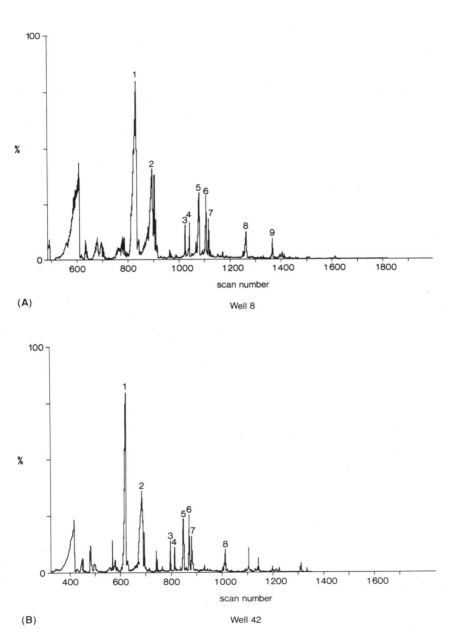

Figure 1 Mass chromatograms of samples taken from six monitoring wells located downgradient from an RCRA landfill.

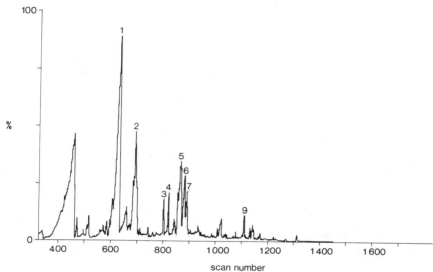

(C)

Well 19SR

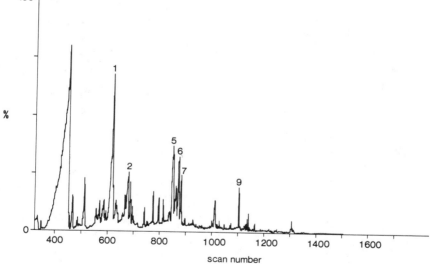

(D)

Well 26CR

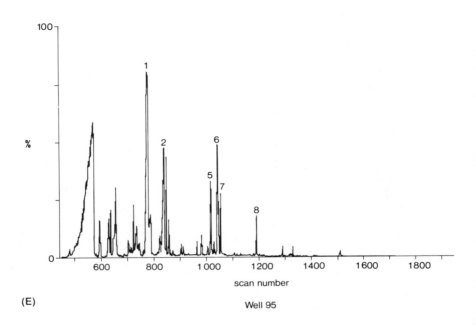

(E)   Well 95

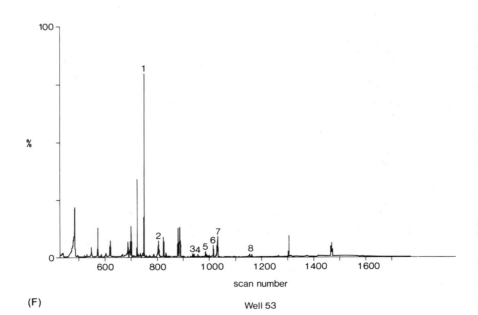

(F)   Well 53

Figure 1 Continued

Table 1 Monitoring Well Compound Identifications

Peak	Well A	Well B	Well C	Well D	Well E	Well F
1	Heptanenitrile	Heptanenitrile	6-Cyanohexanoic acid	Heptanenitrile	Hepatnenitrile	1-Azabicyclo-[3.1.0]hexane
2	3-Hydroxy-3-methyl-2-heptanone	3-Hydroxy-3-methyl-2-heptanone	5-Methyl-5-nonanol	3-Hydroxy-3-methyl-2-heptanone	3-Hydroxy-3-methyl-2-heptanone	3-Hydroxy-3-methyl-2-heptanone
3	4-Isopentyl-1,3-cyclopentane-dione	4-Isopentyl-1,3-cyclopentane-dione	4-Isopentyl-1,3-cyclopentane-dione	Not found	Not found	4-Isopentyl-1,3-cyclopentane-dione
4	1-Azabicyclo-[3.1.0]hexane	1-Azabicyclo-[3.1.0]hexane	Not found	Not found	Not found	1-Ethyl-3-methyl-*trans*-cyclopentane
5	3-Oxo-6-heptenoic acid methyl ester	9-Hydroxy-2-nonanone	3-Oxo-6-heptenoic acid methyl ester	3-Oxo-6-heptenoic acid methyl ester	3-Oxo-6-heptenoic acid methyl ester	2-(Cyclohexyloxy)-ethanol
6	Octahydro-1,5-dinitroso-1,5-diazocine	2,3,4A,5,6,7-Hexahydro-1,4-benzodioxin	1,3,4A,5,6,7-Hexahydro-1,4-benzodioxin	2,3,4A,5,6,7-Hexahydro-1,4-benzodioxin	Octahydro-1,5-dinitroso-1,5-diazocine	2,3,4A,5,6,7-Hexahydro-1,4-benzodioxin
7	1-(1-Oxopentyl)pyrrolidine	1-(1-Oxopentyl)pyrrolidine	1-(1-Oxopentyl)pyrrolidine	Not found	1-(1-Oxopentyl)pyrrolidine	1-(1-Oxopentyl)pyrrolidine
8	2-Methyl-hexane-dinitrile	2-Methyl-hexane-dinitrile	Not found	N-Cyclopentyli-dene-ethanamine	Octanenitrile	N-Cyclopentyli-dene-ethanamine
9	(2-Bromoethyl)-cyclohexane	Not found	1,1',1"-(1-Ethanyl-2-ylidene)tris-cyclohexane	1,1',1"-(1-Ethanyl-2-ylidene)tris-cyclohexane	Not found	Not found

Table 2   Well E Duplicate Sample: Comparison of Compound Identifications

Peak	Well E	Well E duplicate
1	Heptanenitrile	6-Cyanohexanoic acid
2	3-Hydroxy-3-methyl-2-heptanone	Not found
3	Not found	Not found
4	Not found	1-Azabicyclo[3.1.0]hexane
5	3-Oxo-6-heptenoic acid methyl ester	3-Oxo-6-heptenoic acid methyl ester
6	Octahydro-1,5-dinitroso-1,5-diazocine	Octahydro-1,5-dinitroso-1,5-diazocine
7	1-(1-Oxopentyl)pyrrolidine	1-(1-Oxopentyl)pyrrolidine
8	Octanenitrile	N-Cyclopentylidine-ethanamine
9	Not found	Not found

wells and also over a period of several years. The data were therefore used like fingerprints, and the unknown compounds were assigned code names. The same process of identifying recurring compounds, verifying that they yielded reproducible mass spectra, and assigning them code names was repeated for samples from the RCRA facility standpipes located in the leachate collection system and for samples taken from monitoring wells inside the old landfill and upgradient from the RCRA landfill. In this way suites of nonpriority pollutants found in each landfill were identified. When the nonpriority pollutant compounds found in the old landfill were compared to those found in the RCRA facility, there was no correlation. Each set of nonpriority pollutants was unique to one landfill.

Before the nonpriority pollutants were used as indicator compounds, an attempt was made to correctly identify as many of them as possible. A group of mass spectroscopists interpreted the mass spectra and identified many of the compounds. Analytical standards were obtained for the identified compounds so that in future monitoring rounds nonpriority pollutant analysis would be as accurate as priority pollutant analysis. Unfortunately, it is not always possible to identify unknown organic compounds from mass spectral data alone, and several important compounds, including the major component in samples from the old landfill, were not identified. GC/MS is a powerful technique, and GC/MS data can provide remarkably clear insights into the complex chemical processes occurring in contaminated water [10]. Broad-spectrum screening tests, however, are only a first step. They must be followed by refinement of the analytical methods and careful evaluation and interpretation of the data.

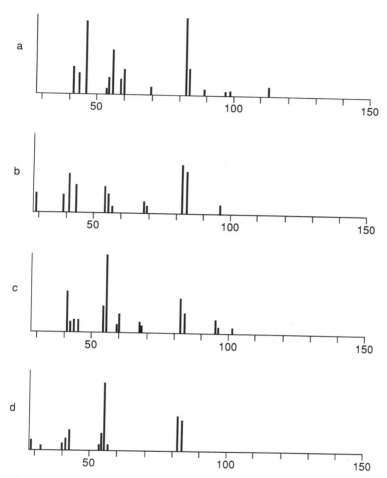

Figure 2 The mass spectra for the unknown compound represented by peak 1 in wells A–F and the three compounds reported as tentative identifications of the unknown compound. (a) Peak 1 in wells A–F; (b) heptanenitrile; (c) 6-cyanohexanoic acid; (d) 1-azabicyclo[3.1.0]hexane.

As the RCRA facility had independent cells, any of which could have been leaking, it was necessary to find an indicator compound for each cell. When the list of the identified nonpriority pollutant indicator compounds was compiled, it did not include a compound from each cell. It was necessary to use some unidentified compounds with only code names as indicator compounds. The retention times of the chromatographic peaks and the characteristic mass spectra were used as fingerprints to identify them in the samples. Semiquantitative data were reported on the basis of an assigned chromatographic response factor of 1 for the unknown compounds.

Four rounds of monitoring were conducted using the modified analytical procedures developed to include the nonpriority pollutant indicator compounds. The compounds found in samples from the old landfill and the RCRA facility were compared to the compounds found in samples from the monitoring wells downgradient from both facilities. Only compounds unique to the old landfill were found downgradient from the two sites. There was no evidence that the RCRA facility had contributed any contaminants to the groundwater downgradient, and the allegation that the RCRA facility was leaking was proven to be false.

In addition to answering the question of whether the RCRA facility was leaking, this program demonstrated that the extent of the plume emanating from the old landfill was greater than had been previously thought. Mapping of the plume using only priority pollutants had underestimated the distance the contaminants had migrated because it had relied on only relatively insoluble compounds. The major component of the old landfill leachate produced a tailing chromatographic peak characteristic of a polar compound such as an organic acid. The same compound was found in downgradient monitoring wells and is responsible for the peaks labeled 1 in Fig. 1. This compound was present in high concentration relative to other components in many of the monitoring wells and was found in wells further downgradient than any of the other indicator compounds, suggesting that it was watersoluble. The inclusion of this indicator compound, even though its identify was unknown, was essential to determining the extent of the contaminant plume from the old landfill.

III. EFFECTIVE NONPRIORITY POLLUTANT ANALYSIS

The approach taken to the identification and analysis of nonpriority pollutants in this case study was extremely labor-intensive and took several years to complete. It was necessary because the only data available for compound identification were broad-screening GC/MS data. A well-designed analytical program is not limited to GC/MS. Because the water-soluble compounds found in groundwater are often nonvolatile, gas chromatography is not always the best available separation technique. New liquid chromatographic methods, which are better suited to these compounds, have been developed and successfully applied to groundwater analysis [2,6]. In addition to mass spectroscopy, other spectroscopic techniques, including infrared and nuclear magnetic resonance spectroscopy, can be used to identify compounds that cannot be identified by mass spectrometry alone. Had information about the nature of wastes deposited at the site been used to help design a comprehensive analytical program including liquid chromatography as well as gas chromatography and infrared and nuclear magnetic resonance spectroscopy in addition to mass spectrometry, the number of identified compounds in the leachate samples would undoubtedly have increased. Although this may sound more complex and time-consuming than using GC/MS alone, better results probably would have been obtained faster and at a lower cost with a well-designed analytical approach.

Regardless of the analytical technique employed, however, successful nonpriority pollutant analysis requires the application of chemical expertise to a complex problem. Too often site assessment data are virtually worthless because standard methods are used without regard to the specific nature of the problem and the skill and judgment of the chemist are replaced by automated instruments and generalized methods.

REFERENCES

1. *Federal Register, 44*:69464–69575, 1979.
2. Bellar, T.A., and Budde, W.L. Determination of nonvolatile organic components in aqueous environmental samples using liquid chromatography/mass spectrometry. *Anal. Chem., 60*:2076, 1988.
3. Plumb, R.H., Jr. A comparison of ground water monitoring data from CERCLA and RCRA sites. *Ground Water Monit. Rev., 7*(4):94, 1987.
4. Johansen, O.J., and Carlson, D.A. Characterization of sanitary landfill leachates. *Water Res., 10*:1129, 1976.
5. Swallow, K.C., Shifrin, N.S., and Doherty, P.J. Hazardous organic compound analysis. *Environ. Sci. Technol., 22*:136, 1988.
6. Finnegan, R.E., and Poppiti, J. Analytical instrumentation: a look to the future. *Environ. Lab., 1*(4):28, 1989.
7. Koehn, J.W., and Stanko, G.H., Jr. Groundwater monitoring. *Environ. Sci. Technol., 22*:1262, 1988.
8. Robertson, J.M., Toussaint, C.R., and Jorque, M.A. *Organic Compounds Entering Ground Water from a Landfill.* U.S. Environmental Protection Agency, Ada, OK, EPA-660/2-74-077, 1974.
9. Grbic-Galic, D., and Vogel, T.M. Transformation of toluene and benzene by mixed methanogenic cultures. *Appl. Environ. Microbiol., 53*(2):254, 1987.
10. Hites, R.A., and Eisenreich, S.J. (eds.). *Sources and Fates of Aquatic Pollutants*, American Chemical Society, Washington, DC, 1987.

3

Establishing an Analytical Protocol for the Measurement of EPA's Appendix IX List of Compounds

Jerry L. Parr, Gary K. Ward, and Gary Walters

Enseco Incorporated, Arvada, Colorado

I. INTRODUCTION

On July 9, 1987, the U.S. Environmental Protection Agency (EPA) promulgated a regulation pertaining to groundwater monitoring at hazardous waste facilities (52 *Federal Register* 25942). This regulation requires the analysis of groundwater for a list of 232 specific constituents, listed in Appendix IX of 40 Code of Federal Regulations (CFR) part 264. This Appendix IX list consists of metals, anions, and a wide variety of organic compounds, including nitrosamines, phenols, polynuclear aromatic hydrocarbons, volatile organics, pesticides, herbicides, and chlorinated dioxins.

Although the EPA provided some guidance in the regulation as to the analytical approach, the guidance did not provide enough detail to clearly specify an approach that could be used by laboratories to generate reliable data. Enseco, as a network of environmental laboratories, needed one approach that could be used consistently by each laboratory to generate comparable data. Thus, a series of activities were initiated to develop a standardized approach for measuring the components in this list.

The process used is illustrative of the processes that must be performed in order to establish a standardized protocol to measure environmental pollutants in compliance with a regulation. In summary, a protocol was established for measuring Appendix IX constituents:

- Review the regulation and related background documents in detail.
- Establish a definitive list of analytes from an analytical chemist's perspective.
- Assess the adequacy of analytical technologies relative to these analytes.
- Select methods based on general performance characteristics.
- Document performance characteristics for each analyte by laboratory studies.
- Establish quality control activities for each method.
- Define sample containers, preservatives, and holding time.

II. REGULATORY OVERVIEW

In 1976, the U.S. Congress enacted legislation to remediate problems associated with the mismanagement of hazardous wastes. This legislation, the Resource and Conservation Recovery Act (RCRA), requires that generators, transporters, and disposers of hazardous waste comply with regulations developed by the U.S. Environmental Protection Agency (EPA). As part of this comprehensive program, the EPA in July 1982 promulgated a regulation for the protection of groundwater from releases of hazardous wastes at treatment, storage, and disposal units (TSDs).

This regulation, contained in the Code of Federal Regulations (CFR), Title 40, Part 264, requires that groundwater be assessed periodically to determine if releases from the TSD units are occurring. This assessment is performed by analyzing for a specified list of hazardous constituents. The 1982 regulation required that analyses be conducted for approximately 400 constituents listed in the regulation (47 FR 32274, July 26, 1982). This list, contained in Appendix VIII of 40 CFR Part 261, contained metallic salts, mixtures, gases, exotic organic compounds, and other materials. In 1987, because of the difficulties in measuring the constituents on this list [1,2], the EPA developed a new list (52 FR 25942, July 9, 1987). This list, codified as Appendix IX to 40 CFR Part 264, contained the Appendix VIII constituents that could be measured using existing technology.

The 1987 rulemaking established a list of 222 entries: 17 metals, 2 anions, 199 specific organic compounds, and 4 entries pertaining to groups of compounds. For each entry, the list contains

- Common names
- *Chemical Abstracts* registry numbers
- *Chemical Abstracts* names
- Suggested EPA methods
- Practical quantitation limits (PQL)

In the regulation, the EPA stated, "The methods . . . may not always be the most suitable method(s) for monitoring an analyte" and "the PQL values in many cases are based only on a general estimate for the method and not on a determination for individual compounds; PQLs are not part of the regulation." In the preamble to the rulemaking, EPA provided additional guidance to laboratories and indicated that additional information was available in the public docket associated with the regulation.

Staff within Enseco had been involved in this regulatory process since 1982, providing comments to EPA for the American Petroleum Institute [3–5], conducting research projects concerning the measurement of Appendix VIII and Appendix IX analytes [6], and participating in an EPA Work Group.

Thus, at the time of the final rule, extensive information existed to support a capability to measure the Appendix IX constituents, but this information had not been standardized to the degree necessary to conduct routine analyses.

III. ESTABLISHMENT OF AN ANALYTE LIST

As discussed previously, Appendix IX consists of 222 entries, including four that are groups of compounds. Virtually all entries contain two or more names. For example, entry 35 contains the names bromoform and tribromomethane as common names and the name methane, tribromo as the CAS name; entry 5 contains the names acetonitrile and methyl cyanide as common names; and entry 172 contains the name polychlorinated biphenyls with a footnote that states, "This category contains congener chemicals, including constituents of Aroclor-1016 . . . ."

Thus, the initial goal was to define a list of analytes for reporting purposes that would (1) contain analyte names understandable by the typical data user, (2) contain only one name for a given compound, and (3) result in specific analytes for the four generic groups. For example, we selected the name bromoform because it is more recognizable than the name tribromomethane. We also selected the name bis(2-chloroisopropyl)ether for entry 32, even though EPA did not list this name among their three names—(bis)2-chloro-1-methylethyl)ether; 2,2'-dichlorodiisopropylether; and propane, 2,2'-ox-bis(1-chloro-)].

The four generic groups required a different solution. For each group, we selected analyte(s) based on accepted practices of the environmental testing industry as follows:

- *Polychlorinated biphenyls*: Measurement of Aroclors 1016, 1222, 1232, 1242, 1248, 1254, and 1260
- *Polychlorinated dibenzo-p-dioxins*: Measurement of tetrachlorodibenzo-p-dioxins, pentachlorodibenzo-p-dioxins, and hexachlorodibenzo-p-dioxins

- *Polychlorinated dibenzofuran*: Measurement of tetrachlorodibenzofur-ans, pentachlorodibenzofurans, and hexachlorodibenzofurans
- *Xylenes* (*total*): Sum of the measurement of *o*-xylene, *p*-xylene, and *m*-xylene

As a result of this exercise, a total of 232 specific analytes were obtained. [Note: As a result of the analytical method selection process, further changes were made in the list. Specifically, *o*-cresol and *p*-cresol were combined into one entry (3/4-methyl phenol) owing to coelution under the analytical conditions, and chlordane was changed to α-chlordane and γ-chlordane (two entires) to reflect current analytical practices for this multicomponent mixture under other EPA programs.]

IV. SELECTION OF ANALYTICAL METHODS

For each entry on the Appendix IX list, EPA suggested the use of one or more SW-846 methods. SW-846 is the methods manual developed by EPA's Office of Solid Waste for compliance monitoring under RCRA. A total of 40 different methods were cited in the regulation as methods that would "provide acceptable analytical results." Furthermore, EPA stated that these methods were not required, nor were they part of the regulation, and also stated that "some of the methods described in SW-846 are less appropriate for screening and are better suited to routine monitoring where the analyst is sure which compounds are present." Thus, laboratories were left to their own as to whether or not to use SW-846 methods and which SW-846 methods to use.

For most Appendix IX entries, two or more methods were cited. For example, Methods 6010 (ICP), 7040 (Flame AA), and 7041 (Furnace AA) were all listed for antimony, and Methods 8010 (GC/Hall), 8020 (GC/PID), 8120 (GC/ECD), and 8270 (GC/MS) were all listed for *o*-dichlorobenzene. Although this array of suggested methods may provide "acceptable" results, the methods obviously do not provide equivalent results in terms of reliability, sensitivity, precision, or cost.

The dilemma of selecting the appropriate methods was further complicated by information contained in SW-846. As stated in a report prepared by Enseco for API [5],

For some analytes, e.g., allyl chloride, the analyte is not listed in the suggested methods (8010 and 8240), nor is it found listed in any other published method, although it can be analyzed by these methods. For others, e.g., acetonitrile, the suggested method is 8015, but it is not listed in Method 8015. Acetonitrile is listed in Method 8030 (but not suggested in Appendix IX); however, there are no performance data for acetonitrile in Method

8030 to support the method's applicability to the analyte. Finally, for some analytes, e.g., tin, the suggested method (7870) has a very high PQL. Tin can be measured with much greater sensitivity by Method 6010, but this method is not suggested in Appendix IX for this analyte. Moreover, tin is not listed in Method 6010.

Thus, the information contained in SW-846 is not directly transferable to establishing a protocol. Enseco's approach to selecting appropriate methods was based on a compound-by-compound survey in which potential methods were selected on the basis of well-known performance characteristics. For example, because of the high degree of reliability in identication and the environmental community's acceptance of GC/MS methods, GC/MS was selected as a potential method for many organic compounds. Other performance characteristics evaluated included sensitivity, freedom from interferences, cost, and development status of the method. For example, Method 8240 (GC/MS) was selected for acetonitrile instead of EPA's recommendation of Method 8015 (GC/FID) because of the interference problems associated with flame ionization detectors.

A primary goal was to limit the total number of methods employed so that the overall analytical costs would be reduced to the greatest extent pos-

Table 1 Analytical Methods for Appendix IX Analytes

SW-846 method	General category/analyte	Technique[a]	Number of analytes measured
6010	Metals	ICP	12
7060	Arsenic	GFAA	1
7421	Lead	GFAA	1
7470	Mercury	CVAA	1
7740	Selenium	GFAA	1
7841	Thallium	GFAA	1
8080	Organochlorine pesticides and PCBs	GC/ECD	26
8150	Herbicides	GC/ECD	3
8240	Volatile organics	GC/MS	52
8270	Semivolatile organics	GC/MS	125
8280	Dioxins and furans	GC/MS	7
9010	Cyanide	Colorimetric	1
9030	Sulfide	Titrimetric	1
		Total	232

[a]ICP, inductively coupled plasma/atomic emission; GFAA, graphite furnace atomic absorption; CVAA, cold vapor atomic absorption; GC/ECD, gas chromatography; GC/MS, gas chromatography/mass spectrometry.

sible. We believed that since this regulation was designed to detect groundwater contamination, general survey methods capable of measuring many compounds were preferable to methods that may offer somewhat better performance for a limited number of compounds in a pristane matrix. If compounds were detected, then alternative methods could be utilized for ongoing monitoring efforts.

As a result of this process, 13 analytical methods were selected to establish a protocol for measuring all of the Appendix IX analytes. These methods and the number of analytes measured by each are listed in Table 1.

V. METHOD PERFORMANCE DATA

Once a potential method was selected, performance data were generated for each analyte that could be measured by the method. The approach was as follows:

Table 2 Specification for Metals and Inorganics

Analyte	Method	Reporting limit (mg/L)	PQL[a] (mg/L)
Antimony	6010	0.05	0.3
Arsenic	7060	0.005	0.01
Barium	6010	0.01	0.02
Beryllium	6010	0.002	0.003
Cadmium	6010	0.005	0.04
Chromium	6010	0.01	0.07
Cobalt	6010	0.01	0.07
Copper	6010	0.02	0.06
Lead	7421	0.005	0.01
Mercury	7470	0.0002	0.002
Nickel	6010	0.04	0.05
Selenium	7740	0.005	0.02
Silver	6010	0.01	0.07
Thallium	7841	0.005	0.01
Tin	6010	0.05	8
Vanadium	6010	0.01	0.08
Zinc	6010	0.02	0.02
Cyanide	9010	0.01	0.04
Sulfide	9030	0.05	10

[a]PQL, practical quantitation limit contained in the Appendix IX rulemaking.

Table 3 Specification for Chlorinated Pesticides and PCBs by Method 8080

CAS no.	Analyte	Reporting limit (μg/L)	PQL[a] (μg/L)
309-00-2	Aldrin	0.05	0.05
12674-11-2	Aroclor 1016	1	50
11104-28-2	Aroclor 1221	1	50
11141-16-5	Aroclor 1232	1	50
53469-21-9	Aroclor 1242	1	50
12672-29-6	Aroclor 1248	1	50
11097-69-1	Aroclor 1254	1	50
11096-82-5	Aroclor 1260	1	50
319-84-6	α-BHC	0.05	0.05
319-85-7	β-BHC	0.05	0.05
319-86-8	δ-BHC	0.05	0.1
58-89-9	γ-BHC (Lindane)	0.05	0.05
5103-71-9	α-Chlordane	0.05	0.1
5103-74-2	γ-Chlordane	0.05	0.1
510-15-6	Chlorobenzilate	0.1	10
72-54-8	4,4'-DDD	0.1	0.1
72-55-9	4,4'-DDE	0.1	0.05
50-29-3	4,4'-DDT	0.1	0.1
2303-16-4	Diallate	1	10
60-57-1	Dieldrin	0.1	0.05
959-98-8	Endosulfan I	0.05	0.1
33213-65-9	Endosulfan II	0.1	0.05
1031-07-8	Endosulfan sulfate	0.1	0.5
72-20-8	Endrin	0.1	0.1
7421-93-4	Endrin aldehyde	0.1	0.2
76-44-8	Heptachlor	0.05	0.05
1024-57-3	Heptachlor epoxide	0.05	1
465-73-6	Isodrin	0.1	10
143-50-0	Kepone	1	10
72-43-5	Methoxychlor	0.5	2
8001-35-2	Toxaphene	5	2

[a]PQL, practical quantitation limit contained in the Appendix IX rulemaking.

1. Generate basic analytical information for identification and quantitation (retention times, mass spectra, linear range, etc.).
2. Evaluate matrix bias (measured as percent recovery) in an aqueous matrix.
3. Determine a detection limit for reporting purposes.

In some cases much of this information existed. For example, data of this type were available for over 100 compounds measurable in EPA's Contract Laboratory Program. Other compounds had been studied in previous work performed by Enseco [6].

As a result of this process, a method specification was established for Appendix IX analyses of groundwater. This specification defined the analyte list by method, the reporting limit, and footnotes that could affect interpretation of the results.

The information developed in this process is presented in Tables 2–7. Each table lists the analytes measured by a specific method, grouped into metals, organochlorine pesticides and PCBs, phenoxy acid herbicides, volatile organics, semivolatile organics, and chlorinated dioxins and furans. For each analyte, a reporting limit is listed and compared to EPA's practical quantitation limit (PQL) contained in the final rule.

As stated in the rule, "The PQL values are in many cases based only on a general estimate for the method and not on a determination for individual compounds. PQLs are not a part of the regulation." By comparison, the reporting limits are based on MDL studies performed at our laboratories. The reporting limit was set equal to EPA's PQL except where our data indicated that EPA's PQL was not achievable (i.e., the Method Detection Limit or MDL is above the EPA's PQL). In these cases, the PQL is set at 5–10 times the MDL. The procedure for determination of the MDL is defined in the *Federal Register*, 40 CFR Part 8136, Appendix B.

Note that, as shown in Table 6, four Appendix IX compounds are not recoverable using EPA Method 8270. These compounds are measurable as

Table 4 Specification for Phenoxy Acid Herbicides by Method 8150

CAS no.	Analyte	Reporting limit (μg/L)	PQL[a] (μg/L)
94-75-7	2,4-D	1	10
93-76-5	2,4,5-T	0.2	2
93-72-1	Silvex	0.2	2

[a]Practical quantitation limit contained in the Appendix IX rulemaking.

Table 5 Specification for Volatile Organics by Method 8240

CAS no.	Analyte	Reporting limit (μg/L)	PQL[a] (μg/L)
67-64-1	Acetone	10	100
75-05-8	Acetonitrile	200	—
107-02-8	Acrolein	100	5
107-13-1	Acrylonitrile	100	5
0107-05-1	Allyl chloride	10	100
71-43-2	Benzene	5	5
75-27-4	Bromodichloro-methane	5	5
75-25-2	Bromoform	5	5
74-83-9	Bromomethane	10	10
78-93-3	2-Butanone	10	100
75-15-0	Carbon disulfide	5	5
56-23-5	Carbon tetrachloride	5	5
108-90-7	Chlorobenzene	5	5
75-00-3	Chloroethane	10	10
67-66-3	Chloroform	5	5
74-87-3	Chloromethane	10	5
0126-99-8	Chloroprene	5	10
124-48-1	Dibromochloro-methane	5	5
96-12-8	1,2-Dibromo-3-chloropropane	10	5
106-93-4	1,2-Dibromoethane	10	5
74-95-3	Dibromomethane	5	5
110-57-6	*trans*-1,4-Dichloro-2-butene	5	5
75-71-8	Dichlorodifluoro-methane (Freon 12)	20	5
75-34-3	1,1-Dichloroethane	5	5
107-06-2	1,2-Dichloroethane	5	5
75-35-4	1,1-Dichloroethene	5	5
540-59-0	1,2-Dichloroethene	5	5
78-87-5	1,2-Dichloropropane	5	5
10061-01-5	*cis*-1,3-Dichloro-propene	5	5
10061-02-6	*trans*-1,3-Dichloro-propene	5	5

Table 5 Continued

CAS no.	Analyte	Reporting limit (μg/L)	PQL[a] (μg/L)
123-91-1	1,4-Dioxane	500	—[b]
100-41-4	Ethylbenzene	5	5
97-63-2	Ethyl methacrylate	20	5
591-78-6	2-Hexanone	10	50
74-88-4	Iodomethane	5	5
078-83-1	Isobutanol	200	—[b]
126-98-7	Methylacrylonitrile	5	5
75-09-2	Methylene chloride	5	5
80-62-6	Methyl methacrylate	20	5
108-10-1	4-Methyl-2-penta-none	10	50
107-12-0	Propionitrile	5	5
100-42-5	Styrene	5	5
630-20-6	1,1,1,2-Tetrachloro-ethane	5	5
79-34-5	1,1,2,2-Tetrachloro-ethane	5	5
127-18-4	Tetrachloroethene	5	5
108-88-3	Toluene	5	5
71-55-6	1,1,1-Trichloro-ethane	5	5
79-00-5	1,1,2-Trichloro-ethane	5	5
79-01-6	Trichloroethene	5	5
75-69-4	Trichlorofluoro-methane	5	5
96-18-4	1,2,3-Trichloro-propane	5	5
108-05-4	Vinyl acetate	10	5
75-01-4	Vinyl chloride	10	10
1330-20-7	Xylenes (total)	5	5

[a]Practical quantitation limit for Method 8240 contained in the Appendix IX rulemaking.
[b]PQL of 150 μg/L in the regulation references Method 8015; Method 8015, a GC/FID method, does not list these compounds as analytes.

Table 6 Specification for Semivolatile Organics by Method 8270

CAS no.	Analyte	Reporting limit (μg/L)	PQL[a] (μg/L)
83-32-9	Acenaphthene	10	10
208-96-8	Acenaphthylene	10	10
98-86-1	Acetophenone	10	10
53-96-3	2-Acetylamino-fluorene	100	10
92-67-1	4-Aminobiphenyl	10	10
62-53-3	Aniline	10	10
120-12-7	Anthracene[b]	10	10
140-57-8	Aramite	10	10
56-55-3	Benzo[a]anthracene	10	10
50-32-8	Benzo[a]pyrene	10	10
205-99-2	Benzo[b]fluoran-thene	10	10
191-24-2	Benzo[g,h,i]perylene	10	10
207-08-9	Benzo[k]fluoran-thene	10	10
100-51-6	Benzyl alcohol	10	20
101-55-3	4-Bromophenyl phenyl ether	10	10
85-68-7	Butylbenzylphthalate	10	10
88-85-7	2-sec-Butyl-4,6-dinitrophenol	10	10
106-47-8	4-Chloroaniline	10	20
111-91-1	bis(2-Chloroeth-oxy)methane	10	10
111-44-4	bis(2-Chloroethyl) ether	10	10
108-60-1	bis(2-Chloroiso-propyl) ether	10	10
59-50-7	4-Chloro-3-methyl-phenol	10	20
91-58-7	2-Chloronaphthalene	10	10
95-57-8	2-Chlorophenol	10	10
7005-72-3	4-Chlorophenyl phenyl ether	10	10
218-01-9	Chrysene	10	10
53-70-3	Dibenz[a,h]anthra-cene	10	10
132-64-9	Dibenzofuran	10	10
84-74-2	Di-n-butylphthalate	10	10
95-50-1	1,2-Dichlorobenzene	10	10

Table 6   Continued

CAS no.	Analyte	Reporting limit (μg/L)	PQL[a] (μg/L)
541-73-1	1,3-Dichlorobenzene	10	10
106-46-7	1,4-Dichlorobenzene	10	10
91-94-1	3,3'-Dichlorobenzidine	20	20
120-83-2	2,4-Dichlorophenol	10	10
87-65-0	2,6-Dichlorophenol	10	10
84-66-2	Diethylphthalate	10	10
60-51-5	Dimethoate[c,d]	—	10
60-11-7	p-Dimethylaminoazobenzene	10	10
57-97-662	7,12-Dimethylbenz[a]anthracene	10	10
119-90-4	3,3'-Dimethylbenzidine	10	10
122-09-8	a,a-Dimethylphenethylamine	10	10
105-67-9	2,4-Dimethylphenol	10	10
131-11-3	Dimethylphthalate	10	10
99-65-0	1,3-Dinitrobenzene	10	10
534-52-1	4,6-Dinitro-2-methylphenol	50	50
51-28-5	2,4-Dinitrophenol	50	50
121-14-2	2,4-Dinitrotoluene	10	10
606-20-2	2,6-Dinitrotoluene	10	10
117-84-0	Di-n-octylphthalate	10	10
122-39-4	Diphenylamine[e]	10	10
298-04-4	Disulfton[d]	50	10
117-81-7	Bis(2-ethylhexyl)phthalate	10	10
62-50-0	Ethyl methanesulfonate	10	10
52-85-7	Famphur[c,d]	—	10
206-44-0	Fluoranthene	10	10
86-73-7	Fluorene	10	10
118-74-1	Hexachlorobenzene	10	10
87-68-3	Hexachlorobutadiene	10	10
77-47-4	Hexachlorocyclopentadiene	10	10

Table 6 Continued

CAS no.	Analyte	Reporting limit (μg/L)	PQL[a] (μg/L)
67-72-1	Hexachloroethane	10	10
70-30-4	Hexachlorophene[c]	—	10
1888-71-7	Hexachloropropene	10	10
193-39-5	Indeno[1,2,3-c,d]pyrene	10	10
78-59-1	Isophorone	10	10
120-58-1	Isosafrole	20	10
91-80-5	Methapyrilene	10	10
56-49-5	3-Methylcholanthrene	10	10
66-27-3	Methyl methanesulfonate	10	10
91-57-6	2-Methylnaphthalene	10	10
298-00-0	Methylparathion[d]	50	10
95-48-7	2-Methylphenol	10	10
106-44-5	3/4-Methylphenol[f]	10	10
91-20-3	Naphthalene	10	10
130-15-4	1,4-Naphthoquinone	10	10
134-32-7	1-Naphthylamine	10	10
91-59-8	2-Naphthylamine	10	10
88-74-4	2-Nitroaniline	50	50
99-09-2	3-Nitroaniline	50	50
100-01-6	4-Nitroaniline	50	50
98-95-3	Nitrobenzene	10	10
88-75-5	2-Nitrophenol	10	10
100-02-7	4-Nitrophenol	50	10
56-57-5	4-Nitroquinoline-1-oxide[c]	—	10
924-92-2	N-Nitrosodi-n-butylamine	10	10
55-18-5	N-Nitrosodiethylamine	10	10
62-75-9	N-Nitrosodimethylamine	10	10
86-30-6	N-Nitrosodiphenylamine[e]	10	10

Table 6 Continued

CAS no.	Analyte	Reporting limit (μg/L)	PQL[a] (μg/L)
621-64-7	N-Nitrosodi-n-propylamine	10	10
10595-95-6	N-Nitrosomethyl-ethylamine	10	10
59-89-2	N-Nitrosomorpho-line	10	10
100-75-4	N-Nitrosopiperidine	10	10
930-55-2	N-Nitrosopyrrolidine	10	10
9-55-8	5-Nitro-o-toluidine	10	10
56-38-2	Parathion[d]	10	10
608-93-5	Pentachlorobenzene	10	10
76-01-7	Pentachloroethane	10	10
82-68-8	Pentachloronitro-benzene	50	10
87-86-5	Pentachlorophenol	50	50
62-44-2	Phenacetin	10	10
85-01-8	Phenanthrene	10	10
108-95-2	Phenol	10	10
106-50-3	4-Phenylenedi-amine[c]	—	10
298-02-2	Phorate[d]	100	10
109-06-8	2-Picoline	10	10
23950-58-5	Pronamide	10	10
129-00-0	Pyrene	10	10
110-86-1	Pyridine	20	10
94-59-7	Safrole[b]	10	10
3689-24-5	Sulfotepp[d]	10	10
95-94-3	1,2,4,5-Tetrachloro-benzene	10	10
58-90-2	2,3,4,6-Tetrachloro-phenol	50	10
297-97-2	Thionazine[d]	50	10
95-53-4	2-Toluidine	10	10
120-82-1	1,2,4-Trichloroben-zene	10	10
95-95-4	2,4,5-Trichloro-phenol	50	10
88-06-2	2,4,6-Trichloro-phenol	10	10

Table 6 Continued

CAS no.	Analyte	Reporting limit (μg/L)	PQL[a] (μg/L)
126-68-1	O,O,O-Triethyl phosphorothioate[d]	10	10
99-35-4	1,3,5-Trinitrobenzene	10	10

Surrogates

4165-60-0	S1	Nitrobenzene-d$_5$
321-60-8	S2	2-Fluorobiphenyl
1718-51-0	S3	Terphenyl-d$_{14}$
4165-62-2	S4	Phenol-d$_5$
367-12-4	S5	2-Fluorophenol
118-79-6	S6	2,4,6-Tribromophenol

[a]Practical quantitation limit for Method 8270 or 8250 contained in the Appendix IX rulemaking.
[b]Measured as two isomers.
[c]Not consistently recoverable using standard analytical method, and consequently method detection limits cannot be established.
[d]The Appendix IX semivolatiles contain organophosphorus pesticides measurable by Method 8140 at lower detection limits. If Method 8140 is included in an Appendix IX package, the following compounds are deleted from the 8270 list:

Phorate	Thionazine
Disulfoton	Dimethoate
Parathion	Famphur
Sulfotepp	Methyl parathion
O,O,O-Triethylphosphorothioate	

[e]Cannot distinguish diphenylamine from N-nitrosodiphenylamine due to decomposition in the injection part.
[f]3-Methylphenol and 4-methylphenol cannot be differentiated on the basis of their mass spectra, and retention times are identical.

Table 7 Specification for Chlorinated Dioxins and Furans by Method 8280

CAS no.	Analyte	Reporting limit[a] (μg/L)	PQL[b] (μg/L)
1-331	Tetrachlorodibenzodioxins (total)	0.010	0.010
1746-01-6	2,3,7,8-Tetrachlorodibenzodioxin	0.005	0.005
1-289	Pentachlorodibenzodioxins	0.010	0.010
1-200	Hexachlorodibenzodioxins	0.010	0.010
1-332	Tetrachlorodibenzofurans	0.010	0.010
1-290	Pentachlorodibenzofurans	0.010	0.010
1-201	Hexachlorodibenzofurans	0.010	0.010

[a]As specified in Method 8280, reporting limits for dioxins and furans are established for each sample. The reporting limits shown are typical values.
[b]Practical quantitation limit contained in the Appendix IX rulemaking.

calibration standards indicating that the sample preparation method is not appropriate. Enseco is pursuing an alternative approach to the measurement of these compounds using LC/MS and has submitted data to the EPA related to these compounds [7].

VI. ESTABLISHMENT OF AN APPENDIX IX PROTOCOL

As discussed in more detail by Taylor [8], a protocol is a set of definitive instructions that encompass activities in addition to the analytical technique. Although EPA attempted to establish analytical procedures, there was no guidance from EPA relating to these other activities. Enseco established a protocol for Appendix IX analyses by defining all other activities associated with the analyses. In addition to generating performance data as discussed in Section V, these activities included defining sample containers, preservatives, and holding times and establishing spike components, spike levels, and control limits for quality control purposes.

Table 8   Recommended Containers and Preservatives for Appendix IX Groundwater Monitoring

Sample container	Preservation	Filtration	Minimum sample size	Methods/ parameters	Recommended holding time
3 × 40-mL glass (VOA)	4°C, HCl to pH <2	No	40 mL ea.	Volatile organics	14 days
5 × 1-L glass	4°C	No	1000 mL ea.	Semivolatile organics, pesticides, herbicides, dioxins	7 days until extraction, 40 days after extraction
Polyethylene	2 mL 50% HNO₃, to pH <2	Yes	500 mL	Metals	6 months
Plastic	2 mL 50% NaOH to pH >12, 4°C	Yes	500 mL	Cyanide	14 days
Plastic	1 mL Zn acetate, 1 mL 50% NaOH to pH >9, 4°C	Yes	250 mL	Sulfide	7 days

Recommended sample containers, preservatives, and holding times are listed in Table 8. These recommendations are based on standard industry practices as initially established by EPA under the Clean Water Act and promulgated in 40 CFR Part 136.

The quality control specifications presented in Tables 9–13 list the specific Appendix IX analytes selected for spiking, appropriate spike levels, and control criteria. The components and spike levels were patterned after EPA's Contract Laboratory Program. For organic methods, both target and surrogate analytes were selected. Target analytes are spiked into reagent water for laboratory controls and into sample matrices for data assessment. Surrogate analytes are spiked into each sample for data assessment.

Depending on the sample type (sample, laboratory control, matrix spike, etc.), the control limits are used to either control the laboratory operations (laboratory controls) or assess the data (samples, matrix spikes) [9]. The control limits were based on limits in EPA's Contract Laboratory Program and on internal data.

Table 9 Quality Control Components, Spike Levels, and Control Limits for Metals and Inorganic Elements

Compound	Method	Spike (mg/L)	Accuracy objective (% recovery)	Precision objective (RPD)
Antimony	6010	0.5	75–125	< 20
Arsenic	7060	0.04	75–125	< 20
Barium	6010	2.0	75–125	< 20
Beryllium	6010	0.05	75–125	< 20
Cadmium	6010	0.05	75–125	< 20
Chromium	6010	0.2	75–125	< 20
Cobalt	6010	0.5	75–125	< 20
Copper	6010	0.25	75–125	< 20
Cyanide	9010	0.12	75–125	< 20
Lead	7421	0.02	75–125	< 20
Mercury	7470	0.001	75–125	< 20
Nickel	6010	0.5	75–125	< 20
Selenium	7740	0.01	75–125	< 20
Silver	6010	0.05	75–125	< 20
Sulfide	9030	0.5	80–120	< 20
Thallium	7841	0.05	75–125	< 20
Tin	6010	0.4	75–125	< 20
Vanadium	6010	0.5	75–125	< 20
Zinc	6010	0.5	75–125	< 20

RPD, relative percent difference.

Table 10 Quality Control Components, Spike Levels, and Control Limits for Organo-chlorine Pesticides and PCBs

Compound	Spike (μg/L)	Accuracy objective (% recovery)	Precision objective (RPD)
Spike components			
γ-BHC (Lindane)	0.2	56–123	< 15
Heptachlor	0.2	40–131	< 20
Aldrin	0.2	40–120	< 22
Dieldrin	0.5	52–126	< 18
Endrin	0.5	56–121	< 21
4,4'-DDT	0.5	38–127	< 27
Surrogate spike			
Dibutylchlorendate	1	48–136	N.A.

N.A., not applicable.

Table 11 Quality Control Components, Spike Levels, and Control Limits for Phenoxyacid Herbicides

Compound	Spike (μg/L)	Accuracy objective (% recovery)	Precision objective (RPD)
Spike components			
2,4-D	5.0	49–103	< 29
2,4,5-TP (Silvex)	1.0	31–121	< 29
Surrogate spike			
DCAA	5	60–120	N.A.

DCAA, 2,4-Dichlorophenylacetic acid.

Table 12 Quality Control Components, Spike Levels, and Control Limits for Volatile Organics

Compound	Spike (μg/L)	Accuracy objective (% recovery)	Precision objective (RPD)
Spike components			
1,1-Dichloroethene	50	61–145	< 14
Trichloroethene	50	71–120	< 14
Benzene	50	76–127	< 11
Toluene	50	76–125	< 13
Chlorobenzene	50	75–130	< 13
Surrogate spike			
1,2-Dichloroethane-d_4	50	76–114	N.A.
4-Bromofluorobenzene (BFB)	50	86–115	N.A.
Toluene-d_8	50	88–110	N.A.

Table 13 Quality Control Components, Spike Levels, and Control Limits for Semivolatile Organics

Compound	Spike (μg/L)	Accuracy objective (% recovery)	Precision objective (RPD)
Spike components			
Phenol	200	12–89	< 42
2-Chlorophenol	200	27–123	< 40
1,4-Dichlorobenzene	100	36–97	< 28
N-Nitrosodi-*n*-propylamine	100	41–116	< 38
1,2,4-Trichlorobenzene	100	39–98	< 28
4-Chloro-3-methylphenol	200	23–97	< 42
Acenaphthene	100	46–118	< 31
4-Nitrophenol	200	10–80	< 50
2,4-Dinitrotoluene	100	24–96	< 38
Pentachlorophenol	200	9–103	< 50
Pyrene	100	26–127	< 31
Surrogate spike			
Phenol-d_5	200	10–94	N.A.
2-Fluorophenol	200	21–100	N.A.
2,4,6-Tribromophenol	200	10–123	N.A.
Nitrobenzene-d_5	100	35–114	N.A.
2-Fluorobiphenyl	100	43–116	N.A.
Terphenyl-d_{14}	100	33–141	N.A.

In summary, the implementation of a protocol to perform routine analyses of groundwater for EPA's list of Appendix IX constituents was performed by

- Reviewing the regulation and related background documents in detail
- Establishing a definitive list of analytes from an analytical chemist's perspective
- Assessing the adequacy of analytical technologies relative to these analytes
- Selecting methods based on general performance characteristics
- Documenting performance characteristics for each analyte by laboratory studies
- Establishing quality control activities for each method
- Defining sample containers, preservatives, and holding time.

REFERENCES

1. Memos from the EPA: 12/4/84 by Bob Kayser Appendix VIII issues and 12/27/84 by John Skinner RCRA ground water monitoring equipment.

2. Enseco—Rocky Mountain Analytical Laboratory. Evaluation of ground water monitoring requirements for Appendix VIII constituents, Mar. 15, 1986. Prepared for the American Petroleum Institute.
3. Evaluation of the feasibility of analyzing groundwater for the Appendix VIII hazardous constituents. Prepared for the American Petroleum Institute, May 1983.
4. Enseco—Rocky Mountain Analytical Laboratory. Evaluation of the applicability of the SW-846 manual to support all RCRA Subtitle C testing, Dec. 20, 1984. Prepared for American Petroleum Institute.
5. Enseco—Rocky Mountain Analytical Laboratory. Evaluation of analytical methods for measuring Appendix IX constituents in groundwater. Prepared for the American Petroleum Institute. API Publication 4499, July 1989.
6. Enseco—Rocky Mountain Analytical Laboratory. Capability of EPA Methods 624 and 625 to measure Appendix IX compounds. Prepared for the American Petroleum Institute.
7. Cornell, J. (Enseco), and Payne, D. (EPA Region V). Internal communication.
8. Taylor, K. *Quality Assurance of Chemical Measurements*, Lewis, Chelsea, MI, 1987.
9. Carlberg, K.A., Hanisch, R.C., and Parr, J.L. Quality assurance in the environmental laboratory. Fourth Annual Waste Testing and Quality Assurance Symposium, Washington, DC, July 1988.

4

Hierarchical Analytical Approach to Evaluating the Transport and Biogeochemical Fate of Organic Compounds in Sewage-Contaminated Groundwater, Cape Cod, Masachusetts

Larry B. Barber II

Geological Survey, U.S. Department of the Interior, Boulder, Colorado

This chapter discusses the nature of organic and inorganic substances in a plume of sewage-contaminated groundwater at Cape Cod, Massachusetts. The hydrogeology of the site is characterized by an unconfined aquifer consisting of highly permeable sand and gravel. Disposal of secondary-treated sewage effluent through rapid-infiltration beds for more than 50 years has resulted in a plume of contaminated groundwater approximately 4000 m long, 1000 m wide, and 30 m thick. Distributions of nonreactive inorganic constituents such as chloride, and boron and parameters such as specific conductance were used to establish the extent of the contamination plume. A variety of measurements were used to characterize organic contamination. Dissolved organic carbon (DOC) concentrations in the sewage effluent were about 12 mg/L, and DOC concentrations in the contaminated groundwater ranged from 1 to 5 mg/L. Dissolved organic carbon fractionation (DOC_f) data indicate that there is considerable variability in the character of DOC within the plume. The hydrophobic-neutral DOC fraction is particularly important because it is related to anionic and nonionic surfactants and their metabolites, which are major contaminants in the plume. Anionic surfactants were determined as methylene-blue-active substances. Total volatile

$n = 10^{-9}$

$m = 10^{-3}$

$C = 10^{-2}$

chlorinated hydrocarbons, another important class of organic contaminants in the plume, were determined as purgeable organic chloride. Trace concentrations of volatile and semivolatile organic compounds were measured by closed-loop stripping, purge-and-trap, liquid–liquid extraction, and solid-phase extraction followed by gas chromatography/mass spectrometry (GC/MS) analysis. Surfactant-derived compounds were further characterized by liquid–liquid and solid-phase extraction followed by infrared spectroscopy, proton and carbon-13 nuclear magnetic resonance spectroscopy, and derivatization GC/MS analysis.

The types of organic compounds determined by these methods were complex, and more than 100 contaminants were identified, including di-, tri-, and tetrachloroethene; dichlorobenzene (ortho, meta, and para isomers); alkylbenzenes (C_1–C_{14} isomers); di-tert-butylbenzoquinone; alkylphenol isomers; alkylphenol polyethoxylate and ethoxycarboxylate isomers; linear and branched-chain alkylbenzenesulfonates and their carboxylated metabolites; phthalate esters; aldehydes; ketones; and aromatic and aliphatic hydrocarbons. Concentrations of individual compounds ranged from less than 10 ng/L to more than 1 mg/L.

The dominant controls on contaminant distributions within the aquifer appear to be variation in the sewage-effluent composition over time, biological degradation, and sorption to the aquifer sediments. Most of the DOC in the secondary-sewage effluent is removed during treatment and infiltration, but the residual components that reach the water table are not significantly attenuated during subsurface transport.

I. INTRODUCTION

This chapter summarizes organic geochemical results from a comprehensive study of a plume of sewage-contaminated groundwater on Cape Cod, Massachusetts. The investigation has been ongoing since 1978 as part of a multidisciplinary research program involving organic and inorganic geochemistry, geology, hydrogeology, and microbiology. The topics considered in this chapter are limited to (1) The types of organic compounds present in the contaminated groundwater, (2) the most suitable methods for analyzing the range of organic compounds present, and (3) the geochemical and biochemical processes that control the behavior and transport of organic contaminants in the aquifer.

Investigation of groundwater contamination by trace-level organic compounds involves detecting and identifying the contaminants, determining their spatial distributions, and interpreting their fate and transport processes within the hydrogeochemical framework of the aquifer. Although there are numerous sources of contamination that may introduce organic compounds into the subsurface, disposal of sewage effluent by rapid-infiltration land treatment (municipal) and septic tanks (domestic) is of particular significance owing to the volume of wastewater disposed of (approximately 57×10^6 m^3/day produced in the United States) and the number of sites (19.5 million septic tanks) located throughout the country [1–3]. Areas with high population densities that dispose of wastewater into shallow unconfined aquifers are particularly vulnerable to groundwater contamination [1,4]. Although domestic septic systems dispose of small quantities compared to municipal disposal facilities, the abundance and disperse nature of septic systems can significantly affect groundwater quality. Rapid infiltration requires permeable sediments in the unsaturated and saturated zones to accommodate large quantities of wastewater [5–7], a characteristic that increases the susceptibility of an aquifer to contamination. Commonly, the only permeable sediments suitable for wastewater disposal also are major water supply aquifers.

Wastewater disposed of by rapid infiltration usually undergoes advanced treatment that removes the bulk of the organic matter; however, some compounds, such as chlorinated hydrocarbons, are only partially removed [8,9]. Organic compound concentrations are further decreased by biological, chemical, and physical processes during infiltration through the unsaturated zone [10,11]. Depending on wastewater application rates and depth to the water table, it is possible that pollutants can pass directly into the saturated zone. Persistent water-soluble organic compounds not removed during infiltration can adversely affect groundwater quality because they may not be attenuated in the aquifer where biological and chemical activity is diminished [12–18].

Because of the range of chemical properties of organic substances in sewage-contaminated groundwater, multiple analytical techniques are needed to identify the major components present. Sensitive analytical methods are necessary because of the complexity of the organic matrix and because of the small concentrations of individual substances (often 100 ng/l or less). Keith [19,20], Leenheer [21], Suffet and Malaiyandi [22], and Pellizzari et al. [23] discuss comprehensive methods for determining organic compounds in water. The most widely use methods for isolating specific compounds are (1) extraction with organic solvents, (2) solid-phase extraction followed by solvent or thermal desorption, and (3) gas purging followed by trapping on solid-phase adsorbents and solvent or thermal desorption. Once isolated and concentrated, compounds are often identified and quantified by gas chromatography combined with mass spectrometry (GC/MS). In this study, dissolved organic carbon (DOC), DOC fractionation (DOC$_f$), methylene-blue-active substances (MBAS), and purgeable organic chloride (POCl) were used to characterize the bulk organic chemistry of the groundwater. Closed-loop stripping (CLS), purge-and-trap (PT), liquid–liquid extraction (LLE), and solid-phase extraction (SPE) combined with GC/MS analysis were used to determine specific contaminants. High-performance liquid chromatography (HPLC), and carbon-13 nuclear magnetic resonance (^{13}C-NMR), proton (^{1}H) NMR, and infrared (IR) spectroscopy also were used to characterize contaminants.

II. FIELD INVESTIGATION

A. Site Location and Description

The field study was conducted at the U.S. Geological Survey's Cape Cod Toxic-Substances Hydrology Research Site (referred to hereafter as the Cape Cod site), a rapid-infiltration wastewater disposal facility located at Otis Air Base, near Sandwich, Massachusetts (Fig. 1). The Cape Cod site has an extensive plume of contaminated groundwater resulting from disposal

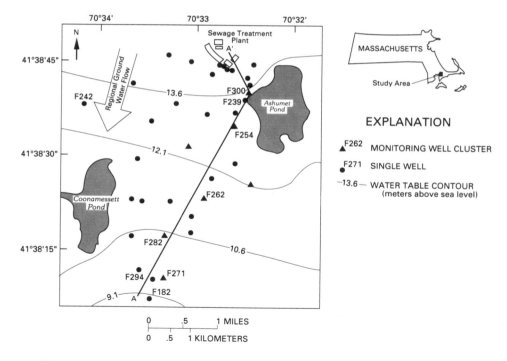

Figure 1   Map of the study area showing locations of wells sampled. Wells described in this chapter are labeled with site identification numbers. Line of section (A-A') is used in subsequent figures. Water-table contours from LeBlanc [4].

of secondary-treated sewage effluent into sand infiltration beds [4,18]. The plume of contaminated water is approximately 4000 m long, 1000 m wide, and 30 m thick. The secondary-treatment facility has been in operation since 1941 and consists of a bar screen, aerated grease-removal unit, Imhoff tanks for anaerobic digestion, trickling filters for aeration, and settling tanks for clarification. Between 1936 and 1941 the wastewater generated on the military base was probably treated by primary digestor. The secondary-treated sewage effluent is discharged to sand infiltration beds, where it percolates through 6 m of unsaturated sediment to the water table. Because the present population at the military base is much smaller than it was during the period from 1936 to 1975 and the volume of wastewater generated is much less than the sewage treatment plant was designed to process, treated effluent is recycled through the treatment plant in order to maintain operational efficiency.

The surficial geology at the Cape Cod site is characterized by Pleistocene to Holocene Age glacial outwash [24,25]. The sediments are stratified sand and gravel consisting of igneous and metamorphic rock fragments and discrete mineral grains. Most of the sediment is medium to coarse sand, and the silt and clay fractions typically comprise less than 1% [26]. Zones of coarse gravel and cobbles are common. Predominant minerals are quartz and feldspar, with minor amounts of glauconite, iron oxides, micas, tourmalines, garnets, amphiboles, and pyroxenes [26]. Kaolinite and illite are the only clay minerals present and occur at trace levels. The amount of sediment organic carbon is typically less than 0.05% but is higher ($>0.1\%$) near the infiltration beds [26]. Many of the grains have oxide coatings, and extractable iron and manganese concentrations are typically 0.2% and 0.02%, respectively [26]. Surface area varies as a function of particle size and ranges from 10 m^2/g for silt to 0.5 m^2/g for medium to coarse sand [26]. Cation-exchange capacity is about 10 meq/100 g [27].

The aquifer is part of the Cape Cod regional aquifer system and is the sole source of drinking water for the area. Annual recharge from precipitation has been estimated to be about 50 cm (45% of mean annual precipitation), and seasonal variations in recharge (as the result of variations in evaporation and transpiration rates) cause the altitude of the water table to fluctuate by 0.3–1 m/yr [4]. The groundwater occurs under water-table conditions, and the water table slopes to the south at 1.5 m/km [4,28,29]. The direction of groundwater flow is generally to the south, but flow patterns are distorted near large kettle-hole ponds. Porosity of the sediments is about 30–40%, the horizontal hydraulic conductivity is about 60–110 m/day, and the flow velocity ranges from 0.2 to 0.7 m/day [4,28–31].

The shallow native groundwater has a pH of 5–6, a specific conductance of 50–80 μS, and a temperature of 10°C and is near oxygen saturation [4,

27,32]. In contrast, the contamination plume is characterized by a pH of 6–7, a specific conductance of 100 to more than 400 μS, a temperature of 10–14°C, high alkalinity, and elevated concentrations of sodium, magnesium, calcium, potassium, ammonium, nitrate, chloride, sulfate, and boron [4,27,32]. The oxidation state of the plume is complex and varies depending on location. Near the infiltration beds, the groundwater is typically anoxic, although microaerophilic conditions probably exist (concentrations of oxygen near the limit of detection, 0.05 mg/L, have been measured in most wells). In the midportion of the plume, oxygen levels increase to about 0.5 mg/L. In the downgradient part of the plume, oxygen concentrations are greater than 1.0 mg/L. Along the length of the plume, contaminated groundwater is overlain by uncontaminated, oxygenated water derived from recharge. Other dissolved gases detected in the plume include nitrogen, nitrous oxide, and carbon dioxide. However, no sulfide or methane has been detected. Denitrification appears to be the major electron-accepting process in the anoxic part of the plume [33].

B. Monitoring-Well Installation and Construction

A network of more than 160 monitoring wells has been installed at the site for collection of water-level and water-quality data. Figure 1 shows the location of all wells sampled in this investigation. The wells discussed in this chapter are numbered. Each well has a unique descriptor; the letter prefix indicates town (F = Falmouth, S = Sandwich), the three-digit number indicates site location, and the number following the hyphen indicates depth of the bottom of the screened interval in feet below land surface (metric equivalents are not used because depth in feet is part of the well identification system). Two methods were used to install the wells (1) hollow-stem auger and (2) drive and wash. Results of geophysical evaluation of the two drilling methods indicate that augering disturbs the aquifer more than drive and wash [34]. Wells installed prior to 1978 are cased with 3.2- or 5.1-cm diameter, glued-joint PVC. Wells installed after 1978 are constructed of 5.1-cm diameter, flush-joint PVC. Most wells have 0.5–1-m screens (0.1-mm slots that are 0.6–1.0 m long). The well installation methods and construction materials are consistent with recommendations reported in the literature [35–38].

Most locations have well clusters consisting of several wells screened at different depths. Close-interval multilevel samplers [39] also are installed at several locations. Studies by LeBlanc et al. [40] and Smith et al. [41] describe the steepness of geochemical gradients in the aquifer and indicate the need for detailed sampling scales when evaluating geochemical processes.

C. Groundwater Sampling

The sampling strategies used in this study are consistent with the principles of sampling groundwater for organic compounds that have been reviewed elsewhere [42–44]. Groundwater samples were collected by two methods:

1. Wells with 5.1-cm casing were sampled using a stainless-steel submersible pump (Model SP81, Keck Geophysical Instruments) fitted with Teflon tubing and having a discharge rate of approximately 1 L/min.
2. Wells with 3.2-cm casing were first evacuated with a gasoline-powered centrifugal-suction pump and then sampled using a peristaltic pump (Model 760, Geotech Environmental Equipment) fitted with Teflon tubing and having a discharge rate of 0.1–0.5 L/min.

The pumps were flushed with distilled water between sample collections. Evaluations of the effect of sampling devices similar to those used in this study on the measurement of volatile organic compounds report that submersible pumps yield slightly higher concentrations than peristaltic pumps [45–49].

To ensure a representative groundwater sample (and provide a native groundwater rinse of the pump), at least three casing volumes of water were removed prior to sample collection. Specific conductance (SC), pH, and temperature were monitored during well evacuation, and samples were collected after field values had stabilized. The relation between stabilization of field parameters, purge volume, and water chemistry is discussed by Gibs and Imbrigiotta [50].

Samples for DOC, DOC_f, and MBAS analysis were filtered (0.45 μm silver filters) and collected in glass bottles with Teflon-lined caps. All glass bottles were thoroughly cleaned and then baked at 300°C for 8 hr. Samples for CLS, PT, POCl, LLE, and SPE analysis were collected without headspace in glass bottles (40 mL to 4 L) with Teflon-lined silicon septa. Samples for inorganic analysis (SC, chloride, boron) were filtered (0.45 μm Millipore filters) and collected in clean plastic bottles. After collection, samples were stored at 4°C until analysis.

III. ANALYTICAL METHODS

The hierarchical approach used in this investigation for the analysis of organic compounds is summarized in Fig. 2. The methods cover a broad range of sensitivity and selectivity and when used together provide a comprehensive characterization of organic contamination. The individual methods will be described in detail in the following sections.

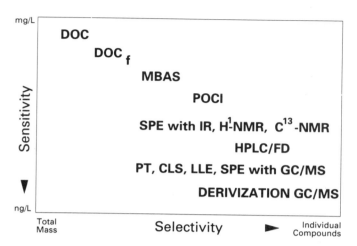

Figure 2 Summary of organic analytical methods used in this study ranked as a function of sensitivity and selectivity. [DOC, dissolved organic carbon; DOC_f, dissolved organic carbon fractionation; MBAS, methylene-blue-active substances; POCl, purgeable organic chloride; HPLC/FD, high-performance liquid chromatography with fluorescence detection; IR, infrared spectroscopy; H^1-NMR, proton nuclear magnetic resonance spectroscopy; C^{13}-NMR, carbon-13 nuclear magnetic resonance spectroscopy; PT, purge and trap; CLS, closed-loop stripping; LLE, liquid–liquid extraction; SPE, solid-phase extraction; GC/MS, gas chromatography/mass spectrometry.]

A. Inorganic Contaminants

Specific conductance was measured in the field with a digital meter (Extech Model 440), and values were corrected for the temperature of the groundwater. Chloride was determined by ion chromatography and colorimetric methods, and boron was determined by directly coupled plasma-emission spectroscopy [51,52].

B. Bulk Organic Contaminants

1. Dissolved Organic Carbon

Dissolved organic carbon (DOC) was determined by high-temperature oxidative combustion and persulfate oxidation [52,53]. In the combustive oxidation method (Coulometric Inc., Model 5011/130) a 10–100-μL sample was injected into an oxygen-swept combustion tube containing barium chromate at 900 °C, and the total carbon was converted to carbon dioxide, which was measured by coulometry. Inorganic carbon was determined by acidification and heating followed by sparging and coulometric detection. Dissolved organic carbon was calculated by difference between total and inorganic carbon. For the persulfate method (Oceanographic International, Model 700) a 1–10-mL sample was acidified with phosphoric acid and sparged with nitrogen, and the liberated carbon dioxide was trapped on molecular sieve adsorbent, thermally desorbed, and measured by infrared detector. After the removal of inorganic carbon, sodium persulfate was added, the sample was heated to 90°C and sparged with nitrogen, and the carbon dioxide produced by oxidation was sparged, trapped, desorbed, and measured by infrared detector.

2. Dissolved Organic Carbon Fractionation

Bulk DOC was separated into functional group classes with differing chemical behavior by DOC_f analysis [54]. This method separates the DOC into six fractions based on sorption of the compounds onto XAD-8, cation-exchange, and anion-exchange resins, adjustment of pH, and the use of different solvents for elution. The various fractions are operationally defined as hydrophobic (adsorbed onto XAD-8) or hydrophilic (not adsorbed onto XAD-8). These two classes are further subdivided into acid, base, and neutral fractions according to the pH at which the compounds are adsorbed or eluted.

3. Methylene-Blue-Active Substances

Anionic surfactants were determined by MBAS analysis [52,55]. Water samples (100 mL) were acidified to pH 2 with sulfuric acid, cationic methylene blue was added, the anionic surfactant/methylene blue complex was extracted into chloroform, and concentrations were measured (using a six-point calibration curve) spectrometrically at a wavelength of 635 nm.

4. Purgeable Organic Chloride

Purgeable organic chloride (POCl) was determined by oxidative combustion/halide generation followed by coulometric detection [56] using a Dohrmann Model DX-25 purgeable organic halide analyzer. Water samples were spraged with oxygen, the gas stream was combusted at 900°C, the hydrogen chloride reaction products were accumulated in a titration cell, and free chloride was titrated with silver ion, which was measured by coulometry. A 7-mL sample volume was used for concentrations greater than 10 μg/L, and a 27-ml sample was used for concentrations less than 10 μg/L.

C. Specific Organic Contaminants

1. Compound Isolation

Trace levels of individual organic compounds were isolated from water samples and concentrated prior to instrumental analysis. The various isolation procedures used are summarized in the following sections.

a. Closed-loop stripping

Trace-level semivolatile organic compounds were isolated by closed-loop stripping (CLS) [27,57–60] using a Tekmar Model CLS-1. A 4-L sample was heated to 40°C and sparged with headspace air (750 mL/min) for 2 hr. The vapor was passed through a 1.5-mg activated carbon trap, and the trap was eluted with 10–20 μL of dichloromethane. Surrogate standards (1-chlorhexane, 1-chlorododecane, and 1-chlorooctadecane in methanol) were added to the field samples at the time of collection, and 1-chlorooctane was added to the dichloromethane extracts as an internal standard.

b. Purge-and-trap analysis

Volatile organic compounds were isolated by purge-and-trap (PT) analysis [55,56,61] using a Tekmar ALS-1. A 5–25-mL water sample was sparged with helium (40 mL/min for 11 min), purged compounds were trapped on Tenax/silica/activated carbon adsorbent (35°C), and the sorbent was thermally desorbed (180°C) into the gas chromatograph (see Section IV.C.2).

c. Liquid–liquid extraction

Semivolatile compounds were concentrated by liquid–liquid extraction (LLE) [55,61–64]. One-liter samples were extracted at pH 7 and pH 2 with either trichloromethane or dichloromethane, the extracts were dried over sodium sulfate, and the solvent was evaporated under a stream of nitrogen. The extracts were then derivatized using boron trifluoride/methanol or diazomethane to form the methyl esters of carboxylic acids.

d. Solid-phase extraction

Semivolatile compounds were isolated by solid-phase extraction (SPE) using cyclohexyl (Analytischem Bondelute, 1000 mg) and n-octyl (Baker SPE-10, 500 mg) surface-modified silica as the adsorbent [65]. A 100–500-mL sample was passed through a preconditioned column, the column was dried under vacuum, and sorbed compounds were eluted with 1 mL of acetonitrile and 4 mL of dichloromethane. The eluant was dried over sodium sulfate and evaporated to a final volume of 100 μL under a stream of nitrogen. Surrogate standards (1,4-dibromobenzene, 2,2'-difluorobiphenyl, 4,4'-dibromobiphenyl, and 1,4-bromofluorobenzene in methanol) were added at the time of collection, and perdeuteronaphthalene was added to the extract as an internal standard.

Large-scale SPE isolations were carried out on samples from select wells using either XAD-8 (reverse-phase) or A-7 (anion-exchange) resins [66–70]. For the XAD-8 isolation, approximately 300 L of water was passed through a 1200-mL column of precleaned resin at a flow rate of 10 bed volumes per hour. The resin was pumped dry and Soxhlet-extracted for 72 hr with methanol (changing the methanol every 24 hr). The combined methanol extracts were rotary-evaporated to dryness and redissolved in D_2O for IR and NMR analysis. A similar volume of water was passed through a 1200-mL column of A-7 anion-exchange resin, and the resin was extracted with 0.1 N sodium hydroxide and 50:50 water/acetonitrile and processed in the same way as the XAD-8 extract.

Specific anionic surfactant compounds and their metabolites were determined in the XAD-8 and A-7 resin isolates using the derivatization procedure of Trehy et al. [71]. This method involves reaction of the resin extracts with phosphorus pentachloride to form the sulfonyl and acyl chlorides of alkylbenzenesulfonates and their carboxylated metabolites, followed by condensation with trifluoroethanol to form trifluoroethyl esters of the sulfonate and carboxylate functionalities. The derivatives were analyzed by GC/MS.

Solid-phase extraction also was used to isolate linear alkylbenzenesulfonate (LAS) surfactants [70,72]. Samples (200 mL) were passed through C_2 surface-modified silica cartridges, followed by elution with water/methanol. The water/methanol eluant was passed through a strong anion-exchange resin cartridge and rinsed with methanol, and LAS was eluted with hydrochloric acid/methanol. Concentrations were determined by addition of C_9 and C_{15} LAS surrogate standards.

2. Compound Analysis and Characterization

a. Gas chromatography/mass spectrometry

The CLS extracts were analyzed in a Hewlett-Packard 5895 GC/MS using a Hewlett-Packard Ultra II 25-m × 0.2-mm i.d. capillary column. One microliter of extract was injected in splitless mode (injector temperature 250°C; split vent opened 30 sec) using helium as the carrier gas (linear flow velocity 28 cm/sec). The gas chromatograph oven was kept at 30°C for 10 min, followed by a 6°C/min linear increase to 300°C, where it was held for 5 min. Mass spectra were acquired from 50 to 450 atomic mass units (amu) at a rate of 2 scans/sec. Concentrations were calculated using total-ion-current peak areas with the 1-chlorododecane surrogate standard as the quantitation reference. This method assumes that target and unknown compounds have a recovery efficiency similar to that of the surrogate standards and an MS relative-response factor of 1. The 1-chlorooctane internal standard was used to evaluate recovery of surrogate standards. Tentative identification of compounds was based on matching of sample mass spectra with those in a computerized mass spectra library. Computerized matches were manually evaluated to ensure quality of identification. Identifications of select target compounds were confirmed by matching mass spectra and retention times with authentic standards.

Purge-and-trap analysis was performed using a Varian 9611 gas chromatograph equipped with a 2.4 m × 2.0 mm glass column packed with 60/80 mesh Carbopack. The oven had an initial temperature of 45°C for 4 min, followed by a 8°C/min ramp to 220°C, where it was held for 21 min. Electron-impact mass spectra were acquired (Finnigan 5100 mass spectrometer) by scanning from 45 to 310 amu. Compound identification was based on matching mass spectra and chromatographic retention times with standards, and quantitation was based on internal standards.

Liquid–liquid and solid-phase extraction extracts and their derivatives were analyzed under the same GC/MS conditions as the CLS samples. Solid-phase extraction samples were analyzed in full-scan and selected ion monitoring (SIM) modes. SIM analysis was carried out for alkylphenols and alkylphenol polyethoxylates by monitoring for base peaks and molecular ions of the compounds [62,63]. Quantitation was based on nonylphenol external standards.

Trifluoroethyl derivatives of anionic surfactants and their metabolites isolated by XAD-8 and A-7 SPE were analyzed by electron-capture negative chemical ionization mass spectrometry using a Finnigan MAT Model TSQ-46 tandem mass spectrometer. The GC conditions were the same as for CLS extract analysis. The mass spectrometer had an ionization energy of 100 eV, a source temperature of 100°C, a source pressure of 0.3 torr, and a scan time of 1 sec and used methane as the reagent gas.

b. Nuclear magnetic resonance and infrared spectroscopy

Isolates from the large-volume XAD-8 and A-7 SPE extractions were characterized by ^{13}C-NMR spectroscopy [66,68,69] using a Varian XL-300 NMR spectrometer. Natural-abundance ^{13}C-NMR spectra were acquired at a carbon resonance frequency of 75.4 MHz. Acquisition parameters for normal broadband decoupled spectra included a 250-ppm (18,850 Hz) spectral window, 45° pulse width, 0.5 sec acquisition time, 1.0 sec pulse delay, continuous broadband decoupling, and line broadening of 1.0 Hz. Samples were analyzed in D_2O using 10-mm NMR tubes, and dioxane was used as an internal reference standard.

Proton NMR spectra were acquired with a Varian FT 80-A NMR spectrometer at a frequency of 79.5 MHz using a 8000-Hz spectral window, 45° pulse angle, 2.0 sec acquisition time, and a 5.0 sec pulse delay [68,69]. Samples were dissolved in D_2O and analyzed in 5-mm NMR tubes.

Infrared absorption spectra of samples prepared as potassium bromide pellets were acquired on a Perkin-Elmer Model 580 IR spectrometer. Samples were scanned from 4000 to 200 cm^{-1}, and the spectra were recorded on a strip chart.

c. High-performance liquid chromatography

High-performance liquid chromatography with fluorescence detection (HPLC/FD) was used to analyze the C_2 SPE isolates for linear alkylbenzenesulfonate [72]. Chromatographic analysis was conducted on a Waters 840 system with dual pumps using a Hitachi Model F1000 dual-grating fluorescence spectrophotometer operated at an excitation wavelength of 225 nm and an emission wavelength of 290 nm with a spectral bandpass of 10 nm. Chromatographic separation was performed in the reverse-phase mode with an Alltech Associates 5-μm C_1 Spherisorb analytical column (25 cm × 4 mm i.d.) using an isocratic binary solvent system (flow rate 1 mL/min) consisting of tetrahydrofuran (45%) and water (55%) with 0.1 M sodium perchlorate added as a phase modifier.

IV. CONTAMINANT OCCURRENCE AND DISTRIBUTION

Groundwater samples were collected between 1978 and 1988 and analyzed for inorganic and organic constituents. Results of these analyses have defined compound distributions within the plume of groundwater contamination [4, 18,27,32,56,58,73,74]. Microbial distributions have also been described [75].

A. Inorganic Contamination

Several dissolved inorganic constituents were used to define the extent of groundwater contamination. Anions such as chloride and bromide do not enter into redox reactions, do not form complexes with other ions, do not significantly adsorb onto mineral surfaces, and do not undergo biological transformations that remove them from solution [76]. As a result, they can be used as tracers of groundwater contamination [77,78]. In contrast to the conservative behavior of chloride and bromide, ions such as nitrate, sulfate, bicarbonate, phosphate, ammonium, calcium, magnesium, and potassium undergo chemical reactions in solution, are adsorbed onto the sediment, or are biologically labile [74,76,79–81]. Boron does not undergo oxidation-reduction reactions, biological degradation, precipitation or significant sorption [82–84].

Dissolved solutes that occur only in sewage effluent are needed to link contaminated groundwater directly and unambiguously to a sewage effluent source. For instance, nitrate is a good indicator of sewage contamination, but it also can originate from a variety of sources such as fertilizer and biological processes in the soil zone. Likewise, there are other soruces of chloride such as road salting and sea spray that can interfere with its use as a tracer of sewage contamination. In contrast, boron is ubiquitous in sewage effluents,

Table 1 Summary of Data for Specific Conductance, Chloride, Boron, Dissolved Organic Carbon, Dissolved Organic Carbon Fractionation, and Methylene-Blue-Active Substances for Samples Collected During 1983

Well no.[a]	SC (μS/cm)	Cl (mg/L)	B (μg/L)	DOC (mg/L)	DOC_f Hydrophobic %T	%A	%B	%N	Hydrophilic %T	%A	%B	%N	MBAS (mg/L)
Effluent	390	33	510	11.9	47	28	1	18	53	40	6	7	0.30
F300-10	138	33	40	1.0	na	na	na	na	na	na	na	na	0.04
F300-30	410	19	530	3.5	43	23	3	17	57	37	10	10	0.27
F239-64	190	11	320	2.0	50	5	0	45	50	35	10	5	0.09
F242-77	51	8	30	0.9	11	11	0	0	89	11	33	45	0.02
F254-26	70	11	40	1.2	36	27	9	0	64	28	0	36	0.03
F254-54	220	8	240	3.2	59	22	3	34	41	25	0	16	0.18
F254-72	225	13	250	2.8	50	18	4	28	50	30	3	17	0.14
F254-107	235	16	290	2.3	48	22	4	22	52	25	2	25	0.25
F254-140	175	15	30	1.9	53	11	11	31	47	0	15	32	0.16
F254-168	115	11	20	1.1	36	0	0	36	64	10	18	36	0.03
F254-216	59	7	20	1.0	36	18	9	9	64	19	0	45	0.03

Well													
F262-41	90	10	30	1.0	na	na	na	na	na	na	na	na	0.04
F262-69	200	16	360	2.8	46	21	4	21	54	15	15	24	0.48
F262-85	255	22	380	3.7	61	14	3	44	39	17	8	14	0.65
F262-109	108	15	40	4.0	na	na	na	na	na	na	na	na	0.10
F262-159	152	4	30	2.0	na	na	na	na	na	na	na	na	0.03
F282-49	100	22	30	1.9	32	16	5	11	68	5	16	47	0.05
F282-70	215	24	380	4.2	79	8	3	68	21	0	7	14	2.0
F282-94	208	20	270	4.0	77	8	2	67	23	4	8	11	2.0
F282-123	143	16	20	3.1	55	2	2	51	45	22	1	22	0.06
F271-41	55	10	20	1.0	na	na	na	na	na	na	na	na	0.05
F271-85	150	21	150	2.0	na	na	na	na	na	na	na	na	1.10
F271-141	132	14	20	2.0	na	na	na	na	na	na	na	na	0.04
F271-165	125	10	20	1.0	na	na	na	na	na	na	na	na	0.03
F294-89	139	16	40	3.1	55	26	3	26	45	39	3	3	0.96
F182-69	80	12	30	1.0	na	na	na	na	na	na	na	na	0.04

SC, specific conductance; Cl, chlorine; B, boron; DOC, dissolved organic carbon; DOC$_f$, dissolved organic carbon fraction; MBAS, methylene-blue-active substances. T, Total; A, acid; B, base; N, neutral; na, not analyzed.
[a]Wells listed in order of increasing distance from the infiltration beds.
Source: Barber [27] and Thurman et al. [32].

has restricted natural and anthropogenic sources, and may behave similarly to chloride in groundwater. In this chapter, specific conductance (SC), chloride, and boron are used to define the extent of sewage contamination and to establish plume boundaries for evaluating the movement and behavior of dissolved organic solutes.

One of the most easily determined inorganic field parameters is SC, a measurement of the total dissolved ionic load. Data for field measurements of SC in the groundwater during 1983 are presented in Table 1 and Fig. 3. There was sufficient contrast in the SC of the uncontaminated groundwater (38–70 μS), the effluent (390 μS), and the contaminated groundwater (90–410 μS) to delineate a plume that extended more than 4000 m downgradient from the infiltration beds, was about 1000 m wide and 30 m thick. Dilution of the sewage effluent by mixing with uncontaminated groundwater beneath the infiltration beds decreased the SC values by 10–15%.

The distribution of chloride is similar to that of SC (Table 1), as would be expected because chloride is the dominant anion in the sewage effluent [4,27,32]. The concentration of chloride was 33 mg/L in the effluent, 4–10 mg/L in the uncontaminated groundwater, and 10–33 mg/L in the contaminated groundwater. The decrease in chloride concentration beneath the infiltration beds was similar to that for SC. Because interferences from sources of chloride other than sewage effluent are present (road salting), and the concentration difference between the contaminated and uncontaminated groundwater is relatively small, chloride was not a definitive indicator of sewage contamination.

The major source of boron in sewage effluents is sodium perborate used as a bleach in detergent powders [82]. Boron concentrations in secondary effluents typically range from 0.3 to 1.5 mg/L [6,7,14,17,18,27,32,82,85, 86]. Boron data for 1983 are summarized in Table 1 and Fig. 4; boron concentration was 510 μg/L in the sewage effluent, <20 to 50 μg/L in the uncontaminated groundwater, and 90–530 μg/L in the contaminated groundwater. Elevated concentrations of boron extended more than 3700 m from the infiltration beds. Although the boundaries of the boron plume were more distinct than those defined by SC and chloride, the general distribution patterns were similar. Boron occurs as orthoboric acid [$B(OH)_3$] or the borate anion [$B(OH)_4^-$], depending on pH. Because its dissociation constant (pK_a) is 9.2 [83], the dominant species in most natural water will be neutral $B(OH)_3$. Potential sites for boron adsorption in soil–sediment systems include (1) iron and aluminum hydroxides (free and associated with clay minerals), (2) micaceous clay minerals, and (3) magnesium hydroxide grain coatings [83, 87–93]. Reports on boron sorption to subsurface sediments suggest that boron is not significantly sorbed during rapid infiltration [6,7,14]. Negligible boron sorption from groundwater would be expected on the basis of the

A

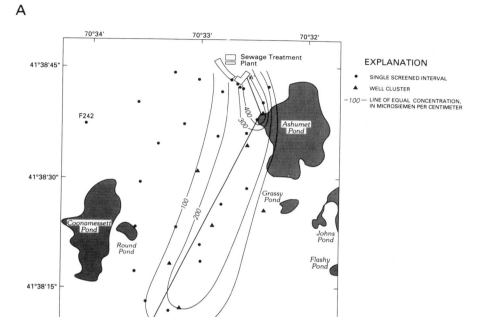

B

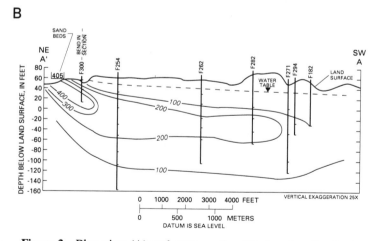

Figure 3 Plan view (A) and cross section (B) showing distribution of specific conductance (in μS/cm) during 1983 sampling. Plan-view contours are plotted using maximum values measured in vertical profiles (see cross section and Table 1).

A

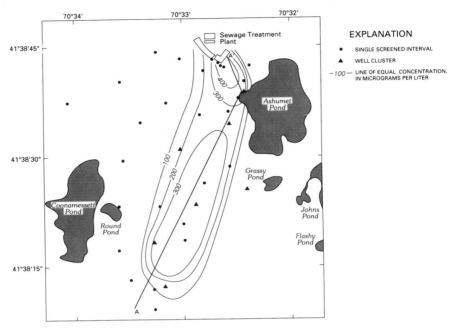

B

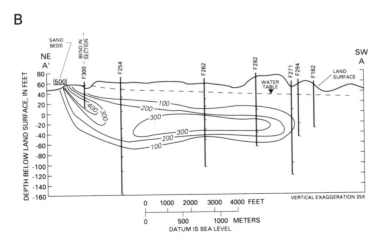

Figure 4 Plan view (A) and cross section (B) showing distribution of boron (in μg/L) during 1983 sampling. Plan-view contours are plotted using maximum concentrations measured in vertical profiles (see cross section and Table 1).

aquifer mineralogy (predominantly quartz and feldspar), the low levels of metallo-hydroxy grain coatings, and the speciation of boron at the pH of the groundwater (5–7).

B. Bulk Organic Contamination

1. Dissolved Organic Carbon

Dissolved organic carbon in sewage effluent is controlled primarily by the level of treatment. Raw sewage can have a DOC as high as 150 mg/L, primary-sewage effluents have a DOC that ranges from 60 to 70 mg/L, and secondary-sewage effluents have a DOC ranging from 10 to 30 mg/L [11,13,14,16,17, 94–96]. The major organic compounds in sewage effluents are sugars, carbo-hydrates, free and bound amino acids, volatile and nonvolatile fatty acids, phenols, sterols, anionic surfactants, nonionic surfactants, cationic surfac-tants, and a variety of neutral compounds such as hydrocarbons. Primary- and secondary-sewage treatment (aerobic and anaerobic biological digestion) reduces the DOC of raw sewage by 80–90%, largely by removal of carbo-hydrates, fatty acids, amino acids, and surfactants. This selective removal results in treated effluents containing relatively recalcitrant and mobile organic compounds. During infiltration, DOC is further reduced in the unsaturated zone by biodegradation and sorption to the sediments. DOC concentrations less than 5 mg/L are typical for contaminated groundwater at rapid infiltra-tion sewage disposal sites.

Dissolved organic carbon data (1983) for groundwater and sewage effluent samples are summarized in Table 1 and Fig. 5. Concentrations of DOC in the uncontaminated groundwater (1–5 mg/L) were higher than in the uncontam-inated groundwater (0.1–0.5 mg/L). The distribution of elevated DOC con-centrations in the groundwater coincides with that of SC, chloride, and boron (Table 1, Figs. 3 and 4). More than 65% of the DOC in the sewage effluent is removed during infiltration, decreasing from 12 mg/L at the infiltration beds to less than 4 mg/L a distance of 30 m downgradient. The decrease in DOC during infiltration is greater than can be explained by simple dilution (based on SC, chloride, and boron data) and indicates that it is being removed by microbial degradation and sorption onto the aquifer sediments. However, DOC removal during infiltration is less than reported for other locations (as much as 90% removal), probably due to (1) the low initial DOC in the sewage effluent and (2) the fact that organic compounds that are not removed during recycling of the effluent through the sewage treatment plant are quite re-sistant to degradation during infiltration. The amount of DOC is further decreased with additional transport in the aquifer, and concentrations are less than 2.0 mg/L (80% removal) a distance of 400 m downgradient from the infiltration beds. The concentration of DOC increases to nearly 4 mg/L

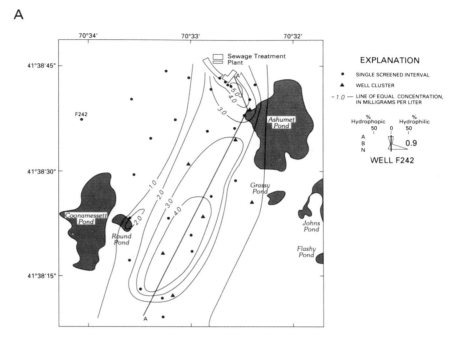

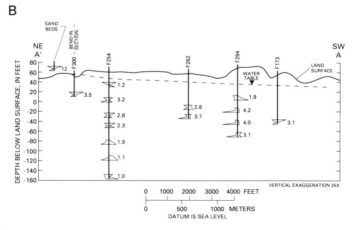

Figure 5 Plan view (A) showing distribution of dissolved organic carbon (mg/L), and cross section (B) showing distribution (in percent) of the various fractions determined by DOC fractionation (hydrophobic and hydrophilic acid (A) base (B) and neutral (N) compounds) during 1983 sampling. Plan-view contours are plotted using maximum concentrations measured in vertical profiles (see cross section and Table 1). Numbers on the cross section are total DOC concentrations (mg/L).

between 1700 and 3300 m from the infiltration beds (Fig. 5). The downgradient zone of elevated DOC is primarily due to nonbiodegradable surfactants introduced prior to 1965 (see Section IV. B. 3). In addition, DOC distributions may reflect higher concentrations of DOC in the sewage efluent during peak activity at the military base (between 1940 and 1970), and the lag time between initiation of sewage disposal and development of microbial populations near the beds capable of effectively removing organic carbon.

The rate of DOC loss decreases with distance from the infiltration beds because (1) microbial populations in the aquifer decrease with distance from the infiltration beds, (2) readily degraded organic compounds are selectively removed near the infiltration beds, leaving a refractory residue, (3) maximum sorption occurs near the infiltration beds, where sediment organic carbon is greatest, and (4) dissolved oxygen is depleted near the infiltration beds and the plume becomes anoxic [18,75,97,98]. The amount of sediment organic carbon at the infiltration beds decreases from more than 1% at the surface to less than 0.05% at a depth of 10 m. Because of the low levels of sediment organic carbon below the infiltration bed surface, sorption probably has less effect on DOC removal in the aquifer than does biodegradation.

Measurements of DOC provide an index of organic contamination [99, 100] and are useful for mass-balance considerations. However, DOC measurements give no indication of the specific compounds present. To acquire such data, class- or compound-specific analyses are required.

2. Dissolved Organic Carbon Fractionation

The various classes of compounds measured by dissolved organic carbon fractionation (DOC_f) analysis behave differently in groundwater [101]. The DOC fractions most important in this study are hydrophobic neutral, hydrophobic acid, hydrophilic neutral, and hydrophilic acid compounds. The hydrophilic fractions are the most mobile because they are highly water-soluble and do not sorb to aquifer sediments. The hydrophobic fractions have a greater affinity for sorbing onto sediments than the hydrophilic fractions, but because sediment organic carbon concentrations are low, the hydrophobic fractions also tend to be mobile.

The DOC_f data collected during 1983 are summarized in Table 1 and Fig. 5B. These data indicate that zones of different bulk DOC characteristics occur within the plume. In background well F242, the DOC was almost 90% hydrophilic, which is typical for natural groundwater [102]. The sewage efluent was approximately 50% hydrophobic and 50% hydrophilic, with the acid fractions being dominant. Near the infiltration beds, DOC_f ratios change with depth from predominantly hydrophilic in the uncontaminated water overlying the plume to almost equal proportions of the hydrophobic and hydrophilic fractions within the plume. With increasing distance from the infiltration beds, DOC_f ratios shift to the hydrophobic neutral fraction, and DOC at well F282 is predominantly hydrophobic neutral. Consistency in DOC_f ratios over the upgradient one-third of the plume indicate that the DOC composition of the contaminated groundwater is similar to the present-day sewage effluent. However, in the downgradient part of the plume, the DOC composition is distinctly different. The zone of maximum DOC matches the zone enriched in the hydrophobic neutral fraction. Anionic and non-ionic surfactants are isolated in the hydrophobic neutral fraction, indicating that the zone of elevated DOC may be of surfactant origin.

3. Methylene-Blue-Active Substances

Anionic surfactants, collectively measured as MBAS, consist of three major types (1) branched-chain alkylbenzenesulfonates (ABS), (2) linear-chain alkylbenzenesulfonates (LAS), and (3) sodium dodecylsulfates (NaLS) [103–106]. The branched-chain alkylbenzenesulfonate surfactants are considered "hard" because they are not easily biologically degraded, whereas LAS and NaLS are considered "soft" because they are more readily biodegraded [106]. Prior to 1965, nonbiodegradable ABS was the most widely used anionic surfactant. In 1965, when it was reported that LAS biodegrades more readily than ABS [107,108], detergent formulations were rapidly switched to LAS.

Concentrations of MBAS in the recent sewage effluent (1983) are low (0.30 mg/L, Table 1) because of (1) the current use of biodegradable surfactant formulations, (2) the relatively small current population at the military base, and (3) the recycling of the sewage effluent to maintain the treatment plant operation efficiency. The plume of elevated MBAS concentrations (Fig. 6) generally coincides with the zones of elevated SC, boron, and DOC (Figs. 3–5). As with DOC, MBAS distributions in the groundwater indicate two distinct zones of elevated concentrations (1) a region of less than 0.3 mg/L that occurs in the upgradient part of the plume and (2) a region of 1–2 mg/L that occurs between 1900 and 4000 m downgradient from the disposal site. Near the infiltration beds, MBAS concentrations decrease from 0.3 mg/L to less than 0.1 mg/L after 30 m of transport as a result of LAS biodegradation [68]. The zone of high MBAS concentrations in the downgradient part of the plume consists of refractory ABS that was introduced into the groundwater prior to 1965 [4,27,66,73], when surfactant concentrations in the effluent were probably higher than they are today.

Experiments with ABS and LAS surfactants indicate that they are retained in the hydrophobic neutral DOC fraction [67]. Only about 50% of LAS and ABS surfactants is carbon; thus, the MBAS of 2.0 mg/L in well F282-94 (Table 1) equates to 1.0 mg/L DOC, and surfactants account for 25% of the DOC. However, using DOC_f data and assuming that the entire

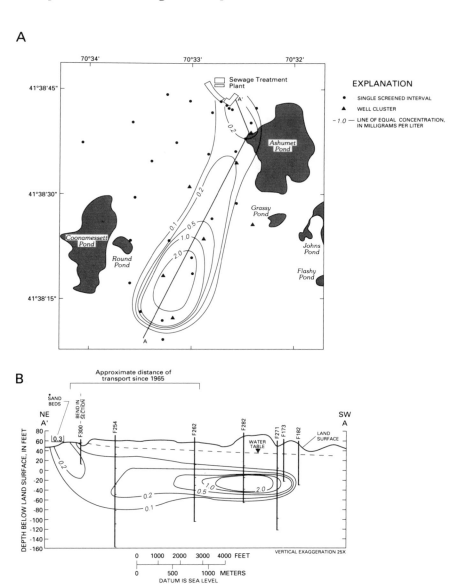

Figure 6 Plan view (A) and cross section (B) showing distribution of anionic surfactants measured as methylene-blue-active substances (mg/L) during 1983 sampling. Plan-view contours are plotted using maximum concentrations measured in
vertical profiles (see cross section and Table 1). Also shown is the estimated transport
distance since 1965 for the tailing edge of branched-chained alkylbenzenesulfonates.

Table 2 Data on Volatile Chlorinated Hydrocarbons Determined by Purge-and-Trap Gas Chromatography/Mass Spectrometry and Total Purgeable Organic Chloride Analysis for Select Groundwater Samples Collected in 1988[a]

Sample[b]	Distance[c]	1,1-DCE	1,2-DCE[d]	TCE	PCE	1,1-DCEa	1,1,1-TCEa	TCM	PT-POCl	POCl
F300-10	500	<0.2	<0.2	<0.2	1.0	<0.2	<0.2	<0.2	1.0	<0.2
F300-30		<0.2	0.3	1.5	0.4	<0.2	<0.2	<0.2	1.8	1.5
F300-50		<0.2	<0.2	<0.2	<0.2	<0.2	<0.2	<0.2	<0.2	0.2
F300-73		<0.2	<0.2	0.2	0.6	<0.2	<0.2	<0.2	0.7	0.3
F300-99		<0.2	<0.2	1.8	4.9	<0.2	<0.2	0.6	6.2	2.8
F254-26	900	<0.2	0.3	<0.2	<0.2	<0.2	<0.2	0.3	0.5	0.8
F254-54		<0.2	5.1	3.7	1.0	<0.2	<0.2	<0.2	7.5	6.2
F254-72		<0.2	240	22	530	5.0	<0.2	<0.2	237	206
F254-107		<0.2	8.7	9.0	14	<0.2	1.2	<0.2	27	19
F254-140		<0.2	2.4	20	0.3	<0.2	0.6	<0.2	19	16
F254-168		<0.2	<0.2	<0.2	<0.2	<0.2	<0.2	<0.2	<0.2	2.0
F254-216		<0.2	<0.2	<0.2	<0.2	<0.2	<0.2	<0.2	<0.2	1.2

Well	Distance									
F262-41	2200	<0.2	<0.2	<0.2	0.3	<0.2	<0.2	0.5	0.7	1.4
F262-69		<0.2	0.5	14	91	0.8	<0.2	<0.2	90	80
F262-85		<0.2	<0.2	5.1	21	<0.2	<0.2	<0.2	22	20
F262-109		0.4	3.9	27	5.8	0.9	1.4	<0.2	32	28
F262-159		<0.2	<0.2	0.6	<0.2	<0.2	<0.2	<0.2	0.5	0.8
F282-49	2900	<0.2	<0.2	<0.2	2.1	<0.2	<0.2	1.8	3.4	1.5
F282-70		<0.2	1.6	53	80	<0.2	<0.2	0.5	198	140
F282-94		0.2	<0.2	8.0	15	<0.2	1.3	<0.2	21	21
F282-123		<0.2	<0.2	<0.2	<0.2	<0.2	<0.2	<0.2	0.2	0.5
F294-64	3500	<0.2	<0.2	<0.2	<0.2	<0.2	<0.2	1.1	1.0	2.2
F294-77		<0.2	<0.2	8.2	10	<0.2	0.5	0.5	16	15
F294-89		<0.2	<0.2	15	8.3	<0.2	0.7	<0.2	20	9.3
F294-109		<0.2	<0.2	<0.2	<0.2	<0.2	<0.2	<0.2	0.0	0.4

DCE, dichloroethene; DCEa, dichloroethane; TCE, trichloroethene; TCEa, trichloroethane; PCE, tetrachloroethene; TCM, trichloromethane; PT-POCl, total purgeable organic chloride determined by PT-GC/Ms; POCl, total purgeable organic chloride.
[a]Concentrations in μg/L for individual compounds and μg Cl/L for POCl.
[b]Value after hyphen is depth below land surface (in feet).
[c]Distance downgradient from the infiltration bed source (in meters).
[d]Combined *cis* and *trans* isomers.

hydrophobic neutral fraction is surfactant related, surfactants account for 65% of the DOC.

4. Purgeable Organic Chloride

Volatile chlorinated hydrocarbons (VCHs) such as chlorinated ethanes and ethenes are among the most commonly detected organic contaminants in groundwater [109,110]. Although these compounds are typically determined by PT-GC/MS analysis, they also can be detected by nonspecific POCl analysis. Analysis by POCl does not provide information on individual compounds, but it does provide a rapid, inexpensive way to screen samples and to track total VCH concentrations [56].

Laboratory POCl spike and recovery experiments for a variety of VCHs indicate that, at concentrations of 0.2–100 µg/L, recoveries average about 95% and relative standard deviations (RSDs) are about 10%. Table 2 and Fig. 7 summarize POCl data for 1988 and indicate a plume of VCH that extends more than 3000 m from the infiltration beds. Concentrations of POCl range from less than 0.2 to more than 900 µg/L (not shown in Table 2). POCl will be discussed further in Section IV.C.2.

C. Specific Organic Compound Contamination

1. Closed-Loop Stripping

More than 200 compounds were detected by CLS-GC/MS analysis in the contaminated groundwater, and many of these were also detected in the sewage effluent. About 60 compounds have been identified by library matches of mass spectra, including chlorinated hydrocarbons (aliphatic and aromatic), alkyl-substituted hydrocarbons (aliphatic and aromatic), alkylphenols, aldehydes, and phthalate esters. These compounds are common in secondary sewage [8,9,11–18,95,96]. Trichloroethene (TCE) and tetrachloroethene (PCE) are the major contaminants and occur at concentrations orders of magnitude higher than those of other compounds. The only other contaminants detected at concentrations greater than 1 µg/L were the combined isomers (primarily *para* and *ortho*) of dichlorobenzene (DCB), the combined

---

Figure 7 Plan view (A) showing distribution of total volatile chlorinated hydrocarbons (in µg/L) measured as purgeable organic chloride, and cross section (B) showing relative proportions of DCE, TCE, and PCE measured by purge-and-trap GC/MS during 1988 sampling. In the cross section, the bar size for wells with only a single compound represents 100% scale, and the numbers represent total concentrations. Plan-view contours are plotted using maximum concentrations measured in vertical profiles (see cross section and Table 2).

A

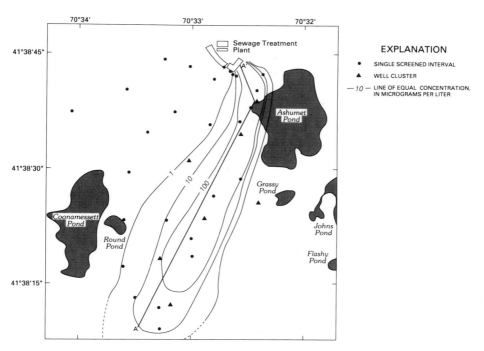

B

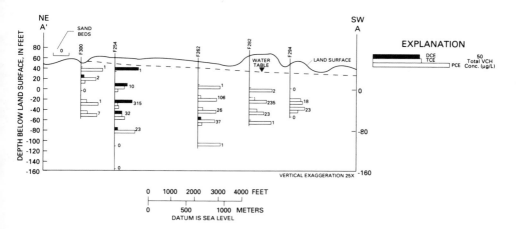

isomers of nonylphenol (NP), 2,6-di-*tert*-butylbenzoquinone (DTBB), alkyl-benzene isomers, and phthalate esters. Tri- and tetrachloroethane are widely used as solvents [109,110]. Dichlorobenzene isomers are common components of cleaning solutions and toilet bowl deodorants [111]. Nonylphenol isomers are degradation products of alkylphenol polyethoxylate (APEO) nonionic surfactants [112]. DTBB is the oxidation product of 2,6-di-*tert*-butylphenol, a widely used antioxidant [113,114]. Alkylbenzene isomers have a variety of sources and are commonly associated with petroleum products [111]. Phthalate esters are ubiquitous contaminants associated with plasticizers [111].

The total mass of semivolatile compounds measured by CLS-GC/MS ranged from 1 to more than 100 μg/L (about 1% of the total DOC). The CLS method is most efficient for a concentration range of 10–1000 ng/L per compound. At lower concentrations (1) there is insufficient mass to provide reliable mass spectra, (2) the background matrix becomes more complex, and (3) system contamination becomes more severe. At concentrations greater than 1000 ng/L, the activated carbon trap becomes saturated, resulting in breakthrough and decreased compound recovery. Average recovery of surrogate standards from distilled water blanks ranged from 50% to 80% with RSDs around 20%, and recovery from groundwater samples ranged from 25% to 50% with RSDs of about 30%. Recovery from groundwater was lower than from distilled water owing to (1) losses resulting from biodegradation and sorption during sample storage, (2) volatilization during sample handling, (3) saturation of the activated carbon trap by coisolated compounds, and (4) enhanced solubility of contaminants resulting from association with dissolved surfactants and humic material. Distilled water spike and recovery experiments for target compounds indicated that recovery was 35–50% for TCE and PCE; more than 80% for DCB, DTBB, and alkyl-benzenes; and 25% for nonylphenol.

A chromatogram for a typical distilled water blank with surrogate and internal standards is shown in Fig. 8A. Compounds considered to be groundwater contaminants were absent from the blanks. Contamination introduced during sampling was evaluated at well clusters that had screens in both the uncontaminated and contaminated parts of the aquifer. Control wells (i.e., F242, Fig. 8B) had only trace levels of organic compounds and were similar to distilled water blanks. In contrast, wells screened within the plume had high levels of contamination (Fig. 8C). Each groundwater sample produced a chromatogram that was a relatively unique "fingerprint" of organic contamination. Samples collected from uncontaminated wells immediately after sampling contaminated wells did not show cross-contamination.

Table 3 presents CLS-GC/MS data from two locations in the plume. Replicate samples were collected from wells F239-64 and F254-72 over a one-year

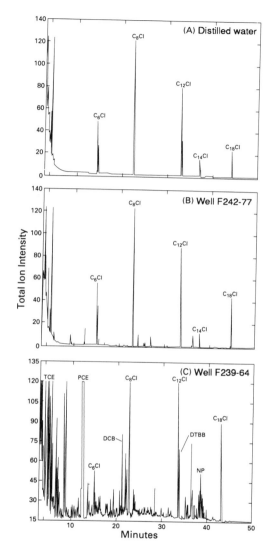

Figure 8 Total ion chromatograms from closed-loop-stripping GC/MS analysis of (A) distilled water blank, (B) uncontaminated groundwater from well F242-77, and (C) contaminated groundwater from well F239-64. Peaks marked C_6Cl, C_8Cl, $C_{12}Cl$, $C_{14}Cl$, and $C_{18}Cl$ are surrogate and internal standards added to the samples at a concentration of 110 ng/L. Also shown are peaks for the target compounds. TCE, trichloroethene; PCE, tetrachloroethene; DCB, dichlorobenzene; DTBB, 2,6-di-*tert*-butylbenzoquinone; NP, nonylphenol.

Table 3 Summary of CLS Data for Time-Series Replicate Analysis of Samples from Wells F254-72 and F239-64 Collected in 1984

	Concentration[a]					
	F254-72			F239-64		
Compound	1	2	3	1	2	3
Trichloroethene	220	1100	540	2840	2610	6440
Methyl cyclohexane	60	150	0	4	10	20
2,3-Dimethyl-2-butanol	5	70	50	7	6	7
Methylbenzene	0	8	0	500	80	120
Tetrachloroethene	9320	16560	13640	18990	22400	11590
Trimethyl cyclohexane isomers	50	50	30	0	90	70
1-Chlorohexane[b]	40	70	20	40	80	60
Dimethylbenzene isomers	20	30	10	10	220	180
Trimethylbenzene isomers	100	70	40	90	180	160
C_{12} Hydrocarbon	40	70	50	20	130	60
1,4-Dichlorobenzene	100	80	70	30	50	30
Decahydronaphthalene	10	20	3	0	20	10
1-Chlorooctane[c]	110	110	110	110	110	110
2,3-Dihydro-1-methyl-1H-indene	30	30	30	10	10	10
1-Chlorododecane[b]	50	60	2	0	130	50
2,6-Di-*tert*-butyl-*p*-benzoquinone	310	150	390	110	100	40
1,1,3,3-Tetramethylbutyl-phenol	350	80	120	30	50	10
Nonylphenol isomers	1100	240	480	290	790	180
Dibutyl phthalate	450	110	190	3	0	60

[a]Concentrations in ng/L calculated against 1-chlorooctane internal standard.

[b]Surrogate standard.

[c]Internal standard.

period to evaluate variability with time. The results from both wells had good analytical and sampling precision over the one-year period for most compounds (exceptions are TCE and PCE). In contrast, concentrations of many compounds increased significantly at well F282 between 1983 and 1984 (data not shown). The increase in concentrations indicates that the plume of contamination had moved into well F282, an observation that is consistent with boron, DOC, and MBAS data.

Distributions of TCE, PCE, DCB, DTBB, and NP determined by CLS-GC/MS (1983 and 1984) delineated several distinct plumes of contamination

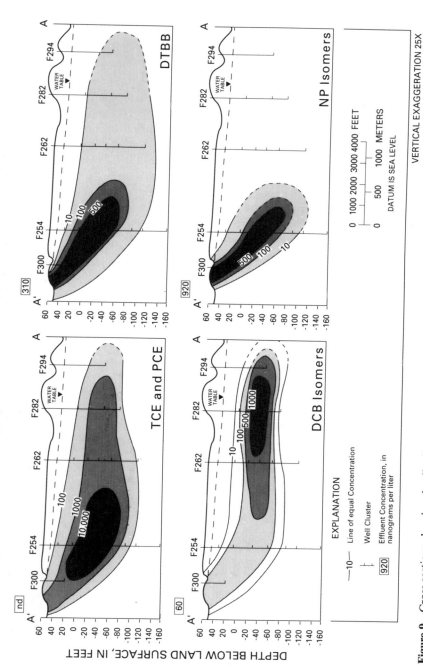

Figure 9 Cross sections showing the distribution of (A) combined trichloroethene (TCE) and tetrachloroethene (PCE), (B) di-*tert*-butylbenzoquinone (DTBB), (C) total dichlorobenzene (DCB) isomers, and (D) total nonylphenol (NP) isomers determined by closed-loop-stripping GC/MS analysis during 1983 and 1985 sampling.

(Fig. 9). TCE, PCE, DCB, and DTBB had maximum downgradient transport distances similar to those of SC, boron, MBAS, and DOC (Figs. 3–6). Assuming the same source histories, the compounds appear to be transported at about the same velocity. Maximum concentrations for TCE and PCE occurred near well F254. Although concentrations of several hundred micrograms per liter of TCE and PCE were measured in the contaminated groundwater, these compounds were not detected in the sewage effluent. The DCB plume had maximum concentrations near well F282. Concentrations of DCB in the sewage effluent and contaminated groundwater near the infiltration beds were similar but were less than DCB concentrations in the downgradient part of the plume. The DTBB plume had a distribution similar to those of the TCE and PCE plumes, with maximum concentrations at well F254. DTBB concentrations in groundwater near the infiltration beds were similar to concentrations in the effluent. The plume of NP was restricted to the area near the infiltration beds, and concentrations in the groundwater were similar to those in the sewage effluent. The distributions of these target compounds were representative of distributions for other trace organic contaminants.

The distributions of TCE and PCE determined by CLS-GC/MS were consistent with those determined by PT-GC/MS. Although concentrations determined by PT-GC/MS were much higher than those determined by CLS-GC/MS (as a result of better recoveries, see Section IV.C.2), the apparent maximum extent of TCE and PCE contamination was farther downgradient for CLS-GC/MS data than for PT-GC/MS data (because of CLS-GC/MS's lower detection limit).

2. Purge-and-Trap Analysis

The major contaminants determined in the groundwater by PT-GC/MS were total (*cis* and *trans* isomers) 1,2-dichloroethene (DCE), TCE, and PCE, all of which occurred at concentrations from 1 μg/L to more than 100 μg/L. DCE was not detected by CLS-GC/MS because it coelutes with the dichloromethane solvent peak. Several other compounds, including 1,1-dichloroethane, 1,1-dichloroethene, 1,2-dichloroethane, 1,1,1-trichloroethane, trichloromethane, and tetrachloromethane, were detected infrequently or at concentrations less than 1 μg/L. Distilled water spike and recovery experiments indicate that at concentrations of 0.2–100 μg/L, recoveries for DCE, TCE, and PCE are greater than 90%, and RSDs are about 10% [56]. The only contaminants detected in distilled water field and laboratory blanks were trichloromethane and dichloromethane, common laboratory solvents.

Results from PT-GC/MS analysis for select wells sampled in 1988 are summarized in Table 2. In some wells, the combined concentrations of DCE, TCE, and PCE were nearly 1000 μg/L; however, concentrations typically were less than 10 μg/L. About 75% of the total samples analyzed (166) had measurable DCE, TCE, or PCE. Figure 7 shows the distribution of total VCH determined by POCl (plan view) and distributions of DCE, TCE, and PCE determined by PT-GC/MS (cross section). Maximum concentrations were restricted to the upgradient part of the plume and did not coincide with maxima in DOC and MBAS. Also, the relative distribution of DCE, TCE, and PCE varies with location in the plume. In many wells, compound concentrations and ratios changed significantly between 1983 and 1988.

Results for PT-GC/MS analysis were in close agreement with POCl data when the organic chloride contents of individual compounds were summed. Because of the similarity in the data, POCl was a complementary method to PT-GC/MS for (1) screening samples for high or low levels of VCH, (2) providing an independent check of PT-GC/MS data, and (3) use as a surrogate measurement for monitoring VCH [56].

3. Liquid–Liquid Extraction

The LLE-GC/MS method measured compounds with low volatility and moderate polarity that were not measured by CLS or PT methods. Organic acids not amenable to GC/MS analysis were derivatized to form the methyl esters. The most important compounds detected in the sewage effluent and groundwater by LLE-GC/MS were APEO nonionic surfactant degradation products, including 1- and 2-alkylphenolethoxylate homologues (1- and 2-APEO), 1-and 2-alkylphenolethoxycarboxylate homologues (1- and 2-APEOC), and NP. Although only 1- and 2-APEO and 1- and 2-APEOC were detected, homologues with longer ethylene oxide chains may be present but were not measured by this method. The predominant alkyl-chain length was C_9, but C_8 isomers also were present. Concentrations of NP determined by LLE-GC/MS (40–50 μg/L) were greater than concentrations determined by CLS-GC/MS because of poor recoveries for NP by CLS.

The distribution of NP, 1- and 2-APEO, and 1- and 2-APEOC homologues in the sewage effluent and contamination plume was complex. Nonylphenol was present in the sewage effluent and in wells near the infiltration beds. The 1- and 2-APEO and 1- and 2-APEOC homologues were present in the midportion of the plume. Only the 1- and 2-APEOC compounds were detected in the downgradient wells.

4. Solid-Phase Extraction

Analysis of the octyl and cyclohexyl SPE extracts by full-scan GC/MS indicated high levels of analytical artifacts in cartridge blanks. As a result, selected ion monitoring (SIM) analysis was needed to eliminate interferences from cartridge bleed. Nonylphenol concentrations measured by SPE-SIM/GC/MS analysis (40–50 μg/L) were higher than concentrations measured by CLS-GC/MS analysis but were approximately the same as concentrations measured by LLE-GC/MS. The SPE method was useful for measuring NP and APEO at low concentrations only in the SIM mode; it was less useful for broad-spectrum analysis because of the interferences.

Anionic surfactants such as LAS are readily determined as MBAS. However, MBAS analysis is subject to interference and does not identify specific compounds. Field and Leenheer [115] reported that carboxylated LAS degradation products (sulfophenylcarboxylates, SPC) also respond to the MBAS test. To better characterize the material giving rise to the groundwater MBAS response, large-volume groundwater samples were collected from wells in the low-MBAS and high-MBAS zones.

The ^{13}C-NMR spectrum of the XAD-8 isolates from the high-MBAS zone confirmed the presence of nonbiodegradable ABS, whereas the spectra for samples from the low-MBAS zone near the infiltration beds show the presence of LAS [66]. The SPC intermediates of LAS were isolated by anion exchange. The ^{13}C-NMR spectra from the A-7 extracts indicate that SPCs are present in the groundwater and can account for about 25% of the MBAS response [68,69]. Although LAS and SPC were identified in the groundwater by ^{13}C-NMR, little information about isomeric or homologue distributions could be determined.

Infrared spectra of the XAD-8 and A-7 isolates contained characteristic absorption bands (*para*-substituted aromatic sulfonates) of LAS and SPC and indicated the presence of linear alkyl chains [68,69]. The ^{1}H-NMR spectra of the isolates were characterized by sharp contaminant-related resonances that indicated that LAS, SPC, APEO, and APEOC were present, superimposed on broad resonances indicative of humic substances [68,69].

Data on individual isomers and homologues of ABS, LAS, and SPC were provided by GC/MS analysis of the trifluoroethyl derivatives of the XAD-8 and A-7 resin isolates. A complex mixture of LAS, SPC, and other related compounds, including dialkyltetralinesulfonates (DATS), were identified in the groundwater by the derivatization technique [68]. Analyses by fast atom bombardment mass spectrometry (FAB/MS) also indicated that LAS and DATS were present at trace levels in the groundwater near the infiltration beds [68,70].

Linear alkylbenzenesulfonate was specifically evaluated using SPE-HPLC/FD analysis of samples collected along a transect through the plume in 1989 [68,70,72]. Concentrations of LAS in the groundwater near the infiltration beds were below the detection limit of the method (10 μg/L), even though MBAS concentrations were 250 μg/L. Concentrations of ABS determined by SPE-HPLC/FD in the downgradient part of the plume were similar to concentrations determined as MBAS.

V. BIOGEOCHEMICAL PROCESSES AND TRANSPORT

The observed distribution of organic contaminants in the groundwater results from (1) hydrogeological factors such as dilution and dispersion, (2) variations in the sewage effluent composition over time, (3) changes in bed-loading location and amount of loading, (4) retardation resulting from sorption onto the aquifer sediments, and (5) biological degradation and transformation.

The sharp boundaries of the various organic and inorganic contaminant plumes (Figs. 3–7,9) show that dilution and dispersion have not substantially decreased concentrations in the center of the plume after 40 years and almost 4000 m of transport. Dilution of inorganic and organic solutes by mixing with groundwater beneath the infiltration beds is only about 10–15%.

The distributions of boron, DOC, MBAS, and DCB, all with maximum concentrations 3000 m downgradient from the infiltration beds, indicate that they were introduced into the aquifer in the late 1940s to mid-1960s, when activity at the military base was greatest. The distributions are consistent with estimated distances of transport based on assumptions of (1) similar source histories, (2) an average transport velocity of 0.3 m/day, and (3) negligible retardation of transport by sorption to the sediments. Lower concentrations in the upgradient portion of the plume for many compounds may be due, in part, to a smaller population and decreased chemical use on the military base in the past 20 years.

The elevated MBAS values in the downgradient part of the plume are due to ABS introduced prior to 1965 [27,66,73]. The observed distribution of ABS is consistent with the expected distance of transport (making the same assumptions as above) if no ABS was introduced after 1965. The transport distance of the tailing edge of the ABS plume was estimated to be about 1880 m from the infiltration beds, which coincides with the tailing edges of the downgradient zones of high concentrations of hydrophobic neutral DOC and MBAS (Figs. 5 and 6).

The distribution of ABS and LAS, which results from changing detergent formulations in 1965, is a good example of the effects of variation in the sewage effluent composition on the nature and distribution of organic contaminants in the groundwater. The distribution of NP, which is restricted to the area near the infiltration beds, also may be due to changing sewage effluent composition over time. If NP was introduced during the switchover from ABS to LAS, it should have been transported 1880 m since 1965 (making the same assumptions about transport given above). Although the distance of NP transport is consistent with its being introduced in the mid-1960s, early detergent literature reports that APEO surfactants (the likely source of NP) have been used as degreasing agents since the 1940s [103]. The distribution of TCE, PCE, and DCB, with maximum concentrations of DCB occurring downgradient from maximum TCE and PCE concentrations, indicates that DCB may have been introduced into the aquifer before TCE and PCE.

Separation of organic mixtures can result from differential rates of movement for individual compounds as the result of sorption to aquifer sediments. An estimate of the relative transport velocity of a given compound (compared to the bulk water) can be obtained by treating transport through the aquifer as a one-dimensional process with constant flow in a homogeneous porous medium [80]. The average retardation factor (R_{fz}) for a sorbing species (z) is related to the equilibrium distribution coefficient (K_{dz} in cm^3/g) between water and sediment. The octanol/water partition coefficient (K_{owz}) and the sediment organic carbon content (f_{oc}) can be used to estimate the value of K_{dz} [116–118]. Assuming that only the <0.125-mm fraction (f) of aquifer material is important for sorption and that sorption follows a linear isotherm, R_{fz} can be calculated from the equation [119]

$$R_{fz} = T_z/T_w = 1 + 3.2 f f_{oc}(K_{owz})^{0.72} p(1-e)/e$$

where T_z is the average residence time of solute z; T_w, the average residence time of water; p, grain density of the aquifer material (g/cm^3); and e, total porosity (decimal fraction). This equation is valid for sediments with greater than 0.1% by weight organic carbon ($f_{oc} = 0.001$). Interactions with the inorganic matrix also may be important for organic-poor sorbents such as aquifer sediments [118,120,121]. Compounds may be transported at a rate faster than predicted assuming equilibrium conditions at high groundwater flow velocities because of slow sorption kinetics.

The log K_{ow} values for TCE, PCE, DCB, DTBB, and NP are 2.29, 2.88, 3.38, 3.70, and 4.20, respectively [18,116]. With the values $f = 0.10$, $f_{oc} = 0.001$, $p = 2.65$ g/cm^3, and $e = 0.35$, R_f values for TCE, PCE, DCB, DTBB, and NP are 1.07, 1.19, 1.43, 1.73, and 2.66, respectively. These R_f values indicate that TCE and PCE should move at similar velocities, but both should move at a faster rate than DCB. The R_f calculations support the hypothesis that high TCE and PCE concentrations upgradient from maximum DCB concentrations (Fig. 9) are due to changing sewage effluent composition rather than sorption processes.

Although R_f calculations predict that TCE should be transported faster than PCE, DCB, and DTBB, all of the compounds were detected near the leading edge of the plume (well F294). Estimated travel distances between 1940 and 1983 (assuming a flow velocity of 0.3 m/day and the above R_f values) would be 4390 m for TCE, 3950 m for PCE, 3280 m for DCB, 2710 m for DTBB, and 1770 m for NP. Although observed transport distances for the apparent leading edges of the various plumes were slightly less than those predicted, the two estimates of transport are consistent and indicate decreasing transport distance with increasing K_{ow}. The calculations do not take into account variability in transport properties of the aquifer at the scale of the plume or the unknown source history with respect to initiation of disposal and initial composition of the effluent.

The observed distributions of the APEO nonionic-surfactant-derived compounds are consistent with a partition-based sorption model. The least soluble compounds (NP) have been transported the shortest distance, the intermediate solubility compounds (APEO) have been transported an intermediate distance, and the most soluble compounds (APEOC) have been transported the greatest distance. On the basis of limited data, there appears to be separation of NP from APEO and APEOC as a result of sorption to the low-carbon aquifer sediments. The similarity in the observed and predicted transport distance for NP, the highest K_{ow} compound considered, indicates that its distribution is controlled by sorption to the aquifer sediments rather than source variability.

The sorption of LAS and ABS by soils increases as SOC increases [104, 122]. However, because of their high water solubility (low K_{ow}), anionic surfactants have little tendency to sorb to low-carbon aquifer sediments. In fact, the presence of anionic surfactants in the groundwater may enhance transport of less soluble organic compounds [73,102,123].

The effect of biological processes on the fate of organic contaminants in the groundwater can be inferred from the chemical data. Most hydrocarbons identified in the groundwater are highly branched and are not readily biodegraded [124,125]. Similarly, DTBB is a highly branched molecule that is not rapidly biodegraded. Nonylphenol is the residual biological degradation product of APEO surfactants and is persistent.

Chlorinated hydrocarbons, such as DCE, TCE, and PCE, can be biologically degraded by reductive dehalogenation under highly reducing (methanogenic) conditions [126–131]; however, under weaker reducing conditions (denitrifying) they are resistant to biodegradation [132–134]. The distribution patterns for DCE, TCE, and PCE (Fig. 7B) follow a trend similar to the reductive dehalogenation pathway, suggesting that biodegradation may be occurring in the aquifer even though the groundwater is not strongly reducing (denitrifying). However, the observed distributions also may be the result of changes in solvent use. Capillary PT-GC/MS analysis was performed on select samples to determine the distribution of *cis*- and *trans*-DCE isomers, in an attempt to distinguish between solvent use and biogeochemical effects. Reductive dehalogenation results in *cis*-DCE being the predominant isomer [126,128]. DCE in the contaminated groundwater is predominantly the *cis* isomer, which is consistent with a reductive dehalogenation process. Vinyl chloride, one of the terminal reaction products of reductive dehalogenation, was not detected in any of the groundwaters. Although available data indicate that biodegradation may be affecting VCH distributions in the aquifer, further investigation is necessary to evaluate more specifically whether reductive dehalogenation is occurring.

Biological degradation of DCB in the aquifer also seems to be negligible, an observation that is not consistent with its reported behavior [118,135, 136]. Barber [137] attributed the apparent lack of DCB degradation in the aquifer to inhibiting factors, including (1) low concentrations of primary and secondary substrates [135,138,139] and (2) denitrifying redox conditions [132–134,136,140]. Because of the long residence times in the aquifer, the role of acclimation is probably not a significant factor contributing to persistency of labile organic compounds [98,141].

Concentrations of LAS were rapidly attenuated during initial infiltration, primarily because of biodegradation. However, further attenuation during transport in the aquifer was significantly less. The decrease in LAS concentration during infiltration is attributed to biodegradation rather than sorption because of the concomitant increase in the concentration of biological metabolites (SPCs). The amount of LAS sorption was negligible because of the small K_{ow} for LAS and the low organic carbon content of the sediments ($<0.1\%$). In contrast, DATS compounds, which are more resistant to biodegradation than LAS, were not removed during infiltration and subsequent transport [68,70]. Although only a minor component of the anionic surfactant mixture in the sewage effluent, DATS becomes the major anionic surfactant derived compound in the groundwater near the infiltration beds. The increased relative abundance of DATS with respect to LAS indicates preferential removal of biologically labile compounds. Biodegradation of ABS in the aquifer does not appear to be significant, as indicated by high concentrations of ABS in the groundwater after more than 25 years.

The presence of the suite of APEO nonionic surfactant biological degradation products in the contaminated groundwater is evidence that biodegradation of compounds of this class is occurring. However, there are presently insufficient data to evaluate the relative importance of biodegradation with respect to sorption in the fate and transport of these compounds in the aquifer.

VI. SUMMARY AND CONCLUSIONS

Evaluation of the biogeochemical fate of organic compounds in sewage-contaminated groundwater requires a comprehensive analytical approach that includes a range of techniques from simple bulk measurements to highly sophisticated procedures. Bulk inorganic and organic measurements, such as specific conductance and dissolved organic carbon (DOC), are useful for determining the total mass of material present and for establishing overall compound distribution patterns. Class-specific measurements such as dissolved organic carbon fractionation (DOC_f), methylene-blue-active substances (MBAS), and purgeable organic chloride (POCl) provide important data on compounds that may not be readily amenable to specific-compound analysis. For determining specific compounds, a variety of methods are necessary because of the complex nature of contaminant DOC. Purge-and-trap gas chromatography/mass spectrometry (GC/MS) provides information on volatile chlorinated hydrocarbons (VCHs), a class of compounds that are likely to occur at concentrations greater than 1 μg/L. Although VCHs are good indicators of organic contamination, they may be only the "tip of the iceberg" and have a complex suite of other trace organic contaminants associated with them. The determination of other types of contaminants requires very sensitive analytical techniques, such as closed-loop stripping GC/MS analysis for trace levels (< 1 μg/L) of semivolatile compounds, or very specific methods, such as solid-phase extraction followed by derivatization GC/MS analysis for anionic and nonionic surfactants.

Comprehensive chemical analysis of contaminated groundwater at the Cape Cod site indicates that most organic substances that are not removed during secondary-sewage treatment and infiltration are of a persistent nature. These compounds have remained in the groundwater for more than 40 years at concentrations that apparently have not been significantly decreased by physical, chemical, and biological processes.

ACKNOWLEDGMENTS

This chapter is the synthesis of studies conducted with many colleagues. The work was funded by the Toxic-Substances Hydrology Program of the U.S. Geological Survey (USGS). Field support for well installation and sample collection was provided by Denis LeBlanc, Richard Quadri, Stephen Garabedian, and Kathryn Hess, USGS Massachusetts District. Much of the research was done in close collaboration with Micheal Thurman, Jennifer Field, Jerry Leenheer, Richard Smith, Ronald Harvey, and Michael Schroeder. I also thank Wilfred Pereria, Jennifer Field, Stephen Garabedian, and Denis LeBlanc for reviewing this chapter.

REFERENCES

1. Miller, D.W. *Waste Disposal Effects on Ground Water: A Comprehensive Survey of the Occurrence and Control of Ground-Water Contamination Resulting from Waste Disposal Practices*, Premier Press, Berkeley, CA, 1980.

2. Pye, V.I., Patrick, R., and Quarles, J. *Groundwater Contamination in the United States*, Univ. Pennsylvania Press, Philadelphia, PA, 1983.

3. Pye, V.I., and Kelley, J. The extent of groundwater contamination in the United States. In *Studies in Geophysics—Groundwater Contamination*. National Academy Press, Washington, DC, 1984, p. 23.

4. LeBlanc, D.R. Sewage plume in a sand and gravel aquifer, Cape Cod, Massachusetts. U.S. Geol. Survey Water-Supply Paper 2218, 1984.

5. Bouwer, H. Design and operation of land treatment systems for minimum contamination of ground water. *Ground Water, 12*:140, 1974.

6. Bouwer, H., Lance, J.C., and Riggs, M.S. High-rate land treatment. II: Water quality and economic aspects of the Flushing Meadows Project. *J. Water Pollut. Control Fed., 46*:844, 1974.

7. Idelovitch, E., and Michail, M. Soil-aquifer treatment—a new approach to an old method of wastewater reuse. *J. Water Pollut. Control Fed., 56*:936, 1984.

8. Reinhard, M., Dolce, C.J., McCarty, P.L., and Argo, D.G. Trace organic removal by advanced waste treatment. *J. Environ. Eng., 105*:675, 1979.

9. McCarty, P.L., and Reinhard, M. Trace organic removal by advanced wastewater treatment. *J. Water Pollut. Control Fed., 52*:1907, 1980.

10. Roberts, P.V., McCarty, P.L., Reinhard, M., and Schreiner, J. Organic contaminant behavior during groundwater recharge. *J. Water Pollut. Control Fed., 52*:161, 1980.

11. Bouwer, E.J., McCarty, P.L., and Lance, J.C. Trace organic behavior in soil columns during rapid infiltration of secondary wastewaster. *Water Res., 15*:151, 1981.

12. Tomson, M.B., Dauchy, J., Hutchins, S., Curran, C., Cook, C.J., and Ward, C.H. Groundwater contamination by trace level organics from a rapid infiltration site. *Water Res., 15*:1109, 1981.

13. Clark, L., and Baxter, K.M. Organic micropollutants in effluent recharge to groundwater. *Water Sci. Technol., 14*:15, 1982.

14. Bouwer, H., and Rice, R.C. Renovation of wastewater at the 23rd Avenue rapid infiltration project. *J. Water Pollut. Control Fed., 56*:76, 1984.

15. Bouwer, E.J., McCarty, P.L., Bouwer, H., and Rice, R.C. Organic contaminant behavior during rapid infiltration of secondary wastewater at the Phoenix 23rd Avenue project. *Water Res., 18*:463, 1984.

16. Hutchins, S.R., Tomson, M.B., Wilson, J.T., and Ward, C.H. Fate of trace organics during rapid infiltration of primary wastewater at Fort Devens, Massachusetts. *Water Res., 18*:1025, 1984.

17. Montgomery, H.A.C., Beard, M.J., and Baxter, K.M. Effects of recharge of sewage effluents upon the quality of Chalk groundwater. *Water Pollut. Control, 83*:349, 1984.

18. Barber, L.B., II, Thurman, E.M., Schroeder, M.P., and LeBlanc, D.R. Long-term fate of organic micropollutants in sewage contaminated groundwater. *Environ. Sci. Technol., 22*:205, 1988.

19. Keith, L.H. (ed.) *Identification and Analysis of Organic Pollutants in Water*, Ann Arbor Sci. Publ., Ann Arbor, MI, 1976.

20. Keith, L.H. (ed.) *Advances in the Identification and Analysis of Organic Pollutants in Water*, Vols. 1 and 2, Ann Arbor Sci. Publ., Ann Arbor, MI, 1981.

21. Leenheer, J.A. Concentration, partitioning, and isolation techniques. In *Water Analysis*, Vol. III, Academic, New York, 1984, p. 83.

22. Suffet, I.H., and Malaiyandi, M. (eds.) *Organic Pollutants in Water: Sampling, Analysis, and Toxicity Testing* (Adv. Chem. Ser. 214), American Chemical Society, Washington, DC, 1987.

23. Pellizzari, E.D., Sheldon, L.S., Bursey, J., Michael, L.C., and Zweidinger, R.A. Master analytical scheme for organic compounds in water. Final Rep. on U.S. Environmental Protection Agency contract 68-03-2704, U.S. Environmental Protection Agency, Washington, DC, 1985.

24. Mather, K.F., Goldthwait, R.P., and Thiesmeyer, L.R. Pleistocene geology of western Cape Cod, Massachusetts. *Bull Geol. Soc. Am., 53*:1127, 1942.

25. Oldale, R.N. Notes on the generalized geologic map of Cape Cod. U.S. Geol. Survey Open-File Rep. 76-765, 1976.

26. Barber, L.B., II. Geochemical heterogeneity in a glacial outwash aquifer: effect of particle size and mineralogy on the sorption of nonionic organic solutes. Ph.D. Dissertation, Univ. Colorado, Boulder, CO, 1990.

27. Barber, L.B., II. Geochemistry of organic and inorganic compounds in a sewage-contaminated aquifer, Cape Cod, Massachusetts. Master Sci. Thesis, Univ. Colorado, Boulder, CO, 1985.

28. LeBlanc, D.R. Digital modeling of solute transport in a plume of sewage-contaminated ground water. In *Movement and Fate of Solutes in a Plume of Sewage-Contaminated Ground Water, Cape Cod, Massachusetts* (D.R. LeBlanc, ed.), U.S. Geol. Survey Toxic Waste Ground-Water Contamination Program, U.S. Geol. Survey Open-File Rep. 84-475, 1984, p. 11.

29. LeBlanc, D.R., Garabedian, S.P., Hess, K.M., Gelhar, L.W. Quadri, R.D., Stollenwerk, K.G., and Wood, W.W. Large-scale natural-gradient tracer test in sand and gravel, Cape Cod, Massachusetts. 1. Experimental design and observed tracer movement. *Water Resources Res., 27*:895, 1991.

30. Garabedian, S.P., LeBlanc, D.R., Gelhar, L.W., and Celia, M.A. Large-scale natural-gradient tracer test in sand and gravel, Cape Cod, Massachusetts. 2. Analysis of spatial moments for a nonreactive tracer. *Water Resources Res., 27*:911, 1991.

31. Wolf, S.H., Celia, M.A., and Hess, K.M. Evaluation of hydraulic conductivities calculated from multiport-permeameter measurements. *Ground Water, 29*: 516, 1991.

32. Thurman, E.M., Barber, L.B., II, Ceazan, M.L., Smith, R.L., Brooks, M.G., Schroeder, M.P., Keck, R.J., Driscoll, A.J., LeBlanc, D.R., and Nichols, W.J., Jr. Sewage contaminants in groundwater. In *Movement and Fate of Solutes in a Plume of Sewage-Contaminated Ground Water, Cape Cod, Massachusetts* (D.R. LeBlanc, ed.), U.S. Geological Survey Toxic Waste Ground-Water Contamination Program, U.S. Geol. Survey Open-File Rep. 84-475, 1984, p. 47.

33. Smith, R.L., and Duff, J.H. Denitrification in a sand and gravel aquifer. *Appl. Environ. Microbiol., 54*:1071, 1988.

34. Morin, R.H., LeBlanc, D.R., and Teasdale, W.E. A statistical evaluation of formation disturbance produced by well-casing installation methods. *Ground Water, 26*:207, 1988.
35. Gibb, J.P., Schuller, R.M., and Griffin, R.A. Procedures for the collection of representative water quality data from monitoring wells. Illinois State Water Survey Cooperative Groundwater Rep. 7, Champaign, IL, 1981.
36. Scalf, M.R., McNabb, J.F., Dunlap, W.J., Cosby, R.L., and Fryberger, J. *Manual of Ground-Water Sampling Procedures,* Natl. Water Well Assoc. NWWA/ EPA Series, Worthington, OH, 1981.
37. Gillham, R.W., Robin, M.J.L., Barker, J.F., and Cherry, J.A. *Groundwater Monitoring and Sample Bias,* Am. Petroleum Inst., 1983.
38. Barcelona, M.J., Gibb, J.P., and Miller, R.A. A guide to the selection of materials for monitoring well construction and ground-water sampling. Illinois State Water Survey Rep. 327(EPA-600/S2-84-024), Champaign, IL, 1983.
39. Cherry, J.A., Gillham, R.W., Anderson, E.G., and Johnson, P.E. Migration of contaminants in groundwater at a landfill: a case study. 2. Groundwater monitoring devices. *J. Hydrol., 63*:31, 1983.
40. LeBlanc, D.R., Quadri, R.D., and Smith, R.L. Delineation of thin contaminated zones in sand and gravel by closely spaced vertical sampling, Cape Cod, Massachusetts. *Eos, 68*:313, 1987.
41. Smith, R.L., Harvey, R.W., and LeBlanc, D.R. Importance of closely-spaced vertical sampling in delineating chemical and microbiological gradients in groundwater studies. *J. Contam. Hydrol., 7*:285, 1990.
42. Pettyjohn, W.A., Dunlap, W.J., Cosby, R., and Keeley, J.W. Sampling ground water for organic contaminants. *Ground Water, 19*:180, 1981.
43. Schuller, R.M., Gibb, J.P., and Griffin, R.A. Recommended sampling procedures for monitoring wells. *Ground Water Monit. Rev., 1*:42, 1981.
44. Gibb, J.P., and Barcelona, M.J. Sampling for organic contaminants in groundwater. *J. Am. Water Works Assoc., 76*:48, 1984.
45. Ho, J.S.Y. Effect of sampling variables on recovery of volatile organics in water. *J. Am. Water Works Assoc., 75*:583, 1983.
46. Barcelona, M.J., Helfrich, J.A., Garske, E.E., and Gibb, J.P. A laboratory evaluation of ground water sampling mechanisms. *Ground Water Monit. Rev., 4*:32, 1984.
47. Barcelona, M.J., Helfrich, J.A., and Garske, E.E. Sampling tubing effects on groundwater samples. *Anal. Chem., 57*:460, 1985.
48. Thurman, E.M., Brooks, M.G., and Barber, L.B., II. Sampling and analysis of volatile organic compounds in a plume of sewage-contaminated ground water. In *U.S. Geological Survey Program on Toxic Waste—Ground-Water Contamination*, Proceedings of the Second Technical Meeting, Cape Cod, Massachusetts, Oct. 21–25, 1985 (S.E. Ragone, ed.), U.S. Geol. Survey Open-File Rep. 86-481:B21, 1988.
49. Imbrigiotta, T.E., Gibs, J., Fusillo, T.V., Kish, G.R., and Hochreiter, J.J. Field evaluation of seven sampling devices for purgeable organic compounds in groundwater. In *Ground-Water Contamination: Field Methods* (A.G. Collins and A.I. Johnson, eds.), Am. Soc. Testing Materials Spec. Publ. 963, Philadelphia, PA 1988, p. 258.

50. Gibs, J., and Imbrigiotta, T.E. Well-purging criteria for sampling purgeable organic compounds. *Ground Water, 28*:68, 1990.

51. Fishman, M.J., and Friedman, L.C. (eds.) *Methods for Determination of Inorganic Constituents in Water and Fluvial Sediments,* U.S. Geol. Survey Tech. Water-Res. Inv., Bk. 5, Chap. A1, 1985.

52. American Public Health Association. *Standard Methods for the Examination of Water and Wastewater,* 15th ed., Am. Public Health Assoc., Washington, DC, 1981.

53. American Society for Testing and Materials. Annual book of ASTM standards, Sec. 11, Water and Environmental Technology, 11.02 (Method D 4129-82), Philadelphia, PA, 1983, p. 98.

54. Leenheer, J.A., and Huffman, E.W., Jr. Analytical method for dissolved organic-carbon fractionation. U.S. Geol. Survey Water-Res. Inv. Rept. 79-4, 1982.

55. Wershaw, R.L., Fishman, M.J., Grabbe, R.R., and Lowe, L.E. (eds.) *Methods for the Determination of Organic Substances in Water and Fluvial Sediments,* U.S. Geol. Survey Tech. Water Res. Inv., Bk. 5, Chapt. A3, 1983.

56. Barber, L.B., II, Thurman, E.M., and Takahashi, Y. Comparison of purge and trap gas chromatography/mass spectrometry and purgeable organic chloride analysis for monitoring volatile chlorinated hydrocarbons in ground water. *Ground Water*, 1992, in review.

57. Grob, K., and Zurcher, F. Stripping of trace organic substances from water—equipment and procedure. *J. Chromatogr., 117*:285, 1976.

58. Barber, L.B., II, Thurman, E.M., and Schroeder, M.P. Use of closed-loop stripping combined with capillary gas chromatography/mass spectrometry analysis, to define a plume of semi-volatile organic contaminants in ground water. In *Movement and Fate of Solutes in a Plume of Sewage-Contaminated Ground Water, Cape Cod, Massachusetts* (D.R. LeBlanc, ed.), U.S. Geol. Survey Toxic Waste Ground-Water Contamination Program, U.S. Geol. Survey Open-File Rep. 84-475, 1984, p. 89.

59. Coleman, W.E., Melton, R.G., Slater, R.W., Kopfler, F.C., Voto, S.J., Allen, W.K., and Aurand, T.A. Determination of organic contaminants by the Grob closed-loop-stripping technique. *J. Am. Water Works Assoc., 73*:119, 1981.

60. Coleman, W.E., Munch, J.W., Slater, R.W., Melton, R.G., and Kopfler, F.C. Optimization of purging efficiency and quantification of organic contaminants from water using a 1-L closed-loop-stripping apparatus and computerized capillary column GC/MS. *Environ. Sci. Technol., 17*:571, 1983.

61. U.S. Environmental Protection Agency. Guidelines establishing test procedures for the analysis of pollutants under the clean water act; Final rule, interim rule, and proposed rule. *Fed. Regist., 49*:209, 1984.

62. Giger, W., Stephanou, E., and Schaffner, C. Persistent organic chemicals in sewage effluents. I. Identifications of nonylphenols and nonylphenol ethoxylates by glass capillary gas chromatography/mass spectrometry. *Chemosphere, 10*:1253, 1981.

63. Stephanou, E., and Giger, W. Persistent organic chemicals in sewage effluents. 2. Quantitative determinations of nonylphenols and nonylphenol ethoxylates by glass capillary gas chromatography. *Environ. Sci. Technol., 16*:800, 1982.

64. Ahel, M., Conrad, T., and Giger, W. Persistent organic chemicals in sewage effluents. 3. Determinations of nonylphenoxy carboxylic acids by high-resolution gas chromatography/mass spectrometry and high-performance liquid chromatography. *Environ. Sci. Technol., 21*:697, 1987.

65. Rostad, C.E., Pereira, W.E., and Ratcliff, S.M. Bonded-phase extraction column isolation of organic compounds in groundwater at a hazardous waste site. *Anal. Chem., 56*:2856, 1984.

66. Thurman, E.M., Willoughby, T., Barber, L.B., II, and Thorn, K.A. Determination of alkylbenzenesulfonate surfactants in groundwater using macroreticular resins and carbon-13 nuclear magnetic resonance spectrometry. *Anal. Chem., 59*:1798, 1987.

67. Thurman, E.M., and Field, J.A. Separation of humic substances and anionic surfactants from ground water by selective adsorption. In *Aquatic Humic Substances: Influence on Fate and Treatment of Pollutants* (I.H. Suffet and P. MacCarthy, eds.), Adv. Chem. Ser. 219, American Chemical Society, Washington, DC, 1989, p. 107.

68. Field, J.A. Fate and transformation of surfactants in sewage contaminated groundwater. Ph.D. Dissertation, Colorado School of Mines, Golden, CO, 1990.

69. Field, J.A., Leenheer, J.A., Thorn, K.A., Barber, L.B., II, Rostad, C., Macalady, D.L., and Daniel, S.R. Comprehensive approach for identifying anionic surfactant-derived chemicals in sewage effluent and groundwater. *J. Contam. Hydrol.*, 1992, in press.

70. Field, J.A., Barber, L.B., II, Thurman, E.M., Moore, B.L., Lawrence, D.L., and Peake, D.A. Fate of alkylbenzenesulfonates and dialkyltetralinsulfonates in sewage contaminated groundwater. *Environ. Sci. Technol.*, 1992, in press.

71. Trehy, M.L., Gledhill, W.E., and Orth, R.G. Determination of linear alkylbenzenesulfonates and dialkyltetralinesulfonates in water and sediments by gas chromatography/mass spectrometry. *Anal. Chem., 62*:2581, 1990.

72. Castles, M.A., Moore, B.L., and Ward, S.R. Measurement of linear alkylbenzenesulfonates in aqueous environmental matrices by liquid chromatography with fluorescence detection. *Anal. Chem., 61*:2534, 1989.

73. Thurman, E.M., Barber, L.B., Jr., and LeBlanc, D.L. Movement and fate of detergents in ground water: a field study. *J. Contam. Hydrol., 1*:143, 1986.

74. Ceazan, M.L., Thurman, E.M., and Smith, R.L. Retardation of ammonium and potassium transport through a contaminated sand and gravel aquifer: the role of cation exchange. *Environ. Sci. Technol., 23*:1402, 1989.

75. Harvey, R.W., Smith, R.L., and George, L. Effect of organic contamination upon microbial distributions and heterotrophic uptake in a Cape Cod, Massachusetts, aquifer. *Appl. Environ. Microbiol., 48*:1197, 1984.

76. Hem, J.D. *Study and Interpretation of the Chemical Characteristics of Natural Waters*, 2nd ed, U.S. Geol. Survey Water-Supply Paper 1473, 1970.

77. Davis, S.N., Thompson, G.M., Bentley, H.W., and Stiles, G. Ground-water tracers: a short review. *Ground Water, 18*:14, 1980.

78. Bowman, R.S. Evaluation of some new tracers for soil water studies. *Soil Sci. Soc. Am. J., 48*:987, 1984.

79. Cherry, J.A., Gillham, R.W., and Barker, J.F. Contaminants in groundwater: chemical processes. In *Studies in Geophysics—Groundwater Contamination*, National Academy Press, Washington, DC, 1984, p. 46.
80. Freeze, R.A., and Cherry, J.A. *Groundwater*, Prentice-Hall, Englewood Cliffs, NJ, 1979.
81. Gschwend, P.M., and Reynolds, M.D. Monodisperse ferrous phosphate colloids in an anoxic groundwater plume. *J. Contam. Hydrol., 1*:309, 1987.
82. Waggott, A. An investigation of the potential problem of increasing boron concentrations in rivers and water courses. *Water Res., 3*:749, 1969.
83. Choi, W.W., and Chen, K.Y. Evaluation of boron removal by adsorption on solids. *Environ. Sci. Technol., 13*:189, 1979.
84. Greenwood, N.N., and Thomas, B.S. *The Chemistry of Boron*, Pergamon, Oxford, 1975.
85. Matthews, P.J. A survey of the boron content of certain waters of the greater London area using a novel analytical method. *Water Res., 8*:1021, 1974.
86. Carriker, N.E., and Brezonik, P.L. Sources, levels, and reactions of boron in Florida waters. *J. Environ. Qual., 7*:516, 1978.
87. Sims, J.R., and Bingham, F.T. Retention of boron by layer silicates, sesquioxides, and soil materials. Pt. 1. Layer silicates. *Soil Sci. Soc. Am. Proc., 31*: 728, 1967.
88. Sims, J.R., and Bingham, F.T. Retention of boron by layer silicates, sesquioxides, and soil materials. Pt. 2. Sesquioxides. *Soil Sci. Soc. Am. Proc., 32*: 364, 1968.
89. Sims, J.R., and Bingham, F.T. Retention of boron by layer silicates, sesquioxides, and soil materials. Pt. 3. Iron- and aluminum-coated layer silicates and soil materials. *Soil Sci. Soc. Am. Proc., 32*:369, 1968.
90. Rhoades, J.D., Ingvalson, R.D., and Hatcher, J.T. Adsorption of boron by ferromagnesian minerals and magnesium hydroxide. *Soil Sci. Soc. Am. Proc., 34*:938, 1970.
91. Bingham, F.T., and Page, A.L. Specific character of boron adsorption by an amorphous soil. *Soil Sci. Soc. Am. Proc., 35*:892, 1971.
92. McPhail, M., Page, A.L., and Bingham, F.T. Adsorption interactions of monosilicic and boric acid on hydrous oxides of iron and aluminum. *Soil Sci. Soc. Am. Proc., 36*:510, 1972.
93. Evans, C.M., and Sparks, D.L. On the chemistry and mineralogy of boron in pure and mixed systems; review. *Commun. Soil Sci. Plant Anal., 14*:827, 1983.
94. Painter, H.A. Chemical, physical, and biological characteristics of wastes and waste effluents. In *Water and Water Pollution Handbook*, Vol. 1 (L.L. Ciaccio, ed.), Marcel Dekker, New York, 1971, p. 329.
95. Garrison, A.W., Pope, J.D., and Allen, F.R. GC/MS analysis of organic compounds in domestic wastewater. In *Identification and Analysis of Organic Pollutants in Water* (L.H. Keith, ed.), Ann Arbor Sci. Publ., Ann Arbor, MI, 1976, p. 517.
96. Baird, R., Selna, M., Haskins, J., and Chappelle, D. Analysis of selected trace organics in advanced wastewater treatment systems. *Water Res., 13*:493, 1979.

97. Harvey, R.W., and George, L. Growth determinations for unattached bacteria in a contaminated aquifer. *Appl. Environ. Microbiol., 53*:2992, 1987.

98. Harvey, R.W., and Barber, L.B., II. Associations of free-living bacteria and dissolved organic compounds in a plume of contaminated groundwater. *J. Contam. Hydrol.*, 1992, in press.

99. Hughes, J.L., Eccles, L.A., and Malcolm, R.L. Dissolved organic carbon (DOC), an index of organic contamination in ground water near Barstow, California. *Ground Water, 12*:283, 1974.

100. Barcelona, M.J. TOC determinations in ground water. *Ground Water, 22*:18, 1984.

101. Leenheer, J.A. Study of sorption of complex organic solute mixtures on sediment by dissolved organic carbon fractionation analysis. In *Contaminants and Sediments—Analysis, Chemistry, and Biology* (R.A. Baker, ed.), Ann Arbor Sci. Publ., Ann Arbor, MI, 1980, p. 267.

102. Thurman, E.M. *Organic Geochemistry of Natural Waters*, Martinus Nijhoff/ Dr W. Junk, Dordrecht, The Netherlands, 1985.

103. Schwartz, A.M., and Perry, J.W. *Surface Active Agents*: *Their Chemistry and Technology*, Vol. I, Interscience, New York, 1949.

104. Wayman, C., Page, H.L., and Robertson, J.B. Behavior of surfactants and other detergent components in water and soil-water environments. Report prepared for the Technical Studies Program, Federal Housing Administration, U.S. Dept. of Housing and Urban Development, FHA Publ. 532, Washington, DC, 1965.

105. Layman, P.L. Detergent report—brisk detergent activity changes picture for chemical suppliers. *Chem. Eng. News, 62*:17, 1984.

106. Swisher, R.D. *Surfactant Biodegradation*, Marcel Dekker, New York, 1987.

107. Weaver, P.J., and Coughlin, F.J. Measurement of biodegradability. *J. Am. Soil Chem. Soc., 41*:738, 1964.

108. Maehler, C.Z., Cripps, J.M., and Greenberg, A.E. Differentiation of LAS and ABS in water. *J. Water Pollut. Control Fed., 39*:R92, 1967.

109. Zoeteman, B.C.J., Harmsen, K., Linders, J.B.H.J., Morra, C.F.H., and Slooff, W. Persistent organic pollutants in river water and ground water of the Netherlands. *Chemosphere, 9*:231, 1980.

110. Council on Environmental Quality. *Contamination of Ground Water by Toxic Organic Chemicals*, U.S. Govt. Printing Office, Washington, DC, 1981.

111. Moore, J.W., and Ramamoorthy, S. *Organic Chemicals in Natural Waters*: *Applied Monitoring and Impact Assessment*, Springer-Verlag, New York, 1984.

112. Giger, W., Brunner, P.H., and Schaffner, C. 4-Nonylp;henol in sewage sludge: accumulation of toxic metabolites from nonionic surfactants. *Science, 225*: 623, 1984.

113. Jungclaus, G.A., Lopez-Avila, V., and Hites, R.A. Organic compounds in an industrial wastewater: a case study of their environmental impact. *Environ. Sci. Technol., 12*:88, 1978.

114. Lopez-Avila, V., and Hites, R.A. Oxidation of phenolic antioxidants in a river system. *Environ. Sci. Technol., 15*:1386, 1981.

115. Field, J.A., and Leenheer, J.A. Response of anionic detergents and their carboxylated degradation products to the methylene-blue-active-substances test. In *U.S. Geol. Survey Toxic Substance Hydrology Program—Proceedings of the Technical Meeting,* Phoenix, AZ, Sept. 26–30, 1988 (G.E. Mallard and S.E. Ragone, eds.), U.S. Geol. Survey Water-Resources Investigation Rep. 88-4220, 1989, p. 614.

116. Chiou, C.T., Porter, P.E., and Schmedding, D.W. Partition equilibria of nonionic organic compounds between soil organic matter and water. *Environ. Sci. Technol., 17*:227, 1983.

117. Karickhoff, S.W., Brown, D.S., and Scott, T.A. Sorption of hydrophobic pollutants on natural sediments. *Water Res., 13*:241, 1979.

118. Schwarzenbach, R.P., and Westall, J. Transport of nonpolar organic compounds from surface water to groundwater: laboratory sorption studies. *Environ. Sci. Technol., 15*:1360, 1981.

119. Schwarzenbach, R.P., Giger, W., Hoehn, E., and Schneider, J.K. Behavior of organic compounds during infiltration of river water to ground water: field studies. *Environ. Sci. Technol., 17*:472, 1983.

120. McCarty, P.L., Reinhard, M., and Rittmann, B.E. Trace organics in groundwater. *Environ. Sci. Technol., 15*:40, 1981.

121. Karickhoff, S.W. Organic pollutant sorption in aquatic systems. *J. Hydraul. Eng., 110*:707, 1984.

122. Hand, V.C., and Williams, G.K. Structure–activity relationships for sorption of linear alkylbenzenesulfonates. *Environ. Sci. Technol., 21*:370, 1987.

123. Kile, D.E., and Chiou, C.T. Water solubility enhancements of DDT and trichlorobenzene by some surfactants below and above the critical micelle concentration. *Environ. Sci. Technol., 23*:832, 1989.

124. Hammond, M.W., and Alexander, M. Effect of chemical structure on microbial degradation of methyl-substituted aliphatic acids. *Environ. Sci. Technol., 6*:732, 1972.

125. Niemi, G.J., Veith, G.D., Regal, R.R., and Vaishnav, D.D. Structural features associated with degradable and persistent chemicals. *Environ. Toxicol. Chem., 6*:515, 1987.

126. Parsons, F., Wood, P.R., and DeMarco, J. Transformations of tetrachloroethene and trichloroethene in microcosms and groundwater. *J. Am. Water Works Assoc., 76*:56, 1984.

127. Kleopfer, R.D., Easley, D.M., Haas, B.B., Jr., Deihl, T.G., Jackson, D.E., and Wurrey, C.J. Anaerobic degradation of trichloroethylene in soil. *Environ. Sci. Technol., 19*:277, 1985.

128. Parsons, F., and Lage, G.B. Chlorinated organics in simulated groundwater environments. *J. Am. Water Works Assoc., 77*:52, 1985.

129. Parsons, F., Lage, G.B., and Rice, R. Biotransformation of chlorinated organic solvents in static microcosms. *Environ. Toxicol. Chem., 4*:739, 1985.

130. Barrio-Lage, G., Parsons, F.Z., Nassar, R.S., and Lorenzo, P.A. Sequential dehalogenation of chlorinated ethenes. *Environ. Sci. Technol., 20*:96, 1986.

131. Barrio-Lage, G., Parsons, F.Z., Nassar, R.S., and Lorenzo, P.A. Biotransformation of trichloroethene in a variety of subsurface materials. *Environ. Toxicol. Chem., 6*:571, 1987.

132. Bouwer, E.J., Rittmann, B.E., and McCarty, P.L. Anaerobic degradation of halogenated 1- and 2-carbon organic compounds. *Environ. Sci. Technol., 15*: 596, 1981.

133. Bouwer, E.J., and McCarty, P.L. Transformations of 1- and 2-carbon halogenated aliphatic organic compounds under methanogenic conditions. *Appl. Environ. Microbiol., 45*:1286, 1983.

134. Bouwer, E.J., and McCarty, P.L. Transformations of halogenated organic compounds under denitrification conditions. *Appl. Environ. Microbiol., 45*: 1295, 1983.

135. Hutchins, S.R., Tomson, M.B., Wilson, J.T., and Ward, C.H. Microbial removal of wastewater organic compounds as a function of input concentration in soil columns. *Appl. Environ. Microbiol., 48*:1039, 1984.

136. Kuhn, E.P., Colberg, P.J., Schnoor, J.L., Wanner, O., Zehnder, A.J.B., and Schwarzenbach, R.P. Microbial transformations of substituted benzenes during infiltration of river water to groundwater: laboratory column studies. *Environ. Sci. Technol., 19*:961, 1985.

137. Barber, L.B., II. Dichlorobenzene in ground water: evidence for long-term persistence. *Ground Water, 26*:696, 1988.

138. Boethling, R.S., and Alexander, M. Microbial degradation of organic compounds at trace levels. *Environ. Sci. Technol., 13*:989, 1979.

139. Bouwer, E.J., and McCarty, P.L. Modeling of trace organics biotransformation in the subsurface. *Ground Water, 22*:433, 1984.

140. Hutchins, S.R., Tomson, M.B., Wilson, J.T., and Ward, C.H. Anaerobic inhibition of trace organic compound removal during rapid infiltration of wastewater. *Appl. Environ. Microbiol., 48*:1046, 1984.

141. Wilson, J.T., McNabb, J.F., Cochran, J.W., Wang, T.H., Tomson, M.B., and Bedient, P.B. Influence of microbial adaptation on the fate of organic pollutants in ground water. *Environ. Toxicol. Chem., 4*:721, 1985.

5

Application of Soil-Gas Sampling Technology to Studies of Trichloroethylene Vapor Transport in the Unsaturated Zone

Bernadette M. Hughes,[*][†] R. David McClellan,[‡] and Robert W. Gillham

Waterloo Centre for Groundwater Research, University of Waterloo, Waterloo, Ontario, Canada

I. INTRODUCTION

The occurrence of surface or subsurface releases of halogenated organic solvents is of great concern because of the potential impact of solvent spills on underlying groundwater reserves. Although the solubilities of these compounds are quite low, they exceed the recommended drinking water concentration limits by several orders of magnitude. As a consequence, vast amounts of groundwater can become contaminated to unacceptable levels by relatively small volumes of solvent liquids. The spread of this contamination within the aquifer itself can result from the migration of the solvent both as an immiscible liquid and as a solute [1–3]. Migration of volatile contaminants can also take place above the water table as a result of vapor-phase transport in the soil gas of the unsaturated zone. This mechanism of transport is a potentially important one for the halogenated organic solvents because they have relatively high vapor pressures.

---

[*]*Present affiliation*: Environmental Project Control, Inc., Grafton, Massachusetts
[†]Now Bernadette Hughes Conant
[‡]*Present affiliation*: Intera Information Technologies Ltd., Toronto, Ontario, Canada

Following the spill of a halogenated organic solvent, some of the product will be retained as a residual liquid within the unsaturated zone by capillary forces. Volatilization of this liquid, followed by migration in the vapor phase, can result in contamination of both the gas and aqueous phases at some distance from the source. Furthermore, infiltration through the contaminated zone can result in the displacement of contaminated water to the underlying groundwater zone. In order to interpret the resulting contaminant distribution at a spill site and design an effective remediation scheme, a clearer understanding of the vapor-phase transport processes is required.

To address this problem, experiments investigating the behavior of trichloroethylene (TCE) vapors in the unsaturated zone [4] and the effectiveness of vapor extraction for the removal of TCE contamination [5] were conducted at an experimental site located at Canadian Forces Base Borden. These experiments simulated the case where residual liquid solvent retained following a spill is the primary source of soil-gas contamination. A soil-gas sampling methodology was developed for use in these experiments for the purpose of monitoring the spatial and temporal variations of TCE vapor concentrations surrounding an emplaced residual liquid source.

Soil-gas sampling for the detection of organic vapors has become an increasingly popular tool in site investigations [6–13]. The use of sampling methods involving the extraction of soil gas from the subsurface, coupled with analyses of the samples at the site, allows for the sampling of many points over a short time period and provides the advantage of "real-time" data. Most of the documented site investigations for which such soil-gas surveys have been used are concerned with the delineation of underlying groundwater plumes or the detection of leaking underground storage tanks. The requirements of these kinds of surveys will differ somewhat from more focused research projects such as our experiments at Borden and others [14, 15]. Detailed investigations of vapor transport processes require the delineation of small concentration changes over short distances and time spans rather than an averaging over large areas and times. Research applications in general will require a greater degree of resolution and data precision than applied site investigations.

The purpose of the following discussion is to provide an example of an application of soil-gas sampling technology to a research-oriented study. The methodology used in conducting the TCE vapor experiments at the Borden field site is described in detail, and some of the practical problems encountered during sampling and factors affecting data quality are discussed. The success of the methodology is assessed by considering the level of data quality attained and its usefulness in interpreting the experimental results. The discussion of the sampling methodology provided here will be limited to the results from the vapor transport experiments, although essentially identical procedures were used for the vapor extraction remediation experiments.

II. DESCRIPTION OF EXPERIMENTS

To investigate the nature of TCE vapor transport behavior in a natural field setting, two experiments were conducted at the Canadian Forces Base Borden site near Alliston, Ontario [16]. A residual liquid contaminant source was introduced into the silty sands of the unsaturated zone by mixing the sediments with liquid TCE at the ground surface and packing the mixture into an excavation. The source was cylindrical, with a vertical axis of 0.8 m and a radius of 0.6 m, and was buried 0.35 m below the ground surface. A network of soil-gas sampling probes was installed at radial distances of 0–9 m from the source and depths of 0.15–3.3 m. Over the course of the experiments the depth to the water table varied from approximately 3.4 to 4.2 m below ground surface. Most of the samplers were concentrated in multilevel nests along one radial axis oblique to the direction of groundwater flow. The configuration of the sampling network along this detailed profile is shown in Fig. 1.

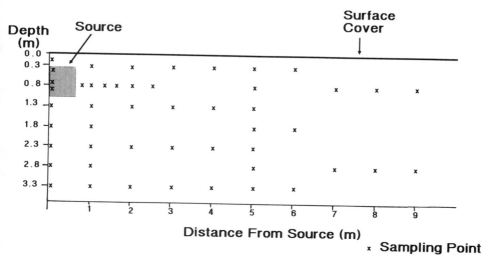

Figure 1   Configuration of the soil-gas sampling network along the detailed depth profile. Individual sampling points are referenced with the labeling system SW*a-b* where *a* is the radial distance of sampling nest from the source (in meters), and *b* is the sampler number in sequence from shallowest to deepest for each sampling nest.

In Experiment 1, conducted in November and December 1988, the ground surface of the 18 m × 18 m site was covered with a sheet of PVC plastic (0.51 mm thick) to prevent vapor losses to the atmosphere and infiltration of precipitation to the soil. Experiment 2 was carried out during July and August 1989, using essentially the same physical setup. However, the surface cover was removed for this experiment, leaving only a small cover directly overlying the source to prevent infiltration of precipitation through the residual liquid TCE. In addition to the altered surface boundary condition, the second experiment was subject to a different subsurface temperature profile. In each case, the migration of the vapor plume emanating from the emplaced source was monitored over time by collecting soil-gas samples from the network of sampling probes. The TCE vapor concentrations were determined by using a portable gas chromatograph located at the site.

III. SAMPLING EQUIPMENT AND METHODOLOGY

The primary requirement of the sampling methodology was to accurately define TCE vapor concentration gradients over short horizontal and vertical distances with minimal physical disturbance of the vapor contaminant plume by the extraction of soil-gas samples. This required the collection of samples of small volumes from well-defined locations within the unsaturated zone as well as the measurement of TCE vapor concentrations over a range of 6 orders of magnitude. The tendency for hydrophobic organic compounds to sorb to synthetic materials had to be considered in equipment design because of the potential for cross-contamination and false concentrations when plastics were used in the construction [17–20].

The resulting design consisted of a network of sampling probes constructed with very small screened intervals and internal volumes, which could be installed with little disturbance of the surrounding sediments. To minimize the problems of sorption and cross-contamination, sampling equipment was constructed of stainless steel and glass where practical. In addition, nests of individual samplers were used in preference to multilevel devices, and the use of a sampling manifold was avoided. A dilution step was required for the majority of the soil-gas samples prior to analysis to prevent overloading of the gas chromatograph detector by high vapor concentrations.

A. Sampling Probes

The probes were constructed at the University of Waterloo according to the design illustrated in Fig. 2. Each probe was constructed of 6.4 mm o.d. × 3 mm i.d. seamless stainless-steel tubing with a 1-cm-long intake screen and a solid stainless-steel drive point at the downhole end of the probe. The screened

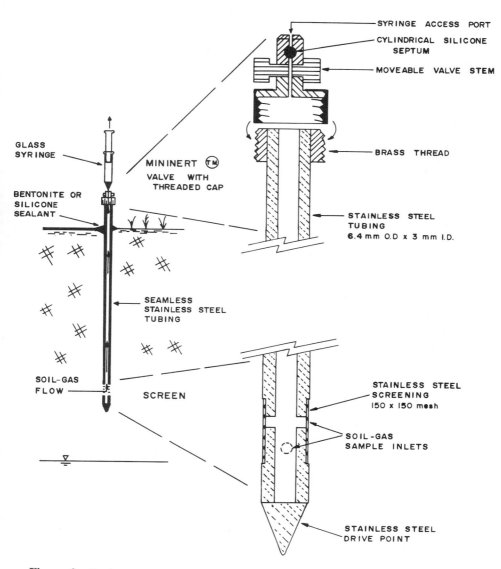

Figure 2 Design of the soil-gas sampling probes.

interval was constructed by machining a recess into a 1-cm section of the wall of the tubing. Three or four holes were drilled through this recessed section and were covered with a 150 × 150 mesh stainless-steel screening, which fit flush with the wall of the tubing. A brass thread was soldered to the top of the probe to accommodate the attachment of a screw-cap Mininert valve (see Fig. 2). The valve served as a reusable seal that could be opened and closed, allowing access to the probe by insertion of a syringe needle. These valves are constructed mainly of Teflon with a silicone rubber spetum core to provide a seal around the syringe needle when it is inserted through the access hole. Sample probes were constructed at a variety of lengths with internal volumes of approximately 5–30 cm³.

The sandy, stone-free nature of the sediments at the site and the narrow diameter of the probes allowed for installation of the samplers by driving them by hand using a clamp and hammer assembly. No drilling or augering of an annular space was required, thereby reducing the possibility of short-circuiting of gas along the outside of the sampler. To further minimize the potential for vertical short-circuiting and access to atmospheric air, the probes were sealed to the surface cover with silicone sealant. In the case of the open surface condition (experiment 2), bentonite powder was placed around the sampler at ground surface.

B. Sample Collection

Prior to the collection of a soil-gas sample, each probe was purged to ensure that the sample would accurately represent the in situ soil gas at the screen location. A purge volume equivalent to 3–4 times the internal volume of the probe was withdrawn using a 60-mL plastic syringe dedicated to the individual sampling probe. The magnitude of the purge volume was based on the results of purge tests carried out at the site on several sample probes of varying lengths. In these tests, samplers were purged incrementally until a relatively constant TCE vapor concentration was obtained for the extracted soil gas.

Following purging, a soil-gas sample was collected using one of two sampling methods. Depending on the anticipated TCE vapor concentrations determined from previous results or initial screening, gas samples were either diluted prior to analysis or injected directly into the gas chromatograph. For TCE vapor concentrations up to approximately 25 ppmv (parts per million per volume), a direct injection procedure was used. A gas sample was extracted from the probe using a 250-μL Hamilton Gastight syringe, and 50–200 μL of sample was injected directly into the gas chromatograph within approximately 1 min of sample collection. When a duplicate sample was required, the probe was resampled within 2–5 min of the original sample, but without repurging.

For soil-gas samples with vapor concentrations greater than 25 ppmv, a dilution step was required. A 0.5–10-mL sample was collected from the probe using a ground-glass syringe with a metal luer tip and a disposable needle. A small drop of distilled water was applied to the upper portion of the plunger to ensure an airtight seal. After an excess of sample was drawn and enough was expelled to give the desired volume, the sample was immediately injected into a 1-L amber glass dilution bottle that had been fitted with a screw cap and Teflon-lined septum (Fig. 3). After a period of at least 10 min was allowed for mixing of the sample and equilibration with the dilution vessel, a sub-sample of this dilution was injected into the gas chromatograph using a 250-μL Hamilton Gastight syringe. For all portions of the sampling procedures involving syringes, the syringes were equilibrated with the gas sample prior to sample collection by drawing the plunger up and down several times to flush the barrel.

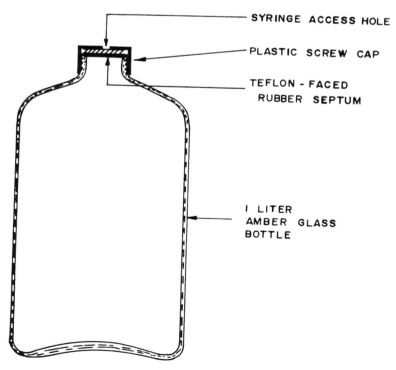

Figure 3 Dilution vessel.

Between analyses, the injection syringes were placed in a cleaning manifold and flushed with prepurified nitrogen. If syringe blanks indicated that significant contamination from exposure to high vapor concentrations persisted after cleaning, the Teflon tip on the end of the plunger was discarded and replaced. The fixed-needle model of the Hamilton syringes was chosen in preference to the removable-needle model because less Teflon is used in its construction. The glass syringes and glass dilution vessels were exposed to the atmosphere between samples and cleaned by flushing with a stream of nitrogen gas prior to each use.

C. Sample Analysis

The TCE vapor concentrations in the soil-gas samples were measured using a Photovac Model 10S70 gas chromatograph equipped with a photoionization detector. This gas chromatograph was chosen on the basis of its portable design for field use and the high sensitivity of the detector to chlorinated alkenes. The chromatograph was fitted with an encapsulated fused silica capillary column (CPSil5CB) and an isothermal column-oven, which was set at 40°C for the analyses. The carrier gas was ultrazero air, and the flow rate was set to 10–12 mL/min. Injections of all samples to the gas chromatograph were performed manually, with injection volumes ranging from 40 to 200 μL. The above configuration resulted in a retention time of approximately 70–80 sec for TCE, depending upon the exact value of the flow rate and the accuracy of the manual injection. A typical chromatogram produced using this configuration is shown in Fig. 4.

The gas chromatograph was calibrated at the start of each monitoring session by direct injection of gas standards. Standards were reanalyzed at the end of each session as a check on the calibration and periodically throughout when large numbers of analyses were conducted. The calibration runs were stored in three different memory locations or "libraries" in the chromatograph computer module. The libraries contained single-point calibrations at TCE concentration levels of approximately 0.5, 5.0, and 15 ppmv, respectively, with different machine settings for each. As a check on accuracy and linearity, a second gas standard at approximately 50% of the calibration concentration was analyzed for each library. Agreement of this second standard to within about 10% of its true concentration was chosen as the criterion for an acceptable calibration. The gas calibration standards were prepared at the site in the same type of glass bottles as were used for the dilution of the soil-gas samples. A known volume (3–30 μL) of a liquid stock solution was injected into a dilution bottle and allowed to fully volatilize and equilibrate with the bottle for a period of approximately 30 min. The liquid stock solutions were prepared in the laboratory by diluting TCE in methanol over a range of concentrations. The stock solutions were stored on ice at the site.

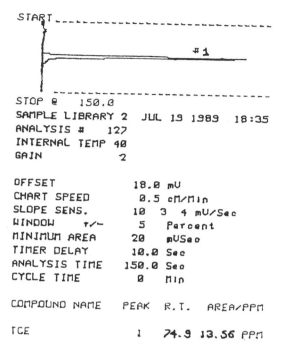

START

#1

STOP @   150.0
SAMPLE LIBRARY 2   JUL 19 1989   18:35
ANALYSIS #   127
INTERNAL TEMP 40
GAIN   2

OFFSET   18.0 mU
CHART SPEED   0.5 cM/Min
SLOPE SENS.   10 3 4 mU/Sec
WINDOW   +/-   5 Percent
MINIMUM AREA   20   mUSec
TIMER DELAY   10.0 Sec
ANALYSIS TIME   150.0 Sec
CYCLE TIME   0   Min

COMPOUND NAME   PEAK R.T.   AREA/PPM

TCE   1   74.9 13.56 PPM

Figure 4   Typical chromatogram for analysis using the Photovac 10S70 gas chromatograph.

IV. EVALUATION OF METHODOLOGY

A. Precision of Sampling Methodology

During monitoring of the vapor transport experiments, an attempt was made to incorporate repeat measurements at various stages of the procedure as an indication of the precision of the methodology. These measurements were not designed to provide detailed statistical information, and therefore the resulting data set does not support a rigorous analysis of variance. However, the following provides an estimate of the variability in the procedure based on the repeat analyses and provides useful information on the methodology.

The majority of the samples analyzed during the two experiments required a dilution step prior to injection into the gas chromatograph. During each monitoring session, duplicates were collected for 10–15% of those samples for which the dilution procedure was used, and the precision of the sampling methodology can be estimated from the many duplicate pairs. The degree of precision for a set of measurements can be presented by the relative standard deviation, also referred to as the coefficient of variation (CV):

$$CV\ (\%)\ =\ 100\ \times\ \frac{\text{standard deviation}}{\text{mean}} \tag{1}$$

for a single pair of measurements ($s1$ and $s2$) the CV can be obtained from

$$CV\ =\ 100\ \times\ \frac{\dfrac{|\ s1 - s2\ |}{\sqrt{2}}}{\dfrac{(s1 + s2)}{2}} \tag{2}$$

The CV value reflects the combined variability of the sample collection procedure and the chromatographic analysis. The statistical resolution for any single pair of measurements is very poor (only one degree of freedom). However, if it is assumed that the variance is constant, the CV measurements of the many duplicate pairs can be pooled to improve the resolution [21]. This pooled or mean CV value is then used as an approximation of the overall precision for the experiment.

The mean CV estimates for experiments 1 and 2 were 9.9% and 5.1%, respectively. These values were obtained as follows. For experiment 1, the CV values for a total of 54 duplicate pairs varied between 0.18% and 81%, with a mean CV of 12%. The mean CV is reduced to 9.9% if the two highest values are removed as outliers (Fig. 5A). For experiment 2 a total of 74 duplicate pairs were analyzed, with CV values ranging from 0.12% to 67%, with a mean of 9.8%. Visual analysis of the data suggested that CV values were not dependent on sample concentration for either experiment. However, if the CV values for experiment 2 are plotted versus the sampling time (Fig. 5B), it can be seen that the poorest agreements between duplicate samples occurred during the first 11 days of the experiment. This corresponds to a period prior to the time (day 12) at which the rubber septa in the valves were replaced in all the probes (see Section V.A). If all those CV values that exceed the maximum CV value observed after replacement of the septa are removed as outliers (see Fig. 5B), the mean CV calculated over the remaining data set is 5.1%.

The difference in the mean CV values for the two experiments arises from several factors including equipment performance, operator experience, and the time of year (and therefore temperatures) at which the experiments were conducted. In addition, the majority of concentration values in experiment 2 were determined by averaging the results of two or three replicate analyses per sample, which reduces the contribution of the analytical procedure error

A)

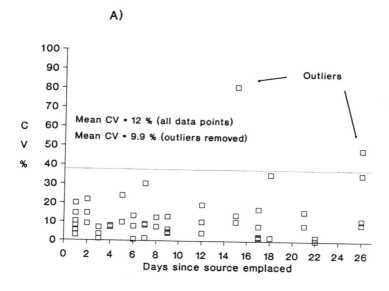

B)

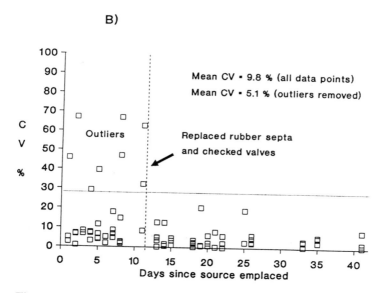

Figure 5 Coefficient of variation (CV) values versus time of sample taking for duplicate analyses of diluted samples for (A) experiment 1 and (B) experiment 2. Each point represents the CV value for a duplicate sample pair.

to the total variation. Replicate analyses were performed for only a small percentage of the samples in experiment 1.

The variability associated with the analytical procedure alone can be estimated from the replicate analyses of single samples. The mean CV value determined for double or triple injections of samples in experiment 1 was 3.6% based on values for 22 samples, and 2.1% for experiment 2 based on 106 samples. The CV values compare well with laboratory values obtained using larger numbers of replicate analyses. The laboratory tests generally gave mean CV values of about 3%, although the use of several different syringes for the analyses of the same sample was found to increase this value to between 4% and 7%. It should be stressed that the variability described by these estimates is included in the variability of the overall sampling procedure described by the comparison of the duplicate samples discussed above.

To partition the overall variability into the contributions from the analytical procedure and sample collection components of the methodology, a nested-design analysis of variance would be appropriate [21,22]. Because the monitoring scheme did not incorporate a balanced design with respect to the schedule of repeat sampling and injection procedures, it can provide only approximate analysis of variance results; nevertheless, it provides good qualitative insight. For both experiments, a greater portion of the variability appears to be due to the sample collection procedure than to the analytical procedure. Improvements in the sample collection technique for experiment 2, however, resulted in an apparent increase in the overall precision over experiment 1.

To characterize the level of precision associated with the direct injection technique used for low concentration soil-gas samples, duplicate and triplicate samples were collected periodically, although not on a regular schedule during each monitoring session. The CV values based on 34 duplicate and nine triplicate samples collected during experiment 2 varied between 0.19% and 21%, with a mean CV of 6.2%. This is slightly higher than the mean CV values obtained for the replicate injections of single samples from sample bottles as discussed above.

For each of the two experiments, an overall confidence interval was approximated using the estimates of total variation as given by the mean CV values for the duplicate analyses of the diluted samples. Based on the empirical rule for normal populations [23], a 95% confidence interval on concentrations was estimated for each experiment as plus or minus twice the mean CV value. This corresponds to a value of $\pm 20\%$ for experiment 1 and $\pm 10\%$ for experiment 2. A more conservative value of $\pm 20\%$ may be more appropriate for the data obtained in the first 11 days of experiment 2.

B. Accuracy, GC Calibration, and Response

The accuracy of a reported concentration reflects how closely the measured value represents the true in situ concentration. In the case of the soil-gas sampling methodology, this will be a function of the success in obtaining a representative sample and maintaining its integrity during subsequent injections and dilutions as well as the accuracy of the analytical procedure. When sampling natural systems, provided all reasonable precautions are taken, it can only be assumed that the sampler design and sampling procedure provide a representative sample, as it is virtually impossible to test this hypothesis. For the experiments described here, the results of duplicate sample comparisons were assumed to provide some indication of the ability to retain sample integrity during the dilution procedure. Alternatively, field samples can be spiked with another compound as a control to allow detection of sample deterioration prior to analysis. Smith and coworkers [14] used gaseous bromochloromethane in this manner as a check on TCE vapor samples collected in glass sampling bulbs.

The accuracy of the analysis itself is largely a function of the quality of the calibration procedure. The preparation of calibration standards from several different liquid stock solutions provided a check on the deterioration of any individual stock solution, but the accuracy of stock solution concentrations was not tested against independent calibration methods. Laboratory tests indicated that the precision in preparing the gas standards from any given stock solution was very good. Three identical gas calibration standards were prepared for each of six different liquid stock solutions with a mean CV value of 1.0% for the six trials. The preparation of daily calibration standards in the same bottles as were used for the diluted samples and the use of the same manual injection procedure subjected the calibration runs to the same biases as the sample analyses. Such biases could include temporal variation in instrument response, different operator techniques, sorption to analytical equipment, and the effects of variable atmospheric conditions such as temperature or humidity.

The accuracy of an analysis is affected by the degree of linearity of the detector response to the variables in the procedure. For the experiment configuration and manual injection method used in these experiments, the major parameters that varied between analyses included TCE concentration, injection volume, and signal gain. Laboratory tests were conducted in which gas standards were injected for various combinations of these three parameters. The results from these tests were variable but suggest that the detector exhibits a slight decrease in sensitivity with increase in concentration or injection volume and is relatively insensitive to gain. The number of data points and the consistency of trends from these tests were insufficient to allow a

A)

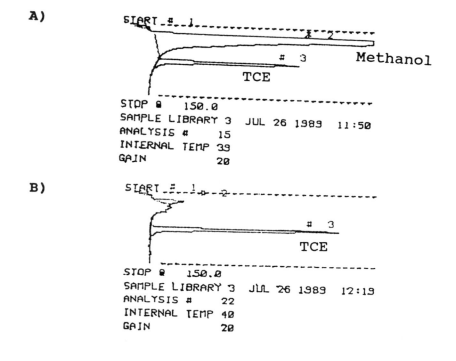

START # 1

2

3   Methanol

TCE

STOP @   150.0
SAMPLE LIBRARY 3   JUL 26 1989   11:50
ANALYSIS #   15
INTERNAL TEMP 39
GAIN   20

B)

START # 1   # 2

3

TCE

STOP @   150.0
SAMPLE LIBRARY 3   JUL 26 1989   12:19
ANALYSIS #   22
INTERNAL TEMP 40
GAIN   20

C)

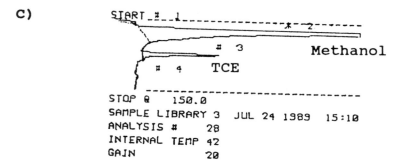

START # 1

2

3   Methanol

4   TCE

STOP @   150.0
SAMPLE LIBRARY 3   JUL 24 1989   15:10
ANALYSIS #   28
INTERNAL TEMP 42
GAIN   20

Figure 6 Chromatograms exhibiting an appropriate TCE peak baseline for (A) a calibration standard analysis with tailing from a methanol peak and (B) a soil-gas sample analysis with no methanol. A poor integration for a calibration standard analysis (C) results from inappropriate choice of input parameters.

regression analysis of the nonlinear relationships. In general, the linearity was found to be good to within about ±10–15% over an order-of-magnitude change in either injection volume or concentration. As there was some uncertainty regarding the magnitude and significance of the nonlinearities, attempts were made to minimize their effects by making use of the multiple calibration memories of the machine and extrapolating the single-point calibration values over small ranges of the concentration or injection volume. For the chromatograph model used, the computer module treats the calibrations from different memory libraries independently. Therefore, three libraries were calibrated at different machine settings, injection volumes, and concentrations. By determining the most appropriate library for each analysis, samples could generally be compared to a calibration standard analyzed within half an order of magnitude of the sample concentration and injection volume.

The methanol used in preparing the standards resulted in a methanol peak on the calibration chromatograms. For the library calibrated at the lowest concentration (about 0.20 ppmv) this presented some difficulty in obtaining good chromatographic results. The higher signal gain required to obtain a reasonable TCE peak area produced such a large methanol peak that the TCE peak was affected by the tailing of the methanol peak. Although good consistency could be obtained between the calibration runs, the soil-gas samples did not contain a methanol peak. This resulted in a subtle change in the shape of the base of the TCE peak. Determination of the baseline used for peak area integration depends on the slope tolerance factors input by the operator. By adjusting these factors, which determine the value of the change in slope of the chromatogram at which the baseline starts and ends for each peak, a satisfactory integration of both the calibration and sample runs could be achieved (Figs. 6A and 6B). Figure 6C is an example of an analysis of a calibration standard for which inappropriate slope factors were chosen, resulting in an underestimation of the true TCE peak area. For the 12 mL/min flow rate and 40°C analysis temperature, the 0.20 ppmv concentration was effectively near the lower limit at which the machine could be reliably calibrated using standards made from the methanol-based stock solutions. Better separation could have been achieved by decreasing the flow rate or column temperature at the expense of longer run times.

V. FACTORS AFFECTING DATA QUALITY

A. Problems with Sample Collection

Among the most problematical factors when dealing with gas samples is leakage. Leaks will generally result in a negative sampling bias caused by either gas losses from a sampling vessel or syringe, or dilution of a sample by atmospheric air as the sample is drawn under negative pressure. When very small gas volumes at negligible flow rates are involved, as is the case when syringes are used, these leaks are almost impossible to detect. The difficulty in dealing with such leakage effects is that they do not always result in a noticeable sample failure but may cause partial dilution of a sample, which will give a low but seemingly reasonable value.

Leakage around the syringe needle as the sample is being drawn from the probes is thought to be one of the major causes of the high CV values obtained for the duplicate analyses in the Borden vapor experiments. It was noted that after many piercings of the rubber septum of a Mininert valve, it no longer maintained a good seal around the syringe needle. This presented problems during purging or drawing of a sample, particularly in probes where the vacuum required to draw a sample was greater due to partial clogging of the screen or the presence of low-permeability sediments. The absence of a proper seal could have permitted atmospheric air to be drawn into the probe, along the annulus of the needle, and into the syringe. On day 12 of experiment 2, because of the poor condition of the rubber septa in several of the samplers, the septa were replaced in all the valves. In addition, the valves were tested for other possible points of leakage by covering each with a soap film and injecting a small volume of air into the probe under positive pressure using a syringe. Those valves that exhibited leaks were either tightened or replaced. For the remainder of the experiment, the condition of the valves was periodically checked, and the rubber septa were replaced whenever it was noted that either the syringe did not fit tightly in the seal or anomalously low TCE concentration values were obtained. The result, as given in Fig. 5B, was a marked improvement in the overall precision of the sampling methodology.

Other potential leaks during the sampling procedure include losses from the dilution bottles, losses from the "gastight" syringes used to collect the undiluted samples, and leaks during injection and analysis. Sample loss resulting from leakage through the septa of the dilution bottles is not thought to be a major cause of error. Several tests of the integrity of the seal on the dilution bottles showed that, even after the septum was punctured as many as 30 times over an 8-hr period, the samples showed very little deterioration. During injection of samples into the gas chromatograph, it was found that the septum on the injection port required regular replacement to avoid sample loss. The addition of a sample loop to the manual injection port could ensure a more reproducible injection volume. After many injections using "gastight" syringes, the small Teflon tips on the plungers, which maintain the seal against the glass barrel, became worn and deformed and had to be replaced. Also, a good seal was more difficult to maintain when the syringes were cold as a result of a slight contraction of the Teflon tips. This may be a contributing factor to the slightly better precision measured for the summer experiment (experiment 2). Kerfoot [24] determined that leaking gastight syringes were the controlling factor for precision in a soil-gas study investigating the distribution of chloroform above a contaminated groundwater plume.

A problem frequently encountered during sample collection and analysis was the plugging of the syringe needles by pieces of rubber septum or Teflon from either the sampler valves or the septa in the dilution vessels. Plugging of the purge syringe needles could generally be detected as a strong resistance to drawing of the purge volume, but it was more difficult to detect for the smaller sampling and injection syringes.

In some cases, failure of the sampling probes occurred as a result of the screen being plugged by fine-grained sediments. In addition, for the deeper samplers, rising water levels placed the screens within or close to the top of the capillary fringe, making it difficult to draw a gas sample. When water-table levels were high enough, water was drawn up the sampler and into the syringe during attempts to purge these samplers.

B. Realistic "Detection" Limits

For experiment 1, an attempt was made to determine the outer limits of the migrating plume, requiring the measurement of "non-detect" levels. The sensitivity of the gas chromatograph detector to TCE is quite high, with detection limits of less than 1 ppbv (parts per billion by volume). It was very difficult to ensure that contamination at such low levels was not introduced by the sampling and analytical equipment, particularly with upwards of 100 analyses per day. Some of the precautions taken during measurement of low-concentration samples included sampling of probes furthest from the source first, reserving some syringes for low-concentration samples only, thoroughly cleaning sampling and analytical equipment, and performing frequent blank runs to check for contamination of the syringes and the gas chromatograph. Although non-detect runs could be achieved in this way, some cross-contamination still occurred. In one case when the gas chromatograph appeared to be "clean," as indicated by an analysis of the carrier gas only, injection of a sample of ultrazero grade air using a new syringe produced a small TCE peak. As a result of potential contamination effects introduced by the procedure, analytical results below about 50 ppbv were not considered to be quantitatively accurate.

C. Temperature Effects

Several properties of gases such as density and vapor pressure are particularly sensitive to temperature variations. The mass of a given volume of soil gas will depend on the temperature of the sample. To determine the correct in situ concentration, some consideration must be given to how the analyzed sample is affected by temperature changes during the sampling procedure, how it relates to the calibration value for the analyses, and what units are used to describe that concentration. A simple application of the ideal gas law can be used to determine the magnitude of the density effect over a given temperature range:

$$\frac{n}{V} = \frac{P}{RT} \tag{3}$$

where P is pressure, V is volume, n is the number of moles of gas, T is temperature (in kelvin), and R is the ideal gas constant (in units consistent with P and V). For example, if the temperature of a soil gas is increased from 0° to 20°C while being maintained at atmospheric pressure, the expected density decrease will be approximately 7%. The significance of temperature effects in a sampling program will be highly dependent on the range of temperature changes encountered and the quality of data required.

In order to contour the results of the Borden field experiments in an unambiguous manner, the concentrations were calculated in units of parts per million by volume (ppmv), which are conservative with temperature change, unlike mass-based units such as μg/L. To properly calibrate the gas chromatograph to these units, the temperature of the calibration bottles was measured and the concentrations of the gas standards were corrected to this temperature using the ideal gas law. The sample dilution bottles were capped at approximately the same temperature as the calibration bottles to ensure that they contained the same molar volume of gas. Even after the septa were punctured, the bottles were capable of withstanding temperature-induced pressure changes without significant sample leakage. Thus, the analysis output concentration multiplied by the appropriate volumetric dilution factor gave a concentration in ppmv that was representative of the in situ sample, even though it was exposed to changing temperatures as it was drawn up the probe. The variation of temperature with depth for the two experiments is shown in Fig. 7.

For the samples analyzed by the direct injection method, however, the measured concentrations are not representative if the temperature of the syringe used to extract and inject the sample was different from that of the bottles and syringes used to calibrate the machine. This did not present a significant problem during the summer (experiment 2), when the temperature difference was generally within 5°C. However, during the winter (experiment 1) the sampling syringes may have been as much as 15–20°C colder. This could result in 7% overestimation of the true ppmv concentration. Maintaining a constant temperature of the calibration, sampling, and analytical equipment in the site trailer was also more difficult during experiment 1.

When dealing with vapor contamination at very high concentrations, an-

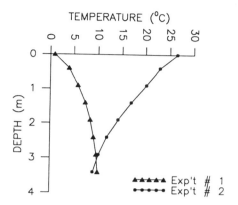

Figure 7 Variation of temperature with depth for experiments 1 and 2.

other property that can affect gas sampling results is vapor pressure. The vapor pressures of most organic compounds are highly sensitive to temperature. If a gas at or near saturation with respect to TCE vapor is moved to an area of lower ambient temperature, the lower equilibrium vapor pressure will result in some condensation of the sample. During experiment 1 the concentrations measured in gas samples collected from probes screened within the source material were consistently lower than predicted on the basis of the equilibrium vapor pressure for the temperature measured at that depth. Conversely, the measured source concentrations for experiment 2 were very close to the theoretically predicted values. This is consistent with the fact that the samples are drawn from depth for experiment 1 would cool and partially condense as they were drawn up the probes, but those collected during experiment 2 would warm, taking them further away from saturation conditions.

VI. DATA INTERPRETATION

To determine if the level of data quality achieved in a soil-gas survey is adequate, consideration must be given to the purpose of the survey and how the data are to be used in the interpretation. For the Borden vapor experiments, the goal was to monitor the migration of a TCE vapor plume over space and time with sufficient resolution to infer how that transport was affected by the spatial and temporal variations encountered in a natural system. Two main methods of data presentation were used in the interpretation of the results of the soil-gas surveys: (1) contoured profiles of TCE vapor concentration across the detailed sampling profile at a given time (concentration "snapshots") and (2) plots of TCE concentration versus time (breakthrough curves) for individual sampling points.

Figure 8 is an example of a contoured profile of TCE vapor concentrations measured along the detailed sampling profile for day 18 of experiment 2. The cross-hatched area in the upper left corner represents the location of the emplaced residual source, and the solid circles indicate the screen locations of the probes that were sampled during that monitoring session. The contours are presented in logarithms of the concentrations in ppmv, with a contour interval of 0.5. The profile is not continued beyond the 1.5 contour (32 ppmv) because of background contamination persisting from the first experiment. Of note, from a data quality perspective, is the exponential nature of the decline in concentration with distance from the source. It can be seen in this example that an error in data precision as large as ±100% for an individual sampling point would not greatly alter the contoured extent of the plume. In this case, the levels of data precision of ±10–20% indicated by the duplicate analyses are more than adequate. In addition to data precision, the degree of resolution of the geometry of the plume will also be a function of the spacing and number of data points.

Plots of the change in TCE concentration over time at individual sampling points were also used to interpret the transport behavior and were the main method used in matching the experimental data to simulations for the validation of a computer model [25]. Figure 9 presents an example of the breakthrough data obtained for the same probe for both experiments with the estimated 95% confidence limit intervals. Including these error bars on the data points provides a more realistic indication of the uniqueness of fit

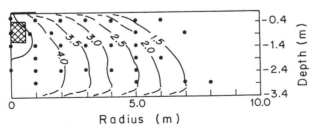

Figure 8 Contour plot of TCE vapor concentrations measured along detailed sampling profile for day 18 of experiment 2. Contours are in logarithm of vapor concentration in ppmv. Cross-hatched area in upper left marks the location of the emplaced source. Base of the domain represents the top of the capillary fringe.

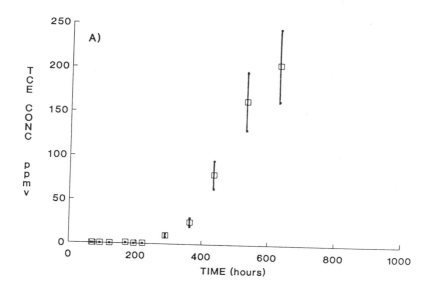

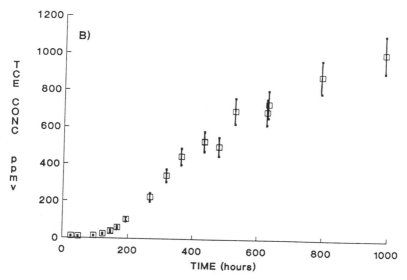

Figure 9 Breakthrough curves for sample probe SW3-1 for (A) experiment 1 and (B) experiment 2. The vertical bars indicate the estimated 95% confidence level intervals. Note that the scales on the Y axes are not equivalent.

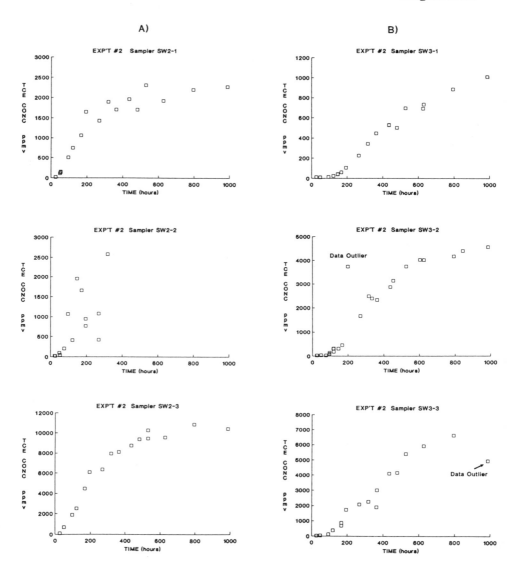

Figure 10   Comparison of breakthrough curves from experiment 2 for two vertical sequences of samplers demonstrating (A) different performance of individual samplers and (B) use of breakthrough curves to detect data outliers.

to be expected when interpreting data trends or curve-matching to model simulations.

A comparison of the variable performance of individual sample probes is provided in Fig. 10A. These three plots compare the results from experiment 2 for three probes from the same sampling nest with a vertical separation distance of 50 cm between samplers (see Fig. 1). The lower plot (SW2-3) is an example of a sample probe that produced a well-defined trend in concentration for the duration of the experiment. In this case the clear trend provides an indication that the data precision is very good, even without the benefit of the error bars. The shallower sample probe shown in the upper plot (SW2-1) also exhibits a well-defined trend in concentration over time but shows slightly more variation. Some of this variation may reflect true variations in vapor concentrations, but it could also have resulted from a slightly leaky valve or valve fitting. The middle plot from the probe located vertically between the other two is an example of a sampler that performed poorly and was abandoned during the experiment. Results from this sampler during experiment 1 were also poor, and it is suspected that the screen was clogged by fine sediments.

Plots of individual breakthrough curves on linear scales also make the presence of data outliers more apparent than on the logarithmic contour plots. The plots in Fig. 10B are good examples of curves with obvious outliers. The plot for sampler SW3-2 exhibits a point that clearly does not fit on the trend defined by the rest of the data. Comparison to the curves for the samplers directly above and below, as well as those probes laterally closest (not shown), suggest that the high value measured for SW3-2 does not reflect a real trend in the vapor plume or the response of the analytical equipment. This point likely represents a case where too large a volume was inadvertently injected into the dilution bottle. Similarly, a data outlier can be seen in the data for the deeper sampler (SW3-3). In this case the unreasonably low value was probably caused by leakage. Unrepresentative points such as these were removed from the depth profile plots before contouring.

VII. SUMMARY AND CONCLUSIONS

The sampling methodology described here represents an attempt to apply soil-gas surveying technology to field experiments investigating TCE vapor transport behavior in the unsaturated zone. This research required the detailed delineation of variations in TCE vapor concentrations over small distances and time intervals with concentrations spanning as many as 6 orders of magnitude. The sampling methodology developed in response to these criteria included a network of small-volume sampling probes with small screen intervals that required little disturbance of the subsurface during installation.

Glass and steel components were used whenever practical because of the noted tendency of organic compounds to sorb to synthetic materials. Where synthetic materials were used, attempts were made to minimize cross-contamination through cleaning procedures and strategies such as the use of dedicated sampling components. Vapor concentrations were determined on-site by use of a portable gas chromatograph. To allow for the analysis of high-concentration samples, a dilution procedure was required.

The two experiments performed using this technology provided sufficient data resolution to construct accurate contour plots along a sampling profile. Breakthrough curves plotted for individual sampling points exhibited reasonable trends and provided a means of identifying some of the sampling errors. Data precision, as indicated by statistical analysis of duplicate samples collected during both experiments, was quite good. The 95% confidence intervals on the concentration values estimated from these analyses were ±20% for experiment 1 and ±10% for experiment 2 (after day 12).

Foremost among the practical problems encountered during sampling were the effects of gas leaks and plugging of syringes and the screens on some of the probes. Some difficulty was also experienced in the attempt to accurately measure very low concentrations. The effects of increased experience with the equipment, refinement of the sampling and calibration methodology, and more favorable seasonal temperature conditions were reflected in the improved level of data precision in the second experiment.

To account for the effects of temperature changes on gas samples, the methodology was designed to determine the concentrations of samples in terms of mole ratio units, which are conservative with temperature. Temperature-induced density changes were not accounted for in the direct injection procedure used for the lower concentration samples, and the maximum resulting error for these data points was estimated at 7%. It is expected that temperature effects caused some condensation of the near-saturation concentration samples in experiment 1.

Our experience suggests that with reasonable care, gas-phase monitoring is a very practical endeavor capable of providing reliable data of the high quality suitable for research applications. On-site analysis of the samples was highly advantageous as they were not subjected to transport or storage and the availability of real-time data allowed for the strategic adjustment of the monitoring schedule and reduced the potential error due to problems such as leaks.

ACKNOWLEDGMENTS

We wish to express our gratitude to M. Robin for his guidance and assistance in conducting the statistical analyses as well as to P. Johnson for his assistance in the design and construction of the sampling probes. This research was conducted with funding from the University Consortium for Solvents-in-Groundwater Research sponsored by Ciba-Geigy, Dow Chemical, Eastman Kodak, General Electric, the University Research Incentive Fund of Ontario, and the Natural Sciences and Engineering Research Council of Canada. Assistance and technical support was also provided by Photovac Inc.

REFERENCES

1. Schwille, F. *Dense Chlorinated Solvents in Porous and Fractured Media—Model Experiments* (J. Pankow, transl.), Lewis Publishers, Chelsea, MI, 1988.
2. Feenstra, S., and Cherry, J.A. Subsurface contamination by dense non-aqueous phase liquid (DNAPL) chemicals. Proceedings of the IAH International Groundwater Symposium, Halifax, Nova Scotia, 1988.
3. Hunt, J.R., Sitar, N., and Udell, K.S. Nonaqueous phase liquid transport and cleanup. 1. Analysis of mechanisms. *Water Resour. Res., 24*(8):1247, 1988.
4. Hughes, B.M., Gillham, R.W., and Mendoza, C.A. Transport of trichloroethylene vapours in the unsaturated zone: a field experiment. Proceedings of the IAH Conference on Subsurface Contamination by Immiscible Fluids, Calgary, April 18–20, 1990.
5. McClellan, R.D., and Gillham, R.W. Vapour extraction of trichloroethylene under controlled field conditions. Proceedings of the IAH Conference on Subsurface Contamination by Immiscible Fluids, Calgary, April 18–20, 1990.
6. Marrin, D.L., and Thompson, G.M. Gaseous behaviour of TCE overlying a contaminated aquifer. *Ground Water, 25*(1):21, 1987.
7. Thompson, G.M., and Marrin, D.L. Soil gas containment investigations: a dynamic approach. *Ground Water Monit. Rev., 7*(3):88, 1987.
8. Kerfoot, H.B. Soil-gas measurement for detection of groundwater contamination by volatile organic compounds. *Environ. Sci. Technol., 21*(10):1022, 1987.
9. Kerfoot, H.B. Is soil-gas analysis an effective means of tracking contaminant plumes in ground water? What are the limitations of the technology currently employed? *Ground Water Monit. Rev., 8*(2):54, 1988.
10. Marrin, D.L. Soil-gas sampling and misinterpretation. *Ground Water Monit. Rev., 8*(2):51, 1988.
11. Marrin, D.L., and Kerfoot, H.B. Soil-gas surveying techniques. *Environ. Sci. Technol., 22*(7):740, 1988.
12. Tillman, N., Ranlet, K., and Meyer, T.J. Soil gas surveys: part I. *Pollut. Eng., 21*(7):86, 1989.
13. Tillman, N., Ranlet, K., and Meyer, T.J. Soil gas surveys: part II/procedures. *Pollut. Eng., 21*(8):79, 1989.
14. Smith, J.A., Chiou, C.T., Kammer, J.A., and Kile, D.E. Effect of soil moisture on the sorption of trichloroethylene vapor to vadose-zone soil at Picatinny Arsenal, New Jersey. *Environ. Sci. Technol., 24*(5):676, 1990.
15. Johnson, R.L., and Perrott, M. Gasoline vapor transport through a high-water-content soil. *J. Contam. Hydrol., 8*:317, 1991.

16. Hughes, B.M. Vapour transport of trichloroethylene in the unsaturated zone. A field experiment. M. Sc. Thesis, Univ. of Waterloo. 1991.

17. Reynolds, G.W., Hoff, J.T., and Gillham, R.W. Sampling bias caused by materials used to monitor halocarbons in groundwater. *Environ. Sci. Technol.*, *24*(1):135, 1990.

18. Gillham, R.W., and O'Hannesin, S.F. Sorption of aromatic hydrocarbons by materials used in construction of ground-water sampling wells. In *Ground Water and Vadose Zone Monitoring* (D.M. Nielsen and A.I. Johnson, eds.), ASTM Std. Tech. Publ. 1053, Philadelphia, 1990, p. 108.

19. Parker, L.V., Hewitt, A.D., and Jenkins, T.F. Influence of casing materials on trace-level chemicals in well water. *Ground Water Monit. Rev., 10*(2):146, 1990.

20. Lion, L.W., Stauffer, T.B., and MacIntyre, W.G. Sorption of hydrophobic compounds on aquifer materials: analysis methods and the effect of organic carbon. *J. Contam. Hydrol.,* 5:215, 1990.

21. Snedecor, G.W., and cochran, W.G. *Statistical Methods*, 7th ed., Iowa State Univ. Press, Ames, IA, 1980.

22. Box, G.E.P., Hunter, W.G., and Hunter, J.S. *Statistics for Experimenters*: *An Introduction to Design, Data Analysis and Model Building*, Wiley, New York, 1978.

23. Mendenhall, W. *Introduction to Probability and Statistics*, 6th ed. Duxbury Press, Boston, MA, 1983.

24. Kerfoot, H.B. Shallow-probe soil-gas sampling for the indication of ground-water contamination by chloroform. *Int. J. Environ. Anal. Chem., 30*:167, 1987.

25. Mendoza, C.A., Hughes, B.M., and Frind, E.O. Transport of trichloroethylene vapours in the unsaturated zone: numerical analysis of a field experiment. Proceedings of the IAH Conference on Subsurface Contamination by Immiscible Fluids, Calgary, April 18–20, 1990.

II
MONITORING STRATEGIES

6

Field Monitoring for Polynuclear Aromatic Hydrocarbon Contamination

Terri L. Bulman

Campbell Group Limited,
West Perth, Australia

I. INTRODUCTION

Polynuclear aromatic hydrocarbons (PAHs) are a group of organic compounds that consist of condensed (joined) aromatic ring structures. They form the basis of a wider range of aromatic compounds, which may be alkylated, hydroxylated, or sulfur-, oxygen-, or nitrogen-substituted and are commonly referred to, collectively, as polynuclear or polycyclic aromatic compounds. The aromatic ring structures may be condensed through biological synthesis or through pyrolytic processes. Thus polycyclic aromatics are frequently found in nature, in geological, soil, and plant matrices and in deposits resulting from forest fires and volcanic activities. They are also produced in high quantities by many industrial processes and thus have been found in some environments at levels that are considered hazardous to human health and to the environment.

Activities and waste types that typically result in significant concentrations of PAHs include those related to fossil fuels (oil, gas, and coal exploration, production, transportation, storage and marketing, and use). Sites that most often require investigation and remediation because of PAH contamination are those that have been used for oil refining, waste oil rerefining and disposal, "coal gasification," pipeline and storage tank leakages, wood preservation, and improperly maintained "landfarms" and waste lagoons. On these sites and others, PAHs will form part of a target list of compounds for site investigation. The list may also include highly volatile hydrocarbons (benzene, toluene, xylene, C_3–C_6 alkanes), very soluble hydrocarbons (phenols), halogenated hydrocarbons, pesticides, and metals.

II. PHYSICAL, CHEMICAL, AND BIOLOGICAL PROPERTIES OF PAHs

Polynuclear aromatic compounds share physical and chemical characteristics based on the nature of the aromatic ring structure. Differences in characteristics arise according to molecular weight, number of rings, and substitutions. The basic aromatic ring structure tends to be hydrophobic, relatively stable, and resistant to biological metabolism. It is, however, a strong adsorber of ultraviolet light, with which it reacts readily to form free radicals and decompose. These characteristics tend to increase in significance with an increasing number of condensed rings and are further affected by the presence of hydrophilic or hydrophobic groups or chemically reactive substituents. These characteristics influence where polynuclear compounds will be found in the environment and the potential for environmental effects, as well as the ways in which the compounds will respond to various forms of chemical analysis.

Table 1 lists the relevant characteristics of the PAH compounds most commonly investigated, based on the USEPA CERCLA Appendix VIII and current Appendix IX list. This list of characteristics has been compiled from Bulman et al. [1,2] and Lyman et al. [3]. Many of these properties have been measured for a wide range of compounds in a series of studies performed for the U.S. Environmental Protection Agency [4–7].

Methods for calculating these properties may be useful if order-of-magnitude estimates for the properties are required. The book by Lyman et al. [3] provides an excellent reference for ascertaining chemical properties as experimentally measured values or estimating them mathematically from other known properties. A review of properties estimated in Lyman et al. [3] through structure–activity models suggests that the following molecular components increase octanol/water partitioning and bioconcentration (associated with lipophilic properties): halogens, oxygen or sulfur between two aromatic rings, $N = N$ between two aromatic rings, H, SH, C, CH_3, and benzene rings. Solubility and biodegradability (associated with hydrophilic properties) are likewise decreased. Hydrophilic properties, however, are generally increased

Table 1 Physical, Chemical, and Biological Characteristics of Commonly Investigated PAHs

PAH	Mol. wt	Number of rings	Aqueous solubility (mg/L)	Log oct/water coeff.	Vapor pressure at 20°C (N/m²)	Bioconc. factor (aquatic)
Acenaphthene	154	2.5	3.93	4.33	2.67	387[a]
Acenaphthylene	152	2.5	3.47	4.07	3.87	510[b]
Anthracene	178	3	0.073	4.45	2.61×10^{-2}	917[a]
Benzo[a]anthracene	228	4	0.04	5.61	6.67×10^{-2}	10100[a]
Benzo[b]fluoranthene	252	4.5	0.0012	6.57	6.67×10^{-5}	58000[b]
Benzo[k]fluoranthene	252	4.5	0.00055	6.84	6.67×10^{-5}	93000[b]
Benzo[g,h,i]perylene	276	6	0.0003	7.23	1.33×10^{-8}	180000[b]
Benzo[a]pyrene	252	5	0.004	6.04	6.67×10^{-5}	23000[b]
Chrysene	228	4	0.002	5.61	8.40×10^{-5}	11000[b]
Dibenzo[a,h]anthracene	278	5	0.0025	5.97	1.33×10^{-8}	20000[b]
Fluoranthene	202	3.5	0.264	5.33	9.11×10^{-5}	2702[a]
Fluorene	166	2.5	1.98	4.12	1.73	1300[a]
Indeno(1,2,3-c,d)pyrene	276	5.5	0.062	7.66	1.33×10^{-8}	390000[b]

[a]Measured.
[b]Calculated.
Source: Bulman et al. [1], Lyman et al. [3], Sims et al. [6].

by incorporating oxygen, sulfur, nitrogen, NO_2, SO_2, NH_2, OH, C(O)OH, or $C(O)NH_2$ into the molecule. The properties discussed above are critical to the detection of PAH compounds in environmental media. The partitioning of compounds between soil, air, water, and oily matrices affects the mobility and biodegradability of the compounds in the environment. Partitioning may also affect the ability to detect compounds in different types of environmental media. Furthermore, it determines the potential for volatilization losses from samples during field or laboratory procedures. Compounds such as PAHs that occur preferentially in the soil, organic carbon, or oil phase cannot be assessed adequately by monitoring groundwater only. Cores of the aquifer solid material must also be analyzed. Similarly, water samples to be analyzed for highly volatile compounds must be taken in containers with zero headspace to reduce losses of volatile compounds from the sample. Accurate, quantitative determinations of the affinity of compounds for environmental matrices require direct, site-specific measurement.

All aspects of a monitoring program for PAH compounds, from selection of the compounds to be monitored, to sampling and on to analysis, are inextricably linked and are based on the same fundamental principles. The planning of a monitoring program must integrate all of these aspects.

III. FIELD MONITORING PROGRAMS

The most important factor governing the details of a monitoring program is the use intended for the data: the data quality objectives. The degree of focus for a monitoring program may range from a qualitative survey of background environmental quality to a program specifically focused on regulatory requirements or specific research objectives. Data quality objectives, as well as the nature of the site and previous or current industrial activities, will guide the selection of target compounds for monitoring, the location of sample points, the types of samples to be taken, and the methods used for sampling and analysis.

The following discussion is divided according to three types of monitoring objectives and strategies. The first is based on routine acquisition of qualitative data that are not defined by specific regulatory requirements. The second is a regulation-driven program that is typically precipitated by a known crisis or contamination situation. The third type is a program with research objectives.

A. Qualitative Surveys

1.   Overview

A monitoring program can be developed to provide a qualitative survey of PAH contamination. Such a survey may be intended to form part of an environmental auditing program for an industry that typically uses or produces products that contain PAH compounds. The effects of long-term storage, transport spillages, and disposal of these products on local groundwater quality can be monitored by routinely measuring PAH concentrations for comparison with a reference value or background (off-site) samples. An example of such a survey would be an internal audit of groundwater quality under a land treatment facility for disposal of oily wastes. Key data requirements in this case would involve the comparison of on-site and background concentrations, typically near detection limits. The survey might include comparison of the relative proportions of a suite of compounds that form a characteristic "fingerprint" of contamination from the site source.

2. Selection of Target Compounds

Compounds targeted for analysis should be characteristic of the product or waste, present in high concentrations, easily and reliably detected analytically, and unlikely to be introduced from other sources. Such monitoring data would be used as a basis for continuing operations. If substantial contamination is found, the data will assist in the design of a more rigorous assessment program. A qualitative survey may also be useful as a preliminary assessment when contamination is suspected but the source is not known.

In addition to the target compounds, samples may be analyzed for a number of gross parameters that are selected to give an indication of impact on soil or water quality other than that provided by analysis of specific contaminants. For example, samples may be analyzed for total organic carbon or total extractable hydrocarbon (i.e., oil and grease). Such parameters can also serve as an indicator of other types of contamination. Total nitrogen, sulfur, and halogens can be used as a gross parameter for nitrogenous, sulfurous, and organochlorine contamination, respectively. Total dissolved solids and electrical conductivity can be used as indicators for contaminant mixtures that are ionic. Other site parameters that are important aids in the interpretation of the results of monitoring include hydraulic conductivity, the gradient and direction of groundwater flow, the groundwater quality (salinity, oxygen content, pH, dissolved ions), and the variability of the aquifer material.

3. Data Requirements

A qualitative survey program is typically carried out over an extended time period—several monitoring periods each year for many years. Because of this, there is little need for replicated or repeated analysis for each sampling period. Decisions made on the basis of these data are not usually subject to rigorous time constraints. A large amount of historical data is collected that will ultimately be evaluated to distinguish true anomalies from seasonal trends. Detailed determination of precision and accuracy for each sample is usually not required. The primary data requirements can generally be expressed as Is the compound present? and Are the compounds present in proportions typical of the waste or product?

To answer the first question, analysis of specific PAH compounds is usually performed at trace levels or the limits of detection. Prevention of sample contamination is critical, therefore, to prevent "false positives." In the field, this means careful handling of samples, using clean sampling equipment and sample containers. This is facilitated through the use of dedicated sampling equipment such as dedicated submersible pumps or, at least, dedicated sample tubes for each well to be sampled. Equipment that is moved from well to well for the purpose of taking samples must be rinsed between wells. The preferred method would include one rinse with a noninterfering (analytically) organic solvent such as methanol followed by several rinses with organic-free water. Using the same source of rinse water, field blanks may be taken to confirm that contamination of the samples is not taking place. If analysis of samples routinely indicates the presence of PAH compounds, it is recommended that the sampling program make extensive use of field blanks to determine if the source of PAH compounds is due to field handling or contaminated equipment or sample containers. It is equally important to ensure that laboratory equipment and glassware is free of contamination. Solvent washing and oven drying are effective means to achieve this. It is important to ensure that the source of water is organic-free and that all chromatographic materials, reagents, solvents, and gases are of chromatographic grade and free of contamination. Regular analysis of these materials, particularly new shipments, and the preparation of laboratory blanks are required to ensure high-quality analysis.

Groundwater samples, in particular, should be relatively "clean," that is, have little in the way of interfering organic species. The analytical program should therefore focus on detection (preventing false negatives). The approach would include concentration of a large-volume groundwater sample to lower the detection limit as much as possible and minimize handling steps such as sample cleanup or transfer between vessels. The specific properties of the target contaminants must be given careful consideration to prevent losses during both sampling and analysis. If volatile species are to be investigated, sampling procedures should minimize the use of vacuum sampling methods and should minimize contact with air. Sample containers should be filled completely (zero headspace) and, if possible, should provide a means of access to the sample that avoids air contact, such as a septum-lined cap. Samples should be chilled as quickly after sampling as possible. The U.S. EPA Method 602 recommends acidification as a preservative. Storage should be below 4°C, and analysis should take place as soon as possible.

Because of the hydrophobic nature of PAH compounds, the choice of sampling and laboratory materials should be considered from the point of view of adsorption of the target PAH compounds onto sampling and analytical containers. Materials should also be considered from the point of view of preventing leaching of interfering compounds out of plastic-based materials. Glass, Teflon, and stainless steel have enjoyed the greatest popularity for PAH sampling and analysis because of their general nonreactivity with organic compounds and their ease of cleaning. PAH compounds do adsorb to these materials, however. Periodic analysis of a solvent rinse of these containers is recommended to ensure that significant losses of the target compounds are not occurring. Sample preservation, for example through pH adjustment (to pH 9) or addition of a small amount of a noninterfering miscible solvent (methanol), will also minimize loss of the compound. Samples should also be stored below 4°C to reduce biodegradation losses and should be analyzed as soon as possible.

If substantial PAH contamination is identified in a groundwater sample, the contamination should be confirmed through additional sampling, including field blanks. The focus of the analytical program, in this case, will be to provide good quantitative analysis for a range of target compounds that are present in the potential source of contamination—the waste, vadose zone, or aquifer solid material.

Data obtained from qualitative groundwater surveys often do not conform to a normal distribution and are therefore interpreted with the assistance of nonparametric statistical techniques. As with many other hydrogeochemical investigations, the survey is intended to identify differences between groundwater units (up- and downgradient from a suspected source of contamination). A review of multivariate statistical methods useful for this interpretation is provided by van Tonder and Hodgson [8] and includes cluster analysis, principal component analysis, factor analysis, and discriminant function analysis.

The ratio of specific compounds or compound groups can be compared to provide a fingerprint of the waste and the suspected contamination. Consideration should be given to selecting compounds for comparison that are not likely to undergo a change in proportion due to expected processes such as retardation in the soil matrix or biodegradation. Suitable comparisons would therefore be made within a group of high molecular weight, persistent compounds or within a group of volatile compounds. A comparison on the basis of molecular weight groups (volatile, biodegradable, and persistent compounds together) does provide a useful estimate of the age, or degree of weathering, of contamination in the sample and thus may be useful in evaluating the source of contamination or distance from such a source.

Information such as the types of contaminant groups in a sample or the ratios of concentrations of different contaminants may be evaluated in a semiquantitative manner using multivariate plots. Multivariate plots are obtained by plotting analytical results on several axes together in one graph. the composite plot provides a picture of the relative concentrations of a suite of chemicals from the same sample. Qualitative evaluations may be no more sophisiticated than visually inspecting these grouped data sets or chromatographic profiles for contaminant "fingerprints."

An example of a qualitative survey for PAH compounds is one performed by the Wastewater Technology Centre, Environment Canada, for the Petroleum Association for Conservation of the Canadian Environment (PACE). The survey is reported in full in Bulman et al. [9]. This survey involved monitoring for the 16 Appendix VIII PAH compounds in soil and groundwater at three operating refinery land treatment sites.

Analysis of soil samples involved extraction of an 8-g sample with acetone using a tissue homogenizer. The acetone extracts were further extracted with hexane. The PAH fraction of each sample extract was isolated by gel permeation chromatography, and analysis was performed by gas chromatography/mass spectrometry.

Of the 16 PAH compounds monitored, eight were consistently found in land treatment soils at levels greater than in background soil. These were phenanthrene, fluoranthene, pyrene, chrysene, benzo[*a*]pyrene, benzo[*a*]anthracene, dibenzo[*a,h*]anthracene, and benzo[*g,h,i*]perylene.

Data collected from the sites were evaluated by comparing the concentrations of PAH compounds in groundwater with concentrations in soil from the land treatment areas and other possible sources of contamination. This comparison was facilitated by constructing fingerprint diagrams for the samples based on groupings of PAH compounds by molecular weight. Multivariate plots in Fig. 1 illustrate the PAH groupings in soil from the three land treatment sites. Although the sites had been used for land treatment of waste for different periods of time (3 years for site A and over 10 years for sites B and C) and had different concentrations of residual contamination in the soil, the PAH compounds at all the sites were predominantly in the high molecular weight range (202–278).

Groundwater samples were taken from preexisting wells on the sites that were constructed with 15-cm PVC casing. In all cases, sample wells were

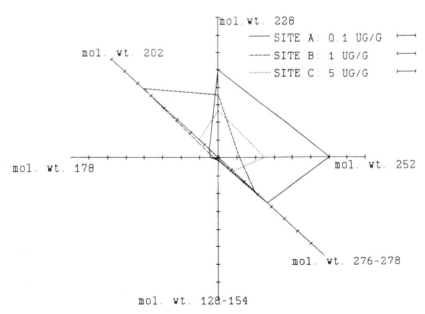

Figure 1 Multivariate plots of PAHs in soil from three oil refinery landfarms. (Data from Bulman et al. [9].)

available upgradient and downgradient from the land application area. Water samples were analyzed for PAH compounds, oil and grease, dissolved organic carbon, pH, and electrical conductivity. For PAH analysis, a 1-L sample was adjusted to pH 10 with NaOH, then extracted three times with distilled-in-glass hexane. The extracts were combined, evaporated, and exchanged into an acetonitrile–water mixture for subsequent HPLC analysis. Heavily contaminated samples were fractionated by gel permeation chromatography prior to solvent exchange. Field and analytical method blanks were analyzed to provide a check on sample contamination. In this case HPLC was used for analysis, rather than GC/MS as performed for the soil samples, to achieve an emphasis on detection (i.e., lower detection limits) for samples that were expected to contain few interfering analytes.

Concentrations of PAH compounds in groundwater near some landfarm sites were found to be elevated over background concentrations. On site C, however, "fingerprinting" was able to distinguish between contaminants from a diesel spill, from the land treatment area, and from a sewage trunk line downgradient from the land treatment area.

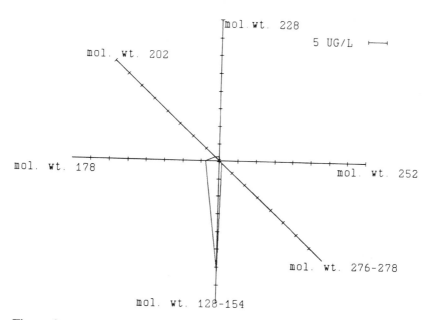

Figure 2 Multivariate plot of PAH compounds in a well contaminated with diesel fuel. (Data from Bulman et al. [9].)

Figures 2–4 are multivariate plots that illustrate the concentrations of PAH compounds in groundwater from individual wells at site C according to the same molecular weight groupings as those in Fig. 1. Figure 2 illustrates the fingerprint of contamination from a well that was known to be directly contaminated with diesel fuel. The PAH compounds in these samples are predominantly in the low molecular weight end (molecular weights 128–154). Figure 3 illustrates PAH contamination in a well downgradient from both the diesel-contaminated well and the land treatment area. The PAH concentrations are lower than that in the diesel-contaminated well but contain the characteristic low molecular weight PAH compounds.

In contrast, Fig. 4 illustrates the fingerprint of PAH compounds from another well on the site with significant PAH contamination. In this case the fingerprint is not that of the diesel fuel contamination but may be due to leakage from a nearby sewage line.

The PAH compounds most commonly measured in excess of background concentrations in groundwater samples were anthracene and chrysene, although phenanthrene, fluoranthene, pyrene, benzo[a]anthracene, benzo[a]pyrene,

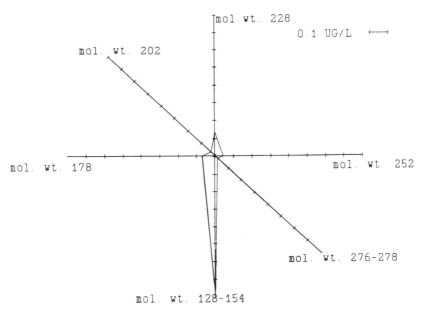

Figure 3   Multivariate plot of PAH compounds from a well located downgradient from oil refinery landfarm. (Data from Bulman et al. [9].)

dibenzo[*a,h*]anthracene, and benzo[*g,h,i*]perylene were also identified in samples at some sites.

As was the case in the above example, statistical analysis is often complicated by a large number of missing or *censored* data. Wells that do not provide adequate sample volume and contaminant levels below detection limits are common problems in this type of sampling program. How these problems are handled depends very much on the objective of monitoring.

Results that are below detection limits are not a problem if the aim of monitoring is to determine whether the concentrations of contaminants are below detection or below a detectable background level. In such a situation, data handling may involve determining the proportion of samples that exceed the established level. The number of samples exceeding the set level may be restricted to a certain percentile (for example, no more than 10%), over which further investigation is triggered. The appropriate percentile would be determined by background variability and sampling and analytical variability, which can be determined from historical data. Plots of the percentile exceeding the level can also be used in evaluating seasonal and annual trends.

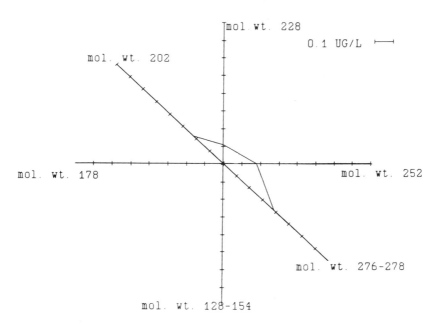

Figure 4 Multivariate plot of PAH compounds in a well located next to sewage trunk line. (Data from Bulman et al. [9].)

In situations where statistical procedures must be used to discriminate between sample populations (for example, various levels of contamination), the absence of quantitative data can be handled by using *censored data* techniques. These methods involve estimating data values based on the probability distribution of values for samples that are considered representative of the missing or unquantifiable sample. A large number of quantified data points are required, and the confidence that can be placed in the statistical treatment decreases as the number of *censored data* estimates increases.

Gleit [10] considered the problem of estimating *censored* data for small data sets with normal distributions and compared the methods of using maximum likelihood estimation, constants such as detection limits, linear estimation, and expected values of order statistics. Replacement of missing values by expected values of order statistics was the recommended method. Gilliom and Helsel [11] evaluated estimation techniques using a range of underlying distributions, with varying proportions of missing values, and compared the reliability of estimates of the mean, standard deviation, median, and interquartile range. The log-probability regression method, in which censored observations are assumed to follow a lognormal distribution fit to uncensored data by least squares regression, was the most robust method and resulted in the smallest errors for the mean and standard deviation. The lognormal maximum likelihood method resulted in the smallest errors for the median and interquartile range.

B. Regulatory Compliance

1. Overview

Another type of monitoring program may be designed in which it is desired to compare environmental quality criteria with certain regulatory standards. This type of assessment may make use of legislative or guideline reference values for soil or groundwater quality. Examples are the so-called, A, B, C criteria initiated by the Dutch government [12] and adopted and modified by other agencies, including the governments of France [13] and the Province of Quebec, Canada [14]. Reference values may, alternatively, be established as part of a site assessment procedure through investigation of local background concentrations or site-specific risk assessment. Whatever the source of the reference values, this type of program is typically carried out on a site where severe contamination is expected or known, and the purpose of the program is to delineate areas of the site according to the level of contamination. In contrast with an industry-initiated site audit type of survey, samples from a contaminated site investigation will contain high concentrations of contaminants, often in mixtures and difficult oily or tarry matrices, which present a special set of problems for the analyst.

2. Selection of Target Compounds

A site investigation to be used to assess regulatory compliance must provide information specifically on compounds for which regulations exists (i.e., compounds for which reference values exist). The investigation may not be limited to those compounds, however, if the history of site use, current visible contamination, or public concerns suggest that contamination with other, nonregulated, compounds may be a problem. In most jurisdictions there is currently a general lack of *cleanup criteria* for many site contaminants. An example of guidelines that do exist for PAH compounds are the Dutch criteria provided in Table 2.

Determination of the contaminants of concern and the appropriate cleanup criteria may form part of the site assessment process. If site-specific cleanup criteria are to be developed for a site, substantial information must also be collected on site soil and groundwater characteristics. A comprehensive pathways risk assessment model, such as the AERIS model (Aid for Evaluating the Redevelopment of Industrial Sites) developed for Environment Canada [15], requires determination of the site characteristics listed in Table 3. These parameters are useful, in a general sense, for interpreting the results of the site investigation. In cases of complex and extensive site remediation, risk assessment models can provide valuable assistance in planning the site investigation.

Table 2 The Dutch Guidelines[a] for PAH Concentrations in Soil and Groundwater

Compound	Soil criterion (mg/kg)	Water criterion (mg/L)
Naphthalene	5	0.007
Phenanthrene	10	0.002
Anthracene	10	0.002
Fluoranthene	10	0.001
Chrysene	5	0.0005
Benzo[a]anthracene	5	0.0005
Benzo[a]pyrene	1	0.0002
Benzo[k]fluoranthene	5	0.0005
Indeno(1,2,3-c,d)pyrene	5	0.0005
Benzo[g,h,i]perylene	10	0.001
Total PAH	20	0.010

[a]B level criteria (indicates additional investigation required).
Source: Rijksinstituut voor Volksgezondheid en Milieuhygiene [12].

Table 3 Examples of Soil and Aquifer Characteristics Required for Determining Site-Specific Cleanup Criteria

Soil characteristic	Aquifer characteristic
Porosity	Hydraulic conductivity
Particle density	Porosity
Saturated water content	Bulk density
Organic carbon content	Longitudinal dispersivity
pH	Thickness
Partition coefficients	Hydraulic gradient
Depth of contamination	Depth to water table

Industrial operations may continue on a contaminated site, or they may have recently ceased. In either case, company records and personnel will be able to provide valuable information regarding the types of contaminants that may be found; locations of old storage, transfer, or disposal areas; and changes in processes or feedstocks. In addition to assisting in selecting a primary target list of contaminants to monitor, this information may alert the analyst to potentially interfering matrices, coexisting contaminants (solvents), or other special analytical problems. In cases where contamination occurs at abandoned sites or landfills that have received a variety of wastes, the selection of contaminant species and analytical methodologies will be based on much less information and will involve substantial preliminary work and method development.

3. Data Requirements

The monitoring program for regulatory compliance will be designed to answer the questions What is the concentration in the sample? and Does the concentration differ significantly from the reference value? This information is generally required quickly and for a large number of samples. Field staff, chemists, and program managers are all under pressure to provide high-quality quantitative results in a short period of time. The cost of analyzing a large number of samples represents a major component of the project budget. The efficiency of the sampling and monitoring program is therefore critical. Site investigation is generally performed in stages so that the major effort can be concentrated on delineating the areas of highest contamination. Methods of investigation include selection and on-site measurement of gross parameters that provide a general indication of contaminant levels. A large number of soil samples may be collected and archived at $-40°C$. In this way, a small number of samples may be analyzed at first, and additional samples selected from those archived may be analyzed at a later date if further data are required.

Preliminary site investigation on contaminated sites normally begins with a rough evaluation of contamination based on coarse grid sampling, soil-gas sampling, and geophysical surveys. These methods are intended to provide a general indication of the presence of contamination, buried drums and other structures, which will be used in defining a more detailed delineation of the contaminated areas. A "coarse" laboratory counterpart, a broad analytical scan, is also required at this point, to qualitatively identify as many potential contaminant groups as may be found on the site. This preliminary analysis will, of necessity, be lacking in sensitivity, however broad it is in scope. A typical starting point for analysis is the list of Appendix VIII or Appendix IX compounds. Analysis for polynuclear aromatics, along with other base-neutral aromatic species, would be performed by GC/MS. If substantial numbers of PAH compounds are identified, methodologies must be worked up that will be suitable for cleaning up and concentrating the PAH compounds in the samples and preparing them for more sensitive resolution through analytical methods such as HPLC.

Because PAH compounds typically occur in oily and tarry products such as refined and heavy oil products and coal tar wastes, sample cleanup is usually a critical and time-consuming part of the analysis. It is equally unlikely that methods, once developed, can be routinely applied to samples in other situations. Although the basic steps may be standardized, considerable innovation is required to handle the samples in a way that will meet rigorous precision and accuracy requirements. The effort required to adapt protocols to specific samples from a site, in the preliminary site investigation stage, is critical to the entire subsequent site investigation. When rigorous cleanup methods are required, contaminant detection limits are often increased. Specific compounds that are not easily resolved in complex matrices (such as anthracene from phenanthrene and benzo[a]anthracene from chrysene) may be dropped from the list of target compounds to be quantified if it is unlikely that the desired quality control parameters can be met.

Site investigation is usually an initial step in site cleanup, and site cleanup inevitably involves substantial cost. The issue of the degree and extent of site contamination therefore carries an onerous financial, and sometimes legal, weight. This determines what volume of soil or aquifer must be subjected to remediation. Analysis of these samples requires substantial documentation of precision and accuracy (a quality assurance/quality control program) and may have to conform to preset standards of precision and accuracy. A reasonable QA/QC standard for PAH compounds, for example, may be 80–120% recovery with 90% of samples falling within that range. If it is apparent during preliminary site assessment that the complexity of the sample matrix precludes attainment of the preset standards, revised standards should be agreed upon.

In some cases, the liability for site contamination and cleanup comes into question, and the results of analyses for site investigations may be used in legal proceedings. In these situations a record of who handled the samples is required, from collection on site through to analysis. It is generally advisable to keep such records, often called a *chain of custody*, on a routine basis for all site investigations.

In the Environment Canada qualitative survey described in Section III.A.3 of this chapter [9], a QA/QC program was used that provides an example of the type of program that could be used for regulatory monitoring. Soil sample variability and analytical recovery were assessed by taking duplicate 8-g subsamples from selected replicate samples. In addition, some duplicate subsamples received additions of surrogate compounds (a suite of 10 PAH compounds) at known concentrations. Deuterated surrogate compounds were added to all soil extracts prior to cleanup and analysis by GC/MS to evaluate analytical recovery and variability due to interference with other contaminants.

The results of the QA/QC program are summarized in Table 4. A range of variability is provided for the suite of 16 PAH compounds analyzed. The greatest source of variability was in the collection of field samples. Field samples consisted of composites of four to ten soil samples (approximately 15 samples per hectare). Analytical variability was also significant and was not consistent among samples or time periods. Sample variability can be reduced by increasing the number of samples analyzed per unit soil area. If an estimate can be made of the spatial variability of contaminants on the site, the required number of samples needed to provide a specific level of precision can be estimated in advance. Increasing the number of samples can be extremely expensive, however, for the incremental increase in precision obtained. Subsample variability can be reduced much more easily by increasing the size of the subsample. If possible, it is desirable to extract the

Table 4   Variability in Landfarm Samples

Source of variability	Method of measurement	Measured variability
Sampling	Triplicate samples	Coefficient of variation 5–160%
Subsampling	Duplicate subsamples	Coefficient of variation 40–100%
Extraction efficiency	"Spiked" duplicates	74–111% recovery
Cleanup and detection	Deuterated "spike" duplicates	100–125% recovery

Source: Data from Bulman et al. [9].

entire field sample for analysis. Analytical variability, such as losses during cleanup and other sample preparation, should be minimized as much as possible by reducing sample handling. The effectiveness of sample extraction procedures and the operation of analytical equipment should be routinely monitored through the use of sample spikes and standards. Steps should be taken to routinely achieve set standards of analytical performance. In the example presented, the results of the QA/QC program provided important assistance in interpreting trends in the data.

Delineation of areas of a site on the basis of contaminant concentration can be assessed initially by plotting concentrations at corresponding sample locations on a site plan. Contour lines can then be drawn on the plan that roughly divide the site into zones of concentration ranges. Various contour smoothing routines are available as computer programs that estimate concentrations between sample locations based on some weighting of the concentrations at nearby sample locations. These smoothing routines typically weight sample concentrations as an inverse function of the distance between each sample location and the point being estimated. It is an inherent assumption, however, that the relationship between the concentrations at two points and the distance between them is the same regardless of the orientation of that distance on the site plan, that is, the direction between sample locations. It is possible that samples some distance apart along a certain orientation—for example, the direction of groundwater flow—are more similar (more "continuous") than samples the same distance apart but along a different orientation. This effect of the direction between sample points on their similarity is known as *anisotropy*. Anisotropy is usually significant in real-world situations.

Variogram modeling and kriging are statistical tools often applied to soil and groundwater data that take into account the effects of both distance and direction. Variogram analysis involves an assessment of the spatial distribution of a parameter (such as the concentration of a contaminant) by comparing measurements between individual sample locations. The variability between locations is assessed according to the distance between them in selected directions. Similarly, cross-variogram analysis assesses the relationship between two parameters as a function of separation distance and direction.

The relationship between the variance and the separation distance is the variogram model. This analysis can be used to interpolate concentrations for other, unsampled locations. The use of variogram models as a predictive tool is known as *kriging*. Kriging estimates concentrations in unsampled locations by using variogram models to account for the distance between sample locations, clustering of locations, and anisotropy. Variogram modeling and kriging were initially developed for use in the mining industry, for the purpose of estimating ore grade. The techniques are very useful for any parameter with a spatially continuous distribution providing sufficient numbers of samples have been measured. A description of the techniques involved using a fully worked example followed through several approaches to kriging has been published by Isaaks and Srivastava [16].

C. Research Programs

1. Overview

A research program differs from other monitoring programs in that it has a much narrower focus and requires not only a high level of precision but also an accurate estimate of the level of precision and confidence in the data. Research programs might involve the evaluation of sampling or analytical methods or the evaluation of fundamental processes such as the fate and transport of compounds. In these situations, the monitoring program is developed to answer questions such as What is the concentration in the sample?, How does this concentration differ from other samples or predicted concentrations?, and What degree of confidence do I have in the measurements?

2. Selection of Target Compounds

A monitoring program for research may deal with a range of concentrations, from trace to very high levels. The method selected for analysis must provide good detection, precision, and accuracy over a wide range of concentrations. The compounds selected for study must therefore respond well to the same analytical technique, yet be easily distinguished from each other. Most semivolatile PAH compounds are readily determined by HPLC. Certain pairs, such as anthracene/phenanthrene and chrysene/benzo[a]anthracene, are not readily distinguished by routine chromatographic procedures. It is therefore important to consider analytical capabilities when defining the target compounds and other research objectives.

3. Data Requirements

All of the cautionary notes regarding sampling and analysis for the qualitative survey and the regulatory monitoring program apply to the research program. As the purposes of research programs vary widely, it is difficult to make general comments. The most important consideration to keep in mind is that the effort involved in sampling and analysis is considerable and generally costly. Good quality results are not easily obtained. The utmost care must therefore be given to all sample handling.

An example of a research program that involved PAH monitoring is a field evaluation study performed by Environment Canada for validation of the RITZ (Regulatory and Investigative Treatment Zone) and VIP (Variable Input Parameter) models [17]. These two related models were developed for the U.S. Environmental Protection Agency to assess the fate of contaminants during land treatment of an oily waste. *Land treatment* refers to the incorporation of a waste into soil to allow soil microorganisms to degrade the organic constituents. These models are intended as field management tools rather than exact descriptions of the mechanisms governing the transport of contaminants in soil. The site used in this study was an operating petroleum refinery waste land treatment area.

Concentrations of PAH compounds were measured on test plots over two field seasons, including measurements at three depths in soil, at two depths in soil pore water, in leachate collected in tile drains, and in groundwater. Concentrations in water were generally at or below detection limits. This prevented a quantitative assessment of the model leachate predictions. Concentrations in soil with time and depth were compared, however, to assess model predictions.

To evaluate the model predictions, it was necessary to estimate the variability in model input parameters and the effect of this uncertainty on model output. A large number of input parameters were required that were estimated on the basis of field sampling of soil and tests in the laboratory. The sensitivity of the model to these parameters was determined by comparing model output with successive runs of the model using input parameters that varied by 10% of the average field measurement. The most sensitive input parameters were loading rates of the compound in the soil, biodegradation rates, soil bulk density, recharge rate, porosity, and phase partitioning coefficients (between soil, water, air, and oil phases). All of these parameters varied spatially over the site. The concentrations of the compounds were also subject to the sources of analytical variability described in the regulatory monitoring example. Analytical variability for PAH compounds contributed to uncertainty in the estimation of the initial loadings to soil, the biodegradation rates, and the phase partitioning coefficients. Analysis also contributed to variability in the measured PAH concentrations in soil over the course of the study, to which model predictions were compared.

One way to compare model predictions with measured values is to calculate the numbers of overestimates and underestimates, as well as the degree of over- or underestimation (as a ratio or percent of the average observed value). A simple way to present this type of comparison graphically is to plot measured versus predicted values, as has been done in Fig. 5 for one of the compounds studied, toluene. Toluene data were selected for the example because the concentrations of PAH compounds in this soil did not decrease sufficiently over the time period to be distinguishable from sample variability. With a perfect fit of measured and predicted values, the data would all lie on the

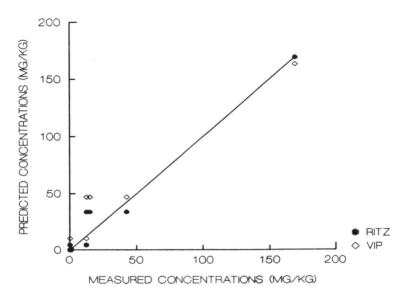

Figure 5   Measured versus predicted concentrations of toluene in soil. (From Wastewater Technology Centre et al. [17].)

solid line in Fig. 5. Most users of a "management" model such as VIP or RITZ would be happy with data that lie close to the line and are evenly distributed on either side of it (i.e., an equal number of over- and underestimates).

Another method for evaluating model predictions is illustrated using VIP predictions and measured toluene concentrations with depth and with time in Figs. 6 and 7. Model predictions are illustrated, including changes due to measured variability (plus or minus one standard deviation) in a single sensitive input parameter.

These plots are used to determine whether model predictions differ from observations in a systematic manner. For example, although the model tended to overpredict concentrations in the upper soil zone (Fig. 6) and underpredict those in the lower soil zone, the location of the contaminant "slug" was approximated reasonably well by the model. Measured concentrations in Fig. 6 were determined in soil taken at intervals from a 75-mm-diameter intact 2-m soil core (split-spoon auger core). In this study [17], the intact core method of sampling provided the most precise results for assessment of contaminant movement vertically in the soil. Differences between model predictions and observed values may have resulted from sampling and analytical variability as well as uncertainty in estimating input parameters such as recharge rate, bulk density, porosity, and partitioning coefficients.

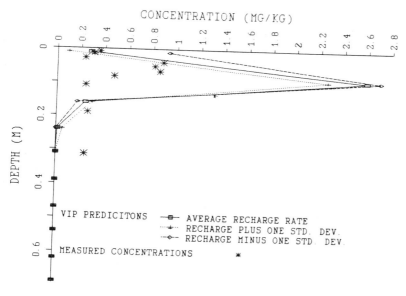

Figure 6 Concentration of toluene with depth. (Data from Wastewater Technology Centre et al. [17].)

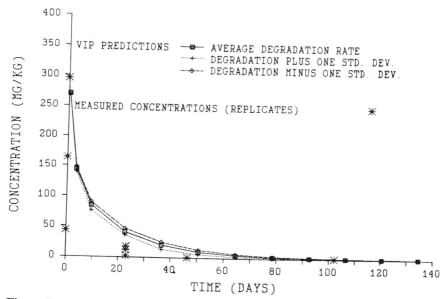

Figure 7 Concentration of toluene with time. (Data from Wastewater Technology Centre et al. [17].)

Model predictions were also compared with measured soil concentrations over time (Fig. 7). In the same study [17], samples were taken in the surface 0–0.15-m depth area using a 20-mm-diameter hand coring tool according to a random grid pattern on the site. Six samples per replicate plot (18 m²) were composited for PAH analysis. Sampling for volatile compounds such as toluene involved the collection of three separate samples, which were placed in septa-capped vials and covered to zero headspace with sodium hexametaphosphate. The model predicted a higher concentration of toluene remaining in soil than was actually measured in early sample periods. This may have been due to underestimation of the biodegradation rate or overestimation of the initial concentration in the soil.

More sophisticated statistical techniques can be used in model evaluation if the research program is designed to collect the appropriate data. Loague et al. [18] provide an excellent review of types of transport models and techniques for interpretation of model evaluation data.

IV. SUMMARY AND RECOMMENDATIONS

A number of problems plague the monitoring of PAH compounds in the field. These problems stem from the characteristics of the PAH compounds themselves (hydrophobicity, volatility of some species), the nature of the environmental media (interferences, variability), and problems related to analysis (detection limits, costs). The problems require imaginative solutions in both sampling and analysis. The appropriate solution will depend on the intended use of the monitoring results. Factors to consider in developing a monitoring program for PAH compounds include the following:

- Detection limits (needs will depend on use of data, effects of sampling methods, and analytical approach)
- Sources of variability (soil/aquifer system, seasonal and spatial variability, sample collection, sample handing, analytical method)
- Sources of contamination (type of equipment, construction materials in equipment, method of sampling)
- Costs (relative cost associated with sampling and analytical methods, acceptable compromises)

Other soil and water characteristics that may be monitored in order to assist in the interpretation of PAH data include total hydrocarbons, total organic carbon, total nitrogen/sulfur, oxygen content, redox state, pH, conductivity, phase partitioning coefficients, hydraulic conductivity, dispersion, gradient, and, if appropriate, specific model parameters.

There is considerable variability in sampling and analytical efficiencies for PAH, particularly in the complex oily matrices with which they are usually associated. A QA/QC program may therefore be critical in evaluation and interpretation of the data.

A recommended approach to developing a monitoring program include the following steps:

1. Review of monitoring objectives with field and laboratory staff
2. Initial characterization of soil, sediment, aquifer
3. (a) Selection of target compounds for monitoring
   (b) Selection of sampling/analytical methodologies
   (c) Establishment of QA/QC program, standards to be met
4. Staging of sampling and analysis, establishment of sampling and analytical priorities
5. Sampling, storage and archiving, analysis of samples
6. Database management
7. Regular data review/progress review, repeat analysis if necessary.

REFERENCES

1. Bulman, T.L., Lesage, S., Fowlie, P.J.A., and Webber, M.D. *The Persistence of Polynuclear Aromatic Hydrocarbons in Soil*, Petroleum Association for Conservation of the Canadian Environment, PACE 85-2, Ottawa, 1985.
2. Bulman, T.L., Lesage, S., Fowlie, P.J.A., and Webber, M.D. The fate of polynuclear aromatic hydrocarbons in soil. In *Oil in Freshwater: Chemistry, Biology, Countermeasure Technology* (J.H. Vandermeulen and S.E. Hrudey, eds.), Pergamon, Toronto, 1987.
3. Lyman, W.J., Reehl, W.F., and Rosenblatt, D.H. (eds.). *Handbook of Chemical Property Estimation Methods: Environmental Behavior of Organic Compounds*, McGraw-Hill, New York, 1982.
4. Loehr, R.C. *Treatability Potential for EPA Listed Hazardous Wastes in Soil*, EPA/600/2-89/011, U.S. Environmental Protection Agency, Washington, DC, 1989.
5. McGinnis, G.D., Borazjani, H., McFarland, L.K., Pope, D.F., and Strobel, D.A. *Characterization and Laboratory Soil Treatability Studies for Creosote and Pentachlorophenol Sludges and Contaminated Soil*, EPA/600/2-88/055, 1988.
6. Sims, R.C., Sims, J.L., Sorensen, D.L., and Hastings, L.L. *Waste/Soil Treatability Studies for Four Complex Industrial Wastes: Methodologies and Results*, Vol. 1, *Literature Assessment, Waste/Soil Characterization, Loading Rate Selection*, EPA/600/6-86/003a, U.S. EPA, Washington, DC, 1986.
7. Walton, B.T., Hendricks, M.S., Anderson, T.A., and Talmage, S.S. *Treatability of Hazardous Chemicals in Soils: Volatile and Semi-volatile Organics*, ORNL-6451, Oak Ridge Natl. Laboratory, Oak Ridge, TN, 1989.
8. van Tonder, G.J., and Hodgson, F.D.I. Interpretation of hydrogeochemical facies by multivariate statistical methods. *Water S. Africa, 12*(1):1–6, 1986.
9. Bulman, T.L., Hosler, K.R., Fowlie, P.J.A., Lesage, S., and Camilleri, S. *Fate of Polynuclear Aromatic Hydrocarbons in Refinery Waste Applied to Soil*, Petroleum Association for Conservation of the Canadian Environment, PACE 88-1, Ottawa, 1988.

10. Gleit, A. Estimation for small normal data sets with detection limits. *Environ. Sci. Technol., 19*(12):1201–1206, 1985.

11. Gilliom, R.J., and Helsel, D.R. Estimation of distributional parameters for censored trace level water quality data. 1. Estimation techniques. *Water Resour. Res., 22*(2):135–146, 1986.

12. Rijksinstituut voor Volksgezondheid en Milieuhygiene. *Leidraad bodemsanering*, afl. 4, November, The Netherlands, 1988.

13. Secrétariat D'Etat de France. *Analyse et Traitement des Sols Pollués*, Bureau du Président du Premier Ministre Chargé de l'Environnement et de la Qualité de la Vie, Paris, France, 1984.

14. Ministère de l'Environnement. *Contaminated Sites Rehabilitation Policy*, Gouvernement du Québec, QEN/SD-8/1, Quebec City, PQ, Canada, 1988.

15. SENES. *Contaminated Soil Cleanup in Canada*, Vol. 6, *User's Guide for the AERIS Model* (*An Aid for Evaluating the Redevelopment of Industrial Sites*), Environment Canada, Decommissioning Steering Committee, Ottawa, 1989.

16. Isaaks, E.H., and Srivastava, R.M. *An Introduction to Applied Geostatistics*, Oxford Univ. Press, Toronto, 1989.

17. Wastewater Technology Centre, Monenco Consultants Ltd., and KRH Environmental Inc. *Field Evaluation of the Regulatory and Investigative Treatment zone (RITZ) Model*, Vol. 2, Field Studies, Environment Canada, Decommissioning Steering Committee, Ottawa, 1992.

18. Loague, K.M., Green, R.E., and Mulkey, L.A. Evaluation of mathematical models of solute migration and transformation: an overview and an example, Proceedings of the International Conference and Workshop on the Validation of Flow and Transport Models for the Unsaturated Zone, Ruidoso, New Mexico, May 23–26, 1988.

7

The Importance of Volatile Organic Compounds as a Disposal Site Monitoring Parameter

Russell H. Plumb, Jr.

Lockheed Engineering & Sciences Company, Las Vegas, Nevada

Groundwater monitoring data from 500 disposal site investigations in the United States were evaluated in order to identify recurring chemical contamination profiles that could provide the basis for an effective monitoring strategy. This effort identified volatile organic compounds as the single most abundant class of organic contaminants in disposal site groundwater. Although no individual volatile compound was universally predominant, this class of contaminants represented 75% of all the detectable events involving organic contaminants. Furthermore, volatile compounds were the most frequently detected class of contaminants in each regional subset of the database, the CERCLA site subset of the database, the RCRA hazardous waste site subset of the database, and the municipal landfill subset of the database. By comparison, pesticide compounds were detected less than 2% of the time in each subset, and nonpriority pollutant organic contaminants were detected only sporadically.

Data from both the CERCLA and RCRA subsets defined an identical mathematical relationship between the number of volatile compounds and the number of organic priority pollutants that were detected. The development of this pragmatic relationship suggests that routine analysis for volatile organic contaminants can be used as a screening technique to establish the need for more extensive organic analysis of groundwater samples during a site investigation. An evaluation of the proposed hierarchical monitoring strategy indicates that the approach will function correctly more than 90% of the time.

I. INTRODUCTION

One of the major environmental issues to emerge during the past decade is an increased awareness of the potential impact of waste disposal sites on the quality of groundwater. In the United States, public recognition of the problem due to the identification of sites such as Love Canal in New York, the Valley of the Drums in Kentucky, and Stringfellow Acid Pits in California provided the social and political impetus to develop regulatory programs in response to these concerns. One of these programs was established under the auspices of the Resource Conservation and Recovery Act (RCRA). The RCRA approach specified a minimum groundwater monitoring program including the number of wells, the location of the wells, the analytes that had to be monitored, and even the specific analytical procedures that had to be used [1,2]. This highly structured monitoring program was to be used by owners/operators of active hazardous waste disposal facilities in order to detect the occurrence of leakage events and to characterize the specific contaminants involved. The second regulatory program to address groundwater contamination in the vicinity of disposal sites was developed under the auspices of the Comprehensive Environmental Response, Compensation, and Liability Act (CERCLA or Superfund) [3]. This program was generally used by government agencies to characterize and prioritize abandoned waste disposal sites for eventual cleanup. However, unlike the RCRA program, the CERCLA strategy did not establish any mandatory monitoring requirements.

The RCRA and CERCLA monitoring strategies have a common goal: to characterize groundwater contamination in the vicinity of waste disposal sites. Despite their similar objectives, these programs have established different monitoring requirements. The purpose of this paper is to compare and contrast composite groundwater monitoring data sets obtained with the approaches outlined in the RCRA and CERCLA programs. The results of these comparisons will provide a pragmatic basis for the development of a more effective strategy to monitor for the presence of organic contaminants in the vicinity of waste disposal sites.

II. DATABASE COMPILATION

The groundwater monitoring data that are summarized and discussed in this paper were obtained from ongoing site investigations across the United States.

The data set consists of groundwater monitoring results obtained at approximately 200 CERCLA site investigations, 150 RCRA Subtitle C detection monitoring facilities, 25 RCRA Subtitle C compliance monitoring facilities, and 125 RCRA Subtitle D municipal landfills. The type of information that is tabulated here includes the regulatory classification of the site (RCRA or CERCLA), the geographic location of the site (EPA region, state), the location of the monitoring well (on-site, off-site; upgradient, downgradient), the groundwater contaminants for which analyses have been reported, the groundwater contaminants that have been detected, and the concentration of each contaminant detected in groundwater.

The resultant project database contains the analytical records for approximately 1200 contaminants for which analyses have been reported. Since the data were derived from different programs, the extent of sample characterization varied widely from site to site. For example, some site investigations contributed data only for the RCRA detection monitoring parameters, and other site investigations contributed data for the complete list of Appendix VIII or Appendix IX parameters (a list of 359 or 245 specific contaminants, respectively, for which monitoring has been or is required during RCRA assessment monitoring) [1,4]. Also, some of the site summaries contain only the results for a single sampling event, whereas others contain monthly sampling surveys for several years. All of the data were generated as part of routine site investigations completed between 1981 and 1986.

III. CONTAMINANTS DETECTED

Prior to the initiation of the research effort discussed in this paper, the largest number of organic contaminants previously reported to be present in groundwater in the vicinity of waste disposal sites was 175 compounds [5]. However, after compilation and review of groundwater monitoring results from more than 500 site investigations conducted in the United States, analytical results for approximately 1000 organic compounds have been tabulated and 425 different compounds have been detected in groundwater during at least one site monitoring program. Throughout the text, the terms *detected event* and *detectable event* refer to situations in which the reported concentrations of specific contaminants in the groundwater exceed the analytical detection limit of the method used in a particular site investigation. Since the compiled data were originally generated by multiple laboratories using both GC and GC/MS methodologies, it is not possible to provide an explicit detection limit value. However, as a point of reference, GC procedures are capable of detecting selected contaminants at concentrations below 1 ppb (provided sufficient sample volume is available and matrix interferences are absent), and GC/MS procedures are capable of detecting contaminant concentrations in the low parts per billion range (provided sufficient sample volume is available and matrix interferences are absent).

Although a complete listing of the hundreds of compounds that have been detected is beyond the scope of this paper, the composite monitoring results have been tabulated as a function of the chemical/analytical classification of the individual contaminants. For the purpose of discussing the data, the following chemical classifications and definitions are used throughout the text:

Volatiles—a standard list of 31 volatile organic compounds routinely determined with SW-846 Method 8240 [6].

Nonstandard volatiles—eight volatile compounds capable of being analyzed with SW-846 Method 8240 [6] but not routinely reported.

Base/neutrals—a standard list of 45 organic compounds extracted from a sample under basic conditions and analyzed with SW-846 Method 8270 [6].

Nonstandard base/neutrals—eight base/neutral compounds capable of being analyzed with Sw-846 Method 8270 [6] but not routinely reported.

Acid-extractables—11 phenolic compounds extracted from an acidified sample and analyzed with SW-846 Method 8270 [6].

Nonstandard acid-extractables—four acidic compounds capable of being detected with SW-846 Method 8270 [6] but not routinely reported.

Pesticides—26 compounds capable of being analyzed with SW-846 Method 8250 [6].

RCRA pesticides—three pesticides required in RCRA Subtitle C detection monitoring [1] and capable of being detected with SW-846 Method 8150 [6].

Priority pollutants—a list of 113 organic compounds that consist of the volatile, base/neutral, pesticide, and acid-extractable classifications identified above [7].

Appendix IX—a list of 208 organic compounds presently required during the compliance monitoring phase of RCRA Subtitle C site monitoring activities [4]. This list includes the 113 priority pollutants plus 95 nonpriority pollutant organic compounds.

Appendix VIII—a list of 359 organic compounds previously required (1985–1987) during the assessment monitoring phase of RCRA Subtitle C site monitoring activities [1]. This list includes the 113 priority pollutants plus 226 nonpriority pollutant organic compounds.

Miscellaneous organics—individual organic compounds that do not fall into any of the previous categories.

Organic monitoring data from abandoned waste disposal sites (CERCLA sites), active waste disposal sites in detection monitoring (RCRA Subtitle C sites), and municipal landfills are summarized in Table 1. An important point illustrated with this information is that the volatile class of compounds represents the single most abundant class of organic contaminants in disposal site groundwater. These compounds account for 75% of all the detectable events involving organic contaminants. The priority pollutant, Appendix IX, and Appendix VIII classifications represent 87–90% of the detectable events, but each of these categories includes the volatile organic compounds. The second most abundant classification of organic contaminants in Table 1 is the miscellaneous compounds that represent 9.6% of the detectable events. However, when one considers the large number of compounds in this category and the relatively small number of explicit detectable events (less than 10 events per compound), very little can be said regarding the occurrence and distribution of these compounds except that they have been detected on a sporadic basis.

The composite data set can also be used to identify the specific organic contaminants that predominate in disposal site groundwater. However, as shown in Table 2, the rank order of the individual constituents will vary depending on the factor chosen to rank the contaminants. For example, when the organic contaminants are rank-ordered according to the number

Table 1 Classifications of Organic Contaminants that Have Been Detected in the Groundwater at 500 Waste Disposal Site Investigations in the United States

Organic classification	No. of compounds	Detectable events	Percent of events
Volatile compounds	31	24,900	75.3
Nonstandard volatile compounds	8	579	1.8
Base/neutral compounds	45	2,041	6.2
Nonstandard B/N compounds	8	19	0.06
Acid-extractable compounds	11	979	3.0
Nonstandard AE compounds	4	50	0.2
Pesticides	26	815	2.5
RCRA pesticides	3	261	0.8
Priority pollutants	113	28,735	86.9
Appendix IX constituents	208	29,760	90.1
Appendix IX nonpriority pollutant constituents	95	1,025	3.1
Appendix VIII constituents	359	29,864	90.4
Appendix VIII nonpriority pollutant constituents	246	1,129	3.4
Miscellaneous organics	784	3,188	9.6

Table 2 Rank-Ordering of the 30 Most Frequently Detected Organic Compounds in Disposal Site Groundwater Based on Detectable Events and the Number of Sites in Which the Contaminant Was Detected

Compound	Detectable events	Compound	Number of sites
Dichloromethane[a]	4558	Dichloromethane[a]	157
Trichloroethene[a]	4001	Trichloroethene[a]	132
Tetrachlorethene[a]	2913	Toluene[a]	131
trans-1,2-Dichloroethene[a]	2357	Benzene[a]	120
Trichloromethane	2137	trans-1,2-Dichloroethene[a]	116
1,1-Dichloroethane[a]	1706	Tetrachloromethane[a]	111
1,1-Dichloroethene[a]	1653	Ethylbenzene[a]	109
1,1,1-Trichloroethane[a]	1609	1,1-Dichloroethane[a]	108
Toluene[a]	1430	1,1,1-Trichloroethane[a]	101
1,2-Dichloroethane[a]	1339	Trichloromethane[a]	89
Benzene[a]	1169	Bis(2-ethylhexyl) phthalate	89
Ethylbenzene[a]	733	Chlorobenzene[a]	86
Phenol	679	1,2-Dichloroethane[a]	82
Chlorobenzene[a]	662	Vinyl chloride[a]	79
Vinyl chloride[a]	580	Phenol	79
Tetrachloromethane[a]	484	1,1-Dichloroethane[a]	76
Bis(2-ethylhexyl) phthalate	383	Chloroethane[a]	61
Naphthalene	369	Naphthalene	61
1,1,2-Trichloroethane[a]	270	Di-N-butyl phthalate	56
Chloroethane[a]	269	Fluorotrichloromethane[a]	45
Acetone[a]	254	Lindane	41
1,2-Dichlorobenzene	240	2,4-Dimethylphenol	38
Isophorone	211	1,1,2-Trichloroethane[a]	37
Fluorotrichloromethane[a]	203	Diethyl phthalate	37
1,4-Dichlorobenzene	191	Isophorone	37
2-Butanone	171	2,4-D	36
1,2,4-Trichlorobenzene	164	1,2-Dichlorobenzene	36
2,4-Dimethylphenol	159	Acetone[a]	34
1,2-Dichloropropane[a]	158	1,4-Dichlorobenzene	34
Dichlorodifluoromethane[a]	154	1,2-Dichloropropane[a]	33

[a]Volatile compound.

of detectable events, the top 12 compounds, and 17 of the top 20 compounds, are volatile organic contaminants. When the contaminants are rank-ordered according to the number of site investigations that reported the presence of the compound in groundwater, the top 10 compounds, and 16 of the top 20 compounds, are volatile organic compounds. Despite the differences in ordering, 25 contaminants (19 of them volatiles) are in both lists in Table 2.

In addition to the abundance of individual volatile organic compounds, two further observations on the data in Table 2 are possible. First, each of the rank-ordering factors decreases exponentially (the 30th ranked compound was detected one-thirtieth as often as the most abundant organic contaminant; the 30th ranked compound was detected at one-fifth the number of sites of the highest ranked contaminant). This observation reinforces the abundance of the volatile compounds in disposal site groundwater. Second, it should be noted that the highest ranked compound in Table 2 is dichloromethane. As the data are not censored on the basis of quality assurance/quality control considerations, the ranking for this contaminant is undoubtedly influenced by the fact that it is a common laboratory contaminant. However, because of the high use of this compound and the high concentrations reported in the groundwater at some of the sites, it would be presumptuous to discard all of the data for this contaminant. Therefore, whereas a strict quality assurance evaluation of the data may alter the ranking of a few contaminants [specifically dichloromethane and bis(2-ethylhexyl) phthalate], this refinement would not alter the overall abundance of volatile organic compounds in groundwater relative to other detected contaminants. Support for this statement is available from a study of 92 disposal sites in western Germany [8]. These results showed that the 15 most frequently detected organic contaminants in disposal site groundwater were all volatile compounds, and dichloromethane was included in the list (12th).

The general predominance of volatile organic compounds in the rank-ordered list of groundwater contaminants is consistent with other published studies. It has been demonstrated [9–11] that volatile organic compounds are more abundant than other organic contaminants when the disposal site monitoring results are ranked according to frequency of detection, number of sites at which a contaminant is detected in groundwater, average contaminant concentration, and maximum detected concentration of each compound. Also, as stated above, studies have shown that the 15 most prevalent organic contaminants in disposal site groundwater in the former Federal Republic of Germany are all volatile organic compounds [8,12].

IV. REGIONAL DISTRIBUTION OF CONTAMINANTS

The initial tabulation of disposal site groundwater monitoring data demonstrated that volatile organic compounds are the most abundant class of organic contaminants and the most frequently detected individual contaminants in the national database. In order to evaluate whether the predominance of volatile compounds could be attributed to a small number of sites or a specific geographic region or was characteristic of the national problem, the project database was geographically sorted into 10 subsets corresponding to the 10 established EPA regions in the United States. The numbers of organic priority pollutants reported for site investigations conducted in each EPA region are tabulated in Table 3. These results demonstrate that volatile compounds constitute the single most abundant class of groundwater contaminants in each of the geographic subsets of the project database [9].

A review of the information in Table 3 indicates that there are differences in the occurrence of volatile compounds and other organic contaminants. For example, the average number of detected volatile compounds per site investigation exceeded 10 in three regions (1, 8, and 9) but was three or less in two regions (5 and 7). This variation is probably due to factors such as regional differences in chemical use and differences in site characterization activities. However, despite these fluctuations, the data clearly establish the need to include volatile analyses in a groundwater monitoring strategy for waste disposal sites in each geographic region.

The frequency with which individual organic priority pollutants and non-standard priority pollutants in the CERCLA portion of the database have been detected on a regional basis is summarized in Table 4. This information illustrates the variability that has been observed between groundwater contaminants within a region and between regions for each contaminant. First, no single contaminant is universally abundant in each of the regional data sets [13]. Although trichloroethene had the highest frequency of detection in five of the regional data sets, it was only the tenth most frequently detected volatile compound in one region and the fifteenth in another. Second, a volatile compound was the most frequently detected compound in eight of the ten regions [phenol was ranked first in one region, and bis(2-ethylhexyl) phthalate, possibly due to sample contamination, was ranked first in another]. In fact, volatile compounds represented four of the top five compounds in all regions except one. In region 7, volatile compounds

Table 3 Average Number of Organic Priority Pollutants Detected per Site in Each EPA Region

Class of contaminant	EPA region									
	1	2	3	4	5	6	7	8	9	10
Volatiles	10.0	6.8	6.4	5.6	3.0	4.9	2.1	11.0	11.2	6.8
Acid-extractables	0.9	1.3	0.8	0.4	0.4	0.5	0.2	1.0	0.3	1.4
Base/neutrals	2.5	3.3	3.1	1.8	1.1	2.8	0.7	4.2	3.0	2.8
Pesticides	0.0	1.1	0.6	1.3	0.5	1.4	1.1	3.6	1.8	1.3

Source: Plumb and Pitchford [9].

Table 4 Frequency of Detection[a] of Organic Priority Pollutants and Nonstandard Priority Pollutants in Groundwater by EPA Region

Groundwater contaminant	EPA region									
	1	2	3	4	5	6	7	8	9	10
Standard volatile priority pollutants										
Acrolein	0.0	0.0	2.4	0.2	0.0	1.0	0.0	0.3	0.0	0.4
Acrylonitrile	0.0	0.0	0.0	0.0	0.0	0.0	0.0	0.0	0.1	0.0
Benzene	19.9	20.3	16.7	9.6	3.8	34.3	2.4	10.7	8.2	6.2
Tetrachloromethane	3.1	0.0	2.4	9.8	1.1	10.3	0.0	0.0	9.4	4.6
Chlorobenzene	2.9	8.7	4.9	18.7	0.4	12.4	0.0	2.0	1.8	3.5
1,2-Dichloroethane	7.9	10.7	8.6	2.3	2.3	28.0	2.4	11.6	19.4	6.6
1,1,1-Trichloroethane	21.3	1.8	14.5	5.1	10.9	23.5	2.4	22.2	22.8	21.6
1,1-Dichloroethane	18.3	9.3	16.5	10.2	5.3	24.8	0.0	17.1	21.9	11.4
1,1,2-Trichloroethane	1.5	0.9	3.3	0.0	1.5	27.6	0.0	5.6	0.5	3.4
1,1,2,2-Tetrachloroethane	1.7	0.9	2.5	1.1	0.0	5.3	0.0	0.3	0.2	4.9
Chloroethane	1.5	1.9	1.6	3.9	1.9	1.1	2.4	0.6	0.8	2.0
2-Chloroethyl vinyl ether[b]	0.0	0.5	0.0	0.0	0.4	0.0	0.0	0.0	0.3	0.0
Trichloromethane	15.3	9.3	17.1	5.7	1.5	25.5	7.3	8.2	43.5	25.3
1,1-Dichloroethene	6.9	4.7	12.2	5.5	4.2	24.5	0.0	15.5	38.5	11.6
1,2-*trans*-Dichloroethene	13.4	12.4	8.9	20.1	6.8	23.2	4.9	13.3	40.2	27.3
1,2-Dichloropropane	2.5	1.4	4.8	0.2	1.1	6.3	0.0	10.7	0.3	0.0
1,3-*trans*-Dichloropropene	0.4	0.0	0.0	0.0	0.0	2.1	0.0	0.0	0.0	0.4
Ethylbenzene	14.9	11.7	18.3	2.8	3.8	12.2	2.4	3.7	2.6	6.2
Dichloromethane	20.8	26.5	8.4	7.4	11.6	30.7	4.9	22.9	21.2	29.8
Chloromethane	0.8	1.9	1.6	0.4	0.0	0.0	2.4	0.6	0.8	1.2
Bromomethane	0.0	0.0	3.2	0.0	0.0	0.0	0.0	0.0	0.8	0.0
Tribromomethane	0.2	0.0	3.3	0.0	0.0	1.1	0.0	0.0	0.4	1.0
Bromodichloromethane	0.6	1.3	0.8	0.0	0.0	2.1	2.4	0.0	1.1	2.6
Fluorotrichloromethane	4.2	0.9	3.0	0.4	1.5	0.0	0.0	0.9	2.5	0.4
Dichlorodifluoromethane	0.4	1.1	4.1	0.6	0.0	0.0	0.0	1.1	0.0	0.0
Chlorodibromomethane	0.4	0.0	0.0	0.2	0.0	1.1	2.4	0.0	0.8	1.3
Tetrachloroethene	16.4	4.3	16.7	8.5	8.3	25.5	4.8	20.7	51.8	21.3
Toluene	20.8	16.4	22.3	7.4	10.9	33.0	2.4	7.7	8.7	11.8
Trichloroethene	26.0	11.2	29.7	4.7	17.5	26.0	7.3	21.7	76.1	31.9
Vinyl chloride	8.6	9.2	7.8	17.3	3.4	7.3	0.0	2.6	0.0	9.8
Nonstandard volatile priority pollutants										
Acetone	9.3	27.3	6.6	15.9	26.7	0.0	—	5.3	0.0	11.3
2-Butanone	4.3	19.4	3.3	12.7	0.7	0.0	—	38.5	0.0	8.5
Carbon disulfide	0.0	0.0	0.0	4.2	0.0	0.0	—	0.0	3.6	0.0
2-Hexanone	0.3	11.1	0.0	0.0	0.0	0.0	—	0.0	3.6	1.4
4-Methyl-2-pentanone	0.8	23.1	0.0	1.4	1.3	0.0	—	5.9	0.0	0.0

Table 4 Continued

Groundwater contaminant	EPA region									
	1	2	3	4	5	6	7	8	9	10
Styrene	0.8	3.1	0.0	8.0	1.3	29.5	—	0.0	3.8	2.9
Vinyl acetate	0.0	0.0	0.0	0.0	0.0	0.0	—	0.0	0.0	0.0
o-Xylene	15.6	26.9	0.0	0.0	1.3	0.0	—	10.0	3.6	10.0
Standard base/neutral priority pollutants										
Acenaphthene	0.0	0.0	0.0	0.5	0.0	2.7	0.0	0.0	0.0	3.8
Benzidine[b]	0.0	0.0	2.4	0.0	0.4	2.7	0.0	0.0	0.0	0.0
1,2,4-Trichlorobenzene	0.0	0.0	1.5	1.5	0.4	10.5	0.0	0.0	0.0	0.5
Hexachlorobenzene	0.0	0.0	0.0	1.5	0.0	1.3	0.0	0.0	0.7	0.0
Hexachloroethane	0.0	0.0	0.0	9.7	0.0	0.0	0.0	0.0	0.0	0.5
Bis(2-chloroethyl) ether	0.0	2.6	0.0	0.0	1.2	12.2	0.0	0.0	1.0	0.0
2-Chloronaphthalene	0.0	0.0	0.0	0.0	0.0	1.4	0.0	0.0	0.0	0.0
1,2-Dichlorobenzene	0.0	1.3	3.7	0.0	0.4	4.0	0.0	1.8	0.0	7.1
1,3-Dichlorobenzene	0.0	0.0	0.0	0.0	0.0	2.7	0.0	0.3	0.0	2.7
1,4-Dichlorobenzene	1.4	1.3	2.4	0.0	0.0	3.9	0.0	0.0	2.9	3.8
3,3'-Dichlorobenzidine	0.0	0.0	0.0	0.0	1.2	0.0	0.0	0.0	0.0	0.0
2,4-Dinitrotoluene	0.0	0.0	0.0	0.0	0.0	1.4	0.0	0.3	0.0	0.0
2,6-Dinitrotoluene	0.0	0.0	1.2	0.0	0.0	1.4	0.0	0.0	0.0	0.0
1,2-Diphenyl hydrazine[b]	0.0	0.0	0.0	0.0	0.0	0.0	0.0	0.3	0.0	0.0
Fluoranthene	1.4	0.6	2.4	0.0	0.4	0.0	0.0	0.6	2.0	0.5
4-Chlorophenyl phenyl ether	0.0	0.0	3.7	0.0	0.4	0.0	0.0	0.0	0.0	0.0
4-Bromophenyl phenyl ether	0.0	0.0	0.0	0.0	0.0	0.0	0.0	0.0	1.0	0.0
Bis(2-chloroisopropyl) ether[b]	0.0	0.0	0.0	0.0	0.0	10.8	0.0	0.0	0.0	0.0
Bis(2-chloroethoxy) methane	0.0	0.7	0.0	0.0	0.0	1.4	0.0	0.0	0.0	0.0
Hexachlorobutadiene	0.0	0.0	0.0	7.8	0.0	0.0	0.0	0.0	0.0	0.5
Hexachlorocyclopentadiene	0.0	0.0	0.0	0.8	0.0	0.0	0.0	0.0	0.0	0.0
Isophorone	0.0	2.0	4.9	0.5	2.7	0.0	0.0	0.6	0.0	1.1
Naphthalene	0.0	2.5	4.5	8.9	1.1	20.7	0.0	1.5	0.0	3.1
Nitrobenzene	0.0	0.0	1.2	0.5	0.0	0.0	0.0	0.3	0.0	0.0
N-Nitrosodiphenylamine	0.0	0.0	2.4	0.0	0.0	0.0	0.0	0.0	0.0	0.6
N-Nitroso-n-propylamine	0.0	0.0	1.2	0.5	0.4	0.0	0.0	0.0	0.0	0.5
Bis(2-ethylhexyl) phthalate	10.1	11.8	10.8	3.6	16.0	23.7	18.9	13.5	2.7	9.2
Butyl benzyl phthalate	5.7	1.3	0.0	1.0	0.8	2.7	0.0	7.1	3.0	2.2
Di-N-butyl phthalate	2.9	5.5	0.0	2.6	3.0	10.7	0.0	4.3	9.1	7.6
Di-N-octyl phthalate	0.0	3.9	4.9	0.5	2.7	0.0	0.0	0.9	0.0	1.1
Diethyl phthalate	2.9	3.9	3.2	4.6	0.0	8.0	0.0	1.5	1.0	2.2
Dimethyl phthalate	0.0	0.7	4.8	1.0	0.0	1.4	0.0	0.0	0.0	0.0
Benzo[a]anthracene	0.0	0.0	2.4	1.0	0.0	0.0	0.0	0.0	0.0	0.0
Benzo[a]pyrene	0.0	0.0	2.4	0.0	0.0	0.0	0.0	0.0	1.6	0.0
Benzo[b]fluoranthene	0.0	0.0	0.0	0.0	0.0	0.0	0.0	0.0	0.0	0.0
Benzo[k]fluoranthene	0.0	0.0	2.4	0.0	0.0	0.0	0.0	0.0	0.0	0.0

Table 4 Continued

Groundwater contaminant	EPA region									
	1	2	3	4	5	6	7	8	9	10
Chrysene	1.4	0.0	2.4	0.5	0.0	0.0	0.0	0.0	0.0	0.0
Acenaphthylene	0.0	0.0	0.0	0.0	0.0	5.4	0.0	0.0	0.0	0.5
Anthracene	0.0	0.0	3.0	0.0	0.4	1.4	0.0	0.6	0.0	0.0
Benzo[g,h,i]perylene	0.0	0.0	0.0	0.0	0.0	0.0	0.0	0.0	0.0	0.5
Fluorene	0.0	0.0	0.0	0.0	0.0	4.1	0.0	0.0	0.0	0.0
Phenanthrene	1.4	0.0	3.7	0.0	0.0	0.0	0.0	0.0	1.0	0.0
Dibenzo[a,h]anthracene	0.0	0.7	0.0	0.0	0.0	0.0	0.0	0.0	0.0	0.0
Indeno[1,2,3-cd]pyrene	0.0	0.0	0.0	0.0	0.0	0.0	0.0	0.0	0.0	0.0
Pyrene	1.4	1.3	3.0	1.0	0.0	0.0	0.0	0.6	0.0	0.0
Nonstandard base/neutral priority pollutants										
Aniline	0.0	0.0	0.0	—	0.0	0.0	—	0.0	0.0	0.0
Benzyl alcohol	0.0	0.0	0.0	—	0.0	0.0	—	0.0	0.0	1.6
4-Chloroaniline	0.0	0.0	0.0	—	0.0	0.0	—	0.0	0.0	0.0
Dibenzofuran	0.0	0.0	0.0	100.0	0.0	0.0	—	0.0	0.0	0.0
2-Methyl naphthalene	0.0	0.0	0.0	23.5	0.0	0.0	—	0.0	0.0	0.0
2-Nitroaniline	0.0	0.0	0.0	—	0.0	0.0	—	0.0	0.0	0.0
3-Nitroaniline	0.0	0.0	0.0	—	0.0	0.0	—	0.0	0.0	0.0
4-Nitroaniline	0.0	0.0	0.0	—	0.0	0.0	—	6.2	0.0	0.0
Standard acid-extractable priority pollutants										
2,4,6-Trichlorophenol	0.0	0.0	0.0	0.0	0.0	14.8	0.0	0.0	0.0	2.7
p-Chloro-m-cresol	0.0	0.0	0.0	0.0	0.0	0.0	0.0	0.0	0.0	4.4
2-Chlorophenol	0.0	0.7	1.4	0.0	0.0	5.1	0.0	0.0	0.0	4.9
2,4-Cichlorophenol	0.0	0.0	0.0	0.6	0.0	7.2	0.0	0.0	0.0	6.6
2,4-Dimethylphenol	1.4	2.0	1.3	0.6	2.3	1.0	0.0	1.2	0.0	5.5
2-Nitrophenol	0.0	0.0	3.9	0.0	0.0	0.0	0.0	0.0	1.0	1.6
4-Nitrophenol[b]	0.0	0.7	2.6	0.6	0.0	1.1	0.0	0.0	0.0	0.5
2,4-Dinitrophenol	1.4	0.0	0.0	0.6	0.0	0.0	0.0	0.0	0.0	0.0
4,6-Dinitro-2-methylphenol	0.0	0.0	0.0	0.0	0.0	0.0	0.0	0.0	2.0	0.0
Pentachlorophenol	1.4	11.5	17.6	0.0	3.5	0.0	0.0	1.5	0.0	4.9
Phenol	14.3	27.4	5.0	3.0	2.7	15.6	11.7	18.2	0.0	13.1
Nonstandard acid-extractable priority pollutants										
Benzoic acid[b]	0.0	3.7	0.0	50.0	0.0	5.9	—	0.0	0.0	3.2
2-Methylphenol	0.0	9.7	0.0	100.0	2.0	0.0	—	0.0	0.0	14.3
4-Methylphenol	0.0	14.3	0.0	100.0	3.2	0.0	—	5.9	0.0	17.5
2,4,5-Trichlorophenol	0.0	0.0	0.0	100.0	0.0	0.0	—	0.0	0.0	1.6

Table 4 Continued

Groundwater contaminant	EPA region									
	1	2	3	4	5	6	7	8	9	10
Standard pesticide priority pollutants										
Aldrin	0.0	0.0	0.0	0.0	0.0	0.0	0.0	0.0	0.0	0.0
Dieldrin	0.0	0.5	0.0	3.5	0.8	12.3	0.0	1.0	0.8	0.5
Chlordane	0.0	0.5	1.8	2.4	0.0	5.0	2.1	0.7	1.6	0.0
4,4'-DDT	0.0	0.0	0.0	0.6	0.4	7.6	0.0	1.7	0.0	0.0
4,4'-DDE	0.0	0.0	0.0	0.0	0.8	9.3	0.0	0.7	0.0	1.6
4,4'-DDD	0.0	0.0	0.0	0.0	0.4	8.5	0.0	1.0	1.4	0.5
Endosulfan I	0.0	1.1	0.0	0.0	0.4	1.0	0.0	0.0	0.0	1.1
Endosulfan II	0.0	0.0	0.0	0.0	0.0	0.0	0.0	0.0	0.0	0.0
Endosulfan sulfate	0.0	0.0	0.0	0.4	0.8	1.0	0.0	0.0	0.0	0.0
Endrin	0.0	2.6	0.8	1.5	0.0	1.4	0.0	0.3	0.0	0.5
Endrin aldehyde	0.0	0.0	0.0	0.0	0.0	1.0	0.0	0.0	0.0	0.0
Heptachlor	0.0	3.2	0.0	0.5	0.8	2.0	1.9	0.0	0.8	1.1
Heptachlor epoxide	0.0	1.1	0.0	0.9	0.4	0.0	1.9	0.0	0.0	0.5
α-BHC	0.0	0.0	1.8	2.4	0.4	24.8	6.2	0.3	17.9	3.2
β-BHC	0.0	1.1	0.0	2.1	0.0	23.9	4.3	2.6	0.8	1.6
δ-BHC	0.0	0.5	0.0	1.2	0.8	23.1	2.1	0.7	34.7	1.6
Lindane	0.0	2.1	0.0	2.7	0.8	15.9	2.0	2.7	18.2	3.4
PCB-1242	0.0	0.0	0.0	0.3	0.4	0.0	0.0	1.0	0.0	1.0
PCB-1254	0.0	0.0	17.9	0.7	0.0	0.0	0.0	0.3	0.0	0.5
PCB-1221	0.0	0.0	0.0	0.0	0.0	0.0	0.0	0.0	0.0	0.0
PCB-1232	0.0	0.0	0.0	0.0	0.0	0.0	0.0	0.0	0.0	0.0
PCB-1248	0.0	0.0	3.4	0.0	0.0	0.0	0.0	0.0	0.0	0.0
PCB-1260	0.0	0.0	0.0	0.3	0.0	1.0	0.0	0.7	0.0	4.2
PCB-1016	0.0	0.5	0.0	0.0	0.0	0.0	2.1	0.0	0.0	0.0
Toxaphene	0.0	0.0	0.0	0.0	0.0	0.9	0.0	0.0	0.7	0.0
2,3,7,8-p-Dioxin	0.0	0.0	0.0	0.0	0.0	0.0	0.0	0.0	0.0	0.0
RCRA pesticides										
Methoxychlor	0.0	—	0.0	0.0	—	1.7	0.0	2.2	0.0	0.0
2,4-D	20.0	—	0.0	50.0	—	2.6	—	4.6	22.5	29.7
Silvex	0.0	—	0.0	0.0	—	2.6	—	1.4	0.0	15.4

[a]Frequency of detection expressed as the percentage of reported analyses for which the contaminant concentration exceeded analytical detection limits.
[b]Compound not included in the list of Appendix IX compounds.
— No analysis reported.
Source: Plumb and Parolini [13].

ranked only third and fourth, but a total of 14 volatile compounds were detected compared to 8 pesticides, 1 base/neutral, and 1 acid-extractable compound. Third, only 12 organic priority pollutants—benzene, 1,2-dichloroethane, 1,1,1-trichloroethane, chloroethane, trichloromethane, *trans*-1,2-dichloroethene, ethylbenzene, dichloromethane, tetrachloroethene, toluene, trichloroethene, and bis(2-ethylhexyl) phthalate—were reported in the groundwater in all 10 regions, and 11 of these compounds are volatile organic constituents. In reviewing the data, it must be remembered that the number of analytical attempts are not the same for all substances and, in particular, the explicit number of attempts are substantially lower for the nonstandard priority pollutants. Thus, some of the high detection frequencies for the nonstandard priority pollutants are artifacts of the small number of analytical attempts that have been reported. [For example, several substituted phenols in the nonstandard acid-extractable category have calculated detection frequencies of 100% in region 4 (Table 4) when only explicit analytical results are considered. However, when it is considered that each of the 158 acid scans reported in the region are acceptable alternative analyses for nonstandard acid-extractable compounds [4], a more appropriate detection frequency for these compounds would be $(1+0) \times 100/(1+158)$, or 0.62%, when implicit analytical results are considered. This situation holds for most of the nonstandard priority pollutant compounds listed in Table 4 because the number of implicit analyses is much greater than the number of explicit analyses for each of these compounds.]

The results obtained when the composite database was sorted geographically clearly show that the high frequency of occurrence of volatile organic compounds is not an artifact due to a small number of sites or regions. Rather, the predominance of volatile organic compounds is indicative of the groundwater contamination that can occur in the vicinity of waste disposal sites. The absolute abundance of individual organic compounds may vary geographically, but the presence of volatile organic compounds is a dominant characteristic of the groundwater contamination profile associated with mixed waste disposal sites.

V. CONTAMINANT OCCURRENCE BY FACILITY TYPE

The project database was also sorted according to the regulatory classification of the site. The categories that were used for this purpose are CERCLA sites (unmanaged hazardous waste disposal sites), RCRA Subtitle C detection monitoring sites (active hazardous waste disposal sites following the detection monitoring/interim status requirements of the RCRA groundwater monitoring program), RCRA Subtitle C compliance monitoring sites (active

hazardous waste disposal sites following the compliance monitoring requirements of the RCRA groundwater monitoring program), and municipal landfills. The results of this sorting effort are summarized in Table 5.

Data from the CERCLA sites indicate that the volatile compounds account
for the largest percentage of detectable events at this type of facility. Volatile
compounds represent 90% of all the priority pollutant events and 78% of
all the organic detectable events (Table 5). A similar trend is apparent in the
RCRA compliance monitoring sites data set and the municipal landfill data
set. At the RCRA compliance monitoring sites, volatile compounds represent
70% of the priority pollutant detectable events and 69% of all the reported
organic detectable events. The corresponding figures for volatile compounds
at municipal landfills are 82% and 75%, respectively. Although the relative
abundance of volatile contaminants is slightly reduced at RCRA compliance
monitoring sites and municipal landfills (compared to CERCLA sites), this
group still represents the single most abundant classification of organic contaminants that have been detected in site groundwater and is more abundant
than the combined total for all other groups of organic contaminants.

Table 5 Occurrence of Organic Contaminants in Groundwater as a Function of Chemical Classification and Type of Waste Disposal Facility[a]

Chemical classification	CERCLA sites		RCRA-C detection monitoring		RCRA-C assessment monitoring		RCRA municipal landfills	
	AA	DE	AA	DE	AA	DE	AA	DE
Volatiles	134,108	15,670	0	0	5,436	3,514	165,799	9,020
Nonstandard volatiles	7,133	307	0	0	308	36	2,971	272
Base/neutrals	66,362	605	15	0	20,400	941	55,608	1,278
Nonstandard base/ neutrals	2,916	7	0	0	195	0	324	12
Acid-extractables	16,229	401	0	0	5,099	281	13,641	541
Nonstandard acid- extractables	1,521	45	0	0	177	5	236	5
Pesticides	50,767	580	7,821	110	17,406	276	38,089	195
RCRA pesticides	9,068	162	7,781	118	3,770	20	5,735	72
Appendix IX	—	—	—	—	103,266	5,088	—	—
Nonpriority pollu- tant, App. IX	—	—	—	—	6,000	76	—	—
Miscellaneous organics	13,089	2,318	—	—	—	—	20,864	887

[a]AA, number of reported analyses for compounds in the classification; DE, number of detectable events for compounds in the classification.

The one data set in Table 5 that does not demonstrate the abundance of volatile organic compounds is the composite data set from 125 RCRA Subtitle C detection monitoring facilities. However, this is due to the fact that no analytical attempts have been reported at these facilities. Currently (1990), the RCRA Subtitle C detection monitoring program does not require any monitoring for volatile compounds.

Each type of facility summarized in Table 5 follows a particular groundwater monitoring strategy. The RCRA Subtitle C program specifies four indicator parameters that are to be used to detect groundwater contamination near a facility (detection monitoring or interim status) followed by an extensive list of analyses that must be completed to identify and characterize the contaminants that may be present (compliance monitoring) [1,2]. The CERCLA program is completely discretionary. Finally, there are no formal national requirements to monitor groundwater conditions in the vicinity of municipal landfills. (A monitoring program has been proposed for landfills but has not been finalized at this time [14].) However, despite these divergent strategies, the composite data in Table 5 suggest that there is a strong similarity between the type and abundance of organic contaminants that can be expected in the groundwater near these different types of facilities. Specifically, volatile compounds are the most abundant organic contaminants at each type of facility. Also, the frequency of detection of pesticides (the class of compounds routinely determined with SW-846 Method 8250 [6]) is uniformly low in the data sets for each classification (1.1% at CERCLA sites, 1.4% at detection monitoring sites, 1.6% at compliance monitoring sites; and 0.5% at municipal landfills). Furthermore, the frequency of detection of base/neutral compounds and acid-extractable compounds is similar in the three facility classifications for which there are data. In addition, it has previously been shown that the frequency of detection of individual inorganic contaminants is very similar in the RCRA and CERCLA data sets [11]. Because of the comparability of the groundwater contamination profiles that have been observed for a broad spectrum of chemicals (volatiles, base/neutrals, acid-extractables, pesticides, and inorganics) at these different facilities, it is suggested that a universal groundwater monitoring program can be developed for use at these different facilities [11].

VI. EVOLUTION OF A GROUNDWATER MONITORING STRATEGY

A. Need for Volatile Monitoring

A review of composite monitoring data from 500 abandoned waste disposal sites, active hazardous waste disposal sites, and municipal landfills in the

United States has demonstrated the high relative abundance of volatile compounds compared to other classifications or organic contaminants in groundwater near these facilities. The volatile compounds are the most frequently detected class of compounds in the total project database, each geographic subset of the database, and each facility subset of the database. These observations establish a prima facie need for routine volatile monitoring as part of waste disposal site investigations [9–13,15,16]. Furthermore, a consideration of the general properties of volatile compounds, such as their high solubility in water and high octanol–water partitioning coefficients (K_{ow}), would suggest that these compounds would be expected to be more mobile than other organic contaminants when they are associated with disposal site leakage events. For example, 22 of the top 25 most frequently detected organic contaminants have log K_{ow} values of -3 or greater. Thus, the physical and chemical characteristics of volatile compounds both explain and reinforce the need to monitor for these contaminants.

B. Assessment of a Volatile Screening Strategy

Recognition of the qualitative abundance of volatile organic compounds in disposal site groundwater led to the development of an empirical relationship between the number of volatile contaminants and the number of organic priority pollutants detected during a site investigation [9]. Based on composite data from 114 CERCLA sites, the resultant linear expression between these parameters was defined by the equation

Priority pollutants $= 1.394 \times$ (No. of VOCs) $+ 1.072$

where "priority pollutants" is the number of organic priority pollutants detected in the groundwater; "No. of VOCs" is the number of volatile organic priority pollutants detected in the groundwater. The calculated linear correlation coefficient (r^2) between these two parameters was 0.955 ($N = 114$).

Subsequent to the derivation of the mathematical expression above, monitoring data were obtained from an additional 42 disposal site investigations [16]. These data are plotted in Fig. 1 along with a plot of the originally derived volatile organic–priority pollutant relationship. As can be seen, the second set of data produced a nearly identical relationship defined by the equation

Priority pollutants $= 1.348 \times$ (No. of VOCs) $+ 1.380$

The linear correlation coefficient for the second equation is 0.908 ($N = 42$).

There are two important observations that relate to the information presented in Fig. 1. First, the fact that the two data sets can be defined by essentially the same mathematical relationship reinforces an earlier suggestion [9] that a volatile scan can be used as an analytical screening procedure to

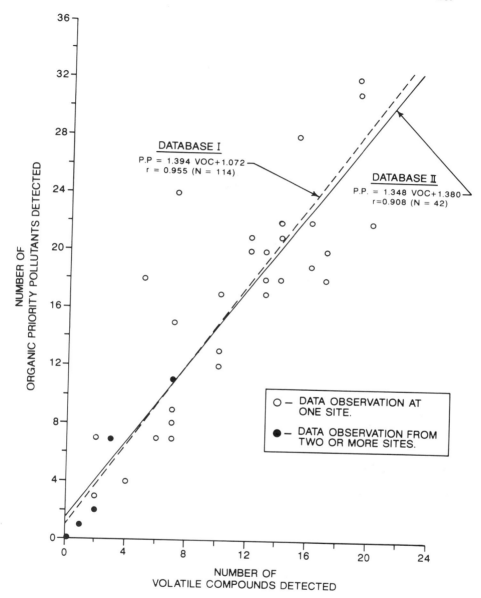

Figure 1 Comparison of volatile organic–priority pollutant relationship derived from CERCLA monitoring data (database I, dashed line) and RCRA monitoring data (database II, solid line). (From Plumb [16].)

establish the need for more extensive organic characterization of ground-water samples collected during waste disposal site monitoring activities. Thus, there would be no need for more extensive and expensive organic character-ization of routine monitoring samples unless the number of detected volatile compounds exceeded a predetermined limit, in which case there would be a strong probability that the additional analytical effort would be useful in characterizing existing site conditions. Second, the two data sets summarized in Fig. 1 represent two distinct categories of waste disposal sites (database I consists of composite data from 114 CERCLA sites, and database II con-sists of composite data from 42 RCRA Subtitle C compliance monitoring sites and municipal landfills). The implication of this observation is that the proposed volatile screening strategy would have general applicability to all classifications of mixed waste sites (i.e., RCRA sites, CERCLA sites, and municipal landfills).

A preliminary evaluation of the type of performance that can be expected with the proposed volatile screening strategy has been initiated [16]. In order to achieve this objective, the observed distribution of organic contaminants in disposal site groundwater has been summarized as shown in Fig. 2. For this purpose, each site was entered into the matrix based on the number of volatile organic compounds detected in groundwater samples and the total number of organic priority pollutants detected in the monitoring samples. Thus, there were five sites in the data subset that reported no volatile com-pounds and no organic priority pollutants in the collected groundwater sam-ples, five sites that reported either one or two volatile compounds and one to three organic priority pollutants in the collected groundwater samples, etc. There is no fundamental reason to select a specific volatile action limit. However, the type of data summary shown in Fig. 2 provides a mechanism to evaluate all possible volatile compound action limits. For example, the empirical relationship shown in Fig. 1 suggests that six volatile compounds would be indicative of approximately nine organic priority pollutants being present. If these values (volatile compounds 6; priority pollutant compounds 9) are superimposed on the data distribution shown in Fig. 2, the data distribu-tion reduces to the simplified format shown at the bottom of Fig. 2. This distribution provides an estimate of the effectiveness of a volatile screening strategy if it is decided to establish six volatile compounds as the action limit:

1. The approach is effective (by definition) when the number of detected volatile compounds is 6 or less and the number of detected priority pollu-tants is 9 or less (box 1) or when the number of detected volatile compounds is 7 or more and the number of detected priority pollutants is 10 or more (box 4). Based on the actual observed distribution at 49 sites, a desired response occurred 45 times (92%).

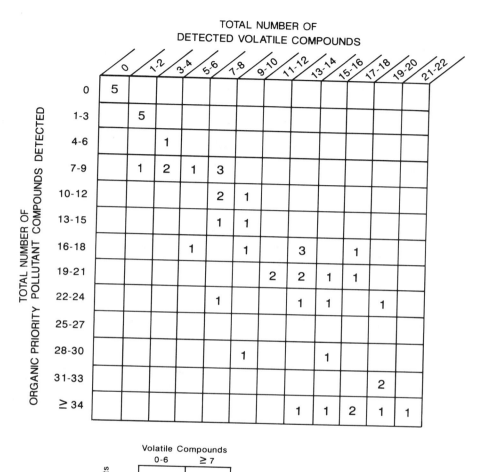

Figure 2 Reported distribution of volatile organic compounds and organic priority pollutant compounds in groundwater at 49 RCRA waste disposal sites. (From Plumb [16].)

2. The possibility of the volatile screening approach triggering a false positive response indicating the need for more analyses when they are not necessary can be estimated by the cases where the number of detected volatile compounds is 7 or more (indicating the need for more analyses by definition) but the number of detected priority pollutant compounds is less than 9 (box 2). This occurred in 3 of the 49 investigations (6%).

3. The possibility of the volatile screening approach producing a false negative response (failing to indicate the need for more organic analyses when it is really necessary) can be estimated by the number of cases when the number of detected volatile compounds is 6 or less but the number of detected priority pollutants is 10 or more (Box 3). This occurred once in 49 investigations (2%).

The process of evaluating possible volatile action limits was repeated for values ranging from 3 to 15. These results are presented in Table 6. This information suggests that the volatile screening strategy would be 92–98% effective for an action level of 11 volatile compounds or less [16]. The estimated occurrence of both false positive and false negative responses would be 8% or less with this approach. On the basis of these performance estimates, any volatile action limit between 3 and 11 would be equally effective. Although an action limit of 11 volatile compounds would be functionally effective, it is obviously too high for practical purposes. Conversely, although an action limit of three volatile compounds would be effective and practical, it would defeat the purpose of a screening strategy because it would trigger the need for expensive organic analyses even though the number of priority pollutants likely to be detected is very small. A consideration of these factors suggests that six volatile compounds would be a reasonable action limit for

Table 6 Estimated Performance of Volatile Scans as a Groundwater Screening Technique as a Function of Potential Volatile Action Limits

Response	Potential volatile action limits expressed as number of detectable volatile compounds						
	>2	>5	>7	>9	>11	>13	>15
Desirable response	98	94	92	92	92	84	86
Overanalyze	0	0	6	2	0	8	6
Underanalyze	2	6	2	6	8	8	8

[a]Data represent the estimated frequency of occurrence of the indicated response for each potential volatile action limit based on observed contaminant distributions at 49 waste disposal sites.
Source: Plumb [16].

the volatile screening strategy: the technique would be 92–94% effective, the level is sufficiently low to be practical, and the additional organic analyses would be justified by the contaminants that are likely to be present. [If the number of detected volatile compounds is 6 or less, the site assessment would be conducted with the results of the volatile scan plus the inorganic analyses included in the monitoring program. If the number of detected volatile compounds is 7 or more, the site assessment would be conducted with the results of the volatile scan plus the supplemental organic analysis (the nonvolatile priority pollutants can be recommended for this purpose) plus the inorganic analyses included in the monitoring program.]

C. Cost–Benefit Evaluation of Volatile Monitoring Strategy

Two methods were used to evaluate the economic feasibility of the proposed volatile screening strategy. The first method was a simple cost–benefit analysis of the analytical procedure required for volatile analyses. The second method was a comparison of costs for the volatile screening procedure and the organic monitoring requirements of the RCRA detection monitoring program.

The cost–benefit analysis was developed by comparing the relative abundance of organic contaminants with the relative costs of the appropriate analytical procedure [17]. As shown in Table 7, the reported abundance of

Table 7 Estimated Cost/Benefit Ratio for the Individual Organic Analytical Procedures Needed to Conduct Complete Appendix IX Analyses on Groundwater Samples

Analytical determination	Cost per analysis[a] ($)	Percent of costs[b]	Percent of detectable events	Cost/ benefit ratio[c]
Volatiles	225	7.5	84.15	11.22
Base/neutrals	450	15.0	8.55	0.57
Acid-extractables	450	15.0	3.46	0.23
Pesticides	225	7.5	2.95	0.39
Nonpriority pollutants	2000	66.7	0.25	0.004
Priority pollutants	950	31.7	99.15	3.13
Appendix IX	3000	100.0	100.00	1.00

[a]Based on 1987 costs in the EPA Contract Laboratory Program. Individual laboratory costs may vary from these estimates.
[b]Percent of Appendix IX costs.
[c]Defined as percent of detectable events divided by percent of analytical costs. The larger the ratio, the more cost-effective the analytical determination.
Source: Plumb [17].

the organic contaminants varied widely, with volatile compounds representing more than 84% of all detectable events involving organic Appendix IX constituents and the nonpriority pollutant fraction representing only 0.25% of all detectable events involving organic Appendix IX constituents. Similarly, the costs of each analytical procedure can be expressed as a percentage of the total Appendix IX analytical costs. Thus, at $225, a volatile scan represents 7.5% of the Appendix IX analytical costs, the base/neutral and acid-extractable fractions represent approximately 15% of the costs, and the nonpriority pollutant fraction represents approximately 66% of the costs. If each of the analytical procedures were equally useful and cost-effective for characterizing disposal site groundwater conditions, the calculated cost/benefit ratio (defined as the percentage of detectable events divided by the percentage of analytical costs) for each class of contaminants should be 1.0. However, the information in Table 7 clearly demonstrates that the only individual organic analytical procedure with a cost/benefit ratio greater than 1.0 is the volatile fraction. Thus, the proposed use of volatile analyses as a screening technique is also the most economically efficient.

The second method chosen for an economic evaluation of the proposed volatile screening strategy was to compare costs for volatile analysis with costs for the organic indicator parameters (TOC and TOX) specified in the RCRA Subtitle C detection monitoring program [1,2]. As indicated in Table 7, the approximate cost for a volatile analysis is $225. The costs for TOC and TOX analyses are approximately $25 and $100, respectively. However, the RCRA program requires quadruplicate analysis for each of the specified indicator parameters. Thus, the proposed volatile screening strategy would cost less than half ($225 vs. $500 per sample) of the current RCRA detection monitoring strategy if a leakage event has not occurred. Should a leakage event occur, the economic advantage of the volatile screening strategy would be increased because the volatile strategy directly targets the contaminants that have the greatest possibility of being present but the RCRA strategy would have to be supplemented with additional data (at additional cost) to identify the contaminants causing the organic indicator parameters to increase.

D. Exclusion from the Proposed Monitoring Strategy

The previous discussion of composite monitoring data showed that a volatile screening strategy can be technically and economically effective for broad classifications of waste disposal sites such as abandoned waste disposal sites (CERCLA sites), active waste disposal sites (RCRA facilities), and municipal landfills. However, a review of industry-specific data sets has identified several situations where a volatile screening strategy would not be appropriate. Two examples are the following.

1. *Incinerated municipal waste ash monofills.* As the volatile compounds are lost due to volatilization or thermal combustion, there is no reason to monitor groundwater in the vicinity of ash monofills for volatile contaminants. An inorganic monitoring strategy is appropriate for this industry [18].
2. *Wood preservation facilities.* Monitoring data for sites in this industrial classification have repeatedly produced a strong fingerprint for volatile compounds (principally benzene, toluene, and xylene) and base/neutral compounds (principally polynuclear aromatic hydrocarbons) [19,20]. Therefore, a monitoring strategy that emphasizes both volatile compounds and base/neutral compounds would be effective for this industry and permit differentiation from typical mixed waste disposal facilities that may be present in the area.

Sites where the volatile screening strategy would not be appropriate are generally restricted to a single, well-defined waste product (monofills handle incinerator ash; wood treatment plants handle creosote). By taking this factor into account and adjusting the monitoring strategy when a single waste product can be identified, it is possible that the performance of the volatile screening strategy can be improved over the estimates presented in this paper.

VII. CONCLUSIONS

Groundwater monitoring data from approximately 500 waste disposal site investigations conducted in the United States were collected and composited for the purpose of defining possible trends that could provide the basis for an effective monitoring strategy. This effort demonstrated that the class of volatile compounds routinely determined with SW-846 Method 8240 [6] is the single most abundant class of organic contaminants likely to be present in disposal site groundwater. Even though more than 425 individual organic contaminants have been identified in groundwater monitoring samples, the 31 compounds that make up the volatile classification (as defined by SW-846 Method 8240) represent 75% of the detectable events for organic contaminants.

The project database was subsequently sorted according to the geographic location of the individual sites, the regulatory classification of the individual sites, and the industrial classification of the sites. The dominant class of organic contaminant in each of the resultant data subsets was that of volatile organic compounds. As the presence of volatile organic compounds is a preponderant characteristic of the groundwater contamination profile associated with CERCLA sites, RCRA sites, and municipal landfills, the evaluation of composite monitoring data suggests that analyses for volatile organic compounds should be an essential component of any waste disposal site groundwater monitoring program.

The observed occurrence and distribution of volatile organic compounds and priority pollutants in RCRA site groundwater and CERCLA site groundwater can be defined by an empirically derived mathematical expression. Because this relationship has been verified using two large independent data sets, it has been proposed that analysis for volatile organic compounds can serve as an effective screening technique to assess the need for more extensive organic analysis of groundwater samples during waste disposal site monitoring activities. The first phase of this technique would consist of mandatory monitoring for volatile organic compounds, which would directly target the most abundant classification of organic contaminants that can be expected to be present in disposal site groundwater. When the occurrence of volatile compounds exceeds an established threshold, more extensive organic analysis for nonvolatile contaminants would be triggered in the second phase of this hierarchical technique. A preliminary evaluation of this strategy with composite monitoring data suggests that it can correctly assess the need for more extensive organic analysis of groundwater samples at least 92% of the time. Judging by the observed occurrence of organic contaminants in the composite database, it can be recommended that the additional organic analyses be limited to the nonvolatile priority pollutants. This monitoring strategy is significantly more effective and economical than the present strategy embodied in the RCRA Subtitle C program for hazardous waste disposal sites. The only waste disposal sites that can be identified at this time for which the proposed volatile screening strategy would be inappropriate are those such as incinerated waste ash monofills and creosote operations that handle a single, well-defined waste product.

REFERENCES

1. U.S. Environmental Protection Agency. Hazardous waste management system. Part VII. Standards applicable to owners and operators of hazardous waste treatment storage and disposal facilities. *Fed. Regist., 45*(98):33154–33258, 1980.
2. U.S. Environmental Protection Agency. RCRA Ground-Water Monitoring Technical Enforcement Guidance Document. U.S. EPA Office of Solid Wastes, Washington, DC, 1986.
3. U.S. Environmental Protection Agency. Comprehensive Environmental Response, Compensation, and Liability Act of 1980. *Environmental Statutes*, Government Institutes, Inc., Rockville, MD, 1985, pp. 619–666.
4. U.S. Environmental Protection Agency. List (Phase I) of hazardous constituents for ground water monitoring. *Fed. Regis., 52*(131):25942–25953, July 9, 1987.
5. Office of Technology Assessment. Protecting the nation's groundwater from contamination. OTA-0-233. Office of Technology Assessment, U.S. Congress, Washington, DC, 1984.

6. U.S. Environmental Protection Agency. Test Methods for Evaluating Solid Wastes, SW-846, 3rd ed. U.S. EPA Office of Solid Waste and Emergency Response, Washington, DC, 1986.
7. Keith, L.H., and Telliard, W.A. Priority pollutants. I. A perspective view. *Environ. Sci. Technol., 12*(4):416–423, 1979.
8. Arneth, J.D., Milde, G., Kerndorff, H., and Schleyer, R. Waste deposit influences on ground water quality as a tool for waste type and site selection for final storage quality. Proceedings of the Swiss Workshop on Land Disposal, Gerzensee, Switzerland, Mar. 14–18, 1988.
9. Plumb, R.H., Jr. and Pitchford, A.M. Volatile organic scans: implications for ground-water monitoring. Proceedings of the Conference on Petroleum Hydrocarbons and Organic Chemicals in Ground Water—Prevention, Detection, and Restoration, American Petroleum Institute and National Water Well Association, Houston, TX, Nov. 13–15, 1985, pp. 207–222.
10. Plumb, R.H., Jr. Practical alternative to the RCRA organic indicator parameters. Proceedings of HAZMACON 87—Hazardous Materials Management Conference and Exhibition, Santa Clara, CA, Association of Bay Area Government, Oakland, CA, Apr. 21–23, 1987, pp. 135–150.
11. Plumb, R.H., Jr. A comparison of ground water monitoring data from CERCLA and RCRA sites. *Ground Water Monit. Rev., VII*:94–100, 1987.
12. Schleyer, R., Arneth, J.D., Kerndorff, H., and Milde, G. Main contaminants and priority pollutants from waste sites: criteria for selection with the aim of assessment on the gorundwater path. In *Contaminated Soil '88* (K. Wolf, W.J. Van Den Brink, and F.J. Colon, eds.), Kluwer, Dordrecht, 1988, pp. 247–251.
13. Plumb, R.H., Jr. and Parolini, J.R. Organic contamination of ground water near hazardous waste disposal sites: a synoptic overview. Presented at the Geological Society of America Annual Meeting, Phoenix, AZ, September 1987.
14. U.S. Environmental Protection Agency. Solid waste disposal facility criteria; proposed rule. *Fed. Regist., 53*(168):33314–33422, 1988.
15. Koehn, J.W., and Stanko, G.H., Jr. Groundwater monitoring. Accurate assessment and reasonable economy are achievable. *Environ. Sci. Technol., 22*(11): 1262–1264, 1988.
16. Plumb, R.H., Jr. Assessment of volatile organic scans as an alternative RCRA indicator parameter. *Ground Water Monit. Rev.,* submitted 1990.
17. Plumb, R.H., Jr. The occurrence of organic Appendix IX constituents in disposal site ground water. *Ground Water Monit. Rev., 1*:157–164, 1991.
18. Plumb, R.H., Jr. Development of a ground-water monitoring strategy for incinerated municipal waste landfills. Internal report prepared for the U.S. Environmental Protection Agency Environmental Monitoring and Support Laboratory, Las Vegas, NV, EPA/600/X-90/110, 1990.
19. Rosenfeld, J.K. Industry-specific ground-water contamination. In *Minimizing Risk to the Hydrologic Environment* (A. Zaporozec, ed.), Kendall/Hunt, Dubuque, IA, 93–111, 1990.
20. Rosenfeld, J.K., and Plumb, R.H., Jr. Ground-water contamination at wood treatment facilities. *Ground Water Monit. Rev., XI*:133–140, 1991.

8

An Overview of Statistical Methods for Groundwater Detection Monitoring at Waste Disposal Facilities

Robert D. Gibbons

The University of Illinois at Chicago
Chicago, Illinois

I. INTRODUCTION

The objective of a groundwater monitoring program at a waste disposal facility is to determine if the facility is affecting groundwater. To accomplish this objective, owner/operators are required to place detection monitoring wells in both upgradient and downgradient locations around the facility and to monitor those wells at regular intervals, typically quarterly, for a series of "indicator parameters." The logic of this sampling strategy is that upgradient water quality represents the background conditions for that particular region, and downgradient water quality represents background water quality plus any influence produced by the facility. Although the logic of this sampling strategy is sometimes questionable, we will proceed by assuming that it is indeed appropriate, and describe statistical methods that can be used in detection monitoring programs of this kind. In some cases, particularly when background data are available prior to the installation of the facility, intrawell comparisons may be used (i.e., each well compared to its own history). The major advantage of this approach is that it eliminates the spatial component of variability from the comparison. The statistical methodology is illustrated in Figure 1, which graphically portrays the variable bases for statistical comparisons between wells and for intrawell statistical comparisons.*

It should be noted that the U.S. Environmental Protection Agency (US EPA) has paid considerable attention to the problem of statistical methods for monitoring waste disposal facilities as part of its RCRA regulations. Unfortunately, much of this work has been severely misguided. For example, the original regulation that required the use of Cochran's approximation to the Behrens–Fisher t statistic (CABF t statistic) also required that each sample be split into four aliquots to be treated as if they were independent measurements. The result, of course, is a tremendous underestimate of the true standard error of the mean difference between upgradient and downgradient locations and a false positive rate that rapidly approached 100% for all sites regulated by RCRA. US EPA [1] provided alternative methods to the CABF t test that included prediction limits, tolerance limits, control charts, and parametric and nonparametric analysis of variance procedures. Unfortunately, little or no guidance was initially provided as a part of the statistical regulation, and when draft guidance was made available it was riddled with inconsistencies and both statistical and conceptual errors. A major problem with these regulations is that they attempt to minimize false negative rates (i.e., the probability that you conclude that there is no contamination when there is) by increasing false positive rates (i.e., the probability that you conclude that there is contamination when there is not). By requiring huge numbers of monitoring wells and "laundry lists" of constituents, the probability of failing at least one statistical test by chance alone is near certainty, regardless of the particular statistical method that is used to evaluate the data. It is only through a careful monitoring program in which verification resampling plays a key role that false positive and false negative rates can achieve a reasonable balance.

The focus of this chapter is on the use of prediction limits and tolerance limits for groundwater detection monitoring. The term *detection monitoring* refers to the early stage in which monitoring data are evaluated to determine if a statistically significant increase over background water quality levels has occurred. Such an increase does not mean the site is leaking! It only means that the result is not consistent with chance expectations given the assumptions underlying the statistical model. A statistically significant detection monitoring result indicates that further and perhaps more detailed study is

---

*Special thanks to Marty Sara, Principal Hydrogeologist for Waste Management Inc., for review of this work and preparation of the figures.

required. This further study may take on many different forms depending on site hydrogeology, sampling protocol, type of constituent for which the exceedance occurred, etc. The details of this process are clearly beyond the scope of this chapter.

In my opinion, the dynamic nature of groundwater systems rules out the use of any methods that attempt to average downgradient measurements of any kind (i.e., either within or across downgradient wells). This eliminates methods such as Student's t statistic, the CABF t statistic, and analysis of variance (ANOVA). Of the methods remaining, prediction limits, tolerance limits, and control charts all provide viable alternatives, but they have very different statistical properties and answer very different questions. This chapter focuses on the use of prediction limits and tolerance limits used individually and in conjunction with one another. In the latter case, very reasonable statistical detection monitoring decisions can be made even at facilities with fairly large numbers of monitoring wells and constituents. These methods are now described in some detail.

II. ASSUMPTIONS

The statistical methods described here are suitable for data that are normally distributed, rare-event data that have a Poisson distribution, or data for which the distribution is unknown and/or atypical (i.e., nonparametric). In addition, normally distributed data that have some proportion (less than 90%) below a detection limit (i.e., nondetects) can also be accommodated. I will refer to such data as "censored" and to the corresponding distribution as a censored normal distribution. Given these four choices (i.e., normal, censored normal, Poisson, or nonparametric), the only case that is not covered is when nothing is detected in the background water quality samples (see Fig. 1). For this case, I present a method by which method detection limits can be obtained based on an analyte present calibration study. These limits may, in turn, be used as criteria for decisions in detection monitoring (i.e., limits).

In addition to selecting the proper sampling distribution, the most critical assumption underlying all of the statistical methods presented here is that of independence. These models strictly assume that observations are the result of a random sampling and that each observation is an independent random sample from the parent population. In the context of groundwater monitoring, this assumption rules out the use of replicate samples, daily samples, and perhaps even monthly samples, given how slowly groundwater moves as well as the fact that the act of sampling affects the local groundwater regimen. Therefore, the adoption of a quarterly sampling program is strongly recommended.

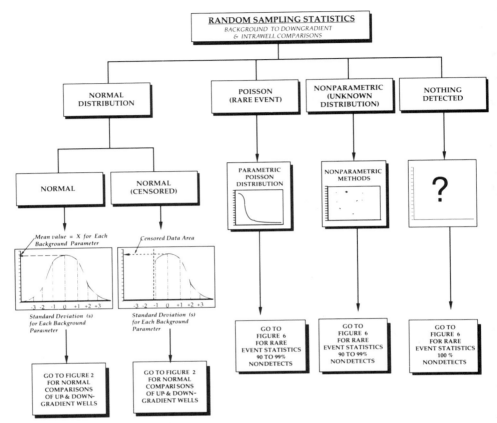

Figure 1 Overview of distributional choices for constituents with various detection frequencies.

III. STATISTICAL OVERVIEW

In detection monitoring programs, the investigator obtains a sample from a monitoring well and must decide whether the facility has had an impact on concentrations of a series of indicator parameters. It is critically important to realize that each new measurement is not a mean value but rather a single, new observation in a supposedly dynamic (flowing) groundwater system. Statistical methods for the comparison of mean values (e.g., Student's *t* test and analysis of variance) do not apply. From a statistical perspective, therefore, the problem is to estimate the probability that each new datum was drawn from the population of pristine, background water quality, for which we have only estimates of mean and variance obtained from a limited number of upgradient measurements.

If the investigator knew that a particular indicator parameter was normally distributed and somehow had the privileged information of knowing the population mean μ and variance σ^2, then he/she could construct the interval $\mu \pm 1.96\sigma$, which would contain 95% of all individual measurements (not means) that were drawn from that population, and the job would be finished. However, in practice, the investigator never knows the values of μ and σ but has only the sample-based estimates ($\bar{x}$ and s) obtained from n independent upgradient measurements. As such, uncertainty is twofold. First, a range of possible values exists when sampling from a normal distribution with known parameters; second, there is a range of possible means ($\bar{x}$) and standard deviations (s) that could be obtained from drawing a sample of size n from a normally distributed population with mean μ and variance σ^2. This latter source of uncertainty will require a multiplier that is larger than 1.96 if one requires reasonable confidence that 95% of the population is contained within the interval. As the number of background water quality measurements approaches infinity, however, the multiplier once again approaches 1.96.

When the sample size is small, say $n = 8$, the required multiplier for 95% confidence that 95% of the population is contained is 3.732 (i.e., $\bar{x} \pm 3.732s$). For a sample size of $n = 30$, 95% confidence is achieved by using a multiplier of 2.549; and for $n = 100$, the multiplier is 2.233. For the purpose of this presentation, a type I error rate of 5% (i.e., 95% confidence) and a coverage proportion of 95% will be used throughout in keeping with US EPA [1] guidelines. There is, of course, nothing magical about these critical values, and either more or less conservative values could be used depending on the application.

A. Tolerance Intervals

The previously described intervals are known as two-sided tolerance intervals in the statistical literature and are largely due to the work of Wald and Wolfowitz [2]. When one is concerned only with values that are too large, one-sided tolerance limits can be constructed as $\bar{x} + ks$, where the multiplier k is somewhat smaller than the previous two-sided tolerance limit factors.

Figure 2 shows the statistical procedure for evaluation of background and downgradient water quality for parameters that normally have detectable values. For example, for $n = 8$, the one-sided tolerance limit is obtained as $\bar{x} + 3.188s$; for $n = 30$, $k = 2.220$; and for $n = 100$, $k = 1.927$. Table 1 presents values of the multiplier k for $n = 4$–100 that are required to provide 95% confidence that 95% of a normally distributed population is contained

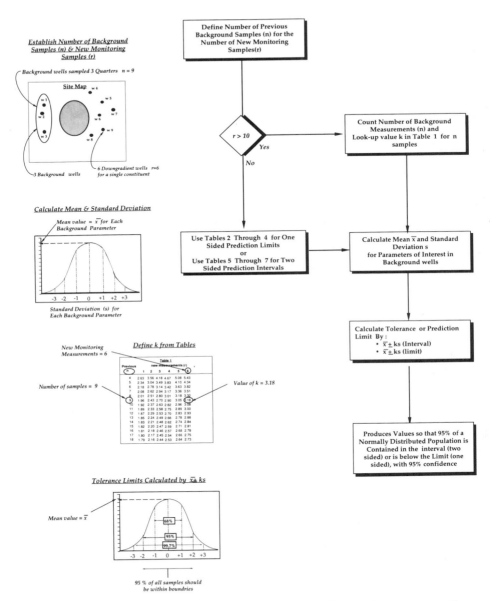

Figure 2 Statistical procedure for evaluation of constituents that are normally distributed and always detected.

Table 1 Factors (k) for Constructing Two-Sided and One-Sided Normal Tolerance Limits ($\bar{x} \pm ks$ and $\bar{x} + ks$) with 95% Confidence that 95% of the Distribution is Covered

n	Two-sided	One-sided	n	Two-sided	One-sided
4	6.370	5.144	20	2.752	2.396
5	5.079	4.210	21	2.723	2.371
6	4.414	3.711	22	2.697	2.350
7	4.007	3.401	23	2.673	2.329
8	3.732	3.188	24	2.651	2.309
9	3.532	3.032	25	2.631	2.292
10	3.379	2.911	30	2.549	2.220
11	3.259	2.815	35	2.490	2.166
12	3.169	2.736	40	2.445	2.126
13	3.081	2.670	50	2.379	2.065
14	3.012	2.614	60	2.333	2.022
15	2.954	2.566	80	2.272	1.965
16	2.903	2.523	100	2.233	1.927
17	2.858	2.486			
18	2.819	2.453			
19	2.784	2.423			

in the interval (i.e., two-sided) or is below the limit (i.e., one-sided). These multipliers reflect the number of standard deviation units required to produce the desired confidence and coverage.

B. Prediction Intervals

Although tolerance intervals are generally quite useful in quality control problems, which are similar to groundwater detection monitoring, even more precise probability statements are possible. For example, in the context of groundwater monitoring, one is generally less interested in what can happen in 95% of all possible samples and more interested in what can happen on the next round of sampling, for which measurements are to be obtained from the monitoring wells at the facility. Because we know the number of future comparisons (i.e., monitoring wells), we can construct an interval (two-sided) or limit (one-sided) that will contain the next r measurements with 95% confidence. If r, in this case the number of monitoring wells, is reasonably small, it will provide a more conservative test than the corresponding 95% confidence and 95% coverage tolerance interval.

For example, for a facility with $n = 8$ background measurements and $r = 3$ monitoring wells, the multiplier for a one-sided 95% confidence and

95% coverage tolerance limit is $k = 3.188$, whereas the corresponding factor for a 95% prediction limit is only $k = 2.80$. However, if the facility had $r = 10$ monitoring wells, the tolerance limit factor would, of course, be unchanged, but the prediction limit factor would now be $k = 3.71$, which is considerably larger than the corresponding tolerance limit factor.

In general, for 95% confidence and 95% coverage, tolerance intervals will be more conservative for facilities with $r > 10$ monitoring wells, and prediction limits will be more conservative (i.e., the limits will be smaller) for facilities with $r \leqslant 10$ monitoring wells. Given the large number of detection monitoring wells at most modern waste disposal facilities, tolerance intervals may well be the method of choice.

Tables 2–4 contain factors for computing one-sided 95% prediction limits based on background samples of $n = 4$–100 and number of monitoring wells of $r = 1$–100. Corresponding two-sided 95% prediction interval factors are provided in Tables 5–7. As in the case of tolerance intervals and limits, these factors are applied as $\bar{x} \pm ks$ (interval) and $\bar{x} + ks$ (limit).

A detailed description of prediction limits in the context of groundwater monitoring problems is provided by Gibbons [3]. Briefly, the equation for a normal upper 95% prediction limit for the next r measurements is

$$\bar{x} + t_{[n-1,.05/r]}s\sqrt{1 + 1/n}$$

Therefore, the factors in Tables 2–4 are actually

$$k = t_{[n-1,.05/r]}\sqrt{1 + 1/n}$$

for the one-sided limits, and in Tables 5–7,

$$k = t_{[n-1,.05/2r]}\sqrt{1 + 1/n}$$

for the two-sided limits, where $t_{[n-1,\alpha]}$ is the $(1 - \alpha)100\%$ point of Student's t distribution on $n - 1$ degrees of freedom.

C. A Two-Stage Approach

As the number of future comparisons gets large (i.e., as r increases), perhaps due to large numbers of monitoring wells or indicator parameters or both, the prediction limit (which includes all r future measurements) becomes extremely large and is no longer protective of the environment. Conversely, as r increases, the tolerance limit remains constant; however, the number of expected failures becomes greater than zero, and the facility will be continuously out of compliance. For example, the 95% upper normal prediction limit for a facility with $r = 100$ future comparisons and eight background measurements is $\bar{x} + 5.74s$, which is enormous. The false negative rate for such a decision rule would be extremely high (i.e., failure to detect contamination when it existed). Conversely, the 95% confidence, 95% coverage upper normal tolerance limit is only $\bar{x} + 3.19s$, which is far more conservative. It is critically important to note, however, that the 95% confidence, 95% coverage tolerance limit will yield as many as five statistical failures per sampling event (out of $r = 100$ comparisons) with 95% confidence by chance alone. A regulatory agency looking at the results of the tolerance limit would incorrectly conclude that the facility had affected groundwater quality. How can we obtain a statistical solution that balances the false positive and false negative rates at reasonable levels, particularly for those facilities that make large numbers of statistical comparisons on each and every sampling event?

A simple solution to this problem can be obtained by using a two-stage sampling approach. In the first stage, groundwater samples are taken from each of the r monitoring wells and compared to a 95% coverage, 95% confidence upper normal tolerance limit, as a screening device. By chance alone, we would expect as many of $0.05r$ exceedances. For example, with $r = 100$, we would expect as many as five significant results by chance alone. The second stage involves a resampling of only those wells that failed the initial screening test. Because we would expect as many as $0.05r$ failures by chance alone, a natural choice is to construct a 95% upper prediction limit for the next $0.05r$ measurements. For the $r = 100$ example, this would amount to a 95% prediction limit for the next five measurements. It is important to emphasize that the computation of the prediction limit is based on $0.05r$ future measurements regardless of the actual number of wells that failed the initial screening test. Certainly, if contamination exists, we would expect more than the $0.05r$ chance failures.

The advantage of this approach over a direct adjustment for $r = 100$ future measurements, or simply applying a 95% confidence, 95% coverage tolerance limit, is that it controls the overall sitewide false positive rate, in this case at 5% (which is clearly not the case for the tolerance limit alone), but at the same time dramatically reduces the false negative rate. For example, with $n = 16$ background measurements, the prediction limit for the next $r = 100$ measurements is $\bar{x} + 4.20s$, which is quite large (i.e., only sensitive to a 4.2 SD unit difference). In contrast, the two-stage procedure would involve the use of an initial screening tolerance limit of $\bar{x} + 2.52s$, followed by a verification prediction limit (applied only to those wells that did not pass the initial screening test) of $\bar{x} + 2.68s$, both of which are sensitive to contamination at considerably lower levels than 4.2 SD units (i.e., approximately 2.5 SD units).

Table 2 Factors for Obtaining One-Sided 95% Prediction Limits for r Additional Samples Given a Background Sample of Size n

Previous n	Number of new measurements (r)														
	1	2	3	4	5	6	7	8	9	10	11	12	13	14	15
4	2.63	3.56	4.18	4.67	5.08	5.43	5.74	6.03	6.29	6.53	6.76	6.97	7.17	7.36	7.54
5	2.34	3.04	3.49	3.83	4.10	4.34	4.54	4.73	4.89	5.04	5.18	5.31	5.44	5.55	5.66
6	2.18	2.78	3.14	3.42	3.63	3.82	3.97	4.11	4.24	4.35	4.46	4.56	4.65	4.73	4.81
7	2.08	2.62	2.94	3.17	3.36	3.51	3.65	3.76	3.87	3.96	4.05	4.13	4.20	4.27	4.34
8	2.01	2.51	2.80	3.01	3.18	3.32	3.43	3.54	3.63	3.71	3.79	3.86	3.92	3.98	4.04
9	1.96	2.43	2.70	2.90	3.05	3.18	3.29	3.38	3.46	3.54	3.60	3.67	3.72	3.78	3.83
10	1.92	2.37	2.63	2.82	2.96	3.08	3.18	3.26	3.34	3.41	3.47	3.53	3.58	3.63	3.68
11	1.89	2.33	2.58	2.75	2.89	3.00	3.09	3.17	3.25	3.31	3.37	3.42	3.47	3.52	3.56
12	1.87	2.29	2.53	2.70	2.83	2.93	3.02	3.10	3.17	3.23	3.29	3.34	3.39	3.43	3.47
13	1.85	2.26	2.49	2.66	2.78	2.88	2.97	3.04	3.11	3.17	3.22	3.27	3.32	3.36	3.40
14	1.83	2.24	2.46	2.62	2.74	2.84	2.93	3.00	3.06	3.12	3.17	3.22	3.26	3.30	3.34
15	1.82	2.21	2.44	2.59	2.71	2.81	2.89	2.96	3.02	3.07	3.12	3.17	3.21	3.25	3.28
16	1.81	2.20	2.41	2.57	2.68	2.78	2.85	2.92	2.98	3.04	3.09	3.13	3.17	3.21	3.24
17	1.80	2.18	2.40	2.54	2.66	2.75	2.83	2.89	2.95	3.01	3.05	3.09	3.13	3.17	3.21
18	1.79	2.17	2.38	2.53	2.64	2.73	2.80	2.87	2.93	2.98	3.02	3.07	3.10	3.14	3.17
19	1.78	2.16	2.36	2.51	2.62	2.71	2.78	2.85	2.90	2.95	3.00	3.04	3.08	3.11	3.14
20	1.77	2.14	2.35	2.49	2.60	2.69	2.76	2.83	2.88	2.93	2.98	3.02	3.05	3.09	3.12
21	1.77	2.13	2.34	2.48	2.59	2.67	2.75	2.81	2.86	2.91	2.96	3.00	3.03	3.07	3.10
22	1.76	2.13	2.33	2.47	2.57	2.66	2.73	2.79	2.85	2.89	2.94	2.98	3.01	3.05	3.08
23	1.75	2.12	2.32	2.46	2.56	2.65	2.72	2.78	2.83	2.88	2.92	2.96	3.00	3.03	3.06
24	1.75	2.11	2.31	2.45	2.55	2.63	2.70	2.76	2.82	2.87	2.91	2.95	2.98	3.01	3.04
25	1.74	2.10	2.30	2.44	2.54	2.62	2.69	2.75	2.81	2.85	2.89	2.93	2.97	3.00	3.03
26	1.74	2.10	2.29	2.43	2.53	2.61	2.68	2.74	2.79	2.84	2.88	2.92	2.95	2.99	3.01
27	1.74	2.09	2.29	2.42	2.52	2.61	2.67	2.73	2.78	2.83	2.87	2.91	2.94	2.97	3.00
28	1.73	2.09	2.28	2.42	2.52	2.60	2.66	2.72	2.77	2.82	2.86	2.90	2.93	2.96	2.99

29	2.98	2.95	2.92	2.89	2.85	2.81	2.77	2.71	2.66	2.59	2.51	2.41	2.28	2.08	1.73
30	2.97	2.94	2.91	2.88	2.84	2.80	2.76	2.71	2.65	2.58	2.50	2.40	2.27	2.08	1.73
31	2.96	2.93	2.90	2.87	2.83	2.79	2.75	2.70	2.64	2.58	2.50	2.40	2.27	2.07	1.72
32	2.95	2.92	2.89	2.86	2.83	2.79	2.74	2.69	2.64	2.57	2.49	2.39	2.26	2.07	1.72
33	2.94	2.92	2.89	2.85	2.82	2.78	2.74	2.69	2.63	2.56	2.49	2.39	2.26	2.07	1.72
34	2.94	2.91	2.88	2.85	2.81	2.77	2.73	2.68	2.62	2.56	2.48	2.38	2.25	2.06	1.72
35	2.93	2.90	2.87	2.84	2.81	2.77	2.72	2.67	2.62	2.55	2.48	2.38	2.25	2.06	1.71
36	2.92	2.90	2.87	2.83	2.80	2.76	2.72	2.67	2.61	2.55	2.47	2.37	2.25	2.06	1.71
37	2.92	2.89	2.86	2.83	2.79	2.76	2.71	2.66	2.61	2.54	2.47	2.37	2.24	2.06	1.71
38	2.91	2.88	2.86	2.82	2.79	2.75	2.71	2.66	2.60	2.54	2.46	2.37	2.24	2.05	1.71
39	2.91	2.88	2.85	2.82	2.78	2.75	2.69	2.66	2.60	2.54	2.46	2.36	2.24	2.05	1.71
40	2.90	2.87	2.85	2.81	2.78	2.74	2.70	2.65	2.60	2.53	2.46	2.36	2.23	2.05	1.71
41	2.90	2.87	2.84	2.81	2.77	2.74	2.69	2.65	2.59	2.53	2.45	2.36	2.23	2.05	1.70
42	2.89	2.87	2.84	2.80	2.77	2.73	2.69	2.64	2.59	2.53	2.45	2.35	2.23	2.04	1.70
43	2.89	2.86	2.83	2.80	2.77	2.73	2.69	2.64	2.59	2.52	2.45	2.35	2.23	2.04	1.70
44	2.88	2.86	2.83	2.80	2.76	2.73	2.68	2.64	2.58	2.52	2.44	2.35	2.22	2.04	1.70
45	2.88	2.85	2.82	2.79	2.76	2.72	2.68	2.63	2.58	2.52	2.44	2.35	2.22	2.04	1.70
46	2.88	2.85	2.82	2.79	2.76	2.72	2.68	2.63	2.58	2.51	2.44	2.34	2.22	2.04	1.70
47	2.87	2.85	2.82	2.79	2.75	2.72	2.67	2.63	2.57	2.51	2.44	2.34	2.22	2.03	1.70
48	2.87	2.84	2.81	2.78	2.75	2.71	2.67	2.62	2.57	2.51	2.43	2.34	2.22	2.03	1.70
49	2.86	2.84	2.81	2.78	2.75	2.71	2.67	2.62	2.57	2.51	2.43	2.34	2.21	2.03	1.69
50	2.86	2.84	2.81	2.78	2.74	2.71	2.67	2.62	2.57	2.50	2.43	2.34	2.21	2.03	1.69
60	2.84	2.81	2.78	2.75	2.72	2.68	2.64	2.60	2.55	2.48	2.41	2.32	2.20	2.02	1.68
70	2.82	2.79	2.77	2.74	2.70	2.67	2.63	2.58	2.53	2.46	2.40	2.31	2.19	2.01	1.68
80	2.80	2.78	2.75	2.72	2.69	2.66	2.62	2.57	2.52	2.46	2.39	2.30	2.18	2.00	1.67
90	2.79	2.77	2.74	2.71	2.68	2.65	2.61	2.46	2.51	2.45	2.38	2.29	2.17	2.00	1.67
100	2.79	2.76	2.73	2.71	2.67	2.64	2.60	2.56	2.51	2.45	2.38	2.29	2.17	1.99	1.67

Factor $= t_{(n-1,1-\alpha/r)}\sqrt{1+1/n}$.

Table 3 Factors for Obtaining One-Sided 95% Prediction Limits for *r* Additional Samples Given a Background Sample of Size *n*

Previous *n*	Number of new measurements (*r*)														
	16	17	18	19	20	21	22	23	24	25	26	27	28	29	30
4	7.71	7.87	8.03	8.19	8.33	8.48	8.61	8.75	8.88	9.00	9.13	9.25	9.36	9.48	9.59
5	5.76	5.86	5.95	6.05	6.13	6.21	6.29	6.37	6.45	6.52	6.59	6.66	6.72	6.79	6.85
6	4.89	4.96	5.03	5.09	5.15	5.22	5.27	5.33	5.38	5.43	5.48	5.53	5.58	5.62	5.67
7	4.40	4.46	4.51	4.56	4.61	4.66	4.71	4.75	4.80	4.84	4.88	4.91	4.95	4.99	5.02
8	4.09	4.14	4.19	4.23	4.27	4.31	4.35	4.39	4.43	4.46	4.50	4.53	4.56	4.59	4.62
9	3.87	3.92	3.96	4.00	4.04	4.08	4.11	4.14	4.18	4.21	4.24	4.26	4.29	4.32	4.34
10	3.72	3.76	3.80	3.83	3.87	3.90	3.93	3.96	3.99	4.02	4.05	4.07	4.10	4.12	4.15
11	3.60	3.64	3.67	3.71	3.74	3.77	3.80	3.83	3.85	3.88	3.91	3.93	3.95	3.98	4.00
12	3.51	3.54	3.58	3.61	3.64	3.67	3.70	3.72	3.75	3.77	3.79	3.82	3.84	3.86	3.88
13	3.43	3.47	3.50	3.53	3.56	3.58	3.61	3.64	3.66	3.68	3.71	3.73	3.75	3.77	3.79
14	3.37	3.40	3.43	3.46	3.49	3.52	3.54	3.57	3.59	3.61	3.63	3.65	3.67	3.69	3.71
15	3.32	3.35	3.38	3.41	3.43	3.46	3.48	3.51	3.53	3.55	3.57	3.59	3.61	3.63	3.64
16	3.27	3.30	3.33	3.36	3.39	3.41	3.43	3.46	3.48	3.50	3.52	3.54	3.56	3.57	3.59
17	3.24	3.27	3.29	3.32	3.35	3.37	3.39	3.41	3.44	3.46	3.47	3.49	3.51	3.53	3.54
18	3.20	3.23	3.26	3.29	3.31	3.33	3.36	3.38	3.40	3.42	3.44	3.45	3.47	3.49	3.50
19	3.18	3.20	3.23	3.26	3.28	3.30	3.32	3.34	3.36	3.38	3.40	3.42	3.44	3.45	3.47
20	3.15	3.18	3.20	3.23	3.25	3.27	3.30	3.32	3.33	3.35	3.37	3.39	3.40	3.42	3.44
21	3.13	3.15	3.18	3.20	3.23	3.25	3.27	3.29	3.31	3.33	3.34	3.36	3.38	3.39	3.41
22	3.11	3.13	3.16	3.18	3.21	3.23	3.25	3.27	3.29	3.30	3.32	3.34	3.35	3.37	3.38
23	3.09	3.11	3.14	3.16	3.19	3.21	3.23	3.25	3.26	3.28	3.30	3.32	3.33	3.35	3.36
24	3.07	3.10	3.12	3.15	3.17	3.19	3.21	3.23	3.25	3.26	3.28	3.30	3.31	3.33	3.34
25	3.06	3.08	3.11	3.13	3.15	3.17	3.19	3.21	3.23	3.25	3.26	3.28	3.29	3.31	3.32
26	3.04	3.07	3.09	3.11	3.14	3.16	3.18	3.19	3.21	3.23	3.25	3.26	3.28	3.29	3.31
27	3.03	3.06	3.08	3.10	3.12	3.14	3.16	3.18	3.20	3.22	3.23	3.25	3.26	3.28	3.29
28	3.02	3.04	3.07	3.09	3.11	3.13	3.15	3.17	3.19	3.20	3.22	3.23	3.25	3.26	3.28

29	3.01	3.03	3.05	3.08	3.10	3.12	3.14	3.16	3.17	3.19	3.21	3.22	3.24	3.25	3.26
30	3.00	3.02	3.04	3.07	3.09	3.11	3.13	3.14	3.16	3.18	3.19	3.21	3.22	3.24	3.25
31	2.99	3.01	3.04	3.06	3.08	3.10	3.12	3.13	3.15	3.17	3.18	3.20	3.21	3.23	3.24
32	2.98	3.00	3.03	3.05	3.07	3.09	3.11	3.12	3.14	3.16	3.17	3.19	3.20	3.22	3.23
33	2.97	2.99	3.02	3.04	3.06	3.08	3.10	3.11	3.13	3.15	3.16	3.18	3.19	3.21	3.22
34	2.96	2.99	3.01	3.03	3.05	3.07	3.09	3.11	3.12	3.14	3.15	3.17	3.18	3.20	3.21
35	2.96	2.98	3.00	3.02	3.04	3.06	3.08	3.10	3.12	3.13	3.15	3.16	3.17	3.19	3.20
36	2.95	2.97	3.00	3.02	3.04	3.06	3.07	3.09	3.11	3.12	3.14	3.15	3.17	3.18	3.19
37	2.94	2.97	2.99	3.01	3.03	3.05	3.07	3.08	3.10	3.12	3.13	3.15	3.16	3.17	3.19
38	2.94	2.96	2.98	3.00	3.02	3.04	3.06	3.08	3.10	3.11	3.12	3.14	3.15	3.17	3.18
39	2.93	2.96	2.98	3.00	3.02	3.04	3.05	3.07	3.09	3.10	3.12	3.13	3.15	3.16	3.17
40	2.93	2.95	2.97	2.99	3.01	3.03	3.05	3.07	3.09	3.10	3.11	3.13	3.14	3.15	3.17
41	2.92	2.94	2.97	2.99	3.01	3.03	3.04	3.06	3.08	3.09	3.11	3.13	3.13	3.15	3.16
42	2.92	2.94	2.96	2.98	3.00	3.02	3.04	3.05	3.08	3.09	3.10	3.12	3.13	3.14	3.15
43	2.91	2.94	2.96	2.98	3.00	3.02	3.03	3.05	3.07	3.08	3.10	3.11	3.12	3.14	3.15
44	2.91	2.93	2.95	2.97	2.99	3.01	3.03	3.04	3.07	3.08	3.10	3.11	3.12	3.13	3.14
45	2.90	2.93	2.95	2.97	2.99	3.01	3.02	3.04	3.06	3.07	3.09	3.10	3.11	3.13	3.14
46	2.90	2.92	2.94	2.96	2.98	3.00	3.02	3.04	3.06	3.07	3.09	3.10	3.11	3.12	3.13
47	2.90	2.92	2.94	2.96	2.98	3.00	3.02	3.03	3.05	3.06	3.08	3.10	3.10	3.12	3.13
48	2.89	2.92	2.94	2.96	2.98	2.99	3.01	3.03	3.05	3.06	3.08	3.09	3.10	3.11	3.12
49	2.89	2.91	2.93	2.95	2.97	2.99	3.01	3.02	3.04	3.05	3.07	3.09	3.10	3.11	3.12
50	2.89	2.91	2.93	2.95	2.97	2.99	3.00	3.02	3.04	3.05	3.07	3.08	3.09	3.10	3.12
60	2.86	2.88	2.90	2.92	2.94	2.96	2.97	2.99	3.01	3.02	3.03	3.05	3.06	3.07	3.08
70	2.84	2.86	2.88	2.90	2.92	2.94	2.95	2.97	2.98	3.00	3.01	3.03	3.04	3.05	3.06
80	2.83	2.85	2.87	2.89	2.91	2.92	2.94	2.95	2.97	2.98	3.00	3.01	3.02	3.03	3.05
90	2.82	2.84	2.86	2.88	2.89	2.91	2.93	2.94	2.96	2.97	2.98	3.00	3.01	3.02	3.03
100	2.81	2.83	2.85	2.87	2.89	2.90	2.92	2.93	2.95	2.96	2.98	2.99	3.00	3.01	3.02

Factor $= t_{(n-1,\,1-\alpha/r)}\sqrt{1 + 1/n}$.

Table 4 Factors for Obtaining One-Sided 95% Prediction Limits for *r* Additional Samples Given a Background Sample of Size *n*

Previous Size *n*	Number of new measurements (*r*)														
	30	35	40	45	50	55	60	65	70	75	80	85	90	95	100
4	9.59	10.11	10.58	11.02	11.42	11.80	12.15	12.49	12.80	13.11	13.40	13.68	13.94	14.20	14.45
5	6.85	7.14	7.40	7.64	7.86	8.06	8.25	8.42	8.59	8.75	8.90	9.04	9.18	9.31	9.43
6	5.67	5.87	6.05	6.22	6.37	6.50	6.63	6.75	6.86	6.97	7.06	7.16	7.25	7.34	7.42
7	5.02	5.18	5.32	5.45	5.57	5.67	5.77	5.86	5.95	6.03	6.10	6.17	6.24	6.31	6.37
8	4.62	4.75	4.87	4.98	5.07	5.16	5.24	5.32	5.39	5.46	5.52	5.58	5.63	5.68	5.74
9	4.34	4.46	4.57	4.66	4.74	4.82	4.89	4.96	5.02	5.07	5.13	5.18	5.23	5.27	5.31
10	4.16	4.25	4.35	4.43	4.51	4.57	4.64	4.69	4.75	4.80	4.85	4.89	4.93	4.98	5.01
11	4.00	4.10	4.18	4.26	4.33	4.39	4.45	4.50	4.55	4.60	4.64	4.68	4.72	4.76	4.79
12	3.88	3.97	4.05	4.12	4.19	4.25	4.30	4.35	4.40	4.44	4.48	4.52	4.55	4.59	4.62
13	3.79	3.87	3.95	4.02	4.08	4.13	4.18	4.23	4.27	4.31	4.35	4.39	4.42	4.45	4.48
14	3.71	3.79	3.86	3.93	3.99	4.04	4.09	4.13	4.17	4.21	4.25	4.28	4.31	4.34	4.37
15	3.64	3.72	3.79	3.86	3.91	3.96	4.01	4.05	4.09	4.12	4.16	4.19	4.22	4.25	4.28
16	3.59	3.67	3.74	3.79	3.85	3.90	3.94	3.98	4.02	4.05	4.08	4.11	4.14	4.17	4.20
17	3.54	3.62	3.68	3.74	3.79	3.84	3.88	3.92	3.96	3.99	4.02	4.05	4.08	4.11	4.13
18	3.50	3.58	3.64	3.70	3.75	3.79	3.83	3.87	3.90	3.94	3.97	4.00	4.02	4.05	4.07
19	3.47	3.54	3.60	3.66	3.71	3.75	3.79	3.82	3.86	3.89	3.92	3.95	3.97	4.00	4.02
20	3.44	3.51	3.57	3.62	3.67	3.71	3.75	3.79	3.82	3.85	3.88	3.91	3.93	3.96	3.98
21	3.41	3.48	3.54	3.59	3.63	3.68	3.72	3.75	3.78	3.81	3.84	3.87	3.89	3.92	3.94
22	3.38	3.45	3.51	3.56	3.61	3.65	3.68	3.72	3.75	3.78	3.81	3.83	3.86	3.88	3.90
23	3.36	3.43	3.49	3.54	3.58	3.62	3.66	3.69	3.72	3.75	3.78	3.80	3.83	3.85	3.87
24	3.34	3.41	3.46	3.51	3.56	3.60	3.64	3.67	3.70	3.73	3.75	3.78	3.80	3.82	3.84
25	3.32	3.39	3.44	3.49	3.53	3.57	3.61	3.64	3.67	3.70	3.73	3.75	3.78	3.80	3.82
26	3.31	3.37	3.42	3.47	3.52	3.55	3.59	3.62	3.65	3.68	3.71	3.73	3.75	3.77	3.80
27	3.29	3.35	3.41	3.46	3.50	3.54	3.57	3.60	3.63	3.66	3.69	3.71	3.73	3.75	3.77
28	3.82	3.34	3.39	3.44	3.48	3.52	3.55	3.59	3.61	3.64	3.67	3.69	3.71	3.73	3.75

n															
29	3.74	3.72	3.70	3.67	3.65	3.62	3.60	3.57	3.54	3.50	3.47	3.42	3.38	3.32	3.26
30	3.72	3.70	3.68	3.66	3.63	3.61	3.58	3.55	3.52	3.49	3.45	3.41	3.36	3.31	3.25
31	3.70	3.68	3.66	3.64	3.62	3.59	3.57	3.54	3.51	3.48	3.44	3.40	3.35	3.30	3.24
32	3.69	3.67	3.65	3.63	3.61	3.58	3.55	3.53	3.50	3.46	3.43	3.39	3.34	3.29	3.23
33	3.68	3.66	3.64	3.62	3.59	3.57	3.54	3.51	3.48	3.45	3.42	3.38	3.33	3.28	3.22
34	3.66	3.64	3.62	3.60	3.58	3.56	3.53	3.50	3.47	3.44	3.41	3.37	3.32	3.27	3.21
35	3.65	3.63	3.61	3.59	3.57	3.55	3.52	3.49	3.46	3.43	3.39	3.35	3.31	3.26	3.20
36	3.64	3.62	3.60	3.58	3.56	3.54	3.51	3.48	3.45	3.42	3.39	3.35	3.30	3.25	3.19
37	3.63	3.61	3.59	3.57	3.55	3.53	3.50	3.47	3.44	3.41	3.38	3.34	3.29	3.24	3.19
38	3.62	3.60	3.58	3.56	3.54	3.52	3.49	3.46	3.44	3.40	3.37	3.33	3.29	3.24	3.18
39	3.61	3.59	3.57	3.55	3.53	3.51	3.48	3.46	3.43	3.40	3.36	3.32	3.28	3.23	3.17
40	3.60	3.58	3.56	3.54	3.52	3.50	3.47	3.45	3.42	3.39	3.35	3.31	3.27	3.22	3.17
41	3.59	3.58	3.56	3.54	3.51	3.49	3.47	3.44	3.41	3.38	3.35	3.31	3.27	3.22	3.16
42	3.59	3.57	3.55	3.53	3.51	3.48	3.46	3.43	3.41	3.37	3.34	3.30	3.26	3.21	3.16
43	3.58	3.56	3.54	3.52	3.50	3.48	3.45	3.43	3.40	3.37	3.33	3.30	3.25	3.20	3.15
44	3.57	3.55	3.53	3.51	3.49	3.47	3.45	3.42	3.39	3.36	3.33	3.29	3.25	3.20	3.14
45	3.56	3.55	3.53	3.51	3.49	3.46	3.44	3.41	3.39	3.36	3.32	3.28	3.24	3.19	3.14
46	3.56	3.54	3.52	3.50	3.48	3.46	3.43	3.41	3.38	3.35	3.32	3.28	3.24	3.19	3.13
47	3.55	3.53	3.52	3.50	3.48	3.45	3.43	3.40	3.38	3.34	3.31	3.27	3.23	3.18	3.13
48	3.55	3.53	3.51	3.49	3.47	3.45	3.42	3.40	3.37	3.34	3.31	3.27	3.23	3.18	3.12
49	3.54	3.52	3.50	3.48	3.46	3.44	3.42	3.39	3.37	3.34	3.30	3.26	3.22	3.18	3.12
50	3.53	3.52	3.50	3.48	3.46	3.44	3.41	3.39	3.36	3.33	3.30	3.26	3.22	3.17	3.12
60	3.49	3.47	3.46	3.44	3.42	3.40	3.37	3.35	3.32	3.29	3.26	3.22	3.18	3.14	3.08
70	3.46	3.45	3.43	3.41	3.39	3.37	3.35	3.32	3.30	3.27	3.24	3.20	3.16	3.19	3.06
80	3.44	3.42	3.41	3.39	3.37	3.35	3.33	3.30	3.28	3.25	3.22	3.18	3.14	3.10	3.05
90	3.42	3.41	3.39	3.37	3.35	3.33	3.31	3.29	3.26	3.23	3.20	3.17	3.13	3.08	3.03
100	3.41	3.39	3.38	3.36	3.34	3.32	3.30	3.27	3.25	3.22	3.19	3.16	3.12	3.07	3.02

Factor $= t_{(n-1,\,1-\alpha/r)}\sqrt{1 + 1/n}$.

Table 5 Factors for Obtaining Two-Sided 95% Prediction Limits for r Additional Samples Given a Background Sample of Size n

Previous n	Number of new measurements (r)														
	1	2	3	4	5	6	7	8	9	10	11	12	13	14	15
4	3.56	4.67	5.43	6.03	6.53	6.97	7.36	7.71	8.03	8.33	8.61	8.88	9.13	9.36	9.59
5	3.04	3.83	4.34	4.73	5.04	5.31	5.55	5.76	5.95	6.13	6.29	6.45	6.59	6.72	6.85
6	2.78	3.42	3.82	4.11	4.35	4.56	4.73	4.89	5.03	5.15	5.27	5.38	5.48	5.58	5.67
7	2.62	3.17	3.51	3.76	3.96	4.13	4.27	4.40	4.51	4.61	4.71	4.80	4.88	4.95	5.02
8	2.51	3.01	3.32	3.54	3.71	3.86	3.98	4.09	4.19	4.27	4.35	4.43	4.50	4.56	4.62
9	2.43	2.90	3.18	3.38	3.54	3.67	3.78	3.87	3.96	4.04	4.11	4.18	4.24	4.29	4.34
10	2.37	2.82	3.08	3.26	3.41	3.53	3.63	3.72	3.80	3.87	3.93	3.99	4.05	4.10	4.15
11	2.33	2.75	3.00	3.17	3.31	3.42	3.52	3.60	3.67	3.74	3.80	3.85	3.91	3.95	4.00
12	2.29	2.70	2.93	3.10	3.23	3.34	3.43	3.51	3.58	3.64	3.70	3.75	3.79	3.84	3.88
13	2.26	2.66	2.88	3.04	3.17	3.27	3.36	3.43	3.50	3.56	3.61	3.66	3.71	3.75	3.79
14	2.24	2.62	2.84	3.00	3.12	3.22	3.30	3.37	3.43	3.49	3.54	3.59	3.63	3.67	3.71
15	2.21	2.59	2.81	2.96	3.07	3.17	3.25	3.32	3.38	3.43	3.48	3.53	3.57	3.61	3.64
16	2.20	2.57	2.78	2.92	3.04	3.13	3.21	3.27	3.33	3.39	3.43	3.48	3.52	3.56	3.59
17	2.18	2.54	2.75	2.89	3.01	3.09	3.17	3.24	3.29	3.35	3.39	3.44	3.47	3.51	3.54
18	2.17	2.53	2.73	2.87	2.98	3.07	3.14	3.20	3.26	3.31	3.36	3.40	3.44	3.47	3.50
19	2.16	2.51	2.71	2.85	2.95	3.04	3.11	3.18	3.23	3.28	3.32	3.36	3.40	3.44	3.47
20	2.14	2.49	2.69	2.83	2.93	3.02	3.09	3.15	3.20	3.25	3.30	3.33	3.37	3.40	3.44
21	2.13	2.48	2.67	2.81	2.91	3.00	3.07	3.13	3.18	3.23	3.27	3.31	3.34	3.38	3.41
22	2.13	2.47	2.66	2.79	2.89	2.98	3.05	3.11	3.16	3.21	3.25	3.29	3.32	3.35	3.38
23	2.12	2.46	2.65	2.78	2.88	2.96	3.03	3.09	3.14	3.19	3.23	3.26	3.30	3.33	3.36
24	2.11	2.45	2.63	2.76	2.87	2.95	3.01	3.07	3.12	3.17	3.21	3.25	3.28	3.31	3.34
25	2.10	2.44	2.62	2.75	2.85	2.93	3.00	3.06	3.11	3.15	3.19	3.23	3.26	3.29	3.32
26	2.10	2.43	2.61	2.74	2.84	2.92	2.99	3.04	3.09	3.14	3.18	3.21	3.25	3.28	3.31
27	2.09	2.42	2.61	2.73	2.83	2.91	2.97	3.03	3.08	3.12	3.16	3.20	3.23	3.26	3.29
28	2.09	2.42	2.60	2.72	2.82	2.90	2.96	3.02	3.07	3.11	3.15	3.19	3.22	3.25	3.28

29	2.08	2.41	2.59	2.71	2.81	2.89	2.95	3.01	3.05	3.10	3.14	3.17	3.21	3.24	3.26
30	2.08	2.40	2.58	2.71	2.80	2.88	2.94	3.00	3.04	3.09	3.13	3.16	3.19	3.22	3.25
31	2.07	2.40	2.58	2.70	2.79	2.87	2.93	2.99	3.04	3.08	3.12	3.15	3.18	3.21	3.24
32	2.07	2.39	2.57	2.69	2.79	2.86	2.92	2.98	3.03	3.07	3.11	3.14	3.17	3.20	3.23
33	2.07	2.39	2.56	2.69	2.78	2.85	2.92	2.97	3.02	3.06	3.10	3.13	3.16	3.19	3.22
34	2.06	2.38	2.56	2.68	2.77	2.85	2.91	2.96	3.01	3.05	3.09	3.12	3.15	3.18	3.21
35	2.06	2.38	2.55	2.67	2.77	2.84	2.90	2.96	3.00	3.04	3.08	3.12	3.15	3.17	3.20
36	2.06	2.37	2.55	2.67	2.76	2.83	2.90	2.95	3.00	3.04	3.07	3.11	3.14	3.17	3.19
37	2.06	2.37	2.54	2.66	2.76	2.83	2.89	2.94	2.99	3.03	3.07	3.10	3.13	3.16	3.19
38	2.05	2.37	2.54	2.66	2.75	2.82	2.88	2.94	2.98	3.02	3.06	3.09	3.12	3.15	3.18
39	2.05	2.36	2.54	2.66	2.75	2.82	2.88	2.93	2.98	3.02	3.05	3.09	3.12	3.15	3.17
40	2.05	2.36	2.53	2.65	2.74	2.81	2.87	2.93	2.98	3.01	3.05	3.08	3.11	3.14	3.17
41	2.05	2.36	2.53	2.65	2.74	2.81	2.87	2.92	2.97	3.01	3.04	3.08	3.11	3.13	3.16
42	2.04	2.35	2.53	2.64	2.73	2.80	2.87	2.92	2.97	3.00	3.04	3.07	3.10	3.13	3.15
43	2.04	2.35	2.52	2.64	2.73	2.80	2.86	2.91	2.96	3.00	3.03	3.07	3.10	3.13	3.15
44	2.04	2.35	2.52	2.64	2.73	2.80	2.86	2.91	2.96	3.00	3.03	3.06	3.09	3.12	3.14
45	2.04	2.35	2.52	2.63	2.72	2.79	2.85	2.90	2.95	2.99	3.02	3.06	3.09	3.12	3.14
46	2.04	2.34	2.51	2.63	2.72	2.79	2.85	2.90	2.95	2.99	3.02	3.05	3.08	3.11	3.14
47	2.03	2.34	2.51	2.63	2.72	2.79	2.85	2.90	2.94	2.98	3.02	3.05	3.08	3.11	3.13
48	2.03	2.34	2.51	2.62	2.71	2.78	2.84	2.89	2.94	2.98	3.01	3.04	3.07	3.10	3.13
49	2.03	2.34	2.51	2.62	2.71	2.78	2.84	2.89	2.93	2.97	3.01	3.04	3.07	3.10	3.12
50	2.03	2.34	2.50	2.62	2.71	2.78	2.84	2.89	2.93	2.97	3.00	3.04	3.06	3.10	3.12
60	2.02	2.32	2.48	2.60	2.68	2.75	2.81	2.86	2.90	2.94	2.97	3.01	3.03	3.09	3.08
70	2.01	2.31	2.47	2.58	2.67	2.74	2.79	2.84	2.88	2.92	2.95	2.98	3.01	3.06	3.06
80	2.00	2.30	2.46	2.57	2.66	2.72	2.78	2.83	2.87	2.91	2.94	2.97	3.00	3.04	3.05
90	2.00	2.29	2.45	2.56	2.65	2.71	2.77	2.82	2.86	2.89	2.93	2.96	2.98	3.02	3.03
100	1.99	2.29	2.45	2.56	2.64	2.71	2.76	2.81	2.85	2.89	2.92	2.95	2.98	3.00	3.02

Factor $= t_{(n-1,1-\alpha/r)}\sqrt{1+1/n}$.

Table 6   Factors for Obtaining Two-Sided 95% Prediction Limits for r Additional Samples Given a Background Sample of Size n

Previous Size n	Number of new measurements (r)														
	16	17	18	19	20	21	22	23	24	25	26	27	28	29	30
4	9.80	10.01	10.21	10.40	10.58	10.76	10.93	11.10	11.26	11.42	11.57	11.72	11.87	12.01	12.15
5	6.97	7.09	7.20	7.30	7.40	7.50	7.59	7.68	7.77	7.86	7.94	8.02	8.10	8.17	8.25
6	5.75	5.83	5.91	5.98	6.05	6.12	6.19	6.25	6.31	6.37	6.42	6.48	6.53	6.58	6.63
7	5.09	5.15	5.21	5.27	5.32	5.38	5.43	5.48	5.52	5.57	5.61	5.65	5.69	5.73	5.77
8	4.67	.473	4.78	4.83	4.87	4.92	4.96	5.00	5.04	5.07	5.11	5.15	5.18	5.21	5.24
9	4.39	4.44	4.49	4.53	4.57	4.61	4.64	4.68	4.71	4.74	4.78	4.81	4.83	4.86	4.89
10	4.19	4.23	4.27	4.31	4.35	4.38	4.42	4.45	4.48	4.51	4.53	4.56	4.59	4.61	4.64
11	4.04	4.08	4.11	4.15	4.18	4.21	4.24	4.27	4.30	4.33	4.35	4.38	4.40	4.43	4.45
12	3.92	3.96	3.99	4.02	4.05	4.08	4.11	4.14	4.16	4.19	4.21	4.24	4.26	4.28	4.30
13	3.82	3.86	3.89	3.92	3.95	3.98	4.00	4.03	4.05	4.08	4.10	4.12	4.14	4.16	4.18
14	3.74	3.78	3.81	3.84	3.86	3.89	3.92	3.94	3.96	3.99	4.01	4.03	4.05	4.07	4.09
15	3.68	3.71	3.74	3.77	3.79	3.82	3.84	3.87	3.89	3.91	3.93	3.95	3.97	3.99	4.01
16	3.62	3.65	3.68	3.71	3.74	3.76	3.78	3.81	3.83	3.85	3.87	3.89	3.90	3.92	3.94
17	3.58	3.60	3.63	3.66	3.68	3.71	3.73	3.75	3.77	3.79	3.81	3.83	3.85	3.86	3.88
18	3.53	3.56	3.59	3.61	3.64	3.66	3.68	3.71	3.73	3.75	3.76	3.78	3.80	3.82	3.83
19	3.50	3.53	3.55	3.58	3.60	3.62	3.65	3.67	3.68	3.70	3.72	3.74	3.76	3.77	3.79
20	3.47	3.49	3.52	3.54	3.57	3.59	3.61	3.63	3.65	3.67	3.68	3.70	3.72	3.73	3.75
21	3.44	3.46	3.49	3.51	3.54	3.56	3.58	3.60	3.62	3.63	3.65	3.67	3.69	3.70	3.72
22	3.41	3.44	3.46	3.49	3.51	3.53	3.55	3.57	3.59	3.61	3.62	3.64	3.65	3.67	3.68
23	3.39	3.41	3.44	3.46	3.49	3.51	3.53	3.54	3.56	3.58	3.60	3.61	3.63	3.64	3.66
24	3.37	3.39	3.42	3.44	3.46	3.48	3.50	3.52	3.54	3.56	3.57	3.59	3.60	3.62	3.63
25	3.35	3.38	3.40	3.42	3.44	3.46	3.48	3.50	3.52	3.53	3.55	3.57	3.58	3.60	3.61
26	3.33	3.36	3.38	3.40	3.42	3.44	3.46	3.48	3.50	3.52	3.53	3.55	3.56	3.58	3.59
27	3.32	3.34	3.36	3.39	3.41	3.43	3.45	3.46	3.48	3.50	3.51	3.53	3.54	3.56	3.57
28	3.30	3.33	3.35	3.37	3.39	3.41	3.43	3.45	3.46	3.48	3.50	3.51	3.53	3.54	3.55

29	3.54	3.52	3.51	3.50	3.48	3.47	3.45	3.43	3.42	3.40	3.38	3.36	3.34	3.31	3.29
30	3.52	3.51	3.50	3.48	3.47	3.45	3.44	3.42	3.40	3.38	3.36	3.34	3.32	3.30	3.28
31	3.51	3.50	3.48	3.47	3.45	3.44	3.42	3.41	3.39	3.37	3.35	3.33	3.31	3.29	3.27
32	3.50	3.48	3.47	3.46	3.44	3.43	3.41	3.39	3.38	3.36	3.34	3.32	3.30	3.28	3.25
33	3.48	3.47	3.46	3.44	3.43	3.42	3.40	3.38	3.37	3.35	3.33	3.31	3.29	3.27	3.24
34	3.47	3.46	3.45	3.43	3.42	3.41	3.39	3.37	3.36	3.34	3.32	3.30	3.28	3.26	3.23
35	3.46	3.45	3.44	3.42	3.41	3.39	3.38	3.36	3.35	3.33	3.31	3.29	3.27	3.25	3.23
36	3.45	3.44	3.43	3.41	3.40	3.39	3.37	3.35	3.34	3.32	3.30	3.28	3.26	3.24	3.22
37	3.44	3.43	3.42	3.40	3.39	3.38	3.36	3.35	3.33	3.31	3.29	3.27	3.25	3.23	3.21
38	3.44	3.42	3.41	3.40	3.38	3.37	3.35	3.34	3.32	3.30	3.29	3.27	3.25	3.23	3.20
39	3.43	3.41	3.40	3.39	3.37	3.36	3.35	3.33	3.31	3.30	3.28	3.26	3.24	3.22	3.20
40	3.42	3.41	3.39	3.38	3.37	3.35	3.34	3.32	3.31	3.29	3.27	3.25	3.23	3.21	3.19
41	3.41	3.40	3.39	3.37	3.36	3.35	3.33	3.32	3.30	3.28	3.27	3.25	3.23	3.21	3.18
42	3.41	3.39	3.38	3.37	3.35	3.34	3.33	3.31	3.29	3.28	3.26	3.24	3.22	3.20	3.18
43	3.40	3.39	3.37	3.36	3.35	3.33	3.32	3.30	3.29	3.27	3.26	3.23	3.21	3.19	3.17
44	3.39	3.38	3.37	3.36	3.34	3.33	3.31	3.30	3.28	3.27	3.25	3.23	3.21	3.19	3.17
45	3.39	3.37	3.36	3.35	3.34	3.32	3.31	3.29	3.28	3.26	3.24	3.22	3.20	3.18	3.16
46	3.38	3.37	3.36	3.34	3.33	3.32	3.30	3.29	3.27	3.25	3.24	3.22	3.20	3.18	3.16
47	3.38	3.36	3.35	3.34	3.33	3.31	3.30	3.28	3.27	3.25	3.23	3.21	3.19	3.17	3.15
48	3.37	3.36	3.35	3.33	3.32	3.31	3.29	3.28	3.26	3.24	3.23	3.21	3.19	3.17	3.15
49	3.37	3.35	3.34	3.33	3.32	3.30	3.29	3.27	3.26	3.24	3.22	3.20	3.19	3.17	3.14
50	3.36	3.35	3.34	3.32	3.31	3.30	3.28	3.27	3.25	3.24	3.22	3.20	3.18	3.16	3.14
60	3.32	3.31	3.30	3.29	3.27	3.26	3.25	3.23	3.22	3.20	3.18	3.17	3.15	3.13	3.11
70	3.30	3.28	3.27	3.26	3.25	3.24	3.22	3.21	3.19	3.18	3.16	3.14	3.12	3.11	3.08
80	3.28	3.26	3.25	3.24	3.23	3.22	3.20	3.19	3.17	3.16	3.14	3.13	3.11	3.09	3.07
90	3.26	3.25	3.24	3.23	3.21	3.20	3.19	3.17	3.16	3.15	3.13	3.11	3.09	3.08	3.05
100	3.25	3.24	3.23	3.22	3.20	3.19	3.18	3.16	3.15	3.13	3.12	3.10	3.08	3.06	3.04

Factor $= t_{(n-1,1-\alpha/r)}\sqrt{1+1/n}$.

Table 7 Factors for Obtaining Two-Sided 95% Prediction Limits for r Additional Samples Given a Background Sample of Size n

Previous Size n	Number of new measurements (r)														
	30	35	40	45	50	55	60	65	70	75	80	85	90	95	100
4	12.15	12.80	13.40	13.94	14.45	14.92	15.37	15.79	16.19	16.57	16.93	17.28	17.62	17.94	18.25
5	8.25	8.59	5.90	9.18	9.43	9.67	9.89	10.10	10.29	10.48	10.66	10.83	10.99	11.14	11.29
6	6.63	6.86	7.06	7.25	7.42	7.57	7.72	7.85	7.98	8.10	8.21	8.32	8.42	8.52	8.61
7	5.77	5.95	6.10	6.24	6.37	6.49	6.59	6.69	6.79	6.88	6.96	7.04	7.12	7.19	7.26
8	5.24	5.39	5.52	5.63	5.74	5.83	5.92	6.00	6.08	6.15	6.21	6.28	6.34	6.40	6.45
9	4.89	5.02	5.13	5.23	5.31	5.40	5.47	5.54	5.60	5.66	5.72	5.78	5.83	5.87	5.92
10	4.64	4.75	4.85	4.93	5.01	5.09	5.15	5.21	5.27	5.32	5.37	5.42	5.47	5.51	5.55
11	4.45	4.55	4.64	4.72	4.79	4.86	4.92	4.97	5.02	5.07	5.11	5.16	5.20	5.24	5.27
12	4.30	4.40	4.48	4.55	4.62	4.68	4.73	4.78	4.83	4.88	4.92	4.96	4.99	5.03	5.06
13	4.18	4.27	4.35	4.42	4.48	4.54	4.59	4.64	4.68	4.72	4.76	4.80	4.83	4.86	4.89
14	4.09	4.17	4.25	4.31	4.37	4.42	4.47	4.52	4.56	4.60	4.63	4.67	4.70	4.73	4.76
15	4.01	4.09	4.16	4.22	4.28	4.33	4.37	4.41	4.45	4.49	4.53	4.56	4.59	4.62	4.65
16	3.94	4.02	4.08	4.14	4.20	4.25	4.29	4.33	4.37	4.40	4.44	4.47	4.50	4.53	4.55
17	3.88	3.96	4.02	4.08	4.13	4.18	4.22	4.26	4.30	4.33	4.36	4.39	4.42	4.45	4.47
18	3.83	3.90	3.97	4.02	4.07	4.12	4.16	4.20	4.23	4.27	4.30	4.33	4.35	4.38	4.40
19	3.79	3.86	3.92	3.97	4.02	4.07	4.11	4.14	4.18	4.21	4.24	4.27	4.29	4.32	4.34
20	3.75	3.82	3.88	3.93	3.98	4.02	4.06	4.10	4.13	4.16	4.19	4.22	4.24	4.27	4.29
21	3.72	3.78	3.84	3.89	3.94	3.98	4.02	4.05	4.09	4.12	4.15	4.17	4.20	4.22	4.24
22	3.68	3.75	3.81	3.86	3.90	3.95	3.98	4.02	4.05	4.08	4.11	4.13	4.16	4.18	4.20
23	3.66	3.72	3.78	3.83	3.87	3.91	3.95	3.98	4.01	4.04	4.07	4.10	4.12	4.14	4.16
24	3.63	3.70	3.75	3.80	3.84	3.88	3.92	3.95	3.98	4.01	4.04	4.06	4.09	4.11	4.13
25	3.61	3.67	3.73	3.78	3.82	3.86	3.89	3.93	3.96	3.98	4.01	4.03	4.06	4.08	4.10
26	3.59	3.65	3.71	3.75	3.80	3.83	3.87	3.90	3.93	3.96	3.98	4.01	4.03	4.05	4.07
27	3.57	3.63	3.69	3.73	3.77	3.81	3.85	3.88	3.91	3.93	3.96	3.98	4.01	4.03	4.05
28	3.55	3.61	3.67	3.71	3.75	3.79	3.83	3.86	3.89	3.91	3.94	3.96	3.98	4.00	4.02

n															
29	3.54	3.60	3.65	3.70	3.74	3.77	3.81	3.84	3.87	3.89	3.92	3.94	3.96	3.98	4.00
30	3.52	3.58	3.63	3.68	3.72	3.76	3.79	3.82	3.85	3.87	3.90	3.92	3.94	3.96	3.98
31	3.51	3.57	3.62	3.66	3.70	3.74	3.77	3.80	3.83	3.86	3.88	3.90	3.92	3.94	3.96
32	3.50	3.55	3.61	3.65	3.69	3.73	3.76	3.79	3.81	3.84	3.86	3.89	3.91	3.93	3.95
33	3.48	3.54	3.59	3.64	3.68	3.71	3.74	3.77	3.80	3.83	3.85	3.87	3.89	3.91	3.93
34	3.47	3.53	3.58	3.62	3.66	3.70	3.73	3.76	3.79	3.81	3.84	3.86	3.88	3.90	3.92
35	3.46	3.52	3.57	3.61	3.65	3.69	3.72	3.75	3.77	3.80	3.82	3.84	3.86	3.88	3.90
36	3.45	3.51	3.56	3.60	3.64	3.67	3.71	3.73	3.76	3.79	3.81	3.83	3.85	3.87	3.89
37	3.44	3.50	3.55	3.59	3.63	3.66	3.70	3.72	3.75	3.77	3.80	3.82	3.84	3.86	3.88
38	3.44	3.49	3.54	3.58	3.62	3.65	3.69	3.71	3.74	3.76	3.79	3.81	3.83	3.85	3.86
39	3.43	4.48	3.53	3.57	3.61	3.64	3.68	3.70	3.73	3.75	3.78	3.80	3.82	3.84	3.85
40	3.42	3.47	3.52	3.56	3.60	3.64	3.67	3.69	3.72	3.74	3.77	3.79	3.81	3.83	3.84
41	3.41	3.47	3.51	3.56	3.59	3.63	3.66	3.69	3.71	3.74	3.76	3.78	3.80	3.82	3.83
42	3.41	3.46	3.51	3.55	3.59	3.62	3.65	3.68	3.70	3.73	3.75	3.77	3.79	3.81	3.82
43	3.40	3.45	3.50	3.54	3.58	3.61	3.64	3.67	3.69	3.72	3.74	3.76	3.78	3.80	3.82
44	3.39	3.45	3.49	3.53	3.57	3.60	3.63	3.66	3.69	3.71	3.73	3.75	3.77	3.79	3.81
45	3.39	3.44	3.49	3.53	3.56	3.60	3.63	3.65	3.68	3.70	3.73	3.74	3.76	3.78	3.80
46	3.38	3.43	3.48	3.52	3.56	3.59	3.62	3.65	3.67	3.70	3.72	3.74	3.76	3.77	3.79
47	3.38	3.43	3.48	3.52	3.55	3.58	3.61	3.64	3.67	3.69	3.72	3.74	3.76	3.77	3.78
48	3.37	3.42	3.47	3.51	3.55	3.58	3.61	3.63	3.66	3.68	3.71	3.73	3.75	3.77	3.78
49	3.37	3.42	3.46	3.50	3.54	3.57	3.60	3.63	3.65	3.68	3.70	3.72	3.74	3.76	3.77
50	3.36	3.41	3.46	3.50	3.53	3.57	3.60	3.63	3.65	3.68	3.70	3.72	3.74	3.75	3.76
60	3.32	3.37	3.42	3.46	3.49	3.52	3.55	3.58	3.60	3.62	3.64	3.66	3.68	3.70	3.71
70	3.30	3.35	3.39	3.43	3.46	3.49	3.52	3.54	3.57	3.59	3.61	3.63	3.65	3.66	3.68
80	3.28	3.33	3.37	3.41	3.44	3.47	3.50	3.52	3.54	3.57	3.59	3.60	3.62	3.64	3.65
90	3.26	3.31	3.35	3.39	3.42	3.45	3.48	3.50	3.53	3.55	3.57	3.58	3.60	3.62	3.63
100	3.25	3.30	3.34	3.38	3.41	3.44	3.46	3.49	3.51	3.53	3.55	3.57	3.59	3.60	3.62

Factor $= t_{(n-1,\,1-\alpha/r)}\sqrt{1 + 1/n}$.

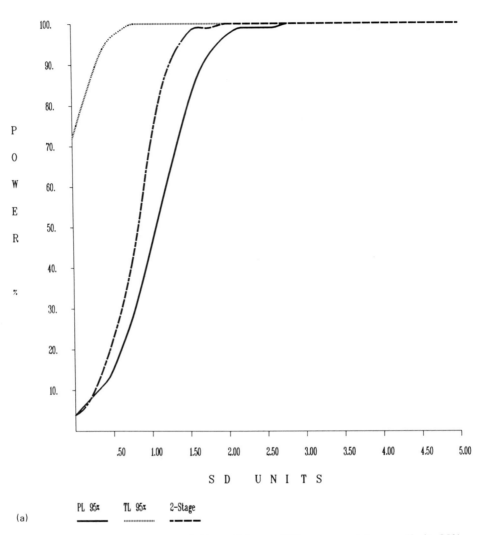

(a)

Figure 3 Power curves for 95% confidence, 95% coverage tolerance limit, 95% confidence prediction limit for the next $r = 100$ measurements, and the two-stage procedure (a) with all 100 new measurements contaminated; (b) with only 1 of 100 new measurements contaminated.

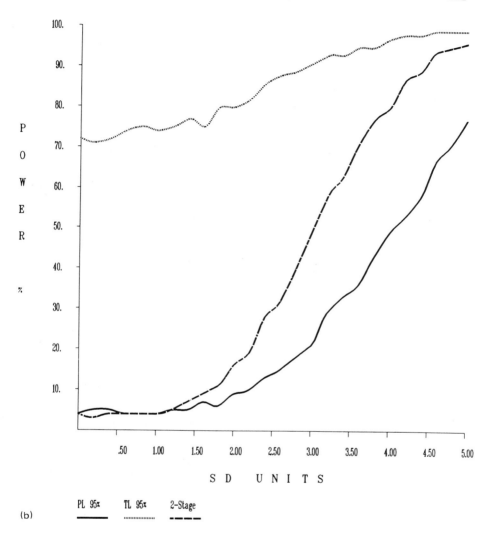

(b)

These points are illustrated in Figs. 3a and 3b. Figure 3a displays the estimated power curves for the 95% confidence, 95% coverage tolerance limit, the 95% prediction limit for the next $r = 100$ measurements, and the two-stage procedure. When the upgradient versus downgradient difference is 0 SD unit, the percentage of statistically significant results represents the test's false positive rate. When the upgradient versus downgradient difference is greater than zero, the percentage of statistically significant results represents the test's power, and 100% − power reveals the test's false negative rate. Power curves are useful because for a given sample they allow the owner/operator and regulatory agencies to select an optimal test in the sense that it achieves its intended false positive rate (e.g., 5%) while simultaneously achieving a reasonable false negative rate (e.g., 5%) for the smallest possible upgradient versus downgradient difference (expressed in standard deviation units, i.e., effect size). It should, of course, be noted, however, that simply selecting a test based on its false negative rate is not at all acceptable because a test with a high false positive rate will necessarily have a low false negative rate. In the extreme case in which the false positive rate is 100%, the false negative rate is 0%. There are far more reasonable and statistically appropriate ways of minimizing false negative rates that do not involve compromising false positive rates. These methods are illustrated in the following.

The power curves in Fig. 3a were obtained by simulating a comparison between a background sample of size $n = 16$, and 100 new monitoring measurements. Twenty-six evenly spaced points between 0 and 5 SD units (i.e., difference between upgradient and downgradient water quality in standard deviation units) were selected to illustrate the effect of contamination spread evenly across the 100 monitoring wells. Each point was replicated 1000 times; therefore, each power curve is the result of 26,000 simulations of $n = 16$ background measurements and $r = 100$ monitoring measurements. The results displayed in Fig. 3 are exactly as predicted. When the difference between background and monitoring well means is 0 SD unit, the false positive rate is over 70% for the tolerance limit (because we would expect up to five failures per sampling event), but both the prediction limit for all $r = 100$ future measurements and the two-stage procedure achieve their intended nominal error rates of 5%. However, the two-stage procedure exhibits a substantially reduced false negative rate throughout the entire range of effect sizes, which is also as expected.

Figure 3a represents an overly optimistic picture of statistical power in that all 100 new monitoring measurements were of equally contaminated samples. Certainly, this is not representative of practice. Figure 3b presents the other extreme, in which only one of the 100 monitoring wells is influenced (i.e., detecting a needle in a haystack). Of course, much larger effect sizes (i.e., SD unit differences) are required to consistently detect contamination, and the prediction limit does not achieve a reasonable level of statistical power even for a difference of 5 SD units, which is enormous. Again, however, the same conclusions can be drawn. The two-stage procedure is the only approach that both achieves its intended false positive rate of 5% and substantially decreases the false negative rate relative to the single-stage alternative.

In practice, the power of the two-stage procedure would typically be expected to be intermediate between these two extremes, achieving 95% power (i.e., a 5% false negative rate) in the 2.5–3.0 SD unit range. Of course, as the number of background samples is increased, the statistical power is also increased, and the 5% false negative rate is achieved for a smaller effect size.

For the purpose of illustration, the number of future measurements r has been equated to the number of monitoring wells, but this is not necessarily the case. A site with 20 monitoring wells that is required to statistically evaluate five chemical constituents will also be making $r = 100$ future comparisons per sampling event. To the extent that the five compounds are redundant (i.e., correlated), the number of independent future comparisons will, in fact, be less than 100; however, groundwater monitoring parameters are usually selected to tap relatively independent sources of contamination, and therefore the number of monitoring parameters (i.e., chemical constituents) is a critical factor and must be considered in any rigorous statistical evaluation of the data that is designed to balance false positive and false negative rates.

The two-stage procedure outlined here (see Fig. 4) can be used even in much larger groundwater monitoring programs. For example, if a facility were required to make 1000 statistical comparisons per monitoring event (e.g., 50 monitoring wells and 20 constituents), a 95% confidence, 99% coverage tolerance limit could be used as the screening test in stage 1, and a 95% prediction limit for the next 10 future measurements [i.e., 0.01(1000) = 10 expected chance failures] could be used for verification in stage 2.

IV. SELECTING THE NUMBER OF BACKGROUND SAMPLES

A common question asked of statisticians is, How many background samples do I need? This question became even more common when, in previous regulations (40 CFR Part 264), owner/operators were required to demonstrate that their alternate statistical procedures balanced false positive and false negative results. Fortunately, this requirement is no longer a part of the new (1989) RCRA statistical regulation. The reason the difficulty arose is that the false negative rate, or 1 − statistical power, that is required to define number of samples, is dependent on three things: the false positive rate (selected *a priori* by the regulation; that is, 5% for the facility as a whole), the number of background measurements, and the effect size. The effect size describes the smallest difference (typically described in standard deviation units) that is environmentally meaningful.

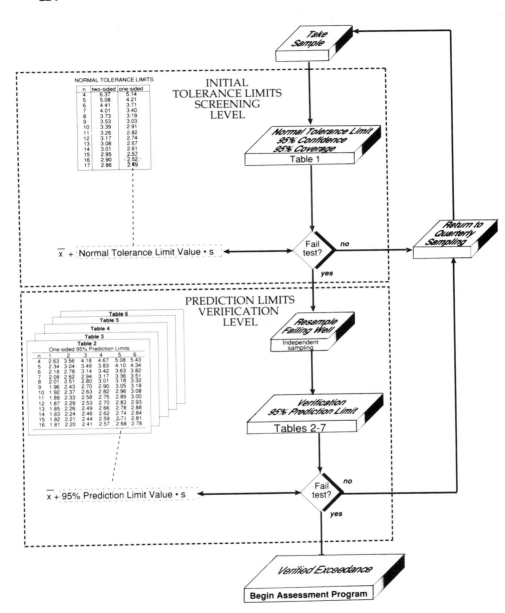

Figure 4 The two-stage statistical procedure.

Since US EPA has not provided such a minimum effect size, it is impossible to derive a sample size that will properly balance false positive and false negative rates of tolerance or prediction limits. As such, it is suggested that the number of samples be selected on the basis of the size of the multiplier. For example, a 95% confidence, 95% coverage one-sided tolerance interval multiplier goes from 7.656 to 5.144 for a change in background sample size of 3 to 4, but only from 2.566 to 2.523 for a change in background sample size of $n = 15$ to $n = 16$. Even doubling the background sample size to $n = 30$ only decreases the multiplier by approximately 1/3 SD unit (i.e., $k = 2.220$). Therefore, a background sample size in the range of 16–32 is recommended. If more observations are available, they should, of course, also be included because they will provide even more precise estimates of μ and σ^2.

V. ALTERNATIVE BACKGROUND SAMPLING

The previous line of reasoning also suggests how the background sample should be selected. The three choices are (1) fixed sample of n time-series historical measurements (e.g., 2-year fixed window), (2) the n most recent measurements (e.g., a 2-year moving window), and (3) all available historical measurements. When using tolerance or prediction limits or intervals, the third option is recommended, as shown in Fig. 5.

Finally, how shall the n background measurements be selected? Should we obtain n measurements from a single background well, a single measurement from n different background wells, or something in between? In the case of upgradient versus downgradient comparisons, there are two possible strategies. For a facility with eight or more upgradient or background wells, the eight measurements could be used to construct a new tolerance or prediction limit specific to each quarterly monitoring event (Fig. 5). With eight background measurements, the one-sided 95% confidence and 95% coverage tolerance limit multiplier would be 3.188. Although the multiplier is somewhat larger than the value of 2.523 obtained for $n = 16$ background samples, this strategy eliminates the temporal component of variability and will therefore yield a smaller standard deviation than if historical measurements are pooled. The result is an effective detection monitoring program that is supported by the new US EPA statistical rule.

For facilities with fewer than eight background water quality wells, at least four upgradient or background wells should be monitored quarterly so that after one year of monitoring, 16 background measurements will be available. Furthermore, four upgradient wells, if widely spaced, will provide a reasonable characterization of the spatial component of variability at the facility. If only a single upgradient well were installed, as required in the

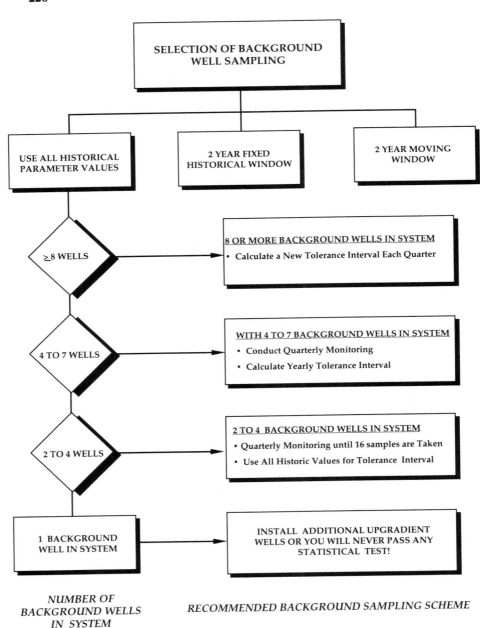

NUMBER OF
BACKGROUND WELLS
IN SYSTEM

RECOMMENDED BACKGROUND SAMPLING SCHEME

Figure 5 Criteria for selecting background well-sampling strategy.

previous RCRA regulations, differences between upgradient and down-gradient water quality would be completely confounded with spatial variability; that is, Is the difference between upgradient and downgradient measurements due to the influence of the facility or simply the difference that one would find by drilling any two holes in the ground? There is, of course, no answer to this question if there is only a single background well.

VI. NONDETECTS

It is very common in groundwater detection monitoring to obtain samples that cannot be properly quantified because of low-level concentrations that are less than the limit of detection of the analytical instrument. This condition can make the direct application of the previously described statistical prediction and tolerance limits and intervals problematic because the usual sample statistics $\bar{x}$ and s are no longer valid estimates of μ and σ. Statistically, these distributions are termed *censored*. In this section, three different levels of "censoring" are considered: (1) up to 90% nondetects, (2) 91–99% nondetects, and (3) nothing detected, as shown in Fig. 6. The use of these methods is proposed for obtaining unbiased estimates of the mean and variance of background (e.g., upgradient) water quality.

A. Case 1. Up to 90% Nondetects

When at least 10% of the groundwater samples have a measurable (detectable) value of a particular indicator parameter, the mean and variance of the distribution can be approximated using a method due to Aitchison [4]. The adjusted mean value is given by

$$\bar{x} = \left(1 - \frac{n_0}{n}\right)\bar{x}'$$

(1)

where $\bar{x}'$ is the average of the n_1 detected values, n_0 is the number of nondetects, and $n = n_0 + n_1$ is the total number of samples. The adjusted standard deviation is

$$s = \left[\left(1 - \frac{n_0}{n}\right)s^{2\prime} + \frac{n_0}{n}\left(1 - \frac{n_0 - 1}{n - 1}\right)\bar{x}^{2\prime}\right]^{1/2}$$

(2)

where s' is the standard deviation of the n_1 detected measurements. The normal tolerance and prediction limits can then be computed as previously described, using the total sample size n to obtain the appropriate tabulated multiplier.

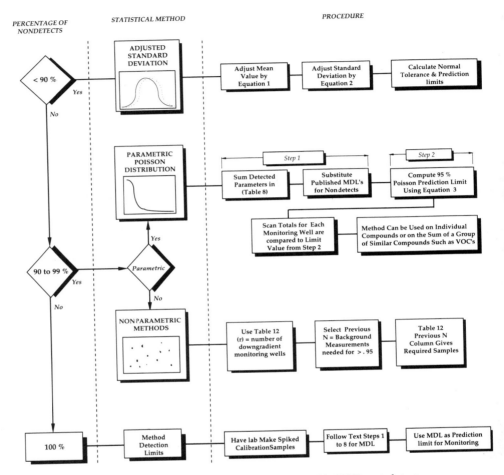

Figure 6 Statistical methods for constituents with 90–100% nondetects.

B. Case 2. More than 90% Nondetects

When the detection frequency is less than 10%, the previous method of obtaining adjusted mean and variance estimates no longer applies. With limited data, it is difficult to know just what to do. What further complicates this problem is that one of the most important classes of detection monitoring compounds are the volatile organic priority pollutants (VOCs), which typically have detection frequencies in this range.

To date, the only applicable statistical approach to setting site-specific limits for these compounds is the one described by Gibbons [5]. This procedure is based on tolerance and prediction limits for the Poisson distribution—a distribution that has been widely used for the analysis of rate events such as suicide, mutation rates, and atomic particle emission. These limits can be applied either to detection frequencies (i.e., number of detected compounds per scan) or to the actual concentrations when recorded in parts per billion (ppb). In the latter case, it is assumed that a measurement of 20 ppb of benzene represents a count of 20 g of benzene for every billion grams of water examined. To the extent that this is an accurate description of the true physical measurement process, Poisson prediction and tolerance limits provide a reasonable approximation that appears to be sufficiently accurate for most practical purposes. This sentiment is echoed in the new US EPA statistical regulation:

> Tolerance intervals and prediction intervals have not been widely used by the Agency to evaluate ground water monitoring data. However, the Agency is aware of recent publications that have employed these statistical methods to evaluate ground water monitoring data, especially in evaluating certain classes of chemical compounds (e.g., volatile organic compounds). Several commentors suggested that the Agency incorporate this research into today's final rule, noting that these procedures may be the best way to evaluate data that is below the limit of analytical detection.

In the case of VOCs, the 95% Poisson prediction limit is computed as follows (see Fig. 6).

1. For each US EPA Method 624 volatile organic priority pollutant scan, sum the detected concentrations of the 27 compounds listed in Table 8, substituting the published method detection limit (MDL) (see Table 8) for those compounds that were not detected. For example, if none of the compounds were detected, the sum for that scan is 154 ppb.
2. Compute the 95% Poisson prediction limit as

$$\frac{y}{n} + \frac{t^2}{2n} + \frac{t}{n}\left[y(1 + n) + \frac{t^2}{4}\right]^{1/2} \tag{3}$$

   where y is the total ppb for all n background scans (i.e., the sum of n individual scan totals), n is the number of background scans, t is the $(1 - 0.05/r)100\%$ point of Student's t distribution on $n - 1$ degrees of freedom (see Tables 9–11), and r is the number of monitoring wells.
3. Compare scan totals for each monitoring well (computed as in step 1) to the limit value computed in step 2. By chance alone, we would expect an exceedance 5% of the time given the assumptions in step 2.

Table 8 Method 624 Volatile Organic Compounds
Published Method Detection Limits

Compound	Reported MDL
Benzene	4.4
Bromodichloromethane	2.2
Bromoform	4.7
Bromomethane	10.0
Carbon tetrachloride	2.8
Chlorobenzene	6.0
Chloroethane	10.0
Chloroform	1.6
Chloromethane	10.0
Dibromochloromethane	3.1
1,1-Dichloroethane	4.7
1,2-Dichloroethane	2.8
1,1-Dichloroethene	2.8
trans-1,2-Dichloroethene	10.0
1,2-Dichloropropane	6.0
cis-1,3-Dichloropropene	5.0
trans-1,3-Dichloropropene	10.0
Ethylbenzene	7.2
Methylene chloride	2.8
1,1,2,2-Tetrachloroethane	6.9
Tetrachloroethene	4.1
Toluene	6.0
1,1,1-Trichloroethane	3.8
1,1,2-Trichloroethane	5.0
Trichloroethene	1.9
Trichlorofluoromethane	10.0
Vinyl chloride	10.0

All values reported in μg/L.

For other compounds that have detection frequencies in the range of 1–10%, the same strategy may be applied, either for individual compounds or on the sum of a group of similar compounds, as in the case of the VOCs.

C. A Nonparametric Approach

The foregoing discussion is based on the assumption that the distribution of the parameter(s) of interest is known and has a parametric form (i.e., normal, censored normal, or Poisson). In some cases, however, this assumption is unreasonable (even following transformation, e.g., logarithmic) and a "distribution-free" statistical method may be required. In the context of groundwater monitoring, Gibbons [6] has adapted the nonparametric prediction limit originally described by Chou and Owen [7]. In contrast to the parametric approach in which we estimate a limit value from the mean and standard deviation of a sample of n previous measurements, the nonparametric approach identifies the required number of samples (n) such that the maximum value of those samples is the 95% prediction limit. Gibbons [6] further generalizes the procedure to include the effects of resampling (i.e., taking a verification sample following a statistically significant groundwater monitoring result).

The 95% nonparametric prediction limit can be easily obtained with the aid of Table 12. For example, assume that we have a facility with $r = 10$ downgradient monitoring wells. Furthermore, let us also assume that if we fail a detection monitoring test we are permitted to resample the well before any further action is taken, and if the repeat sample does not fail the test we return to normal detection monitoring. How many background samples (n) are we required to obtain so that the maximum observed measurement of those n samples will contain the next $r = 10$ monitoring measurements given the possibility of a single resample of any well that fails the initial test? Inspection of Table 12 reveals that a background sample of $n = 18$ measurements provides 94.9% confidence, and a background sample of $n = 19$ provides 95.3% confidence. The answer is, therefore, that $n = 19$ background samples must be taken in order to ensure that the maximum of those 19 samples will not be exceeded by the next 10 monitoring measurements (i.e., one at each downgradient well), given that we can resample any well that fails the initial test.

The major advantage of this approach is that it only assumes that the samples are independent and measured on a continuous scale; no particular distribution is specified. Furthermore, as long as at least one groundwater sample has a measurable value, the limit is defined. When nothing is detected in the n background samples, an alternative approach must be taken.

D. Case 3. What to Do When Nothing Is Detected

Statistical methods are of little use without measurable data. Nevertheless, it is surprisingly common to observe a background collection of 10 or so measurements for which nothing was detected. What do we do? Is the tolerance or prediction limit zero? Is it the method detection limit? The answer to this question can be found only by examining the specifics of the analytical measurement process itself (see Gibbons et al. [8]). Interestingly, the analyst's decision as to whether or not a particular substance is present in a particular

Table 9 Values of t for Obtaining One-Sided 95% Poisson Prediction Limits for r Additional Samples Given a Background Sample of Size n

Previous sample of size n	Number of new measurements (r)														
	1	2	3	4	5	6	7	8	9	10	11	12	13	14	15
4	2.35	3.18	3.74	4.18	4.54	4.86	5.14	5.39	5.62	5.84	6.04	6.23	6.41	6.58	6.74
5	2.13	2.78	3.19	3.50	3.75	3.96	4.15	4.31	4.47	4.60	4.73	4.85	4.96	5.07	5.17
6	2.01	2.57	2.91	3.16	3.36	3.53	3.68	3.81	3.93	4.03	4.13	4.22	4.30	4.38	4.45
7	1.94	2.45	2.75	2.97	3.14	3.29	3.41	3.52	3.62	3.71	3.79	3.86	3.93	4.00	4.06
8	1.89	2.36	2.64	2.84	3.00	3.13	3.24	3.33	3.42	3.50	3.57	3.64	3.70	3.75	3.81
9	1.86	2.31	2.57	2.75	2.90	3.02	3.12	3.21	3.28	3.35	3.42	3.48	3.53	3.58	3.63
10	1.83	2.26	2.51	2.68	2.82	2.93	3.03	3.11	3.18	3.25	3.31	3.36	3.41	3.46	3.50
11	1.81	2.23	2.47	2.63	2.76	2.87	2.96	3.04	3.11	3.17	3.22	3.28	3.32	3.37	3.41
12	1.80	2.20	2.43	2.59	2.72	2.82	2.91	2.98	3.05	3.11	3.16	3.21	3.25	3.29	3.33
13	1.78	2.18	2.40	2.56	2.68	2.78	2.86	2.93	3.00	3.05	3.11	3.15	3.20	3.23	3.27
14	1.77	2.16	2.38	2.53	2.65	2.75	2.83	2.90	2.96	3.01	3.06	3.11	3.15	3.19	3.22
15	1.76	2.14	2.36	2.51	2.62	2.72	2.80	2.86	2.92	2.98	3.02	3.07	3.11	3.15	3.18
16	1.75	2.13	2.34	2.49	2.60	2.69	2.77	2.84	2.89	2.95	2.99	3.04	3.07	3.11	3.15
17	1.75	2.12	2.33	2.47	2.58	2.67	2.75	2.81	2.87	2.92	2.97	3.01	3.05	3.08	3.11
18	1.74	2.11	2.31	2.46	2.57	2.65	2.73	2.79	2.85	2.90	2.94	2.98	3.02	3.06	3.09
19	1.73	2.10	2.30	2.44	2.55	2.64	2.71	2.77	2.82	2.88	2.92	2.96	3.00	3.03	3.07
20	1.73	2.09	2.29	2.43	2.54	2.62	2.70	2.76	2.81	2.86	2.90	2.94	2.98	3.01	3.04
21	1.72	2.09	2.28	2.42	2.53	2.61	2.68	2.74	2.80	2.84	2.89	2.93	2.96	3.00	3.03
22	1.72	2.08	2.28	2.41	2.52	2.60	2.67	2.73	2.78	2.83	2.87	2.91	2.95	2.98	3.01
23	1.72	2.07	2.27	2.40	2.51	2.59	2.66	2.72	2.77	2.82	2.86	2.90	2.93	2.96	2.99
24	1.71	2.07	2.26	2.40	2.50	2.58	2.65	2.71	2.76	2.81	2.85	2.89	2.92	2.95	2.98
25	1.71	2.06	2.26	2.39	2.49	2.57	2.64	2.70	2.75	2.80	2.84	2.88	2.91	2.94	2.97
26	1.71	2.06	2.25	2.38	2.48	2.57	2.63	2.69	2.74	2.79	2.83	2.86	2.90	2.93	2.96
27	1.71	2.06	2.25	2.38	2.48	2.56	2.63	2.68	2.73	2.78	2.82	2.85	2.89	2.92	2.95
28	1.70	2.05	2.24	2.37	2.47	2.55	2.62	2.68	2.73	2.77	2.81	2.85	2.88	2.91	2.94

29	1.70	2.05	2.24	2.37	2.47	2.55	2.61	2.67	2.72	2.76	2.80	2.84	2.87	2.90	2.93
30	1.70	2.05	2.23	2.36	2.46	2.54	2.61	2.66	2.71	2.76	2.80	2.83	2.86	2.89	2.92
31	1.70	2.04	2.23	2.36	2.46	2.54	2.60	2.66	2.71	2.75	2.79	2.82	2.86	2.89	2.91
32	1.70	2.04	2.23	2.36	2.45	2.53	2.60	2.65	2.70	2.74	2.78	2.82	2.85	2.88	2.91
33	1.69	2.04	2.22	2.35	2.46	2.53	2.59	2.65	2.70	2.74	2.78	2.81	2.84	2.87	2.90
34	1.69	2.03	2.22	2.35	2.44	2.52	2.59	2.64	2.69	2.73	2.77	2.81	2.84	2.87	2.90
35	1.69	2.03	2.22	2.34	2.44	2.52	2.58	2.64	2.69	2.73	2.77	2.80	2.83	2.86	2.89
36	1.69	2.03	2.22	2.34	2.44	2.51	2.58	2.63	2.68	2.72	2.76	2.80	2.83	2.86	2.88
37	1.69	2.03	2.21	2.34	2.43	2.51	2.57	2.63	2.68	2.72	2.76	2.79	2.82	2.85	2.88
38	1.69	2.03	2.21	2.34	2.43	2.51	2.57	2.63	2.67	2.72	2.75	2.79	2.82	2.85	2.87
39	1.69	2.02	2.21	2.33	2.43	2.50	2.57	2.62	2.67	2.71	2.75	2.78	2.81	2.84	2.87
40	1.68	2.02	2.21	2.33	2.43	2.50	2.56	2.62	2.67	2.71	2.75	2.78	2.81	2.84	2.87
41	1.68	2.02	2.20	2.33	2.42	2.50	2.56	2.62	2.66	2.70	2.74	2.78	2.81	2.84	2.86
42	1.68	2.02	2.20	2.33	2.42	2.50	2.56	2.61	2.66	2.70	2.74	2.77	2.80	2.83	2.86
43	1.68	2.02	2.20	2.32	2.42	2.49	2.56	2.61	2.66	2.70	2.74	2.77	2.80	2.83	2.85
44	1.68	2.02	2.20	2.32	2.42	2.49	2.55	2.61	2.65	2.69	2.73	2.77	2.80	2.82	2.85
45	1.68	2.02	2.20	2.32	2.41	2.49	2.55	2.60	2.65	2.69	2.73	2.76	2.79	2.82	2.85
46	1.68	2.01	2.20	2.32	2.41	2.48	2.55	2.60	2.65	2.69	2.73	2.76	2.79	2.82	2.84
47	1.68	2.01	2.19	2.32	2.41	2.48	2.55	2.60	2.65	2.69	2.72	2.76	2.79	2.82	2.84
48	1.68	2.01	2.19	2.32	2.41	2.48	2.54	2.60	2.64	2.68	2.72	2.75	2.78	2.81	2.84
49	1.68	2.01	2.19	2.31	2.41	2.48	2.54	2.60	2.64	2.68	2.72	2.75	2.78	2.81	2.84
50	1.68	2.01	2.19	2.31	2.40	2.48	2.54	2.59	2.64	2.68	2.72	2.75	2.78	2.81	2.83
60	1.67	2.00	2.18	2.30	2.39	2.46	2.52	2.58	2.62	2.66	2.70	2.73	2.76	2.79	2.81
70	1.67	1.99	2.17	2.29	2.38	2.45	2.51	2.56	2.61	2.65	2.68	2.72	2.75	2.77	2.80
80	1.66	1.99	2.17	2.28	2.37	2.45	2.51	2.56	2.60	2.64	2.67	2.71	2.74	2.76	2.79
90	1.66	1.99	2.16	2.28	2.37	2.44	2.50	2.55	2.59	2.63	2.67	2.70	2.73	2.75	2.78
100	1.66	1.98	2.16	2.28	2.36	2.44	2.49	2.54	2.59	2.63	2.66	2.69	2.72	2.75	2.77

Factor $= t_{(n-1,\,1-\alpha/r)}$.

Table 10 Values of t for Obtaining One-Sided 95% Poisson Prediction Limits for r Additional Samples Given a Background Sample of Size n

Previous n	Number of new measurements (r)														
	16	17	18	19	20	21	22	23	24	25	26	27	28	29	30
4	6.89	7.04	7.18	7.32	7.45	7.58	7.70	7.82	7.94	8.05	8.16	8.27	8.37	8.47	8.57
5	5.26	5.35	5.44	5.52	5.60	5.67	5.75	5.82	5.88	5.95	6.01	6.08	6.14	6.20	6.25
6	4.53	4.59	4.65	4.72	4.77	4.83	4.88	4.93	4.98	5.03	5.08	5.12	5.16	5.21	5.25
7	4.11	4.17	4.22	4.27	4.32	4.36	4.40	4.45	4.49	4.52	4.56	4.60	4.63	4.66	4.70
8	3.85	3.90	3.95	3.99	4.03	4.07	4.10	4.14	4.17	4.21	4.24	4.27	4.30	4.33	4.35
9	3.68	3.72	3.76	3.80	3.83	3.87	3.90	3.93	3.96	3.99	4.02	4.05	4.07	4.10	4.12
10	3.55	3.59	3.62	3.66	3.69	3.72	3.75	3.78	3.81	3.83	3.86	3.88	3.91	3.93	3.95
11	3.45	3.48	3.52	3.55	3.58	3.61	3.64	3.67	3.69	3.72	3.74	3.76	3.78	3.81	3.83
12	3.37	3.40	3.44	3.47	3.50	3.52	3.55	3.58	3.60	3.62	3.65	3.67	3.69	3.71	3.73
13	3.31	3.34	3.37	3.40	3.43	3.45	3.48	3.50	3.53	3.55	3.57	3.59	3.61	3.63	3.65
14	3.26	3.29	3.32	3.35	3.37	3.40	3.42	3.45	3.47	3.49	3.51	3.53	3.55	3.57	3.58
15	3.21	3.24	3.27	3.30	3.33	3.35	3.37	3.40	3.42	3.44	3.46	3.48	3.49	3.51	3.53
16	3.18	3.21	3.23	3.26	3.29	3.31	3.33	3.35	3.37	3.39	3.41	3.43	3.45	3.47	3.48
17	3.15	3.17	3.20	3.23	3.25	3.27	3.30	3.32	3.34	3.36	3.38	3.39	3.41	3.43	3.44
18	3.12	3.15	3.17	3.20	3.22	3.24	3.27	3.29	3.31	3.33	3.34	3.36	3.38	3.39	3.41
19	3.09	3.12	3.15	3.17	3.20	3.22	3.24	3.26	3.28	3.30	3.32	3.33	3.35	3.36	3.38
20	3.07	3.10	3.13	3.15	3.17	3.20	3.22	3.24	3.25	3.27	3.29	3.31	3.32	3.34	3.35
21	3.05	3.08	3.11	3.13	3.15	3.17	3.19	3.21	3.23	3.25	3.27	3.28	3.30	3.32	3.33
22	3.04	3.06	3.09	3.11	3.13	3.16	3.18	3.20	3.21	3.23	3.25	3.26	3.28	3.29	3.31
23	3.02	3.05	3.07	3.10	3.12	3.14	3.16	3.18	3.20	3.21	3.23	3.25	3.26	3.28	3.29
24	3.01	3.03	3.06	3.08	3.10	3.12	3.14	3.16	3.18	3.20	3.21	3.23	3.24	3.26	3.27
25	3.00	3.02	3.05	3.07	3.09	3.11	3.13	3.15	3.17	3.18	3.20	3.21	3.23	3.24	3.26
26	2.99	3.01	3.03	3.06	3.08	3.10	3.12	3.14	3.15	3.17	3.19	3.20	3.22	3.23	3.24
27	2.98	3.00	3.02	3.05	3.07	3.09	3.11	3.12	3.14	3.16	3.17	3.19	3.20	3.22	3.23
28	2.96	2.99	3.01	3.04	3.06	3.08	3.09	3.11	3.13	3.15	3.16	3.18	3.19	3.21	3.22

29	2.96	2.98	3.00	3.03	3.05	3.07	3.09	3.10	3.12	3.14	3.15	3.17	3.18	3.19	3.21
30	2.95	2.97	3.00	3.02	3.04	3.06	3.08	3.09	3.11	3.13	3.14	3.16	3.17	3.18	3.20
31	2.94	2.96	2.99	3.01	3.03	3.05	3.07	3.08	3.10	3.12	3.13	3.15	3.16	3.18	3.19
32	2.93	2.96	2.98	3.00	3.02	3.04	3.06	3.08	3.09	3.11	3.12	3.14	3.15	3.17	3.18
33	2.93	2.95	2.97	2.99	3.01	3.03	3.05	3.07	3.09	3.10	3.12	3.13	3.14	3.16	3.17
34	2.92	2.94	2.97	2.99	3.01	3.03	3.04	3.06	3.08	3.09	3.11	3.12	3.14	3.15	3.16
35	2.91	2.94	2.96	2.98	3.00	3.02	3.04	3.06	3.07	3.09	3.10	3.12	3.13	3.14	3.16
36	2.91	2.93	2.95	2.98	3.00	3.01	3.03	3.05	3.07	3.08	3.10	3.11	3.12	3.14	3.15
37	2.90	2.93	2.95	2.97	2.99	3.01	3.03	3.04	3.06	3.07	3.09	3.10	3.12	3.13	3.14
38	2.90	2.92	2.94	2.97	2.98	3.00	3.02	3.04	3.05	3.07	3.08	3.10	3.11	3.12	3.14
39	2.89	2.92	2.94	2.96	2.98	3.00	3.02	3.03	3.05	3.06	3.08	3.09	3.11	3.12	3.13
40	2.89	2.91	2.94	2.96	2.98	2.99	3.01	3.03	3.04	3.06	3.07	3.09	3.10	3.11	3.13
41	2.89	2.91	2.93	2.95	2.97	2.99	3.01	3.02	3.04	3.05	3.07	3.08	3.10	3.11	3.12
42	2.88	2.91	2.93	2.95	2.97	2.98	3.00	3.02	3.03	3.05	3.06	3.08	3.09	3.10	3.12
43	2.88	2.90	2.92	2.94	2.96	2.98	3.00	3.01	3.03	3.05	3.06	3.07	3.09	3.10	3.11
44	2.88	2.90	2.92	2.94	2.96	2.98	3.00	3.01	3.03	3.04	3.06	3.07	3.08	3.10	3.11
45	2.87	2.89	2.92	2.94	2.96	2.97	2.99	3.01	3.02	3.04	3.05	3.06	3.08	3.09	3.10
46	2.87	2.89	2.92	2.93	2.95	2.97	2.99	3.00	3.02	3.03	3.05	3.06	3.07	3.09	3.10
47	2.87	2.89	2.91	2.93	2.95	2.97	2.99	3.00	3.02	3.03	3.04	3.05	3.07	3.08	3.10
48	2.86	2.89	2.91	2.93	2.95	2.96	2.98	3.00	3.01	3.03	3.04	3.05	3.07	3.08	3.09
49	2.86	2.88	2.91	2.92	2.94	2.96	2.98	2.99	3.01	3.02	3.04	3.05	3.06	3.08	3.09
50	2.86	2.88	2.90	2.92	2.94	2.96	2.97	2.99	3.01	3.02	3.03	3.04	3.06	3.07	3.09
60	2.84	2.86	2.90	2.90	2.92	2.93	2.95	2.97	2.98	3.00	3.01	3.02	3.04	3.05	3.06
70	2.82	2.84	2.88	2.88	2.90	2.92	2.93	2.95	2.96	2.98	2.99	3.00	3.02	3.03	3.04
80	2.81	2.83	2.86	2.87	2.89	2.90	2.92	2.94	2.95	2.97	2.98	2.99	3.00	3.02	3.03
90	2.80	2.82	2.85	2.86	2.88	2.90	2.91	2.93	2.94	2.96	2.97	2.98	2.99	3.01	3.02
100	2.79	2.82	2.84	2.85	2.87	2.89	2.90	2.92	2.93	2.95	2.96	2.97	2.99	3.00	3.01

Factor = $t_{(n-1,\,1-\alpha/r)}$.

Table 11 Vaues of t for Obtaining One-Sided 95% Poisson Prediction Limits for r Additional Samples Given a Background Sample of Size n

Previous n	Number of new measurements (r)														
	30	35	40	45	50	55	60	65	70	75	80	85	90	95	100
4	8.57	9.04	9.46	9.85	10.21	10.55	10.87	11.17	11.45	11.72	11.98	12.23	12.47	12.70	12.92
5	6.25	6.52	6.76	6.97	7.17	7.36	7.53	7.69	7.84	7.98	8.12	8.25	8.38	8.50	8.61
6	5.25	5.44	5.60	5.76	5.89	6.02	6.14	6.25	6.35	6.45	6.54	6.63	6.71	6.79	6.87
7	4.70	4.85	4.98	5.10	5.21	5.31	5.40	5.48	5.56	5.64	5.71	5.78	5.84	5.90	5.96
8	4.35	4.48	4.59	4.69	4.78	4.87	4.94	5.02	5.08	5.14	5.20	5.26	5.31	5.36	5.41
9	4.12	4.23	4.33	4.42	4.50	4.57	4.64	4.70	4.76	4.81	4.86	4.91	4.96	5.00	5.04
10	3.95	4.06	4.15	4.22	4.30	4.36	4.42	4.48	4.53	4.58	4.62	4.67	4.71	4.74	4.78
11	3.83	3.92	4.00	4.08	4.14	4.20	4.26	4.31	4.36	4.40	4.44	4.48	4.52	4.55	4.59
12	3.73	3.82	3.89	3.96	4.02	4.08	4.13	4.18	4.22	4.26	4.30	4.34	4.37	4.41	4.44
13	3.65	3.73	3.81	3.87	3.93	3.98	4.03	4.08	4.12	4.15	4.19	4.23	4.26	4.29	4.32
14	3.58	3.66	3.73	3.80	3.85	3.90	3.95	3.99	4.03	4.07	4.10	4.13	4.16	4.19	4.22
15	3.53	3.61	3.67	3.73	3.79	3.84	3.88	3.92	3.96	3.99	4.03	4.06	4.09	4.11	4.14
16	3.48	3.56	3.62	3.68	3.73	3.78	3.82	3.86	3.90	3.93	3.96	3.99	4.02	4.05	4.07
17	3.44	3.52	3.58	3.64	3.69	3.73	3.77	3.81	3.84	3.88	3.91	3.94	3.96	3.99	4.01
18	3.41	3.48	3.54	3.60	3.65	3.69	3.73	3.77	3.80	3.83	3.86	3.89	3.92	3.94	3.96
19	3.38	3.45	3.51	3.56	3.61	3.65	3.69	3.73	3.76	3.79	3.82	3.85	3.87	3.90	3.92
20	3.35	3.42	3.48	3.53	3.58	3.62	3.66	3.69	3.73	3.76	3.79	3.81	3.84	3.86	3.88
21	3.33	3.40	3.46	3.51	3.55	3.59	3.63	3.66	3.70	3.73	3.75	3.78	3.80	3.83	3.85
22	3.31	3.38	3.43	3.48	3.53	3.57	3.60	3.64	3.67	3.70	3.73	3.75	3.77	3.80	3.82
23	3.29	3.35	3.41	3.46	3.50	3.54	3.58	3.61	3.64	3.67	3.70	3.72	3.75	3.77	3.79
24	3.27	3.34	3.39	3.44	3.48	3.52	3.56	3.59	3.62	3.65	3.68	3.70	3.72	3.75	3.77
25	3.26	3.32	3.38	3.42	3.47	3.50	3.54	3.57	3.60	3.63	3.66	3.68	3.70	3.72	3.75
26	3.24	3.31	3.36	3.41	3.45	3.49	3.52	3.55	3.58	3.61	3.64	3.66	3.68	3.70	3.72
27	3.23	3.29	3.35	3.39	3.43	3.47	3.51	3.54	3.57	3.59	3.62	3.64	3.67	3.69	3.71
28	3.22	3.28	3.33	3.38	3.42	3.46	3.49	3.52	3.55	3.58	3.60	3.63	3.65	3.67	3.69

29	3.21	3.27	3.32	3.37	3.41	3.44	3.48	3.51	3.54	3.56	3.59	3.61	3.63	3.65	3.67
30	3.20	3.26	3.31	3.36	3.40	3.43	3.47	3.50	3.52	3.55	3.58	3.60	3.62	3.64	3.66
31	3.19	3.25	3.30	3.34	3.38	3.42	3.45	3.48	3.51	3.54	3.56	3.58	3.61	3.63	3.65
32	3.18	3.24	3.29	3.33	3.37	3.41	3.44	3.47	3.50	3.53	3.55	3.57	3.59	3.61	3.63
33	3.17	3.23	3.28	3.33	3.37	3.40	3.43	3.46	3.49	3.52	3.54	3.56	3.58	3.60	3.62
34	3.16	3.22	3.27	3.32	3.36	3.39	3.42	3.45	3.48	3.51	3.53	3.55	3.57	3.59	3.61
35	3.16	3.21	3.26	3.31	3.35	3.38	3.42	3.44	3.47	3.50	3.52	3.54	3.56	3.58	3.60
36	3.15	3.21	3.26	3.30	3.34	3.37	3.41	3.44	3.46	3.49	3.51	3.53	3.55	3.57	3.59
37	3.14	3.20	3.25	3.29	3.34	3.37	3.40	3.43	3.45	3.48	3.50	3.52	3.54	3.56	3.58
38	3.14	3.19	3.24	3.29	3.33	3.37	3.39	3.42	3.45	3.47	3.49	3.52	3.54	3.56	3.57
39	3.13	3.19	3.24	3.28	3.32	3.36	3.38	3.41	3.44	3.46	3.49	3.51	3.53	3.55	3.57
40	3.13	3.18	3.23	3.27	3.31	3.35	3.38	3.41	3.43	3.46	3.48	3.50	3.52	3.54	3.56
41	3.12	3.18	3.23	3.27	3.31	3.34	3.37	3.40	3.43	3.45	3.47	3.49	3.51	3.53	3.55
42	3.12	3.17	3.22	3.26	3.30	3.33	3.37	3.39	3.42	3.44	3.47	3.49	3.51	3.53	3.54
43	3.11	3.17	3.22	3.26	3.30	3.33	3.36	3.39	3.41	3.44	3.46	3.48	3.50	3.52	3.54
44	3.11	3.16	3.21	3.25	3.29	3.32	3.35	3.38	3.41	3.43	3.45	3.48	3.49	3.51	3.53
45	3.10	3.16	3.21	3.25	3.29	3.32	3.35	3.38	3.40	3.43	3.45	3.47	3.49	3.51	3.53
46	3.10	3.15	3.20	3.24	3.28	3.31	3.34	3.37	3.40	3.42	3.44	3.46	3.48	3.50	3.52
47	3.10	3.15	3.20	3.24	3.28	3.31	3.34	3.37	3.39	3.42	3.44	3.46	3.48	3.50	3.51
48	3.09	3.15	3.19	3.24	3.27	3.31	3.34	3.36	3.37	3.41	3.43	3.45	3.47	3.49	3.51
49	3.09	3.14	3.19	3.23	3.27	3.30	3.33	3.36	3.38	3.41	3.43	3.45	3.47	3.49	3.50
50	3.09	3.14	3.19	3.23	3.26	3.30	3.33	3.35	3.38	3.40	3.43	3.45	3.46	3.48	3.50
60	3.06	3.11	3.16	3.20	3.23	3.27	3.30	3.32	3.35	3.37	3.39	3.41	3.43	3.45	3.46
70	3.04	3.09	3.14	3.18	3.21	3.24	3.27	3.30	3.32	3.34	3.37	3.39	3.40	3.42	3.44
80	3.03	3.08	3.12	3.16	3.20	3.23	3.26	3.28	3.31	3.33	3.35	3.37	3.38	3.40	3.42
90	3.02	3.07	3.11	3.15	3.18	3.21	3.24	3.27	3.29	3.31	3.33	3.35	3.37	3.39	3.40
100	3.01	3.06	3.10	3.14	3.17	3.20	3.23	3.26	3.28	3.30	3.32	3.34	3.36	3.38	3.39

Factor $= t_{(n-1,1-\alpha/r)}$.

Table 12   Probability that at Least 1 Out of 2 Samples Will Be Below the Maximum of n Background Measurements at Each of r Monitoring Wells

| Previous n | \multicolumn{15}{c}{Number of monitoring wells (r)} |

Previous n	1	2	3	4	5	6	7	8	9	10	11	12	13	14	15
4	.933	.871	.813	.759	.708	.661	.617	.576	.537	.502	.468	.437	.408	.381	.355
5	.952	.907	.864	.823	.784	.746	.711	.677	.645	.614	.585	.557	.530	.505	.481
6	.964	.930	.897	.865	.834	.804	.775	.748	.721	.695	.670	.646	.623	.601	.580
7	.972	.945	.919	.893	.869	.844	.821	.798	.776	.754	.734	.713	.693	.674	.655
8	.978	.956	.935	.914	.894	.874	.854	.835	.817	.799	.781	.764	.747	.730	.714
9	.982	.964	.946	.929	.912	.896	.879	.863	.848	.832	.817	.802	.788	.773	.759
10	.985	.970	.955	.941	.927	.912	.899	.885	.872	.858	.845	.833	.820	.808	.795
11	.987	.975	.962	.950	.938	.926	.914	.902	.890	.879	.868	.857	.846	.835	.824
12	.989	.978	.967	.957	.946	.936	.926	.915	.905	.895	.886	.876	.866	.857	.847
13	.990	.981	.972	.962	.953	.944	.935	.926	.917	.909	.900	.892	.883	.875	.866
14	.992	.983	.975	.967	.959	.951	.943	.935	.927	.920	.912	.904	.897	.889	.882
15	.993	.985	.978	.971	.964	.957	.950	.943	.936	.930	.922	.915	.909	.902	.895
16	.994	.987	.982	.975	.968	.962	.956	.949	.943	.937	.930	.924	.918	.912	.906
17	.994	.988	.983	.977	.971	.965	.960	.954	.949	.943	.938	.932	.927	.921	.916
18	.995	.990	.984	.979	.974	.969	.964	.959	.954	.949	.944	.939	.934	.929	.924
19	.995	.990	.986	.981	.976	.972	.967	.963	.958	.953	.949	.944	.940	.935	.931
20	.996	.991	.987	.983	.979	.974	.970	.966	.962	.958	.953	.949	.945	.941	.937
25	.997	.994	.991	.989	.986	.983	.980	.977	.975	.972	.969	.966	.964	.961	.958
30	.998	.996	.994	.992	.990	.988	.986	.984	.982	.980	.978	.976	.974	.972	.970
35	.998	.997	.996	.994	.993	.991	.990	.988	.987	.985	.984	.982	.981	.979	.978
40	.999	.998	.997	.995	.994	.993	.992	.991	.990	.988	.987	.986	.985	.984	.983
45	.999	.998	.997	.996	.995	.994	.994	.993	.992	.991	.990	.989	.988	.987	.986
50	.999	.998	.998	.997	.996	.995	.995	.994	.993	.992	.992	.991	.990	.989	.989
60	.999	.999	.998	.998	.997	.997	.996	.996	.995	.995	.994	.994	.993	.993	.992
70	1.00	.999	.999	.998	.998	.998	.997	.997	.996	.996	.996	.996	.995	.995	.994
80	1.00	.999	.999	.999	.998	.998	.998	.998	.997	.997	.997	.997	.996	.996	.995
90	1.00	1.00	.999	.999	.999	.999	.998	.998	.998	.998	.997	.997	.997	.997	.996
100	1.00	1.00	.999	.999	.999	.999	.999	.998	.998	.998	.998	.998	.997	.997	.997

| | Number of monitoring wells (r) | | | | | | | | | | | | | | |
Previous n	20	25	30	35	40	45	50	55	60	65	70	75	80	90	100
4	.252	.178	.126	.089	.063	.045	.032	.022	.016	.011	.008	.006	.004	.002	.001
5	.377	.295	.231	.181	.142	.111	.087	.068	.054	.042	.033	.026	.020	.012	.008
6	.483	.403	.336	.280	.233	.195	.162	.135	.113	.094	.078	.065	.055	.038	.008
7	.569	.494	.430	.373	.324	.281	.244	.212	.184	.160	.139	.121	.105	.079	.060
8	.638	.570	.510	.455	.407	.364	.325	.291	.260	.232	.207	.185	.166	.132	.106
9	.693	.632	.577	.526	.480	.438	.400	.365	.333	.303	.277	.253	.230	.192	.160
10	.737	.683	.633	.586	.543	.503	.466	.432	.400	.371	.343	.318	.295	.253	.217
11	.773	.724	.679	.637	.597	.560	.525	.492	.461	.432	.405	.380	.356	.313	.275
12	.802	.759	.718	.679	.643	.608	.576	.545	.515	.488	.461	.437	.413	.370	.331
13	.826	.787	.750	.715	.682	.650	.620	.591	.563	.537	.512	.488	.465	.423	.384
14	.846	.811	.778	.746	.716	.686	.658	.631	.605	.580	.557	.534	.512	.471	.433
15	.863	.832	.801	.772	.744	.717	.691	.666	.642	.619	.597	.575	.554	.515	.478
16	.877	.849	.821	.795	.769	.744	.720	.697	.675	.653	.632	.612	.592	.554	.519
17	.889	.864	.839	.814	.791	.768	.746	.724	.703	.683	.663	.644	.625	.590	.556
18	.900	.876	.854	.831	.810	.789	.768	.748	.729	.710	.691	.673	.656	.622	.590
19	.909	.888	.867	.846	.826	.807	.788	.769	.751	.733	.716	.699	.683	.651	.620
20	.917	.897	.878	.859	.841	.823	.805	.788	.771	.754	.738	.722	.707	.677	.648
25	.945	.931	.918	.905	.892	.880	.867	.855	.843	.831	.819	.807	.796	.774	.752
30	.960	.951	.941	.932	.922	.913	.904	.895	.886	.877	.868	.860	.851	.834	.817
35	.970	.963	.956	.949	.942	.935	.928	.921	.914	.907	.900	.893	.887	.874	.860
40	.977	.971	.966	.960	.955	.949	.944	.938	.933	.927	.922	.917	.911	.901	.890
45	.982	.977	.973	.968	.964	.959	.955	.950	.946	.942	.937	.933	.929	.920	.912
50	.985	.981	.978	.974	.970	.967	.963	.959	.956	.952	.949	.945	.941	.934	.927
60	.989	.987	.984	.982	.979	.976	.974	.971	.969	.966	.964	.961	.959	.954	.948
70	.992	.990	.988	.986	.984	.983	.981	.979	.977	.975	.973	.971	.969	.965	.962
80	.994	.992	.991	.990	.988	.987	.985	.984	.982	.981	.979	.978	.976	.973	.970
90	.995	.994	.993	.992	.990	.989	.988	.987	.986	.985	.983	.982	.981	.979	.976
100	.996	.995	.994	.993	.992	.991	.990	.989	.988	.987	.987	.986	.985	.983	.981

sample is based on the application of statistical decision rules. Even more remarkable, these decision rules are based on tolerance limits (see Currie [9]) for analyte absent or single concentration detection limit studies and prediction limits for calibration designs (i.e., a series of different spiking concentrations in the range of the method detection limit; see Hubaux and Vos [10] and Clayton et al. [11]).

These method detection limits are defined as the point at which the false positive and false negative rates are both less than 5%, for a test of the null hypothesis that the concentration of the analyte in the sample is zero.

To compute the method detection limit from "spiked" calibration samples, one approach is to use the method of Clayton et al. [11] as follows:

1. Select four concentrations in the range of the hypothesized MDL. For example, for benzene we might select concentrations of 4, 8, 12, and 16 μg/L.
2. Prepare 16 samples, that is, four replicates at each of the four spiking concentrations.
3. Several compounds may be examined simultaneously by including them in the same samples; however, randomize the order of their concentrations so that one sample does not contain all of the lowest concentrations and another samples all of the highest concentrations.
4. Introduce the 16 samples into the usual daily workload of two or more analysts (e.g., two analysts would receive eight samples each). It is essential that the analysts be completely blind to which compounds are present in the samples and their respective spiking concentrations and that they simply be instructed to perform the standard analytic method in question (e.g., Method 624 VOC Scan).
5. Record the results of the analysis as the square root of the ratio of the compound peak area to the internal standard; that is,

$$\text{Response signal} = \left[\frac{\text{peak area count}}{\text{internal standard area count}} \right]^{1/2} = y_i \qquad (4)$$

6. Transform the spiking concentrations as

$$x_i = \sqrt{x_i^* - 0.1} - \sqrt{0.1} \qquad (5)$$

   where x^* is the original spiking concentration.

7. For each compound, compute the slope of the regression line of the instrument response signal (y) against the targeted concentration (x) as

$$b = \frac{\sum_{i=1}^{16} (x_i - \bar{x})(y_i - \bar{y})}{\sum_{i=1}^{16} (x_i - \bar{x})^2} \qquad (6)$$

   where $\bar{x}$ is the average of the four target concentrations and $\bar{y}$ is the average of the 16 instrument response signals as defined above.

8. For each compound, compute the variance of deviations from the regression line as

$$s_{y.x}^2 = \sum_{i=1}^{16} \frac{(y_i - \hat{y}_i)^2}{16 - 2} \tag{7}$$

where $\hat{y}_i = y_i + \hat{b}(x_i - \bar{x})$ is the predicted instrument response for target concentration x_i.

9. Compute the method detection limit for $n = 16$ samples as

$$MDL^* = \frac{3.46 s_{y.x}}{b} \left[1 + \frac{1}{16} + \frac{\bar{x}^2}{\sum_{i=1}^{16} (x_i - \bar{x})^2} \right]^{1/2} \tag{8}$$

where 3.46 is the $\alpha = \beta = 0.95$ percentage point of the noncentral t distribution on $16 - 2 = 14$ degrees of freedom. To express the MDL in the original metric (e.g., μg/L), compute

$$MDL = (MDL^*)^2 + 0.632456(MDL^*) \tag{9}$$

In the absence of any detected values, this estimated MDL can be used as the corresponding prediction limit for detection monitoring. Gibbons et al. [12] have extended this method to the case of multiple future detection decisions and nonconstant variance calibration functions. The interested reader is referred to this work as well as to Clayton and coworkers' original paper [11] for a more detailed treatment of this subject.

We note that there are existing published MDLs for many compounds, including the Method 624 VOCs. More recently, practical quantitation limits (PQLs) have also been published in the Method SW-846 regulation. These national values were established under idealized conditions in which both presence and spiking concentration were known to the analyst and very questionable statistical computations were performed (see Gibbons et al. [8]). Therefore, it is quite reasonable that such levels will not be reached in routine laboratory practice and that the procedure described here will provide more realistic estimates that are consistent with attainable standards in the routine application of these methods. This view is reiterated in the new 40 CFR 264 statistical regulation:

The Appendix IX rule (52 FR 25942, July 9, 1987) listed practical quantification limits (PQLs) that were established from "Test Methods for Evaluating Solid Waste" (SW-846). SW-846 is the general RCRA analytical methods manual, currently in its third edition. The PQLs listed were USEPA's best estimate of the practical sensitivity of the applicable method for RCRA ground water monitoring purposes. However, some of the PQLs may be unattainable because they are based on general estimates for the specific substance. Furthermore, due to site specific factors, these limits may not be reached. For these reasons the Agency feels that the PQLs listed in Appendix IX are not appropriate for establishing a national baseline value for each constituent for determining whether a release to ground water has occurred. Instead, the PQLs are viewed as target levels that chemical laboratories should try to achieve in their analyses of ground water. In the event that a laboratory cannot achieve the suggested PQL, the owner or operator may submit a justification stating the reasons why these values cannot be achieved (e.g., specific instrument limitations). After reviewing this justification, the Regional Administrator may choose to establish facility specific PQLs based on the technical limitations of the contracting laboratory. Thus, US EPA is today clarifying 264.97(h) to allow owners or operators to propose facility specific PQLs. These PQLs may be used with the statistical methods listed in 264.97.

VII. SUMMARY

The statistical methods described here provide a series of general tools by which detection monitoring programs can be designed using indicator parameters that vary from 100% detection to no detection. The methods are completely site-specific with the exception of the case of no detection, for which they are specific to the monitoring laboratory responsible for the routine analysis of the groundwater samples. Even here, use of the actual upgradient groundwater matrix for the laboratory calibration study (in place of reagent water) can lead to the most applicable limits for that particular facility. The statistical procedures are parametric and nonparametric forms of prediction and tolerance intervals and limits and as such are consistent with the new RCRA 40 CFR Part 264 statistical regulation. The facility-wide false positive rate is restricted to 5%; therefore, quarterly monitoring should statistically result in one false positive decision every 5 years. Resampling of the well or wells in question will help minimize false positive results, particularly when used in conjunction with the two-stage procedure previously described. Using the suggested sample sizes should produce false negative rates of less than 5% for monitoring requirements in excess of 2–3 SD units above the background mean. False positive and false negative results are therefore balanced for even modest deviations from background water quality levels.

REFERENCES

1. U.S. Environmental Protection Agency. 40CFR Part 264: Statistical methods for evaluating ground-water monitoring from hazardous waste facilities; final rule. *Fed. Regist., 53*(196):39720–39731, 1988.
2. Wald, A., and Wolfowitz, J. Tolerance limits for a normal distribution. *Ann. Math. Stat., 17*:208–215, 1946.

3. Gibbons, R.D. Statistical prediction intervals for the evaluation of ground-water quality. *Ground Water, 25*:455–465, 1987.
4. Aitchison, J. On the distribution of a positive random variable having a discrete probability mass at the origin. *J. Am. Stat. Assoc., 50*:901–908, 1955.
5. Gibbons, R.D. Statistical models for the analysis of volatile organic compounds in waste disposal facilities. *Ground Water, 25*:572–580, 1987.
6. Gibbons, R.D. A general statistical procedure for ground-water detection monitoring at waste disposal facilities. *Ground Water, 28*:235–243, 1990.
7. Chou, Y.M., and Owen, D.B. One-sided distribution-free simultaneous prediction limits for *p* future samples. *J. Quality Technol., 18*:96–98, 1986.
8. Gibbons, R.D., Jarke, F.H., and Stoub, K.P. Method detection limits. *Proc. fifth Annual USEPA Waste Testing and Quality Assurance Symp., 2*:292–319, 1989.
9. Currie, L.A. Limits for qualitative decision and quantitative determination. *Anal. Chem., 40*:586–593, 1968.
10. Hubaux, A., and Vos, G. Decision and detection limits for linear calibration curves. *Anal. Chem., 42*:849–855, 1970.
11. Clayton, C.A., Hines, J.W., and Elkins, P.D. Detection limits with specified assurance probabilities. *Anal. Chem., 59*:2506–2514, 1987.
12. Gibbons, R.D., Jarke, F.H., and Stoub, K.P. Detection limits: for linear calibration curves with increasing variance and multiple future detection decisions. In *Waste Testing and Quality Assurance*, Vol. 3 (ASTM Monograph, STP 1075) (D. Friedman, ed.), ASTM, Philadelphia, PA, in press.

9

Geochemistry of Groundwater Pollutants at German Waste Disposal Sites

Helmūt Kerndorff, Ruprecht Schleyer, and Gerald Milde[†]

Institute for Water, Soil and Air Hygiene of the Federal Health Office, Langen and Berlin, Germany

Russell H. Plumb, Jr.

Lockheed Engineering & Sciences Company, Las Vegas, Nevada

I. INTRODUCTION

The strong industrial growth over the past decades has resulted in a rapid increase in the number of types and the amount of waste throughout Germany. Municipal solid waste and industrial refuse, construction waste, and production residues have been deposited, often without permission or control, in gravel, sand, and clay pits, rock quarries, bomb craters, and other land depressions or as dumps. Every one of these waste deposit sites can contain hazardous waste along with relatively harmless types of waste such as municipal solid waste or construction waste.

The implementation of the Waste Disposal Act of 1972 in the Federal Republic of Germany (now western Germany) and subsequent waste disposal acts of the federal states (Länder) led to the closing of a number of waste disposal sites. These acts made it legal to deposit waste only in planned and regulated large-scale landfill sites. Unregulated and unauthorized dumps were covered over and landscaped; to the extent that it was detectable, the damage was corrected.

Present affiliations:

[†]University of Bonn, Bonn, Germany

Waste mass is a foreign body in the Earth's crust, which as a rule has neither natural nor artificial seals either at its base or on the surface. Therefore long-term interactions occur between the deposited wastes and the natural elements—groundwater, soil, and air—that can be recognized by the resulting amissions and leakages. These interactions last for decades or centuries, if not longer, until a physicochemical balance (maximum entropy) is obtained. This is also true, to some extent, of the new sealed and regulated disposal sites, because the sealing materials—for example, clay, argilliferous substrates, or even plastic sheeting—frequently leak while the site is in use or after operations are shut down.

II. HYDROGEOCHEMICAL REACTIONS ASSOCIATED WITH WASTE SITES

Large amounts of bio-organic waste are present at practically every waste site, so that each dumping site represents a potential biochemical reactor in which more or less intensive and highly complex processes occur. Aside from chemical, physical, and biological processes within the waste mass, the particular composition of the waste, its density, the hydrologic balance, and the geological properties of the substratum are the primary factors that control the type and duration of the waste transformation and migration processes.

If water from precipitation enters through a waste site, dissolution of waste substances occurs. Chemical reactions, some of which are biologically mediated, can cause the dissolution of components from even relatively insoluble macromolecules. Furthermore, precipitation or coprecipitation of hydrous metal oxides, carbonates, sulfides, humates, and so on, can occur if conditions are favorable.

As the disposal site leachate migrates through the unsaturated zone or aquifer, the concentrations of individual contaminants are altered because of the filtration effects of suspended clay particles and organic colloids. In addition, the leachate concentrations are reduced when seepage water is diluted and distributed throughout an aquifer. At this point, the chemical and biological processes mentioned previously control the fate and behavior of the contaminants in the aquifer, although they can be deminished or intensified or become countervailing, according to the Eh and pH conditions and the "buffer capacity" of the aquifer. The sorption/desorption of dissolved substances on the surface of subterranean materials and the ion-exchange reactions are dependent on the composition of the aquifer material. This is also valid for the ion exchange between components with high exchange capacity, such as humic matter and argilliferous minerals.

The processes within the waste disposal site give rise to the creation of biogeochemical phases. The first of these to occur is the acid fermentation phase, which results in the total reduction of all oxygen owing to increasing microbial activity and the consequent production of fatty acids. Microbial activity then degrades these acids to methane, carbon dioxide, and water during the strictly anaerobic methane phase that follows. Methanogenic bacteria function well only in a very specific Eh and pH ranges and cannot operate in the presence of even small amounts of O_2. Other phases no doubt occur but are not readily observed.

Furthermore, the biogeochemical phases cause the dissolved solids content of the seepage to vary spatially and temporally. Because of the complexity of the source materials within disposal sites, the possibility of their reacting and interacting with their environment makes it impossible to rank the relative importance of the individual processes, effects, and reactions that influence the leakage of contaminants from waste disposal sites. If one wants to identify the relationship between waste disposal sites and groundwater contamination, one is forced to resort to the compilation of statistics, conducting investigations as comprehensively as possible and characterizing the occurrence of substances at a representative number of waste disposal sites.

III. DATA

Analyses of groundwater samples from 236 abandoned waste disposal sites overlying unconsolidated aquifers in western Germany were statistically evaluated. Some of the data are natural background data; the rest relate to contaminated groundwater.

All parameters were tested for statistical normality. Most showed a strongly positively skewed lognormal distribution that can be converted to a normal distribution. A similar data set, consisting of composite groundwater monitoring data from aproximately 500 waste disposal site investigations conducted in the United States [1–3], was made available for this study.

IV. GROUNDWATER CONTAMINATION ASSOCIATED WITH WASTE SITES

A. Hydrochemistry of Uncontaminated, Unconsolidated Aquifers

Generally, the effects of contaminant seepage from waste disposal sites into groundwater can be determined by comparing groundwater downgradient from a waste site with that sampled sufficiently upgradient that it is not influenced by the contamination. This is also primarily valid for naturally occurring inorganic groundwater constituents, the content of which is increased by human influence. It is a different situation with organic contaminants, which are almost exclusively anthropogenic. Their presence in groundwater usually indicates the influence of a waste site. However, this statement must be qualified, as a number of anthropogenic organic compounds may occur in groundwater as a result of contributions from other sources (e.g., agriculture, atmospheric deposition).

The groundwater from upgradient locations in sandy and coarse gravel aquifers that have not been affected by leakage events from waste disposal sites can be characterized as follows. The mean contents of major ionic constituents—calcium, hydrogen carbonate, and sulfate—are natural and indicate that the sediment also contains carbonate and sulfate minerals (Table 1). For the most part, the mean values of potassium, sodium, and chloride content also lie within the "normal range" for this type of aquifer [4]. In several locations there are layers containing organic constituents with high amounts of humic substances, which, by chemical reduction, lead to relatively high concentrations of iron and manganese in the groundwater. The increased nitrate and ammonium content can be traced back to agricultural activities in the general vicinity of abandoned waste disposal sites with a good degree of probability. [For example, there are corn (maize) fields and grain crops planted upgradient from several abandoned sites.] The characteristics noted here are not universal for conditions in Germany because the samples were obtained from petrographically and sedimentarily different aquifers and some sites are represented by only a single sample.

The mean concentrations of trace elements and/or anionic traces can be regarded as natural (Table 1). This also applies to arsenic, the average content of which is 0.5 μg/L.

The cumulative organic parameters that were examined—AOX (adsorbable organic halogens) and DOC (dissolved organic carbon)—proved to have a mean concentration of 2.8 μg/L (AOX) and 1.6 mg/L (DOC). Compounds with low boiling points that were measured (1,1,1-trichloroethane, trichloroethene, tetrachloroethene, dichloromethane, chloroform, carbon tetrachloride, bromoform, bromodichloromethane, and dibromochloromethane) usually occur at concentrations below 1 μg/L in areas not influenced by abandoned waste disposal sites. The highest concentration observed at the upgradient locations included in this study was 0.89 μg/L for bromodichloromethane (Table 2).

B. Hydrochemistry of Contaminated Aquifers

1. Inorganic Substances

Disposal site leakage events can exert a major influence on the abundance and concentration of individual inorganic substances in groundwater. Table 3 lists the statistical parameters, mean, max (maximum value), and 50% (median)

Table 1 Statistical Parameters of Inorganic Constituents from Uncontaminated Groundwater in Western Germany

Parameter	Analytical attempts, total (n)	Detection limit	n >det. lim.[a]	Freq. of detec. (%)	Concentration		Percentile concentration[b]		
					Mean	Max.	50%	75%	90%
Main cations (mg/L)									
Calcium	1287	5.0	1287	100	75.7	425	72.1	103	135
Magnesium	1325	2.0	1319	99.6	11.7	175	8.8	15.8	24.1
Sodium	815	5.0	812	99.6	22.6	140	17.1	29	45.4
Potassium	797	1.0	794	99.6	3.6	67.7	2.4	4.2	6.5
Ammonium	1402	0.02	1004	71.6	0.22	9.5	0.09	0.3	0.55
Iron (total)	1506	0.01	1311	87.1	2.7	63	1.3	2.9	7.8
Manganese	1451	0.01	1139	78.5	0.21	2.2	0.14	0.27	0.5
Main anions (mg/L)									
Hydrogen carbonate	1091	18.0	1091	100	187	567	168	307	366
Chloride	1478	1.0	1477	99.9	38	1000	29	45	71.5
Sulfate	1510	1.0	1420	94	57.5	863	36	80	134
Nitrate	1570	0.5	1228	78.2	10.6	100	1	14.2	36.1
Trace elements (µg/L)									
Aluminum	237	40.0	107	45.1	88	4400	0	13	46
Lead	724	0.05	199	27.5	1.2	500	0	0.1	1.1
Cadmium	715	0.05	159	22.2	1.5	1000	0	0	0.3
Chromium (total)	497	10.0	109	21.9	2.5	1000	0	0	1.0
Copper	79	10.0	48	60.8	4.0	20	1.9	5.9	12.9
Nickel	304	10.0	100	32.9	7.8	1300	0	1.4	9.0

Table 1 Continued

Parameter	Analytical attempts, total (n)	Detection limit	n >det. lim.[a]	Freq. of detec. (%)	Concentration Mean	Concentration Max.	Percentile concentration[b] 50%	Percentile concentration[b] 75%	Percentile concentration[b] 90%
Mercury	404	0.2	27	6.7	0.0	3	0	0	0
Strontium	61	40.0	60	98.3	204	1060	155	270	191
Zinc	165	10.0	98	59.4	95	4250	9.9	50	
Trace anions									
Arsenic (III, V) (μg/L)	472	0.2	115	24.4	0.5	18.3	0	0	1.5
Nitrite (mg/L)	1418	0.1	369	26.0	0.03	28	0	0.01	0.02
Selenium (IV,VI) (μg/L)	149	0.2	6	4.0	0.0	2	0	0	0
Boron (total) (μg/L)	21	20.0	8	86	44	80	45	65	70
Fluoride (mg/L)	498	0.5	382	76.7	0.12	7.5	0.09	0.14	0.19
Cyanide (total) (μg/L)	393	1.0	27	6.9	0.3	10	0	0	0
Phosphate (total) (mg/L)	783	0.1	642	82	0.26	20	0.12	0.4	0.6

[a]Analytical attempts above detection limit.
[b]Percentage of results below or just at the given concentration.

Table 2 Statistical Parameters of Organic Constituents from Groundwater Not Influenced by Waste Sites in Western Germany

Parameter	Analyt. attempts total (n)	$n >$ det. lim.[a]	Detect. freq. (%)	Concentration		Percentile concentration[b]		
				Mean	Max.	50%	75%	90%
Organic sum parameter								
DOC (mg/L)	501	500	99.8	1.6	8.6	1.3	2.0	3.1
AOX (µg/L)	202	75	37.1	2.8	54	0	0.3	9.1
Organic substances (µg/L)								
1,1,1-Trichloroethane	345	51	14.8	0.06	11	0	0	0.05
Trichloroethene	346	62	17.9	0.34	38	0	0	0.22
Tetrachloroethene	345	72	20.9	0.46	39	0	0	0.1
Dichloromethane	302	4	1.3	0.03	7.3	0	0	0
Trichloromethane	280	59	21.1	0.34	60	0	0.01	0.1
Tribromomethane	222	15	6.8	0.1	20	0	0	0
Tribromomethane	144	2	1.4	0.01	0.9	0	0	0
Bromodichloromethane	69	14	20.3	0.89	59	0	0.01	0.18
Dibromochloromethane	126	13	10.3	0.09	10	0	0	0.01

[a]Analytical attempts above detection limit.
[b]Percentage of results below or just at the given concentration.

Table 3 Statistical Parameters of Inorganic Constituents from Contaminated Groundwater in Western Germany

Parameter	Analyt. attempts total (n)	Detection limit	$n >$ det. lim.[a]	Freq. of detec. (%)	Concentration		Percentile conc.[b]	
					Mean	Max.	50%	75%
Main cations (mg/L)								
Calcium	318	5.0	317	99.7	176.6	785.0	155.5	229.0
Magnesium	321	2.0	320	99.7	39.3	436.0	22.0	42.3
Sodium	338	5.0	337	99.7	139.8	3600.0	45.6	130.0
Potassium	217	1.0	217	100.0	34.4	350.0	12.0	36.0
Ammonium	276	0.02	231	83.7	14.4	945.0	0.41	6.2
Iron	335	0.01	330	98.5	10.3	240.0	1.95	11.0
Manganese	329	0.01	311	94.5	1.6	33.1	0.43	1.24
Main anions (mg/L)								
Hydrogen carbonate	195	18.0	195	100.0	526.7	2458.0	407.5	661.0
Chloride	360	1.0	360	100.0	218.0	6020.0	74.2	167.5
Sulfate	352	1.0	350	99.4	218.0	8560.0	122.2	222.8
Nitrate	347	0.5	244	70.3	58.7	11500.0	2.5	24.3

Trace elements (µg/L)							
Aluminum	168	40.0	135	80.4	947.2	59571.0	80.0	573.0
Lead	334	0.05	129	38.6	7.6	450.0	0.0	573.0
Cadmium	335	0.05	126	37.6	40.3	13000.0	0.0	0.3
Chromium	279	10.0	179	64.2	39.5	5123.0	3.0	9.0
Copper	275	10.0	189	68.7	26.5	577.0	7.0	20.0
Nickel	249	10.0	187	75.1	115.8	23168.0	13.0	24.0
Mercury	232	0.2	39	16.8	0.09	3.4	0.0	0.0
Strontium	205	40.0	201	98.0	604.4	5580.0	399.0	763.5
Zinc	328	10.0	276	84.1	921.1	168120.0	64.5	220.0
Trace anions							
Arsenic (III, V) (µg/L)	253	0.2	172	68.0	61.0	4000.0	1.3	6.0
Nitrite (mg/L)	285	0.1	107	37.5	768.1	79000.0	0.0	33.0
Selenium (IV, VI) (µg/L)	112	0.2	20	17.9	0.22	14.0	0.0	0.0
Boron (total) (µg/L)	259	20.0	253	97.7	948.5	31200.0	181.0	529.5
Fluoride (mg/L)	97	0.5	57	58.8	294.1	1700.0	200.0	500.0
Phosphate (total) (mg/L)	207	0.1	113	54.6	2.3	115.0	0.05	0.6

[a]Analytical attempts above detection limit.
[b]Percentage of results (%) below or just at the given concentration.

and 75% percentile concentrations for the parameters measured in groundwater samples collected downgradient from disposal sites.

In order to characterize the impact on groundwater of seepage or leachate migrating from a waste disposal site, the concept of a contamination factor (KF) was developed. The contamination factor represents the ratio of the observed concentration in the downgradient (contaminated) area to the observed concentration in the upgradient (uncontaminated) area of the site. If the site is not leaking or if the substance is not involved in the leakage event, the ratio should be 1.0. However, if the substance is leaking from the site, the ratio will increase to a value greater than 1.0. (The larger the resultant contamination factor, the larger the leakage event.)

Eighty-nine percent of the inorganic parameters examined have a mean contamination factor of >2; 63%, >5; and 30%, >10. However, maximum contamination factors as high as 390 for boron were observed (Table 4). This approach demonstrates the significant changes that disposal site leachate can have on groundwater quality. Furthermore, the approach identifies those specific inorganic substances (with the highest contamination factors) that are likely to be associated with disposal site leachate: arsenic (mean KF 122), ammonium (mean KF 65.5), cadmium (mean KF 26.9), nitrite (mean KF 25.7), boron (mean KF 21.6), chromium (mean KF 15.8) and nickel (mean KF 14.8).

An alternative approach to evaluating the contamination factor data is to rank individual substances according to the frequency with which background concentrations are exceeded. With this approach, the boron concentration in groundwater collected downgradient from waste disposal sites exceeded the background concentration (KF > 1.0) in 222 out of 252 samples (85.7%). Other inorganic substances that exceeded background concentrations, listed in decreasing order, were HCO_3^-, Na, Cl, Mg, Ca, K, Sr, Pb, NO_2, Al, Se, Cd, and Hg (Table 5).

By combining the results of the two approaches, a group (subset) of inorganic parameters that are considered to be characteristic of waste disposal site leachate can be identified (mean contamination factors greater than 10 and background concentrations exceeded more than 50% of the time). These inorganic substances are As (KF 122; 61.3%), NH_4^+ (65.5; 53.6%), B (21.6; 85.7%), Ni (14.8; 64.3%), and Cr (15.8; 56.3%).

In the United States, a different data evaluation technique has been used to identify chemical constituents associated with disposal site leakage events [5]. Rather than examining the data for upgradient–downgradient differences or ratios (which requires an implicit assumption of chemical behavior), this alternative approach is based on changes in monitoring data variance over time. If an inorganic constituent is not involved in a leakage event, the observed concentration for the constituent in all wells in a monitoring network

Table 4   Contamination Factors of Inorganic Groundwater Contaminants

Parameter	Contamination factors (KF)[a]			
	Mean	Max	50%	75%
Main cations				
Calcium	2.3	1.8	2.2	2.2
Magnesium	3.4	2.5	2.5	2.7
Sodium	6.2	25.7	2.7	4.5
Potassium	9.6	5.2	5.0	8.6
Ammonium	65.5	99.5	4.6	20.7
Iron	3.8	3.8	1.5	3.8
Manganese	7.6	15.0	3.1	4.6
Main anions				
Hydrogen carbonate	2.8	4.3	2.4	2.2
Chloride	5.7	6.0	2.6	3.7
Sulfate	3.8	9.9	3.4	2.8
Nitrate	5.5	115.0	2.5	1.7
Trace elements				
Aluminum	10.8	13.5	—[b]	44.1
Lead	6.3	0.9	—[b]	21.0
Cadmium	26.9	13.0	—[b]	—[b]
Chromium	15.8	5.1	—[b]	—[b]
Copper	6.6	28.9	3.7	3.4
Nickel	14.8	17.8	—[b]	33.6
Mercury	—[b]	1.1	—[b]	—[b]
Strontium	3.0	5.3	2.6	2.8
Zinc	9.7	39.6	7.2	4.4
Trace anions				
Arsenic (III, V)	122.0	218.6	—[b]	—[b]
Nitrite	25.7	2.8	—[b]	1.5
Selenium (IV, VI)	—[b]	7.0	—[b]	—[b]
Boron (total)	21.6	390.0	4.0	8.1
Fluoride	1.0	0.2	2.2	3.6
Cyanide (total)				
Phosphate (total)	8.8	5.8	0.4	1.5

[a]KF of mean, max, and 50% and 75% percentile concentrations

$$= \frac{\text{conc. contaminated}}{\text{conc. uncontaminated}}$$

[b]Not calculated because uncontaminated value was below detection limit.

Table 5 Detection Frequencies of Inorganic Groundwater Contaminants[a]

Subst.	Analytical attempts, contaminated wells	No. of samples >backgr. conc.	Detect. freq. >backgr. conc. (%)
B	259	222	85.7
HCO_3^-	233	200	85.7
Na	338	282	83.4
Cl	360	285	79.2
Mg	321	244	76.0
Ca	318	237	74.5
K	217	161	74.2
Sr	205	152	74.1
SO_4^{2-}	352	227	64.5
Ni	249	160	64.3
Mn	329	206	62.6
As	253	155	61.3
Cr	279	157	56.3
NH_4^+	276	148	53.6
F	97	51	52.6
Cu	275	144	52.4
NO_3^-	347	176	50.7
PO_4^{3-}	207	86	41.5
Zn	328	135	41.2
Fe	335	136	40.6
Pb	334	128	38.3
NO_2^-	285	101	35.4
Al	168	57	33.9
Se	112	25	22.3
Cd	335	50	14.9
Hg	232	34	14.7

[a]Detection frequency is the number of values higher than background (= mean uncontaminated).

should be similar (ideally the same). The monitoring data variance for this situation would be minimal and remain low as long as the constituent is not involved in a leakage event. However, when the constituent enters the groundwater because of a leakage event at a waste disposal site, a concentration increase will occur at one or more monitoring locations, and the resultant monitoring data variance will also increase. These changes in monitoring data variance over time can be used to characterize the individual constituents associated with a leakage event [5].

Monitoring data variance was calculated for each of 16 inorganic constituents at 253 waste disposal sites located throughout the United States. The resultant variance values were then classified as follows:

1. The variance is high (with the inference that the constituent is involved in a leakage event at that site).
2. The variance is low (with the inference that the constituent is not involved in leakage event).
3. There were no data or insufficient data to calculate a variance value, and no judgment was made.

The results of this evaluation can be used to identify the waste disposal sites that are apparently leaking, the individual constituents associated with each leakage event, and the number of times an individual constituent is associated with leakage events (Table 6). There are three attributes of the monitoring data variance technique that need to be briefly discussed. First, the distribution

Table 6 Ranking of Inorganic Constituents in Disposal Site Leakage Events Based on High Parameter Variance of the United States Compared to Western Germany Contamination Factors and Detection Frequencies > Background Concentrations

	United States					Western Germany				
Chemical rank	Number of sites		Percentage of sites		(a) Contamination factor (mean values)		(b) Detection frequency (%) > background		Product (a) × (b)	
1	As	28	As	18.7	As	122.0	Na	83.4	As	7479
2	Cd	21	Mg	18.6	Cd	27.0	Cl	79.2	Ni	965
3	Na	18	Zn	16.5	Ni	15.0	Mg	76.0	Na	517
4	Hg	16	Ni	14.8	Al	11.0	Ca	74.5	Mn	476
5	Zn	14	Na	14.3	Zn	9.7	SO$_4$	64.5	Cl	451
6	Pb	13	Cd	14.1	Mn	7.6	Ni	64.3	Cd	402
7	Ni	13	Hg	11.4	Cu	6.6	Mn	62.6	Zn	400
8	Se	12	Ca	10.3	Pb	6.3	As	61.3	Al	373
9	Mg	11	Cu	9.6	Na	6.2	Cu	52.4	Cu	346
10	Mn	11	Se	8.8	Cl	5.7	Zn	41.2	Mg	258
11	Cu	9	Pb	8.2	SO$_4$	3.8	Fe	40.6	SO$_4$	245
12	SO$_4$	9	Mn	7.4	Fe	3.8	Pb	38.3	Pb	241
13	Cl	7	Al	6.7	Mg	3.4	Al	33.0	Ca	171
14	Fe	7	SO$_4$	6.5	Ca	2.3	Se	22.3	Fe	154
15	Ca	6	Cl	5.1	Se	1.0	Cd	14.9	Se	22
16	Al	4	Fe	4.4	Hg	1.0	Hg	14.7	Hg	15

of the inorganic constituents with high variance values does not appear to be random. When one chemical is apparently involved in a leakage event, there are usually several other constituents at the same site that also exceed the variance action limits. This would seem to be a logical, expected observation because leakage events are not planned or controlled. Second, the combination of leaking chemicals at a site can be used to fingerprint the leakage from a particular site or from a particular industry. Third, by simply tabulating the number of times a contaminant is associated with leakage events, the contaminants can be ranked in terms of relative importance.

With regard to the last point, the inorganic contaminants in Table 6 were ranked according to the number of times they were apparently involved in a leakage event and the percentage of sites at which the contaminant was monitored (only those constituents were used for which both West German and American results were available for comparison). The results of this comparison suggest that As, Cd, and Na (plus Mg and Zn) are the specific inorganic constituents that are most frequently involved with disposal site leakage events. When this list is compared with the West German results that identified contaminants associated with leakage events based on high contamination factors (KF) and high frequencies of exceeding background concentrations, the same three contaminants (plus Ni and Cl) are found to be high on both lists. It would seem that the two data evaluation techniques independently verify each other and simultaneously identify the most important inorganic constituents that should be included in site monitoring programs.

2. Organic Substances

In contrast to the relatively small number of inorganic constituents that have been associated with disposal site leakage events, the number of organic substances that can cause possible groundwater contamination downgradient from waste disposal sites is quite large.

The extensive GC/MS analytical data produced during numerous disposal site investigations in the Federal Republic of Germany (236) and the United States (500) resulted in the identification of approximately 1200 organic contaminants that have been reported in disposal site groundwater. However, only a small number of these compounds are present on a frequent basis and at a concentration clearly above established detection limits (approximately 1 μg/L). The majority of the identified contaminants (>1000) have a detection frequency of less than 0.1%. This means that they are detected on average less than once in 1000 samples of groundwater contaminated by waste sites.

A bias for substances or classes of substances due to analysis and the methods used can be almost ruled out, because sample preparation (acid, base/neutral extractable substances, volatile substances, and pesticides) and the analysis

Table 7 Statistical Parameters of Organic Constituents from Contaminated Groundwater in Western Germany

Parameter (μg/L)	Analyt. attempts total (n)	Detection limit (μg/L)	n > detec. limit[a]	Freq. of detec. (%)	Concentration		Perc. conc.[b]	
					Mean	Max.	50%	75%
Tetrachloroethene	227	0.1	195	70.40	56.1	6504.0	1.4	3.7
Trichloroethene	277	0.1	154	55.60	1013.1	128000.0	2.3	11.0
cis-1,2-Dichloroethene	153	4.0	46	30.07	22095.1	411000.0	165.5	1150.0
Benzene	127	1.0	38	29.13	140.5	1795.0	13.5	38.0
1,1,1-Trichloroethane	206	0.1	47	22.82	16.5	270.0	1.0	7.4
m/p-Xylene	92	0.1	21	22.82	39.9	447.0	2.8	4.9
Trichloromethane	236	0.1	52	22.03	76.2	2800.0	1.4	10.1
1,2-Dichloroethane	16	5.0	3	18.75	107.3	210.0	—	—
Chloroethene (VC)	136	1.0	24	17.65	1693.2	12000.0	99.5	1950.0
Toluene	127	0.1	21	16.54	73.2	911.0	3.5	11.0
Dichloromethane	114	10.0	17	14.91	38066.5	499000.0	437.5	16835.0
Tetrachloromethane	201	0.1	29	14.43	1.2	23.0	0.2	0.3
4-Methylphenol (p-cresol)	124	0.1	17	13.71	42.0	283.0	8.6	20.7
Chlorobenzene	93	0.1	12	12.90	52.9	388.0	2.3	6.5
2-Methylphenol (o-cresol)	124	0.1	16	12.90	10.0	63.3	5.9	9.2
1,2-Dichlorobenzene	90	0.1	11	12.22	1.4	6.6	0.9	2.1
1,4-Dichlorobenzene	90	0.1	11	12.22	31.9	265.0	2.2	37.5
Naphthalene	124	0.1	15	12.10	2.2	12.6	1.1	2.1
Ethylbenzene	124	0.1	14	11.29	32.2	160.0	4.4	62.5

Table 7 Continued

Parameter (µg/L)	Analyt. attempts total (n)	Detection limit (µg/L)	n > detec. limit[a]	Freq. of detec. (%)	Concentration Mean	Concentration Max.	Perc. conc.[b] 50%	Perc. conc.[b] 75%
o-Xylene	127	0.1	12	9.45	13.8	69.0	4.7	19.6
2,4,6-Trichlorophenol	124	0.1	11	8.87	3.2	24.1	0.6	1.5
3,5-Dimethylphenol	124	0.1	10	8.06	16.2	61.0	6.9	35.6
Phenol	124	0.1	10	8.06	2.2	5.6	1.5	3.6
1,3-Dichlorobenzene	90	0.1	7	7.78	11.5	74.0	1.1	38.2
trans-1,2-Dichloroethene	134	5.0	10	7.46	57.1	135.0	50.0	95.0
Isopropylbenzene (cumol)	90	0.1	5	5.56	2.4	4.7	3.1	4.5
1,1-Dichloroethane	130	10.0	7	5.38	52.7	110.0	53.0	90.0
Acenaphthene	124	0.1	6	4.84	6.3	32.0	1.3	17.1
2,4-Dichlorophenol	124	0.1	6	4.84	3.5	17.2	0.9	9.6
3-Chlorophenol	124	0.1	6	4.84	12.7	22.8	12.3	20.8
p-Cymol[p-CH$_3$C$_6$H$_4$CH(CH$_3$)$_2$]	90	0.1	4	4.44	1.9	3.5	1.5	2.6
2-Ethyltoluene	90	0.1	4	4.44	0.6	1.0	0.6	0.8
2,4,5-Trichlorophenol	127	0.1	5	3.94	7.1	31.	1.3	16.6
1,3,5-Trimethylbenzene	90	0.1	3	3.33	1.7	4.0	—	—
Phenanthrene	124	0.1	4	3.23	1.5	4.4	0.6	2.6
Tribromomethane	130	1.0	4	3.08	3.0	6.0	2.5	4.5

[a]Number of analytical attempts above detection limit.
[b]Percentage of results below or just at the given concentration.

260

(essentially GC/MS) were geared to determining the largest number of ground-water contaminants possible.

Table 7 lists the most frequently detected organic groundwater contaminants found across western Germany that are attributed to the influence of waste sites. The table shows only those compounds that have a detection frequency of >3%. Among the 19 contaminants with a detection frequency of ⩾10%, 12 (63%) are volatile halogenated compounds (five are alkane derivatives, four are alkene derivatives, and three are substituted benzenes). Benzene and its alkyl derivatives (four compounds) constitute the majority of the seven nonhalogenated contaminants listed in Table 7, which also include two phenolic compounds and naphthalene. The highest mean concentration was obtained for volatile halogenous substances, predominantly for dichloromethane (approximately 38 mg/L), *cis*-1,2-dichloroethene (approximately 22 mg/L), vinyl chloride (approximately 1.7 mg/L), and trichloroethene (approximately 1 mg/L). The high concentrations associated with these volatile organic compounds confirm the significance of this class of substances as major emissions from waste disposal sites [6].

The extensive data from the United States were compared to the West German monitoring data to verify the types and concentrations of contaminants that have been detected in disposal site groundwater. A comparison of the 25 most frequently detected organic contaminants in each data set, including all substances with a (mean) concentration greater than 1 μg/L, showed a high degree of similarity (Fig. 1). Twenty of 25 individual compounds (80%) are identical. Also, a significant similarity between the two data sets is the predominance of halogenated aliphatic compounds in each national data summary (six alkanes and five alkenes). In addition, there are six and five nonhalogenated hydrocarbons and four and seven compounds containing oxygen, that is, phenols and ketones.

V. MULTIVARIATE STATISTICAL EVALUATION OF THE DATA

Generally, numerous contaminants migrate simultaneously from a waste disposal site. Therefore, one can expect strong correlations between waste types and certain parameters.

The largest percentage of waste material deposited in disposal sites in western Germany was classified as "municipal solid waste" and "construction waste" in 1980 (92.9%) and 1984 (88.6%) (Table 8) [7]. It can be assumed that the composition of waste materials was similar in previous years and that most sites contain an equally large percentage of municipal solid waste and construction waste. Therefore, it is highly probable that the types of compounds included in the composite West German data set are representative of these waste classifications. This possibility was investigated using

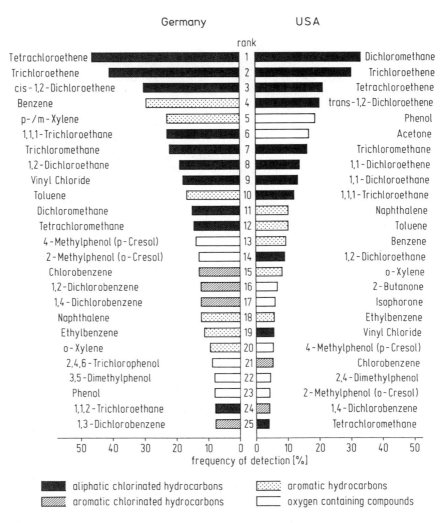

Figure 1 Ranking of frequency detection (⩾1 μg/L) of principal organic contaminants in groundwaters downgradient from waste sites in Germany and the United States.

Table 8 Annual Amounts of Different Types of Waste in Western Germany (Mtonnes)

	1980		1984	
Municipal waste	32.6	(39.4%)	29.6	(34.4%)
Construction waste	44.2	(53.4%)	46.2	(54.2%)
Industrial waste	5.9	(7.1%)	9.8	(11.4%)
Total	82.7	(100%)	86.1	(100%)

factor analysis to determine which individual contaminants or classes of contaminants could be attributed to specific types of waste.

The data used in characterizing groundwater constituents, as mentioned earlier, came from different porous aquifers. Multivariate statistical evaluations based on such heterogeneous data present certain quandaries and can lead to results that are difficult to interpret. Therefore, it was necessary to restrict the assessments to a limited but homogeneous group of data taken from comparable sources of groundwater. These requirements are fulfilled by the samples taken from the Pleistocene region in northern Germany and were the only ones used in the factor analysis.

In addition to the important inorganic groundwater constituents, AOX (halogenated organic compounds that can be adsorbed with activated charcoal) was selected as a variable in the factor analysis along with the two most frequently detected organic contaminants, trichloroethene and tetrachloroethene. Other parameters that were evaluated were temperature, pH, and conductivity measured in the field.

Table 9 shows the results. The five factors mentioned above account for 82.3% of the total variance. Factor 1, with the highest single percentage within the total variance, reflects high values for the parameters sodium, chloride, conductivity, boron, temperature, and AOX, which can be regarded as significant for leaching from municipal solid waste, so this factor is interpreted as a "municipal solid waste" factor. Sodium (0.913) and chloride (0.812) are washed out of the waste mass together and thus enter groundwater dissolved in seepage. The usually high concentration of dissolved substances in contaminated groundwater produces an appropriately high conductivity, to which the high value for conductivity (0.803) in factor 1 can be attributed. The excess boron concentration that is found in all groundwater contaminated by waste sites is represented by a significant value (0.672) in this factor, although a good portion of this boron originates from the sewage sludge deposited along with municipal solid waste [8]. Though the value for AOX is comparatively low it illustrates the fact that not insignificant amounts of contaminants determined with AOX are to be found at the waste sites; AOX, however, is not a typical parameter for municipal solid waste.

Another important class of substances—sulfate, calcium, and strontium—generally show a significant increase in the concentrations downgradient from waste disposal sites whose content mainly (70–90%) consists of construction waste [9]. This is due to the high amounts of these ions in leachates of such waste sites and their similar geochemical behavior in aquifers, which is characterized by a relatively high mobility. For this reason these ions can be found well above background levels at larger distances downgradient from construction waste disposal sites [10]. In these groundwaters their correlation coefficients were found to be 0.53 for SO_4-Ca and 0.54 for SO_4-Sr. Because of this they are also grouped together in factor 4, which for this reason can be regarded as a "construction waste" factor.

Table 9 Factor Matrix After Rotation with Kaiser Normalization of Groundwater Constituents Downgradient from Waste Disposal Sites[a]

	Factor 1	Factor 2	Factor 3	Factor 4	Factor 5	
Eigenvalue	7.1	2.9	2.3	2.1	1.6	
% variance	36.5	14.9	12.1	10.7	8.1	Commonality
Temperature	0.651					0.604
pH value						0.605
Conductivity	0.803					0.976
SO$_4$				0.846		0.744
Cl	0.812					0.899
NO$_3$						0.677
Na	0.913					0.919
Ca				0.618		0.922
Mg						0.849
Fe			0.793			0.736
Mn			0.763			0.686
Sr				0.680		0.743
As						0.467
Pb					0.649	0.877
Cd					0.842	0.787
Ni						0.718
Zn						0.643
Cu						0.855
Cr						0.510
B	0.672					0.638
AOX	0.558					0.756
Trichloroethene		0.951				0.940
Tetrachloroethene		0.609				0.601

[a]Only factor charges > 0.5 are reported.

Factors 2 and 5 are regarded as explanatory variables for refuse from industry and business, respectively. The substances included in these two factors are basically similar in their chemical and/or geochemical characteristics.

Finally, in factor 3, iron and manganese form a significant explanatory variable. Aside from the intensified effect of contamination in this case, the somewhat high natural concentrations of these two elements in groundwater also come to light.

In summary, we can say that both of these principal explanatory factors (1 and 2) are exclusively characterized by substance parameters that can be described as being highly persistent as well as highly mobile.

The parameters in factors 1 and 2 are very well suited for detecting the initial release of leachate (seepage) from a waste disposal site into groundwater (screening parameters). Compared to sodium and chloride, boron has the advantage that it occurs naturally in many loose sedimentary rock deposits in concentrations not exceeding some 30 μg/L. This makes even slight effects caused by seepage from waste more easily recognizable. A comparison of AOX values and volatile halogenated compound content in groundwater samples contaminated by waste disposal sites shows that in the range of 5–10 μg/L and from approximately 300 μg/L upwards the AOX content is below the sum of halogenated compounds. Between approximately 20 and 50 μg/L the AOX content is generally above the sum of halogenated compounds [6]. Despite these differences, the dimensions of both the value for AOX and the sum of volatile halogenated compounds are approximately the same. Different types of results were obtained with groundwater samples that had high AOX content but no volatile compounds. In these cases it was not possible to identify the adsorbed halogenated compounds. Generally one can say that AOX reflects a very wide range of halogenated organic substances and is therefore a particularly useful cumulative organic parameter. Together with GC headspace analysis, used for detecting "low-boiling" substances (especially solvents containing halogens), it provides us with a reliable screening method for groundwater contaminants.

After having used the sulfate value listed in factor 4 along with calcium and strontium at over 100 sites, we consider this to be a suitable screening parameter for determining seepage from construction waste even though it is dependent on aquifer conditions.

VI. THE USE OF PHYSICOCHEMICAL PARAMETERS

Groundwater quality variations in organic constituents that are due to many hydrogeochemical interactions can be recognized by examining a representative number of samples taken from groundwater contaminated by waste sites. The following is meant to list the correlations between the types and quantity of emission and the material composition of the waste mass.

To do this, we use three essential variables: the frequency with which the substance occurs in waste, the amount in which the substance occurs, and the emission behavior of the substance (transfer behavior into groundwater).

Constituents of waste that are present in high amounts, occur frequently, and have a high transfer coefficient would be expected to be detected frequently in groundwater. Substances with low transfer coefficients, including those that are essentially immobile, would not be expected in disposal site groundwater in high concentrations or at a high detection frequency, even though they may be present in waste with a high frequency or in large amounts. Based on this logic a correlation can be recognized between high detection frequencies and high concentrations in disposal site groundwater.

This is clearly shown in Table 7, in which the concentration shows a tendency to decline along with the determined detection frequency. Also, this explanation is consistent with the fact that approximately 1000 substances with a detection frequency of ≤0.1% are usually not detected at concentrations above 1 μg/L. Recognition of this relationship has important consequences with respect to the development of an effective groundwater monitoring strategy (see Schleyer et al., this volume, Chapter 10). Most important is a reduction in the number and types of substances that should be analyzed and evaluated [11,12].

Because the transfer behavior of waste substances governs the behavior of the total seepage to a great extent, this can be calculated in the form of a model by determining those parameters that are essentially responsible for the transfer behavior. Also, it is possible to estimate the transport of a contaminant in groundwater.

The transfer of an organic substance from a waste disposal site into an unconsolidated aquifer is determined more by its physicochemical characteristics than by the chemical properties of the substratum. The organic carbon content of the soil, which is basically responsible for the retention of contaminants, especially of the organic ones, is usually very low ($<0.1\%$), so the migration of a contaminant in groundwater depends almost solely on hydrogeological parameters aside from the properties of the substance itself [13,14].

The most important variables that control the transfer of any waste substance are its persistence and mobility. If a substance is not persistent under the specific conditions, that is, under the path-specific transfer conditions, and additionally has a short half-life, in this case in groundwater, then its mobility is only of subordinate importance. On the other hand, mobility is the decisive variable with persistent substances and/or with those that have a long half-life. If the mobility of a persistent substance is high, then its path-specific transfer behavior will be optimal.

The mobility and persistence of a contaminant along a particular seepage pathway can be estimated by using the substance's relevant characteristic physicochemical data. However, for estimating the mobility of any one substances—for example, one being emitted into the atmosphere from the waste mass—other parameters are needed than those used for transfer from the waste mass into soil and flora. In estimating the mobility of a contaminant that is migrating from the waste mass into groundwater, there are several parameters such as water solubility, vapor pressure, boiling point, evaporation rate, Ostwald's dilution law, the extent to which a substance is volatile in aqueous solution, adsorbability, density, viscosity, dissociation constant, surface tension, liposolubility, and Henry's law that can be used for this purpose.

The fact that many of these parameters are functionally dependent upon one another or correlate and/or overlap in the information they provide makes it necessary to reduce them to the essential ones—those that provide the most specific information for the individual path—and to link them suitably for quantifying the mobility and persistence (see Chapter 10).

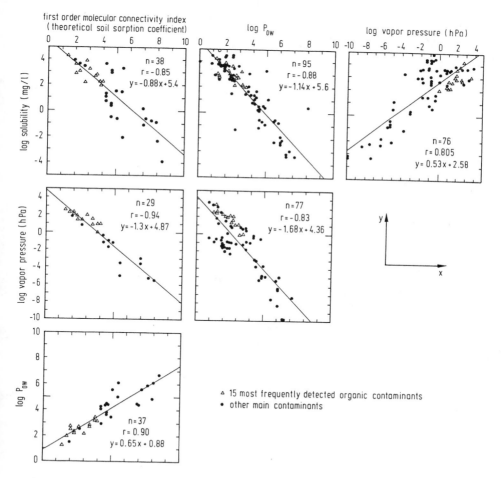

Figure 2 Correlations between four selected physicochemical parameters describing mobility and sorption potential of contaminants in the system waste–soil–groundwater.

The "mobility parameters," water solubility and vapor pressure, correlate negatively with the sorption parameters, K_{ow} and the first-order molecular connectivity index $^1\chi$ [15]. This means, among other things, that substances that are highly soluble in water have a low K_{ow}, and vice versa. This fact is significantly substantiated when one correlates the four parameters that describe mobility (Fig. 2). The available data concerning the important organic groundwater contaminants from waste sites are outlined in the scatter diagrams. All correlations have calculated coefficients greater than 0.8, and, in addition, the contaminants with the highest degrees of detection frequency (open triangles in Fig. 2) are also the ones that characteristically possess high degrees of solubility, high vapor pressure, a low K_{ow}, and a low first-order molecular connectivity index. This substantiates the fact that these four parameters are very well suited for determining the mobility of a substance in groundwater. It is not yet possible to establish a comparable evaluation for persistence, because substantiated measurement results concerning the half-lives of relevant contaminants in groundwater, for example, do not exist. An approach was made to estimate the "persistency potential" of a contaminant by using the COD [16] as a substitute for abiotic degradation and the BOD [17] for biotic degradation. Larger discrepancies between the two values were found due to the different degradation mechanism. For example, trichloromethane has a COD persistence of 92% and a BOD persistence of 1%. To use this information in an evaluation concept as Schleyer et al. do in Chapter 10, it is necessary to take the "worst-case" information.

VII. SUMMARY AND CONCLUSIONS

Downgradient from waste sites, a large variety of inorganic and organic contaminants that have migrated into groundwater have been detected. This investigation shows that the inorganic contaminants that are most frequently associated with disposal site leakage events are arsenic, cadmium, sodium, magnesium, zinc, chloride, ammonium, boron, nickel, and chromium. A significant portion of inorganic leakage can be assigned to the major classes of refuse, such as municipal solid waste, construction waste, and industrial waste in the form of factors, whereas those substances that are geochemically highly persistent and mobile primarily constitute the major explanatory variables.

The most important organic contaminants that are associated with disposal site leakage events belong to the class of halogenated alkanes, alkenes, and aromatic compounds, of which the chlorine-substituted compounds are clearly predominant. These are almost exclusively low-boiling compounds. Other classes of organic contaminants, such as PCBs, dioxins/furans, and pesticides, play a subordinate role in groundwater contamination from waste sites because of their lack of mobility along the groundwater path and/or their usually low hydrogeochemical persistence. The most important non-halogenated contaminants are methyl- and hydroxyl-substituted benzenes.

Because of the gigantic number of waste sites, investigational measures have to be hierarchically structured for determining problematic sites that endanger groundwater, and they have to manage with the least extensive examinations possible. This is necessary to enable the responsible authorities to recognize the dangers threatening groundwater from abandoned waste sites and to take measures to check these dangers. One method to achieve this goal is to develop effective groundwater monitoring strategies for chemical substances that are representative of disposal site leakage events. Such a strategy has been developed by the German Institute for Water, Soil and Air Hygiene. It incorporates the identified main pollutants in a monitoring program to detect and assess leakage events at these sites [9,10,18]. This is discussed in detail in Chapter 10.

Because landfills are still the most frequently practiced type of waste disposal, the dumping of the most relevant groundwater pollutants or waste containing them should be totally avoided at presently operating or future waste sites in order to avoid further damage to groundwater. Some contaminants are certainly metabolites; for example, vinyl chloride is a gas at room temperature and is most probably not dumped as such, but it is a degradation product of tetrachloroethene and trichloroethene. This also applies to *cis*- and *trans*-1,2-dichloroethene, so the parent compounds, tetrachloroethene and trichloroethene, should be excluded from being deposited in landfills.

Basically, one has to assume that total avoidance of substances that endanger groundwater cannot be achieved for several reasons, making it necessary to impose additional requirements for future landfills. First of all, these would include an optimal sealing at the base, particularly for the above-mentioned substances but for other mobile substances as well.

Presently, however, there is only very limited knowledge about the long-term behavior of natural and synthetic sealing materials used in landfills. It seems to be highly improbable that total retention exceeding 100 years will be attained. There are many cases of known groundwater contamination caused by "sealed" landfills. Future landfill sites must be selected extremely carefully and by using scientifically founded criteria. Sites that appear to be suitable are those that either have no groundwater or at which any present or future use of the groundwater, no matter what type, can be excluded.

REFERENCES

1. Plumb, R.H. Jr. Disposal site monitoring data: observations and strategy implications. Proc. Second Canadian/American Conference on Hydrogeology, *Hazardous*

Wastes in Groundwater: A Soluble Dilemma (B. Hitchon and M. Trudell, eds.), Banff, Alberta, Canada, 1985, pp. 66–77.

2. Plumb, R.H., Jr., and Pitchford, A.M. Volatile organic scans: implications for groundwater monitoring. Proc. National Water Well Association/American Petroleum Institute Conference on Petroleum Hydrocarbons and Organic Chemicals in Groundwater, Houston, TX, 1985, pp. 1–15.

3. Plumb, R.H., Jr. A practical alternative to the RCRA organic indicator parameters. Proc. HAZMACON 87, Hazardous Materials Management Conference and Exhibition (T.P.E. Bursztynsky, ed.), Santa Clara, CA, 1987, pp. 135–150.

4. Matthess, G. Die Beschaffenheit des Grundwassers. *Lehrbuch der Hydrogeologie*, Band 2, Gebrüder Bornträger, Berlin, 1973.

5. Plumb, R.H., Jr. Characterizing disposal site leakage events through changes in monitoring data variance. Presented at the ASTM Symposium on Ground Water Monitoring, San Diego, CA, Jan. 31, 1991.

6. Kerndorff, H., Schleyer, R., Arneth, J.-D., and Struppe, T. Vorkommen und Bedeutung halogenorganischer Verbindungen als Grundwasserkontaminanten aus Abfallablagerungen. 1. Langener Kolloquium, Feb. 9–10, 1989, Langen, *Halogenierte Kohlenwasserstoffe in Wasser und Boden* (U. Hagendorf and R. Leschber, eds.), Schriftenreihe des Vereins f. Wasser-, Boden- und Lufthygiene Band 82, Gustav Fischer Verlag, Stuttgart, 1990.

7. Statistisches Bundesamt Wiesbaden (ed.). *Statistisches Jahrbuch 1988 für die Bundesrepublik Deutschland*, Verlag W. Kohlhammer GmbH, Stuttgart, 1988.

8. Scheffer, F., and Schachtschabel, P. *Lehrbuch der Bodenkunde*, 11 Aufl., Ferdinand Enke Verlag, Stuttgart, 1982.

9. Kerndorff, H., Brill, V., Schleyer, R., Friesel, P., and Milde, G. Erfassung grundwassergefährdender Altablagerungen. Ergebnisse hydrogeochemischer Untersuchungen, Institut für Wasser-, Boden- und Lufthygiene des Bundesgesundheitsamtes, WaBoLu-Heft 5/1985.

10. Brill, V., Kerndorff, H., Schleyer, R., Arneth, J.D., Milde, G., and Friesel, P. Fallbeispiele für die Erfassung grundwassergefährdender Altablagerungen aus der Bundesrepublik Deutschland, Institut für Wasser-, Boden- und Lufthygiene des Bundesgesundheitsamtes, WaBoLu-Heft 6/1986.

11. Arneth, J.-D., Schleyer, R., Kerndorff, H., and Milde, G. Standardisierte Bewertung von Grundwasserkontaminationen durch Altlasten. I. Grundlagen sowie Ermittlung von Haupt- und Prioritätskontaminanten. Bundesgesundheitsblatt 31, Heft 4, 1988, pp. 117–123.

12. Schleyer, R., Arneth, J.-D., Kerndorff, H., Milde, G., Dieter, H., and Kaiser, U. Standardisierte Bewertung von Grundwasserkontaminationen durch Altlasten. II. Stoffbewertung, Expositionsbewertung und ihre Verknüpfung. Bundesgesundheitsblatt 31, Heft 5, 1988, pp. 160–168.

13. Kinzelbach, W. *Groundwater Modelling: An Introduction with Sample Programs in Basic*, Elsevier, New York, 1986, p. 333.

14. Friesel, P., and Steiner, B. Interactions of halogenated hydrocarbons with soils. *Fresenius Anal. Chem., 319*:160–164, 1984.

15. Protic, M., and Sabljic, A. Quantitative structure–activity relationships of acute toxicity of commercial chemicals on fathead minnows: effect of molecular size. *Aquat. Toxicol., 14*:47–64, 1989.

16. Janicke, W. *Chemische Oxidierbarkeit organischer Stoffe*, Institut für Wasser-, Boden- und Lufthygiene, WaBoLu-Berichte 1/1983, Dietrich Reimer Verlag, Berlin, 1983.

17. Tabak, H.H., Quave, St. A., Mashni, C.I., and Barth, E.F. Biodegradability studies with organic priority pollutant compounds. *J. WPCF 53*(10):1503–1518, 1981.

18. Arneth, J.D., Kerndorff, H., Brill, V., Schleyer, R., Milde, G., and Friesel, P. Leitfaden für die Aussonderung grundwassergefährdender Problemstandorte bei Altablagerungen. Institut für Wasser-, Boden- und Lufthygiene des Bundesgesundheitsamtes, WaBoLu-Heft 5/1986.

10

Detection and Evaluation of Groundwater Contamination Caused by Waste Sites

Ruprecht Schleyer, Helmūt Kerndorff,   and Gerald Milde[†]

Institute for Water, Soil and Air Hygiene of the Federal Health Office,
Langen and Berlin, Germany

I. INTRODUCTION

The Waste Disposal Act went into effect in the Federal Republic of Germany in 1972. This law regulates the proper disposal of waste in authorized landfills and provides protective measures against possible contaminant seepage. Up until that time, practically every community and many businesses were operating their own dumping sites, at which all of the waste produced was deposited without being sorted and without consideration of the possible consequences for the environment. Individual spectacular instances of soil and groundwater contamination caused by contaminant seepage from such waste sites, as well as the resulting danger for the population, for example via drinking water, made the problem of abandoned waste sites conspicuous in the second half of the 1970s.

In the 1980s, the German federal states began to systematically register abandoned waste sites according to surveys made by using old maps and aerial photographs or questioning the population. Table 1 shows the present status of registering this information with a total of about 50,000 contaminated sites of which some 40,000 are abandoned waste sites [1]. The numbers vary

---

Present affiliations:
[†]University of Bonn, Bonn, Germany

Table 1 Numbers of Closed or Abandoned Waste Disposal Sites and Other Sites Suspected of Being Contaminated in Western Germany

Federal state	Closed or abandoned waste disposal sites	Other sites suspected of being contaminated[a]	Date
Baden-Württemberg	6,500	?	9/1988
Bavaria	482	73	9/1988
Berlin	332	1593	10/1988
Hanseatic City of Bremen	74	169	12/1988
Free Hanseatic City of Hamburg	1,550	290	12/1988
Hesse	5,123	61	9/1988
Lower Saxony	6,200	?	12/1988
North Rhine-Westphalia	8,639	3809	12/1988
Rhineland-Pfalz	7,528	?	1/1989
Saarland	1,728	1868	12/1988
Schleswig-Holstein	2,358	?	9/1988
Total	40,514	7863	

[a]Mainly industrial sites.
Source: SRU [1].

considerably from state to state. To make this point clear, if we were to evenly distribute the entire number over the area known then as the Federal Republic of Germany, which constituted 248,708 km², each square having sides 2.2 km in length would accommodate one contaminated site. Furthermore, taking into consideration that approximately 15–20% of the country's surface area represents a drinking water catchment area, and at present 7.7% has been designated as wellhead protection areas by law (it will be a total of 14% of the country's surface area in the near future) [2], it becomes clear that there will likely be abandoned waste sites within many recharge areas of drinking water wells and even in wellhead protection areas. Assuming that the recharge areas of drinking water wells as well as the abandoned waste sites were statistically equally distributed over this area of western Germany, there would still be between 7500 and 10,000 contaminated sites located within drinking water catchment areas. Specific figures have been supplied for Baden-Württemberg, where 1000 of the known 6000 abandoned waste sites (15.4%) lie within designated wellhead protection areas [3].

II. TYPES OF WASTE AND LEACHATE

The composition of waste that was dumped before the Waste Disposal Act went into effect is not generally known. However, it was probably made up mainly of municipal solid waste and construction waste, as these two types of waste did and still do represent the largest single volumes produced in that area [3]. The possibility, however, can never be excluded that even hazardous industrial waste was dumped at a waste site that today would have to be disposed of as special waste. Moreover, ordinary municipal solid waste and construction waste contain components that can lead to dangerous seepage; examples include paint containers with solvents and batteries with heavy metals.

Waste sites are open systems and therefore leak substances into the environment. Among the various types of emissions—solid (dust, etc.), liquid (represented by gravitational seepage), and gas (e.g., methane)—seepage represents the largest risk potential because of

- The wide range of unknown substances dissolved in it
- The high concentrations of many substances, mainly resulting in high seepage rates leading to extensive pollution of groundwater
- Its entry into groundwater, which represents the most important source of drinking water for western Germany ($>70\%$).

III. RISK POTENTIAL AND ENDANGERED RESOURCES

The risk potential of contaminant seepage into groundwater can be regarded as twofold; there are both acute and latent dangers (see Fig. 1). The acute

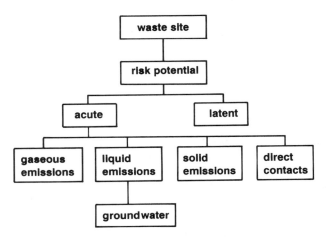

Figure 1 Risk potential of waste sites. The acute liquid seepages into groundwater (leachate) pose the largest risk potential.

risk is a consequence of existing contaminant seepage. The latent risk potential, however, comes from substances encapsulated in the mass of waste, such as in containers, which could be released, for example, if the container rusts, creating an acute risk potential.

Contaminant leachate from waste sites poses a potential danger for people and resources that either are presently exposed to polluted groundwater or could be exposed to it in the future. This primarily refers to human beings who are exposed by the intake of contaminated groundwater in the form of drinking water. Another route of ingestion that poses a danger to people is the consumption of plants that were irrigated with contaminated groundwater or the intake of other links in the food chain. Endangered resources also include surface waters into which contaminated groundwater discharges, the flora and fauna living in them, and buildings that extend into contaminated groundwater and are therefore subjected to a higher degree of corrosion than would normally be the case.

IV. STRATEGY OF EXAMINATION AND EVALUATION

The risk potential of a waste site can be determined only by investigating the emissions and leakages emanating from it. Sample taking and chemical analysis are prerequisites for such determinations. This requires the chemical analysis of samples of groundwater obtained downgradient from the site.

Attempting to determine the risk potential in any other manner, such as by using archives of official agencies or by relying on verbal information furnished by the public, can only be a first step in setting priorities with respect to the sequence of examining the sites. This is due to the fact that such information is usually far from complete, so the uncertainty in evaluating the dangers would be unacceptably large. Also, attempting to determine the risk potential by examining the mass of refuse itself cannot be recommended because of the great degree of inhomogeneity of the mass, which requires that a great number of samples be taken and analyzed. Moreover, there is a chance of increasing the acute risk potential by damaging waste containers in the sampling process.

Unfortunately, contaminant seepage from waste sites into groundwater is still being discovered by chance for the most part—for example, when the sudden emergence of toxic substances in the untreated water of a drinking water well is under investigation. Thus there are no systematic solutions being initiated for identifying which waste sites are a particular danger for groundwater. A solution can be approached after setting priorities, only if the groundwaters downgradient from all waste sites are given highest priority in systematically examining contaminant seepage (criterion example: location of the waste site within the recharge area of a drinking water well).

Taking into consideration the large number of waste sites where the groundwater has to be investigated even after priorities are set, the large number of contaminants that could be present in the groundwater, a rational basis for a phase-oriented examination strategy must constitute the objective. Following primary investigations predominantly concerned with hydrogeology and types of waste, the first analytical step must elucidate whether contaminant seepage has reached the water table. This first analytical step, known as *screening*, must be carried out for all groundwater to be examined (see Fig. 2). Screening must be inexpensive; it must be able to conclusively determine seepage from waste with the least number of test parameters. The second analytical step then has to be taken only at the sites where it has been determined by screening that the groundwater is being affected by seepage. The second step must be more extensive and able to proportionately characterize

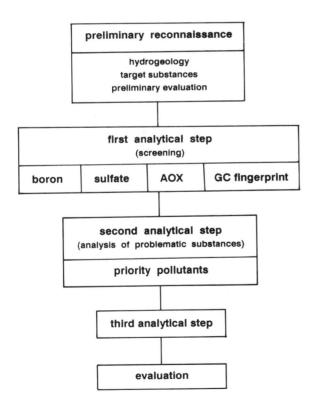

Figure 2 Hierarchical procedure for investigating and evaluating the acute risk potential due to contaminant leachate from waste sites into groundwater.

the seepage influence (analysis of problematic substances or priority pollutants). A third and more comprehensive analytical step, which is ordinarily more expensive (detailed analysis), must be used to register details relevant to the problematic sites identified by the analysis of problematic substances during the second analytical step.

This type of systematic and uniform procedure for registering sites that pose problems for groundwater must be linked to a uniform evaluation strategy for the findings. Remedial action must be carried out, particularly at sites that have been identified as dangerous to humans, the environment, buildings, or other endangered resources. In evaluating contamination, we have to consider substance-specific values, such as the behavior of the substances in groundwater and their toxicity, as well as site-specific values, such as the utilization status of the groundwater and the concentration values to be expected at the site of utilization.

One such self-sufficient comprehensive examination and evaluation concept for waste-site emission will be introduced in the following pages (see Fig. 2). The scheme is divided into

1. Recognition of a possible effect on groundwater
2. Characterization of an established effect on groundwater
3. Evaluation of the groundwater contamination

This scheme should be followed at a large number of waste sites for systematic examination; however, it can be used at any time to examine individual sites. Up to the present time, it has successfully been used at approximately 100 waste sites in western Germany, especially in unconsolidated sediment aquifers.

V. PROCEDURE

A. Recognition of a Possible Effect on Groundwater

1. Preliminary Reconnaissance

The phase of preliminary reconnaissance is the start of groundwater examination downgradient from an identified waste site. Aside from determining the age, extent, volume, contents, and history of the waste site in each case, the evaluation of available hydrogeological documents belongs to this phase. This evaluation must be aimed at registering the groundwater transport conditions and at determining the groundwater's possible upgradient and downgradient ranges and their variability in the vicinity of the waste site. When there are groundwater observation wells suitable for taking representative groundwater samples located downgradient, they should be used for monitoring; however, as a rule it is necessary to demarcate drilling sites and to install one or more monitoring wells.

Establishing so-called target substances

The chemical analyses that are to be conducted on groundwater basically consist of two equally valuable parts:

1. The hierarchically systematized analysis for scientifically confirmed and defined substances
2. Specific analysis for substances that are most likely present, as determined by the results of preliminary reconnaissance, and that can be expected to appear in seepage—the so-called project-specific target substances

Target substances are inorganic and organic substances or classes of substances, such as industrial residues from specific sectors of industry located near the site, that make it possible to conduct situation-oriented, and therefore specific, analyses. Determining the target substances is an essential part of preliminary reconnaissance.

Preliminary evaluation

Preliminary evaluation of waste sites, aimed at setting priorities, can begin following the conclusion of the preliminary reconnaissance phase by taking into consideration information on hydrogeology, utilization status, type of deposited waste, and other data. Any preliminary evaluation that is based exclusively on documents and omits chemical analyses merely provides an indication of risk potential. The actual burden on groundwater can be ascertained only by taking samples directly downgradient from the site and chemically analyzing them.

2. First Analytical Step: Screening

The determination of only four parameters is sufficient for a reliable recognition of possible effects on groundwater quality caused by seepage from a waste site. These are:

- Boron (concentration)
- Sulfate (concentration)
- Adsorbable organic halogen (AOX) (concentration)
- Gas chromatographic fingerprint (number of peaks and total peak area)

If the values obtained for these parameters lie significantly above those of the regional background (see Table 2), then it must be assumed that contaminants are present in the groundwater, and the second analytical step must be initiated. If negative findings are obtained, additional analysis for target substances should be conducted. If these results are negative also, and the groundwater observation well is, in fact, directly downgradient from the site in question, an effect of contaminant seepage can be excluded for

Table 2 Natural Background Concentrations of the Four Screening Parameters and Values Indicating an Influence from Waste Site Leachate for Groundwater in Pleistocene Unconsolidated Sediment Aquifers in Germany

	Boron (μg/L)	Sulfate (mg/L)	AOX (μg/L)	GC fingerprint (total peak area)[a]	
				ECD	FID
Natural back-ground (not influenced)	<20	<70	<10	<30	<5
Influenced by waste site leachate	>50	>150	>20	>60	>10

[a]mV × time.

the time of examination. The groundwater should be checked, however, at reasonable intervals—every 1, 5, or 10 years.

The question arises as to why precisely these four parameters are sufficient for recognizing seepage in groundwater. As mentioned above, municipal solid waste and construction waste are the largest components of the overall waste produced in western Germany, so it can be assumed that they are contained in virtually every waste site. Extensive investigation of groundwater directly downgradient from "unmixed" sites for municipal solid waste found the following inorganic parameters in significantly increased concentrations: K, Na, Mg, NH_4, Mn, B, Cl, NO_3. Comparable investigations at "unmixed" sites for construction waste found significantly higher contents of Ca and SO_4 [4].

Significant increases occur in essentially all parameters germane to municipal solid waste or to construction waste, so that the determination of one parameter is sufficient to recognize any relevant seepage. Among the parameters indicating contaminant seepage from municipal solid waste into groundwater in Pleistocene unconsolidated sediments, one of the most important aquifers in western Germany, boron is the one particularly well suited for this, because it exhibits a low natural background concentration (<20 μg/L) (see Table 2) and functions as a virtually ideal tracer. Sulfate was chosen as the corresponding parameter for indicating seepage from construction waste, although problems may arise, caused by high contents of natural sulfate or in the reduction zones of waste sites, because of possible sulfate reduction to sulfide.

Thus, boron and sulfate encompass the entire range of inorganic leakages from municipal solid waste and construction waste and are representative of many other parameters. When these two parameters were used in almost 100 projects, there was only one case in which the results with respect to them were negative and problematic inorganic substances were found during subsequent detailed examinations. This involved a high lead concentration in the groundwater. However, lead was a target substance here; specific analysis for it had been included in the first analytical step because of the lead glass industry located at the site.

Further extensive examinations have shown that the range of organic substances contained in seepage must be listed according to classes of substances rather than types of waste. The class of halogenated organic compounds can be included quite well in the group parameter AOX (adsorbable organic halogen), so that an increased AOX value indicates seepage containing halogenated organic compounds.

There is no corresponding organic group parameter for nonhalogenated organic compounds. Therefore, still another method must be used that involves the least possible amount of analytical work. Gas chromatographic fingerprints have proved to be a good compromise between the work involved and the quality achieved in detecting this class of substances in groundwater. This method employs appropriate solvents, such as pentane, as extractants for contaminated groundwater. The gas chromatograms obtained for an extract, recorded in parallel by an electron capture detector (ECD) and a flame ionization detector (FID), are evaluated only with respect to the number of peaks and the total peak area. When certain threshold values are exceeded (see Table 2) the second analytical step has to be initiated.

With a total of only four parameters—boron, sulfate, AOX, and GC fingerprints—an influence of waste seepage on groundwater can be determined unambiguously. Moreover, initial conclusions with respect to the composition of waste at the site are possible, an increased boron content indicating municipal solid waste, an increased sulfate content pointing toward construction waste, and an increased AOX content or a strongly structured GC fingerprint signifying problematic organic waste. Furthermore, the results of screening steer the extent of investigation and therefore the costs involved in the second analytical step. If, for example, only the inorganic screening parameters boron and sulfate give positive results and both of the organic parameters prove negative, then the second step has to test exclusively for problematic inorganic substances, forgoing all analysis on organic substances.

There are limits to the use that can be made of both inorganic screening parameters, particularly in cases of high natural background concentration—with boron, for example, in aquifers of marine sedimentary rocks and with sulfate in sedimentary rocks containing gypsum. As for the organic parameters, the GC fingerprint can pose problems, especially when extracts are analyzed with only one selective solvent and only one type of detector. However, contaminant seepage virtually always involves a mixture of many substances, of which several, if present, will be consistently identified by the above-mentioned methods. As experience has shown, the probability is extremely slight that only substances belonging to the class of nonhalogenated organic substances are present that cannot be extracted with pentane, for example, or that cannot be detected by ECD or FID.

Inductively coupled plasma (ICP) has proved to be a modern, reliable, and fast method of analyzing for boron, just as ion chromatography (IC) is for sulfate. Both are multicomponent systems that simultaneously provide measurement values for a number of inorganic parameters, the concentration of which is to be interpreted concurrently during the first analytical step. The first analytical step can therefore also be defined as a combination of methods with which it is possible to unambiguously recognize contaminant seepage from waste sites in groundwater:

ICP/AES (inductively coupled plasma/atomic emission spectroscopy)
IC (ion chromatography)
AOX (adsorbable organic halogen)
GC fingerprint (gas chromatographic fingerprint)

Recent experiences have shown that AOX should be coupled with the headspace method in the first analytical step. On the one hand, it is an inexpensive method that is presently used for routine analysis in many laboratories; on the other hand, the headspace method detects many of the volatile organic compounds that appear most frequently in groundwater downgradient gradient from waste sites. The difference between the AOX results and the sum of the individual volatile halogenated substances detected with the headspace method provides an important indication with respect to the type and extent of groundwater contamination. Furthermore, the headspace method also detects nonhalogenated substances such as benzene, toluene, and xylenes that are overlaid by the solvent peak of the pentane extract and therefore can be missed during screening.

B. Characterization of a Possible Effect on Groundwater

1. Second Analytical Step: Analysis of Priority Pollutants

If the screening results should prove positive with respect to inorganic and/or organic and/or target substances, thus indicating contaminant seepage in groundwater, the subsequent goal has to be the characterization of the contamination. Characterization does not, in this context, refer to the complete analysis of all substances contained in the water. Considering the several hundreds of thousands of substances that theoretically could occur in waste and the resulting seepage, this would not even be possible. Characterization is used here in the sense of the specific analysis of a limited number of substances that are most likely to be found in groundwater downgradient from waste sites (high detection frequency) and that are toxicologically relevant.

Selection of relevant inorganic and organic substances for the second analytical step is based on statistical evaluation or on a large and representative number of extensive groundwater analyses downgradient from waste sites. They are described in detail by Kerndorff et al. in Chapter 9 of this volume. The basic principle can be summarized as follows. Substances that are found more frequently (high detection rate) in a concentration exceeding the relevant concentration of 0.1 μg/L in groundwater downgradient from waste sites are often more water-soluble and more migratory in groundwater than others. Substances with a detection rate of >0.1% are described as *principal contaminants* and are of critical importance for the emission pathway, waste–seepage–groundwater–drinking water. The number of principal contaminants includes about 130 organic and 20 inorganic substances. Those principal contaminants that are also of toxicological relevance are described as *priority pollutants*. Dieter et al. [5] describe a method for evaluating the toxicity of groundwater contaminants in drinking water. This method evaluates toxicity on a scale ranging from 0 to 100 and forms the basis for selecting priority contaminants. At the moment, the toxicological evaluation of each principal contaminant is conducted according to this method. Even now, we can see that the number of priority contaminants that enable us to detect the toxicologically important substances in groundwaters will clearly be less than 50.

On the basis of presently available but still incomplete data, the inorganic and organic substances listed in Table 3 have been declared to be priority pollutants. These are the object of the specific analysis in the second analytical step.

The analytical methods used for detecting the inorganic priority pollutants are generally AAS (atomic absorption spectroscopy) and ICP/AES for metals or IC for anions. Since ICP/AES and IC are usually used in the first analytical step for determining boron and sulfate, many results for inorganic priority pollutants have already been obtained, so we can distinguish between the first and second analytical steps only with respect to the parameters that are to be determined. The same is valid for analyzing the problematic organic substances when the headspace method has been applied in the first analytical step, because many of the organic priority pollutants are determined with this method. The less volatile organic priority pollutants are generally identified and quantified by GC analysis of extracts with ECD and FID running in parallel with reference substances.

Table 3 Inorganic and Organic Priority Pollutants Ranked According to Their Hazard Potential on Groundwater[a]

	Standardized detection frequency[b]	Standardized mean concentration[c]	Toxicological evaluation number[d]	Product
Inorganic				
As	68.9	43.8	100	301 782
Ni	72.7	35.4	61	156 988
Cr	62.5	34.7	55	119 281
NO₂	35.8	61.9	52	115 233
Pb	39.5	33.7	82	109 154
Cu	57.5	37.9	26	56 661
NO₃	55.4	79.0	8.8	38 514
Zn	43.2	54.1	9.6	22 436
Cd	9.7	27.9	58	15 697
Organic				
Benzene	63.3	32.2	100	203 826
Vinyl chloride	38.7	42.0	100	162 540
Trichloroethene	88.4	29.7	53	139 150
Tetrachloroethene	100.0	23.4	39	91 260
Dichloromethane	32.8	56.4	43	79 547
Tetrachloromethane	31.8	21.5	100	68 370
Trichloromethane	48.1	19.2	56	51 717
trans-1,2-Dichloroethene	16.8	37.5	51	32 130
Chlorobenzene	28.5	24.1	43	29 535
1,4-Dichlorobenzene	27.0	22.2	49	29 371
Ethylbenzene	25.0	26.3	43	28 273
p-/m-Xylene	49.8	22.2	17	18 795
1,2-Dichlorobenzene	27.0	13.5	42	15 309
1,3-Dichlorobenzene	17.5	17.9	36	11 277
Phenol	18.1	18.0	32	10 426

[a]The hazard potential on groundwater is defined as the product of the standardized detection frequency, the standardized mean concentration, and the toxicological evaluation number [5]. All substances are listed with a hazard potential on groundwater greater than 10,000.
[b]Maximum detection frequency (tetrachloroethene: 46.21%) is set to 100.
[c]Standardized on a logarithmic scale: 0.1 μg/L is set to 1 and 1 g/L to 100; natural background concentrations were subtracted (inorganics).
[d]According to the toxicological evaluation model from [5].

If no priority pollutants are found in concentrations exceeding the threshold values during the second analytical step, then they can be excluded with a reasonable degree of certainty that no other toxicologically important substances will be found in critical concentration when detailed analysis is conducted. A further, detailed investigation does not have to be carried out in this case. The waste site cannot be listed as endangering groundwater, and its status should be analyzed only at reasonable intervals.

If priority pollutants are found in relevant concentrations during the second analytical step, it must be decided in each individual case whether a third, more specific, analytical step should be initiated. In virtually every case, however, the results of the first and second analytical steps are sufficient for a comprehensive evaluation of the site because the criteria that led to the choice of the determining parameters have been statistically substantiated.

2. Third Analytical Step

A third analytical step usually involves the use of more costly methods, such as GC/MS or HPLC, and is necessary only in individual cases. For instance, a third analysis is indicated when the second analytical step reveals the presence of certain substances or classes of substances that need further elucidation.

C. Evaluation of Groundwater Contamination

When evaluating the contamination of groundwater, the toxic substances contained in the contaminated groundwater and their behavior in the groundwater flow system as well as the specific circumstances related to the site, including the utilization status and possibilities of exposure, all have to be considered. If the groundwater downgradient from a waste site is not being used at present, no utilization is planned, and the groundwater does not threaten any other endangered resource, then, as a rule, the high costs involved in remedial action are not warranted.

Moreover, a clear distinction between the contamination site and the utilization site of the groundwater or the site at which an endangered resource can be damaged by exposure to the pollutants is of primary importance for the evaluation. The contamination site is where the contaminant gains access to the groundwater in high concentration, that is, the actual waste site and possibly the downgradient area directly bordering on it. The groundwater utilization site is, in the majority of cases, a more or less considerable distance from the contamination site; the pollutants are subjected to diverse reversible and irreversible physicochemical processes during subsurface transport, such as dilution, dispersion, adsorption/desorption, absorption, biotic and abiotic degradation, and metabolism.

It is necessary during evaluation to appraise the temporal and spatial effects of these processes and to estimate a potential pollutant concentration at the site at which the groundwater is utilized.

1. Determining a Potential Pollutant Concentration at the Utilization Site

The transport of dissolved substances in groundwater is extremely complex, but it can be estimated to a certain extent by solving the transport equation [6] with the aid of computers. Substance transport in porous media has been described by others [e.g., 7] and will not be the topic of any further discussion here.

In determining the potential pollutant concentration at the utilization site, one should proceed by using numerical models as in the procedure in chemical analysis. One should start with the simplest methods possible and increase the complexity according to the results.

The simplest case of seepage from a waste site can be described as a permanent point source injection into a steady-state groundwater flow system. The following variables must be known to evaluate the two-dimensional distribution of concentration:

Aquifer

Saturated thickness of aquifer (m)
Effective porosity (dimensionless)
Average groundwater velocity (m/day)
Longitudinal dispersivity (m)
Tranverse dispersivity (m)

Substances

Source strength (g/day)
Retardation factor (dimensionless)
Decay constant (day^{-1})

Molecular diffusion $(m^2/H \text{ sec})$ need be taken into consideration only as an additional factor when the velocity of groundwater flow is extremely slow.

With the transport equation and the values for the above-mentioned variables, the potential pollutant concentration can be calculated for every point in time, t (days), at every point $P(x,y)$ (m). Kinzelbach [7] supplies us with a simple computer program written in Basic for doing this.

A major difficulty arises in practice because the values for the variables are frequently unknown. The values of the variables describing the aquifer have to be determined by hydraulic testing. If this is not possible, they must be estimated. This is usually the case for macrodispersion and especially for the ratio of longitudinal to transverse dispersivity. Uncertainties when estimating parameters should be compensated for by worst-case assumptions.

The values for the substance-specific variables are still quite problematic. The source strength can be calculated with the help of the computer program in that a concentration measured at a known distance downgradient from the waste site at point $P(x,y)$ is inserted into the appropriately restated transport equation. The retardation factor and the degradation constant (half-life) are totally unknown for the majority of contaminants, and furthermore, they are dependent on the environment (pH, Eh). To be on the safe side, the worst case is assumed, and a pollutant with unknown data is treated at first as an ideal tracer. That could lead, however, to remedial measures resulting in great expense, for example, to protect a well field although the pollutant may never reach it. Until such a time arrives when comprehensive data concerning mobility, degradation behavior, etc. are available for all the principal contaminants in all the most common geochemical aquifer environments, the behavior of substances in aquifers can be approximated by determining substance-specific potentials.

2. Determining Substance-Specific Potentials with Respect to Behavior in Aquifers

The migration of solutes in groundwater is essentially a result of their mobility, sorption, and persistence, all of which are considered "potentials" that can be determined from various physicochemical data for that substance. Such data for substances are, for example, water solubility and vapor pressure as approximate dominant values for mobility potential just as the octanol/water partition coefficient K_{ow} and the first-order molecular connectivity index $^1\chi$ [8] are for the sorption potential, an opposed correlating value to mobility. A substance's potential for abiotic degradation can be described by using COD (chemical oxygen demand) and BOD (biological oxygen demand) for biotic degradation, thus sufficiently describing the persistence potential.

In order to make the parameters, which have different value ranges, comparable and linkable, the individual values available for the 150 principal contaminants are standardized on a scale ranging from 0 to 100. Should the value ranges encompass some orders of magnitude, the logarithms are used for standardization. Zero corresponds to the next smaller natural number than the lowest value available, and 100 corresponds to the next larger natural number than the largest value available.

Six standardized values between 0 and 100 result for each of the 150 principal contaminants (water solubility, vapor pressure, K_{ow}, $^1\chi$, COD, and BOD). In addition, the two standardized values for the sorption potential

(K_{ow} and $^1\chi$) have to be subtracted from 100 so that an augmentation of these values will signal an increase in the risk potential for groundwater. For the calculation of the three potentials, two of the standardized values are used at a time: The mobility potential of a substance is defined as the product of the standardized water solubility and the standardized vapor pressure, the sorption potential as the product of the K_{ow} and the first-order molecular connectivity index, both standardized and subtracted from 100. The two products are again standardized between 1 and 100. The persistence potential is defined as the mean of the standardized COD and BOD.

A migration potential for a substance in groundwater is calculated by first averaging the mobility and the sorption potentials and then multiplying the resulting mean by the persistence potential, the theoretical maximum being 10,000. If either the persistence potential or the product of mobility and sorption potential is zero, the substance's overall migration potential in groundwater is zero as a result of multiplication. This is reasonable, because a substance that is not persistent, for example, cannot migrate significant distances even though it may be theoretically highly mobile and not very accumulable.

Results of this procedure are shown in Table 4. The resulting figures make the substances comparable with respect to their migration potential in groundwater. Using this migration potential, retardation factors can be assessed and therefore make it possible to calculate the degrees of pollutant concentration at the site of groundwater utilization.

3. Threshold Values Requiring Action

After measured or estimated pollutant concentrations are known for the utilization site, the decisive question still remains: What is the concentration at which remedial action must be initiated? We refer to borderline concentrations as the *threshold values requiring action* to clearly distinguish them from standard values and limits. Threshold values requiring action are dependent on the type of groundwater utilization or on the type of endangered resources that can be damaged by contaminated groundwater. A much higher concentration can be tolerated with respect to expected damage to underground construction components than would be the case with groundwater used as drinking water.

In the case of drinking water, the legal limits cited in the German drinking water regulations correlate with the threshold values requiring action; this means that as soon as a limit is exceeded, remedial action must be initiated. However, in western Germany, limits for drinking water exist for only a few contaminants. This is why typical threshold values have to be deduced predominantly from the substance's toxicity. The above-mentioned toxicological scoring system [5] can be used as an aid in accomplishing this.

Table 4 Organic Priority Pollutants Ranked According to Their Groundwater Migration Potential Which Is Calculated Using Substance-Specific Physicochemical Parameters

	(1) Mobility potential[a]	(2) Sorption potential[b]	(3) Persistence potential[c]	(4) Groundwater migration potential[d]
Trichloroethene	59.4	56.7	77	4470
Tetrachloroethene	47.6	51.3	83	4104
Trichloromethane	71.5	65.0	47	3208
1,4-Dichlorobenzene	35.2	35.7	75	2659
1,3-Dichlorobenzene	42.6	35.3	66	2571
1,2-Dichlorobenzene	39.4	35.6	50	1875
Benzene	62.6	53.6	18	1046
Chlorobenzene	50.7	42.3	22	1023
Ethylbenzene	45.9	37.0	13	539
trans-1,2-Dichloroethene	61.3	69.8	5[e]	328
Dichloromethane	78.5	77.1	1[e]	78
Tetrachloromethane	59.7	54.1	1[e]	57
Phenol	54.4	56.8	1	56

[a]Mobility potential = std. water solubility × std. vapor pressure.
[b]Sorption potential = std. P_{ow} × std. $^1\chi$.
[c]Persistence potential = (std. COD + std. BOD)/2.
[d]Groundwater migration potential = $\{[(1) + (2)]/2\} \times (3)$.
[e]Persistence potential calculated using BOD only.

This system rates the substances on a toxicological scale ranging from 0 to 100. Carcinogenic substances always receive a score of 100 and should not be contained in drinking water, that is, their concentration should be less than the detection level. The lower the toxicity level, the higher the concentration that can be tolerated in drinking water and the higher the threshold value requiring action.

The establishment of threshold values requiring action for other groundwater uses must be given high priority. As a rule, they may be higher than for drinking water.

VI. CONCLUSIONS

Approximately 50,000 waste sites have been registered in western Germany, most of them predating the Waste Disposal Act. Their leachate endangers groundwater that is the primary source of drinking water in the region. The examination and evaluation concept presented here makes it possible to systematically and comparably detect, characterize, and evaluate this large number of waste sites with respect to their groundwater contamination potential.

This analytical method has been particularly valuable in the case of waste sites located in unconsolidated sedimentary and similar aquifers. The conversion of the degrees of concentration mostly measured at the groundwater utilization site by solving the transport equation poses difficulties when data concerning the pollutant's subsurface behavior are missing. A great deal of research is needed in this area; therefore, potentials of the pollutants pertaining to subsurface behavior, calculated from physicochemical data on the substances are used in the interim.

There is also a great need for research in the area of establishing threshold values requiring action for initiating remedial action with respect to the pollutant's toxicity and the greatly varying uses of groundwater. Natural background concentrations also have to be taken into consideration when establishing threshold values requiring action for inorganic substances. If, for example, the natural background concentration of iron is higher than the limit for drinking water, the latter value cannot be the threshold value requiring action. As a goal for the future a threshold value requiring action must be set for each principal contaminant, depending on groundwater use and exposure as well as on the various aquifer types.

When compiling such a catalog, the synergisms of the various pollutants have to be taken into consideration. The pollutant with the highest risk potential should always be used as the triggering criterion for remedial activities. However, it can also occur that the individual substances have concentrations below the threshold values requiring action but their sum poses a danger. Attention must also be paid to the transformation products of pollutants that form during groundwater transport, due to the fact that these can be more dangerous than the original substances. For example, tetrachloroethene in an ambient environment with a low oxygen content may be subjected to reductive metabolism that can lead, via trichloroethene and cis-1,2-dichloroethene, to vinyl chloride, which is a very hazardous substance in terms of human consumption.

All of these important tasks for the future should not cover up the fact that groundwater protection is a topic of primary significance. The goal is the conservation of an all-encompassing natural groundwater quality that is affected to the least possible extent by civilization. Limits for drinking water and threshold values requiring action are always higher than the natural background concentration with only a few exceptions. Contaminated groundwater with a contaminant concentration only slightly lower than these man-made limits can be accepted only in connection with the repair of existing damage. A large-scale attainment of these concentration limits must be avoided at all cost.

REFERENCES

1. SRU (Rat von Sachverständigen für Umweltfragen). Sondergutachten "Altlasten." *Drucksache 11/6191 des Deutschen Bundestags*, Verlag Dr. Hans Heger, Bonn, 1990, p. 304.
2. Müller, J. Wasserversorgung in Niedersachsen, Bestandsaufnahme, Probleme und Lösungsansätze unter besonderer Berücksichtigung des Grundwasserschutzes. *Veröffentlich. Inst. Siedlungswasserwirtsch. Braunschweig, 48*:1, 1990.
3. UBA (Umweltbundesamt). *Daten zur Umwelt 1988/89*, Erich Schmidt Verlag, Berlin, 1989.
4. Kerndorff, H., Brill, V., Schleyer, R., Friesel, P., and Milde, G. Erfassung grundwassergefährdender Altablagerungen—Ergebnisse hydrogeochemischer Untersuchungen. *WaBoLu-Hefte, 5/1985*:175, 1985.
5. Dieter, H.H., Kaiser, U., and Kerndorff, H. Proposal on a standardized toxicological evaluation of chemicals from contaminated sites. *Chemosphere, 20*:75, 1990.
6. Konikow, L.F., and Grove, D.B. Derivation of equations describing solute transport in groundwater. *U.S. Geol. Surv., Water Resour. Invest., 77-19*:30, 1977.
7. Kinzelbach, W. *Groundwater Modelling—An Introduction with Sample Programs in BASIC*, Elsevier, New York, 1986, p. 333.
8. Sabljic, A. On the prediction of soil sorption coefficients of organic pollutants from molecular structure—application of molecular topology model. *Environ. Sci. Technol., 21*:358, 1987.

III
SITE INVESTIGATIONS

11

A Review of Studies of Contaminated Groundwater Conducted by the U.S. Geological Survey Organics Project, Menlo Park, California, 1961–1990

Donald F. Goerlitz

Geological Survey, U.S. Department of the Interior, Menlo Park, California

The U.S. Geological Survey is mandated to evaluate water resources in the United States and has sponsored and actively participated in studies of contaminated groundwater systems. The Organics Project, Menlo Park, California, was begun in 1961 for the purpose of evaluating the quality of water resources. To date, the Project has conducted studies of contaminated groundwater at several locations, six of which are described here. The first, in 1965, was a study of groundwater contamination by chlorinated hydrocarbons from a dump site in Hardeman County, Tennessee. Two studies were of groundwater contamination by explosives in Kitsap County, Washington and Lyon County, Nevada. The three remaining studies were of contamination by wood-preserving chemicals: one at Visalia, California, a second at St. Louis Park, Minnesota, and the third at Pensacola, Florida. The summaries are presented in chronological order, and the salient points of each, including problems, downfalls, and details not included in the formalized reports, are given from a chemist's point of view.

I. INTRODUCTION

Less than two decades ago, groundwater contamination was one of the least recognized environmental problems in the United States. Lack of awareness may have existed because groundwater problems are not readily detected and pathways for contamination are not as noticeable as those affecting surface water. Contamination in groundwater is not only difficult to see but also difficult to evaluate; collecting needed information is very taxing, requiring extensive resources and expertise, and is usually further complicated by the hydrologic setting.

Groundwater contamination by synthetic and processed organic chemicals generally results from the unrestricted use of organic products and the practice of depositing waste on the land surface and in streams. Such point sources are usually associated with domestic and industrial disposal activities. Areal, or non-point, sources of contamination result mainly from applications of pesticides by agricultural workers, transportation authorities, and land management agencies. By their nature and proximity to the surface, porous surficial aquifers are clearly vulnerable to both point and areal sources of contamination.

Until recently, studies of groundwater hydrology and quality in the United States were almost exclusively associated with water supply interests. Today, sophisticated, computer-based groundwater models are widely used as engineering and management tools, not only for supplies but also for evaluating damage to and remediation of contaminated groundwater systems. The U.S. Geological Survey (USGS) is mandated to evaluate the water resources of the United States, including the quality of the water, and has actively sponsored and participated in studies of contaminated groundwater systems. A project to specifically determine the quality of water, to develop methods for the analysis of organic compounds in water resources, and to study the behavior of organic substances in water was established in Menlo Park, California in 1961. The first investigation of groundwater contamination by the Organics Project was at a pesticide manufacturing waste site in Hardeman County, Tennessee in 1965. This study was done in cooperation with the USGS, Tennessee District, and the U.S. Federal Water Pollution Control Agency. This first study was of a contaminated aquifer resulting from the disposal of chlorinated hydrocarbons in a landfill. Since then, the Menlo Park Organics Project has been engaged in similar investigations, six of which are discussed in this chapter. Of the five remaining studies, two were of aquifers contaminated by explosives wastes, and three were of aquifers contaminated by wood preservatives, creosote, and pentachlorophenol (PCP). All of these problems resulted from organic chemicals infiltrating porous surficial aquifers from lagoons and landfills. These studies are summarized herein to recount the evolution of the approaches to site assessment by the Menlo Park Organics

Project. Because the Project was established to study the quality of water with respect to organic substances, which is essentially an organic analytical problem, the experiences are reported from the perspective of the organic chemical analysis.

As might be expected, the early site assessments proceeded with difficulty and uncertainty. With experience, there developed an appreciation for the complexity of the problems associated with these studies. Now it is recognized that a scientific assessment of the fate and transport of organic solutes in groundwater is complex and requires a (complex) multidisciplinary and labor-intensive effort by experienced and knowledgeable investigators. Furthermore, even though there have been a number of intensive studies on the fate and transport of solutes in groundwater, the fundamental processes remain poorly understood and are very complicated, and interpretations still involve a great deal of uncertainty.

Almost all groundwater investigations begin with similar objectives: to define the magnitude and extent of contamination, and to forecast the fate of the pollution. Good, sound chemical data are requisite for meeting these objectives. The correctness of the results is directly and equally affected by the quality of the samples and the chemical data, and attention to both is essential. The data collection plan, including all aspects of sampling and analytical method selection, should be designed and reviewed in consultation with an analytical chemist. There is a tendency not to include chemists in the planning and sampling stages of groundwater investigations. A proclivity to compartmentalize the analytical function as a blackbox data service is naive and problematical. This presumption may explain the existence of burdensome and strict quality assurance and control procedures on chemical analysis and little if any on sampling and field practices. The burden is on the chemist to correct this imbalance, by insisting on reviewing field practices and sampling procedures and by pointing out unnecessary or irrelevant exercises.

In most instances, when environmental contamination is detected, many agencies and jurisdictions become involved, each having differing requirements for dealing with a particular occurrence. Frequently, specific agency objectives complicate any systematic approach to studying the dynamics of transport and the fate of contaminants with respect to a particular situation. Furthermore, the costs of an investigation are often overwhelming and place an urgency on expediting the study, leaving little room for more fundamental examination of the affected aquifer. This scenario occurred in almost every instance described in this chapter, except at the Pensacola site, which was preselected for research. As a consequence, many interesting and important observations were not included in the formal reports produced from these investigations. The purpose of this chapter is to include some of the more interesting details along with a summary of the general study histories at six sites where groundwater has been contaminated. The histories are arranged in chronological order and are presented from a chemist's point of view.

II. PESTICIDE WASTE DUMP, HARDEMAN COUNTY, TENNESSEE

A. Introduction

In October 1964, a pesticide manufacturer located in Memphis, Tennessee began disposing of waste material created during the synthesis of certain chlorinated hydrocarbons (pesticides) by shallow burial at a site in northeastern Hardeman County, Tennessee (Fig. 1). Public health officials expressed concern about the potential hazard for the local population and the potential for leaching of solutes into local surface and groundwater supplies. The manufacturer's representatives held the position that burial of wastes was a time-honored practice and posed no immediate hazard to public health. Presumably, there were no specific laws regulating such dumping practices at the time.

The pesticide manufacturing waste materials dumped at the site were described as still bottoms, spent catalyst, reactor solids, and contaminated trash. The wastes were said to result mainly from the manufacture of endrin, heptachlor, and heptachlor epoxide. No quantitative information was made available by the manufacturer. The accumulated wastes were placed in appropriate 200-L steel and fiber drums. The drums containing liquid were closed with lids to prevent spilling during a 110-km trip to the disposal site. Once at the site, the cargo, in its entirety, was dumped into excavated trenches or ditches approximately 3.5 m deep and 4.5 m wide. In the process of being dumped and covered with earth, many drums were ruptured, and waste was allowed to spill from the temporary containers.

B. Hydrogeologic Setting

The disposal site is located on a 300-acre parcel in northeastern Hardeman County, Tennessee, about 2.5 km northwest of the intersection of State Highways 18 and 100. (See Fig. 1.) At the time of the investigation, the delineated dumping area was on a bulldozed flat situated near the drainage divide along the western side of Pugh Creek Valley. The immediate surface drainage was generally toward the east. The creek passes northward approximately 150 m east of the 1964 property line. Topographically, the parcel is situated on an upland remnant of a fluvial terrace 25–30 m above Pugh Creek.

The terrace is composed of a sequence of nearly horizontal strata consisting of sand, silt, and clay and is illustrated in Fig. 2. Below 30 m, clay is

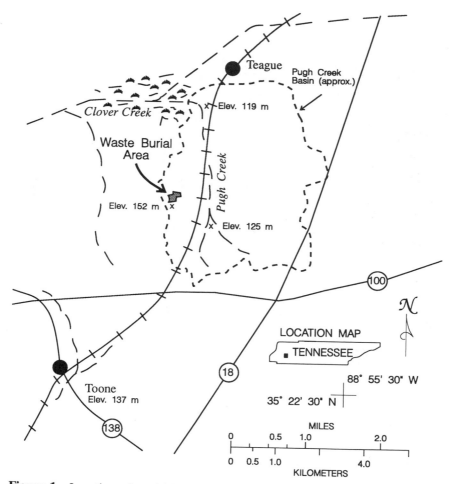

Figure 1 Location of pesticide waste dump, Hardeman County, Tennessee.

predominant. Two distinct water entities occur in a 75-m interval beneath the dump: a water table aquifer and an artesian aquifer. In addition, during the study, perched water was found at about 6 m below the surface throughout the immediate area following heavy precipitation. The water table aquifer is at a depth of about 30 m and slopes to the east. It is recharged by rainfall and discharges through springs and seeps and locally to Pugh Creek, a perennial stream maintained by groundwater during dry periods. The artesian

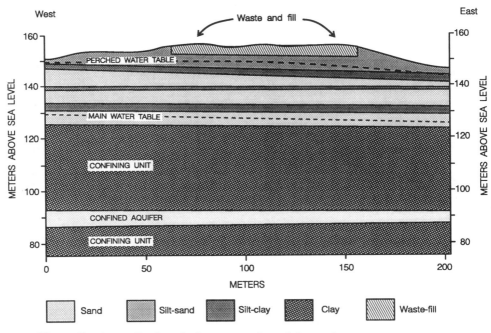

Figure 2   Generalized geologic cross section of dump site.

aquifer is at a depth of 60–70 m beneath the site and is separated from the water table aquifer by a 30-m-thick bed of silt and clay. The thickness and nature of this upper confining bed forestalls direct hydraulic exchange between the aquifers. Recharge of the artesian aquifer takes place mainly at an outcrop area several miles east of the site.

C. Study Plan

In late 1965, the U.S. Geological Survey was asked to investigate the disposal site to identify any potential problems. The request was set within the constraints of the following four questions:

1. Is there now contamination of the environment from disposal pits; if so, what part is now contaminated and to what degree?
2. What is the potential contamination hazard to local groundwater from percolation?
3. What is the potential contamination hazard to contiguous groundwaters?

4. Are there other hydrologic factors related to contamination hazard that have become evident directly or indirectly from this study, such as topography, drainage, or runoff intensity?

A plan of action was developed based mainly on hydrogeologic considerations and information gathered from the manufacturer and was set within the constraints of these four questions.

D. Sample Collection and Analysis

Precipitation and runoff data together with surface and aquifer samples, both water and sediment, were collected in the area of the site and at runoff points, including Pugh Creek. Test holes were drilled at selected locations around the buried waste but not directly through any filled trenches. Several test holes were drilled at inclinations of 30–45° from the vertical to obtain samples from beneath the trenches without penetrating the waste fill. Precautions were taken to avoid significant acceleration of contaminant migration by the test hole drilling. After the aquifer cores were collected, the holes were backfilled with bentonite and Portland cement. None of the test holes were drilled into the confining clay unit above the artesian aquifer. Initially, a 3.5-in. (9-cm, nominal) hollow-stem auger was used for drilling, and a split-spoon sampler was used to take core samples of the aquifer. After the first round of sampling, a Porter piston-drive sampler, also called a retracting-plug sampler [1], was used for the core sampling.

The components of waste in the samples were analyzed by contemporary methods [2]. The compounds were extracted from water using n-hexane followed by a pesticide cleanup procedure on an alumina column. Aquifer samples were kept from drying and were extracted as such, shaking first with acetone and then again after adding n-hexane. This extraction was repeated a total of three times. Gas chromatography (GC) was done using packed columns and both electron capture and combustion/microcoulometric detection. Glass columns 4 mm i.d. × 1.5 m long packed with coated Gas Chrom Q,* 60/80 mesh, were used for gas chromatography. The coatings on two different columns were QF-1 and DC-200 silicone liquid phases. Total chlorine determinations on the extracts were done by programmed temperature chromatography on a 4 mm i.d. × 8 cm long stainless-steel column packed with 60/80 mesh Gas Chrom Q coated with DC200 fluid. The column was heated by direct electrical resistance heating from ambient to 300°C in 1 min. Combustion/microcoulometric titration was used for

---

*The use of brand names in this report is for identification purposes only and does not imply endorsement by the U.S. Geological Survey.

determining the total chlorine in the accumulated combustion products from the column effluent [3].

Samples of waste, surface soil, surface runoff water, and aquifer material were collected and analyzed in May 1967. Gas chromatographic analysis for dieldrin, endrin, heptachlor, and heptachlor epoxide revealed that the surface samples were contaminated at concentrations on the order of a few milligrams per kilogram. Furthermore, all but one of nine grab samples of waste liquor taken from leaking drums were found to contain less than 1% of each of the target compounds. One sample, however, did contain approximately 20% heptachlor by weight. Ten test holes were drilled, but the cores from only one were found to hold concentrations of the four target compounds at levels similar to those found in surface soils. Data obtained from more than 100 analyses disclosed that the top part of each of the split-spoon core samples almost always contained higher amounts of contamination than did the bottom part of the same core. Visual examination of the cores revealed lighter colored, fine-textured material on top of most of the core samples, apparently surface soil that fell down the hole during sampling operations. These samples were judged to be compromised and of marginal use for describing the magnitude of the migration of the waste. Also, it was determined that surface material had been smeared down the sides of each core, further invalidating the core samples and the analytical data.

Examination of gas chromatograms obtained from the above samples using contemporary electron capture gas chromatography (ECGC) on a packed column revealed the presence of a large number, more than 40 distinct GC peaks, of individual components of the waste. Because of the complexity of the mixture, identification of the individual components was not practical using contemporary GC techniques. Gas chromatography coupled with mass spectrometry (GC/MS) was not available at this time. By employing a chloride-specific microcoulometric detector, it was found that nearly all the components displayed by the gas chromatograms were chlorinated. The waste liquor samples contained as much as 40–60% chlorine, as determined by combustion-micoulometric titration.

Because the first attempt to assess the contamination was less than satisfactory, a second round of test hole drilling, sampling, and analysis was done in November 1967. In this second round, very strict precautions were taken to avoid sampling artifacts from contaminated surface material. Drilling platforms and coverings were put in place to generally safeguard the sampling operations. The drilling augers and sampling equipment were carefully cleaned by washing with detergent and water and rinsing with acetone between drilling and sampling. A retractable-piston-drive sampler [1] used for collecting the cores is illustrated in Fig. 3. Cloth packing, placed in the annular space between the hollow auger and the A-rod connected to the core

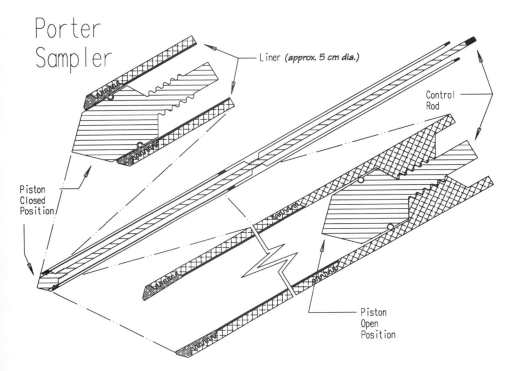

Figure 3 Retractable-piston-drive sampler, for collecting 5×15 cm cores.

sampler, was inserted to reduce the amount of surface material falling inside the apparatus during the drilling and sampling procedure. A processing area was constructed away from the drilling activity to reduce contamination during sample transfer and preparation for shipping.

E. Results and Discussion

Definitive organic analysis was essential in providing a comprehensive response to the questions posed for this investigation. It was soon apparent that selecting or targeting the manufactured products in an attempt to gauge the magnitude and extent of the waste material in the aquifer was an unsatisfactory option. The presence of relatively large amounts of other organic contaminants showed the inadequacy of the preselection of a limited number of target compounds. Another scheme was needed to meet the objectives of the study. The analytical chemists performing the analyses had not been included in the initial planning and sampling phase of the study but, fortunately, were in communication with the hydrologist in charge of the study. This allowed adjustments in the specific approaches and helped salvage some data that might otherwise have been compromised.

Analysis of the water and the aquifer sediments for the target pesticides by ECGC revealed the presence of a large number and amount of unknown "pesticide-like" compounds. By contrast, relatively small amounts of the target pesticides were detected. The difficulty of attempting to identify unknowns by comparison to selected standards is illustrated in Fig. 4. Analysis by microucoulometric GC specifically verified that the "pesticide-like" compounds were highly chlorinated. Later, new information provided by the manufacturer indicated that the waste consisted of aldrin, chlordene, chlorendic acid, chlordenechlorohydrin, hexachlorobicyclohepatdiene, hexachloropentadiene, isodrin, and minor amounts of a large number of other chlorinated

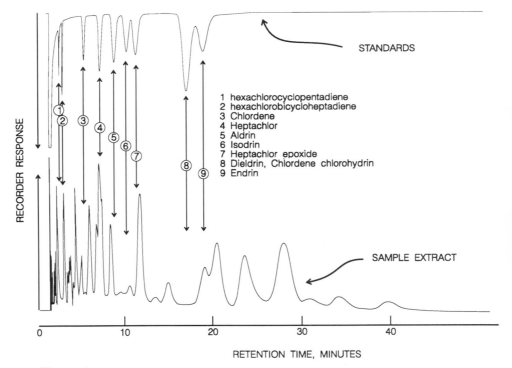

1 hexachlorocyclopentadiene
2 hexachlorobicycloheptadiene
3 Chlordene
4 Heptachlor
5 Aldrin
6 Isodrin
7 Heptachlor epoxide
8 Dieldrin, Chlordene chlorohydrin
9 Endrin

Figure 4 Electron capture gas chromatogram of aquifer sediment extract compared with selected standards.

adducts and products of the Diels–Alder chemical reaction used in the manufacturing process. Owing to the multiplicity of components in the samples, it was not practical to positively identify the individual compounds by contemporary packed-column gas chromatography, even with the element-specific microcoulometric detection. To get around this problem, an analytical method was devised to measure the total amount of organically bound chlorine in water and sediment extracts by oxidative combustion and specific, quantitative determination of the chlorine using microcoulometric titration.

Background levels of organically bound chlorine ranged from 0.1 to 0.5 μg/kg in aquifer sediments collected away from the contamination. In the contaminated sediments, chlorine was found to range as high as 6 g/kg at a depth of 10 m directly beneath the buried waste. A summary of these findings is illustrated in Fig. 5. Aquifer sediments collected away from the dump tested slightly alkaline, pH 7–8, but heavily contaminated samples tested acidic (pH 3.4) [4]. The aquifer porosity was 0.4, and the laboratory-determined hydraulic conductivity ranged between 0.01 and 0.7 m/day. An administrative report to the Federal Water Pollution Control Federation was prepared in 1967, followed by an open-file report published in 1972 [5].

F. Conclusions

Recognition of discrepancies between the original approach of specifying selected compounds and the evidence revealed during chemical analysis changed the conduct of the investigation. By tailoring the sampling techniques and chemical analysis to fit the field situation, the investigation provided much of the information sought in the beginning. This might not have occurred if the analytical chemists had not been in direct, daily communication with the hydrologist in charge of the on-site investigation.

In addition to detailed hydrogeologic and historical information, it is evident in lessons learned from this work that other considerations are important for a rational design to study groundwater contamination. Such considerations include the following.

1. A reconnaissance evaluation is needed to identify any conditions that may interfere with or compromise the objectives. Extent of surface contamination, chemical composition and physical condition of the source material, nature of the aquifer, physical setting of the sampling sites, and remoteness to sample processing facilities are some of the factors that need to be considered.
2. All aspects of the sampling plan should be subjected to interdisciplinary review, preferably by experienced groundwater investigators, and, very important, include analytical chemists—organic analytical chemists in this instance.

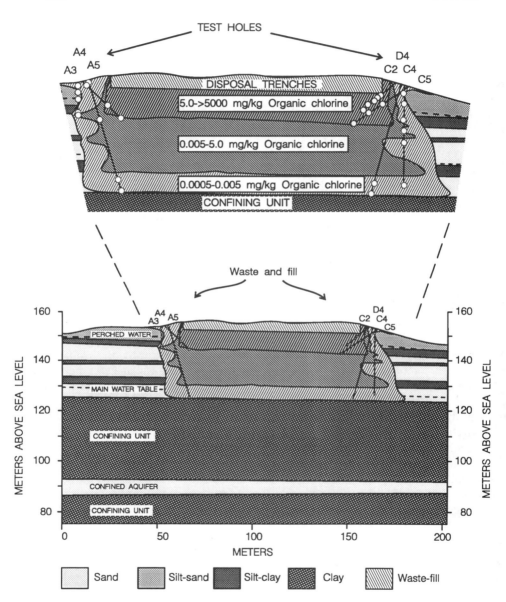

Figure 5 Cross section of test holes showing the distribution of total chlorine concentrations in the sediments.

3. Reliably accurate and specific information on the composition and behavior of contaminants in groundwater is a fundamental requirement if only target compounds are chosen to gauge the magnitude and extent of contamination of the system. An incomplete or poor choice of target compounds may be misleading and yield false negative findings.

4. The direct chemical analysis of the samples for all the components of contamination is the best choice for evaluating the magnitude and extent of contamination in an aquifer.

III. MUNITIONS PROCESSING WASTE DISPOSAL AREA, BANGOR, KITSAP COUNTY, WASHINGTON

A. Introduction

From 1966 to November 1970, the U.S. Navy steam-cleaned emptied surplus projectiles at a facility located at the Bangor Annex, Kitsap County, Washington (Fig. 6). The washings were allowed to drain to a surface depression or pondlike feature. The depression was sloped slightly to the south and was approximately 130 m long and 30 m wide. The wastewater entered by way of a channel or drainage ditch and flowed to a shallow, unlined 30 m² oblong pit area near the south margin of the depression. The ponded wastewater was reported to overflow during storms and occasionally fill a second elongated depression that extended another 100 m to the south. The pond was periodically allowed to dry, after which a layer of surface material was removed for burning at other locations.

Upon learning that the cleaning methods could possibly contaminate the local groundwater, the Navy ceased the operation in November 1970. In February 1972, 1–1.5 m (approximately 380 m³) of the contaminated surface layer of soil was removed to a disposal area. The excavation was backfilled using clean glacial drift obtained from outside the Annex. An estimated 230,000 kg of explosives had been washed into the catchment area, but there were no estimates of how much had been removed from the pond for burning or disposal.

The explosive material washed from the projectiles, designated as Composition B, consisted of a mixture of 60% trinitrotoluene (TNT) and 40% cyclonite (RDX). Both TNT and RDX are moderately soluble in distilled water, approximately 100 mg/L at 20°C. The structural formulas of TNT and RDX are shown in Fig. 7. Other considerations notwithstanding, both of these compounds were judged capable of contaminating the groundwater. In 1971, following preliminary well water testing in the vicinity, the Navy Bureau of Medicine contracted with a commercial laboratory to perform

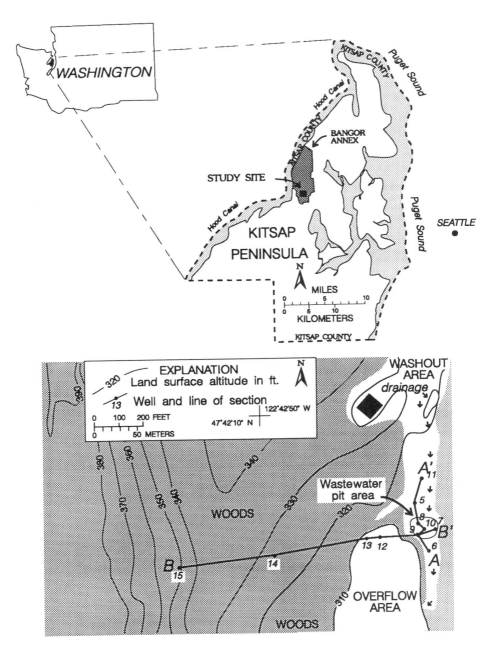

Figure 6 Location map and drawing of the study site, Bangor Annex, Kitsap County, Washington.

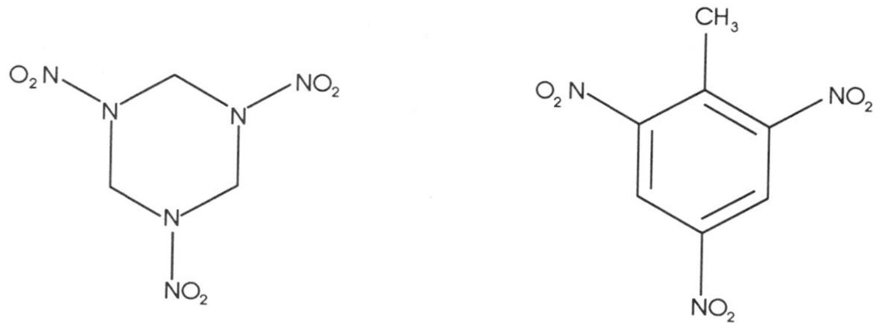

RDX, CYCLONITE, HEXOGEN
Hexahydro-1,3,5-trinitro-1,3,5-triazine
trinitrotriazine
Water solubility, 100 mg/L at 0 °C.

TNT
2,4,6-trinitrotoluene
trinitrotoluene
Water solubility, 100 mg/L at 0 °C.

Figure 7   Structural formulas and chemical and common names for RDX and TNT.

chemical analysis of water samples collected from existing ells. Interpretation of TNT/RDX analytical data from the commercial laboratory was inconclusive, because the subject compounds were detected in well water known to be free of contamination, and the data suggested no sensible relation between the samples and the controls. Subsequently, the Navy drilled a number of test holes beneath the drainage pit area and produced reliable evidence of aquifer contamination.

In 1973 the Navy asked the U.S. Geological Survey to conduct an investigation of the groundwater contamination and hydrology at the Bangor Annex. The study was designed in three phases:

1. To collect and detail existing information and prepare a preliminary work plan for the investigation [6].
2. To obtain additional data to define the extent of contamination and any movement and to identify factors that might impact on the contamination.
3. To determine if a hydrologic or water-quality model should be developed for the system.

B. Hydrogeology

Owing to the limited groundwater development in the vicinity of Bangor Annex, detailed hydrologic information on the study site was not readily available. Geologic and hydrologic information on the general area was summarized in a report by Molenaar [7]. The affected drainage-pond system

(Fig. 6) is underlain by and in direct hydraulic contact with the unconfined aquifer, which consists primarily of glacial drift and unconsolidated interglacial sediments. The unconfined aquifer is tapped by some local domestic wells, but a deeper confined aquifer is the source of water for most domestic wells in the vicinity and for the several large-supply wells at the Bangor Annex. The proximity of domestic and supply wells increased the concern over the possibility of groundwater contamination resulting from the cleaning activities.

Water levels, obtained from piezometers installed near the disposal area, ranged from 15 to 20 m in depth below the disposal pit, and the local water table sloped slightly to the west-northwest. A 2-year record, including the year of the USGS investigation, showed an annual water-table fluctuation of 1–1.5 m. The porosity of the aquifer was 0.3, and an estimate of the hydraulic conductivity was 3 m/day. On the basis of a single aquifer test, the groundwater velocity at the surface was estimated to be 27 m/yr [8].

C. Sampling and Analysis

In 1974, the USGS supervised the drilling of six test wells along a north–south line through the oval waste catchment pit (Fig. 6, A–A') to define the extent of contamination. The line of test holes was nearly perpendicular to the groundwater flow. To gauge contaminant groundwater transport, five test wells for sampling both water and sediment were drilled along a flow line through the former pit area and for a 300-m distance downgradient (Fig. 6, B–B'). One test hole, well 7, was drilled immediately upgradient along the same line. A nearly continuous core was collected from each hole, 0.5 m at a time, down to the water table and directly into the filled collection pit. Well drilling within the pit area was performed by use of a cable tool [1] using drilling jars, drive clamps, and a cleanout tool. The 6-in. (15-cm, nominal) diameter steel casing was driven by the jars and then cleaned out, and the procedure was repeated until the desired depth was reached. No drilling fluids were used. Core samples were taken just below the casing by a split-spoon coring sampler. Another line of wells down the flow-gradient direction were drilled using a hollow-stem auger. Core samples were collected through the hollow stem by means of a similar split-spoon sampler [1]. After core samples were taken, selected wells were cased with 2-in. (5-cm, nominal) steel pipe.

The collected core samples were processed in a protected area away from the drilling. The samples from both coring devices were retained in brass or aluminum core-barrel liners, which expedited removal from the core barrel and facilitated sample handling. Some liners were wrapped in aluminum foil and shipped as such, but usually each liner was emptied from the bottom to avoid artifacts introduced by down-hole contamination. The sediment was transferred to two glass containers, which segregated the material in the bottom from that in the top. The containers were specially cleaned wide-mouth jars with Teflon-lined lids.

Most of the water samples were obtained some days after the wells were developed, although some water was collected as soon as it was observed during the drilling. The water-sampling wells were installed along the flow direction of the principal gradient and set at geometrically increasing distances from the disposal pit. The reasons given for this spacing precaution were that it ensured that the extent of any contaminant migration was sampled in this planned one-time and final round of drilling and satisfied transport-modeling considerations. The well screens were placed close to the ground-water level to skim samples from the very top of the saturated zone. Jet-pump units were permanently installed in each well to limit cross-contamination and to facilitate sampling by means of a portable pump. Before the developed wells were sampled, several casing volumes were removed to ensure that samples were collected from groundwater induced from outside the well casing. The water samples were collected in glass 1-L bottles and sealed with Teflon-lined screw caps.

Generally, samples were shipped to the Menlo Park, California laboratory by air freight, arriving 2 or 3 days after collection. Sediment samples were frozen soon after collection, and the water samples were packed in ice and shipped to the laboratory. Cooling the water samples and freezing the sediment cores were the only means of preservation used for samples collected for the Menlo Park laboratory.

At the outset, the analytical methods used by the contract laboratories in the analysis of the water samples for explosives for the 1971 study were questioned [6]. The unexplained variability in data obtained using the colorimetric analyses suggested the need for another approach. Consequently, alternative methods for the analysis of water and sediment for the subject compounds were developed and tested by the USGS organics laboratory in Menlo Park [9]. Electron capture gas chromatography (ECGC) was the primary technique used for determining low concentrations of TNT and RDX in sample extracts. A Hall detector gas chromatograph (NGC) was used to verify the presence of nitrogen by specific detection. A gas chromatograph coupled with a mass spectrometer (GC/MS), an early (1970) Finnigan System 150, was employed for confirmation when any compounds were found. Two different types of GC columns were used. One was Dexil 300GC, coated 3% by weight on 80/100 mesh Supelcoport and packed into a 1.8 mm i.d. × 1.5 m long Pyrex glass column. The other was OV-17 silicone coated 3% by weight on the same support material and packed in a glass column of similar dimensions.

Both TNT and RDX displayed thermal instability during GC. The presence of benzene as the final solvent for injection of the extracts was found to reduce this problem. Benzene was used as the extraction solvent for water; an acetone–benzene mixture was used to extract the TNT and RDX from wet aquifer sediments.

D. Results and Discussion

Careful evaluation of the analytical data and repetitive analysis of sequentially collected water samples suggested that either the contaminants were varying significantly in concentration from one sample to the next or they were being degraded during the time between collection and analysis. Concentrations of TNT varied by as much as a factor of 3 for "duplicate" water samples. Analytical testing and quality checking of the analytical procedures demonstrated that recoveries for both compounds averaged better than 90% at concentrations producing responses equal to or greater than 10 times the detector signal-to-noise ratio. Recovery tests performed on distilled water samples fortified with TNT and RDX at the 0.5–1.0-μg/L level gave the mean recovery for TNT as 95 $\pm$ 15% and the mean recovery for RDX as 85 $\pm$ 10%. The method limit of detection was 0.1 μg/L. On actual field samples, TNT concentrations in test well water samples were almost always found to be lower in the second sample of a so-called duplicate set if the second sample was analyzed any time after the first. These differences were more pronounced with increased time between analyses. Additional peaks appeared between TNT and RDX on the gas chromatograms (Fig. 8) and became larger in the duplicate companion samples coupled with decreasing TNT concentrations. These additional components were identified by GC/MS as aminodinitrotoluenes (ADNTs), which are products of TNT decomposition. Won et al. [10] demonstrated that TNT is a biologically oxidizable substrate and 2- and 4-aminodinitrotoluenes (2-ADNT and 4-ADNT) were produced as intermediates.

In contrast to the water analysis, data from core sediment analyses compared favorably for sample splits and within the analytical error expected as determined by quality assurance procedures. Recovery tests gave the following results: mean recovery for TNT, 85 $\pm$ 15%; mean recovery for RDX, 93 $\pm$ 10%. The practical method limit of detection was 0.5 μg/kg.

The wastes were reported to consist of TNT and RDX in a ratio of 3:2, and relative concentrations close to a 3:2 ratio were found in water directly beneath the disposal area. The sediments above the water table, however, contained much higher ratios of TNT to RDX, commonly one and sometimes two orders of magnitude more TNT than RDX. Selected data and the spatial relations in the aquifer are illustrated in Figs. 9a and 9b. Water samples taken at the water table directly beneath the disposal area contained 13 mg/L TNT and 5 mg/L RDX. Associated sediments contained 13 mg/kg

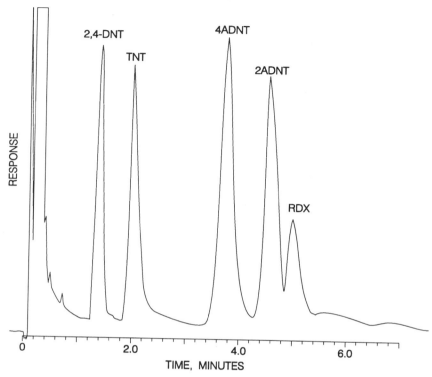

Figure 8 Electron capture chromatogram of extract of water sample with a standard addition of 2,4-dinitrotoluene, TNT, and RDX.

TNT but less than 0.8 mg/kg RDX. At a distance of 60 m from the disposal area, well 13, concentrations in water were 1.2 mg/L TNT and 3.7 mg/L RDX but less than 0.01 mg/kg of each in the sediments.

Two interpretations of the data resulted. The consensus view was that TNT was more strongly sorbed than RDX, and sorption was a major control on the transport behavior of the two compounds. The data from both the water and sediment samples supported this viewpoint. TNT and RDX have approximately equal solubilities in water and were present in the waste at a ratio of 3:2, but TNT was almost always found at much higher concentrations in the sediment than RDX. This suggested that RDX was leached from the sediment faster than TNT. RDX, however, was not found at correspondingly higher concentrations in the water at the water table downgradient.

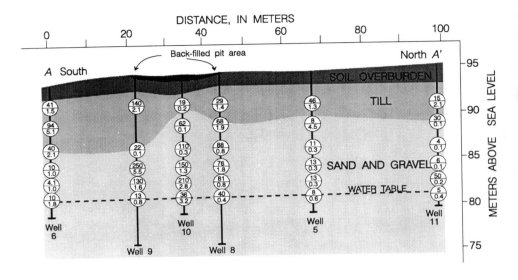

EXPLANATION $\binom{210}{2.8}$ Concentrations of TNT/RDX, mg/kg, on sediments.

(a)

Figure 9   (a) Distribution of TNT and RDX in the aquifer cross section A–A'.
(b) Distribution of TNT and RDX in the aquifer cross section B–B'.

Laboratory observations revealed decreasing concentrations of TNT in samples over time, which suggested that additional or alternative mechanisms—namely, chemical or microbial processes—were influencing the apparent retardation of movement of one or both of the compounds. The possibility of microbial alteration of TNT was consistent with the findings of Won et al. [10] and with those later reported by McCormick et al. [11] with respect to RDX.

E. Conclusions

In an interpretive report to the Navy on this work, Tracy and Dion [8] concluded that the existing contamination was limited to the immediate area of the disposal pit. They estimated that 4300 kg of TNT and 140 kg of RDX remained in the earth materials in 1975. The maximum concentrations of TNT and RDX found in the groundwater directly beneath the disposal area were 13 and 3 mg/L, respectively. The concentrations were less than the detectable limits beyond 165 m downgradient; these low concentrations implied

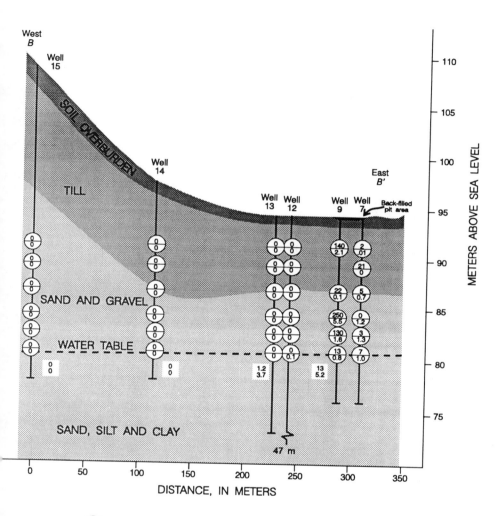

that the wastes had not moved that far since introduction at the source 9 years earlier. The findings and interpretations fulfilled the requirements and objectives of the study, and no further work was immediately planned.

From the chemical perspective, several questions considered beyond the scope of the investigation remained unsolved at the conclusion of the study. Some are listed below.

1. Are the organic solutes being degraded in the aquifer, in the sample bottle, or during the sampling procedure? If so, are these processes abiotic, biotic, aerobic, and/or anaerobic? Also, are the chemical or biochemical processes a factor in the behavior and fate of the compounds?
2. Are sequential samples taken from a well acceptable as "duplicates"? Should a rational procedure be developed to include field sampling and sample handling for quality assurance?
3. Are there any important alteration compounds produced that increase or decrease the endangerment of the groundwater? Do these processes alter the supposed intrinsic behavior of the overall contamination (e.g., sorption, storage)?
4. In addition to the field investigations, are definitive laboratory experiments, as a rule, necessary to establish the physical and chemical relations controlling or resulting from chemical alterations? Should sorption measurements and microbial digester studies be adopted as standard practices?

Clearly, fundamental sorption characteristics of the solutes and the aquifer material were not known. Also, the stability of the solutes were not known. Sequential samples were not comparable to sample splits. The analysis of collecting equipment rinses in an attempt to test for sampling artifacts after the sampling episode increased the confusion and uncertainty because no rational evaluation, either qualitative or quantitative, was forthcoming from positive findings. The implications of these questions go beyond the immediate study and have possible application to other groundwater investigations. These issues were placed on the agenda of subsequent groundwater investigations by the Organics Project. Although 15 years have passed since these initial suspicions, many of these questions remain unanswered.

IV. GROUNDWATER CONTAMINATION BY WOOD-PRESERVING CHEMICALS AT VISALIA, CALIFORNIA

A. Introduction

The groundwater at Visalia, California is contaminated by the wood preservatives creosote and pentachlorophenol (PCP) as a result of wooden power pole treatment operations. Prior to 1968, hot creosote was used as the primary treatment fluid; it then was replaced by a 5% solution of PCP in aromatic diesel oil. Creosote, a coal tar distillate, is a complex mixture of more than 200 compounds that consists of approximately 85% by weight polynuclear aromatic hydrocarbons (PAHs), 12% phenolic compounds, and 3% nitrogen (NH), sulfur (SH), and oxygen (OxH) heterocycles. At the time, creosote was the most extensively used pesticide in the United States, and PCP was the second most used [12]. In 1972, it was discovered that the concrete treatment tanks had been leaking fluid into the underlying shallow aquifer for an unknown length of time. After evaluating the possible extent and impact of the problem, the California Regional Water Quality Control Board ordered the company to stop the leak and remove any existing contamination in the subsurface. Steel liners were installed in the dipping tanks to eliminate the most immediate source of contamination. Samples obtained from test holes drilled adjacent to the treatment tank, however, indicated that large amounts of an oily mixture of both creosote and PCP fluid had already infused into the ground.

B. Hydrogeologic Setting

The wood-treating facility was located in southeastern Visalia, Tulare County, in the southeast part of the San Joaquin Valley of California (Fig. 10). The western foothills of the Sierra Nevada rise a few kilometers to the east. The setting around Visalia is a relatively flat alluvial fan surface that slopes gently west-southwest. The area is drained by tributaries of the Kaweah River and many irrigation canals. Average annual precipitation is about 255 mm and occurs almost entirely during the winter.

The subsurface strata beneath Visalia consist of more than 300 m of unconsolidated sedimentary deposits of Late Pliocene to Holocene Age [13]. These water-bearing sediments consist of thin lenticular layers of limited horizontal extent. The Visalia city water supply is obtained from groundwater, and considerable quantities are taken for irrigation as well as private domestic and industrial use.

The pole-treating site and the area directly downgradient (Fig. 10) are underlain by unconsolidated alluvial deposits. The surficial deposits form an unconfined aquifer to depths of 15–17 m. This aquifer ranges from silts and fine sands near the surface to coarse, poorly sorted sand and pebbles near the base. These surficial deposits are underlain by a discontinuous, less permeable layer of silts and clays that thin out near the southwest corner of the property, then increase in thickness toward the southwest. At the beginning of the study, perched water was found on top of the less permeable layer at a depth of 15–17 m. Below this layer is a lower aquifer, which is

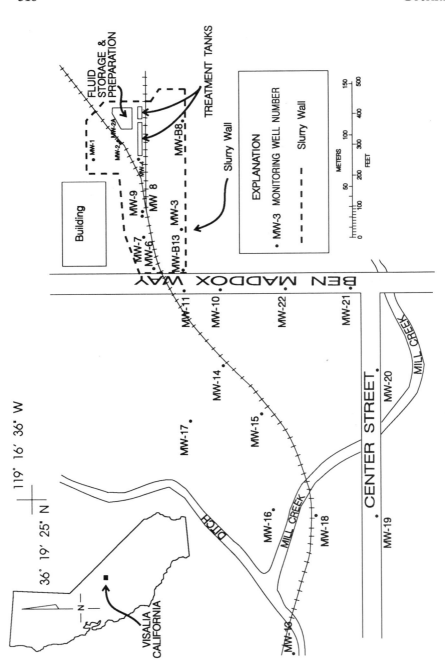

Figure 10 Location of power pole treatment plant, Visalia, California.

the source of water supply for the area and the City of Visalia. The lower aquifer is composed of coarser gravel and appears to be locally confined by the overlying thin silt-clay layer.

Significantly, during the period 1975–1978, drought conditions existed throughout the San Joaquin Valley and groundwater levels declined by tens of feet per year. The upper perched water was not present after June 1977. the water level of the lower aquifer declined at a rate of about 1 ft/month through 1977 and dropped approximately 25 ft over the 3 years of record. Following the drought, by 1981 the water levels of the lower aquifer returned to within 7 ft of the first measurements made in 1974. Coincidentally, the upper or perched water table recurred.

C. Study History

In 1976, the USGS began a field and laboratory program to study the movement and fate of the organic solutes contaminating the groundwater at Visalia. An interdisciplinary study unit was formed that included both organic chemists and groundwater modelers. At a joint meeting, which included representatives of the wood treatment plant and personnel from the California Regional Water Quality Control Board, the USGS team proposed that the Survey conduct an independent study so as not to interfere with any activities of the plant operator or the State of California. It was agreed that data and observations would be freely exchanged. In September 1976, test wells drilled by the plant operator showed that significant amounts of both PCP and creosote had moved beyond the southwest boundary of the property. Subsequently, a subsurface bentonite-cement slurry wall was constructed around the pole-treating facility. It was completed in June 1977 and is shown in Fig. 10. The purpose of constructing the slurry wall was to prevent further lateral migration of contaminated groundwater from the pole yard. The wall extended from the surface to a depth of 1 m into the silt/clay layer approximately 17 m below. The drought conditions and installation of the slurry trench changed the flow pattern of the otherwise simple system in such a way that it was no longer practical as a site for numerical modeling experimentation. Chemical data and other observations were salvaged from the work that was done. These findings revealed the occurrence of several processes not previously reported, and they are summarized here.

D. Methods and Analysis

Chemical analysis of the first samples of groundwater collected from observation and test wells in 1977 showed that PCP and several PAHs, tentatively identified as mainly naphthalene and the methylnaphthalenes, had migrated more than 500 m downgradient in a westerly direction from the suspected point source (Figs. 11 and 12). As a wood preservative and biocide,

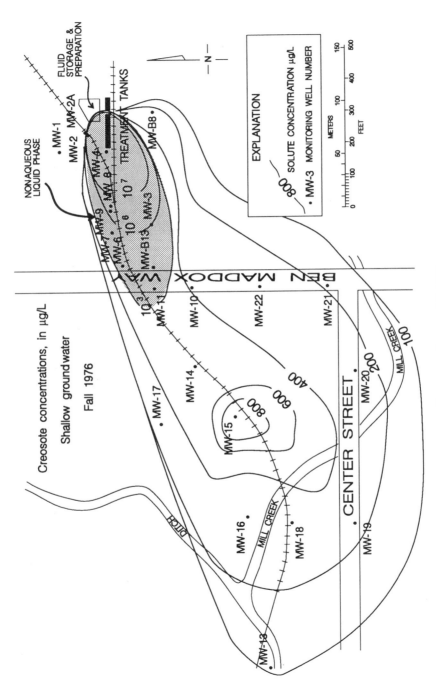

Figure 11 Concentration of creosote solutes in shallow groundwater, fall 1976.

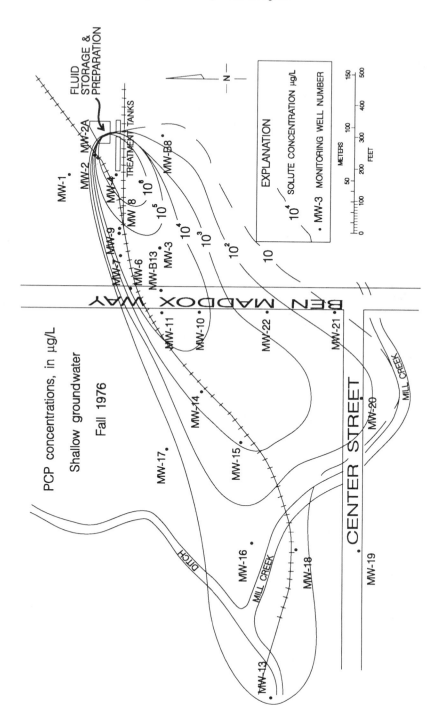

Figure 12 Concentration of PCP in shallow groundwater, fall 1976.

PCP was thought to be resistant to environmental degradation, although it has been shown that PCP is degraded under idealized laboratory conditions by acclimated microbes isolated from soils obtained near PCP-treated fencepost drying racks [14]. PCP appeared to behave in a conservative manner within the aquifer, apparently being neither degraded nor sorbed during transport. Groundwater pH ranged from 7.9 to 8.6 at a temperature of 19°C. Dissolved oxygen, determined colorimetrically (Chemetrics, Inc., Warrenton, VA), ranged from 2 to 6 ppm. The migration of PCP by more than 500 m since the beginning of leakage into the aquifer, a period estimated at 5 years, indicated that the groundwater velocity was about 100 m/yr at the water table. Polynuclear aromatic hydrocarbons were also detected at that distance. Interestingly, although creosote consists of approximately 85% by weight PAHs, 12% phenols, and 3% heterocycles (NH, OxH, and SH), only the PAHs were detected in the well water samples. It was first thought that the source of creosote used at the site either was lacking in phenols and heterocycles or was depleted by leaching and dispersion of these solutes in the aquifer.

Initially, the PAHs and PCP were isolated from the water by extraction into dichloromethane solvent. The PAHs in the extracts were determined by flame-ionization GC (FID/GC). PCP, in a solution of methanol and dichloromethane, was first converted to a methyl ether by reacting with diazomethane, then analyzed by ECGC. Later, water samples were directly analyzed for PCP by high-performance liquid chromatography (HPLC) [15].

To study the transport and fate of the solutes, the USGS developed and performed laboratory experiments to determine the sorption properties of major solutes, specifically PCP and naphthalene. Column studies and shaking-flask experiments were done to ascertain the influence of microbes on the fate of the solutes in the groundwater. Field analytical methods using HPLC were developed for the purpose of identifying and measuring PCP and PAH on-site during test hole drilling and well development. HPLC was performed on both water and extracts of aquifer sediments.

E. Observations

Water samples showed the presence of an immiscible oil phase with a creosote-naphthalene odor in most of the downgradient wells within the boundaries of the treatment yard and also in wells MW10 and MW11 located in the street, Ben Maddox Way. Well MW 4, located some 3 m from the treatment tank, yielded principally oily fluid, and only occasionally was water observed to separate on standing. From this source, the nonaqueous fluid separated into two layers, one lighter than water and the other denser than water. From the start, one of the observation wells, MW 9, which was installed to tap the supply aquifer, was pumped at between 20 and 100 L/min over an extended period and typically produced water samples showing visible contamination. The contamination ranged from a sheen on the surface to a complete covering by a layer of oily fluid. During the well construction, the drillers reported that the oily fluid appeared to occur in former tree-root passages permeating the aquifer. Approximately 1.9×10^7 L was pumped from MW 9 in 1977 and 4.2×10^7 L in 1978. The contaminated water was discharged into the Visalia sewer system, which ultimately flowed to the sewage treatment plant. During 1977, concentrations of PCP in the well water samples ranged from 1 to 5 mg/L, and those of creosote-related compounds, from 1 to 10 mg/L. These wide-ranging concentrations were observed for both the samples taken monthly and those taken during a test by sequential sampling of the well flow. It was found that the concentrations of PCP and PAH remained reasonably constant, however, if the oily phase was removed from the sample prior to analysis.

During laboratory experiments to determine sorption relations of PCP and naphthalene on samples of the aquifer sediments, the standard comparison solutions prepared in PCP-free aquifer water were found to be losing substantial amounts of PCP. These solutions were prepared in the range 2–12 mg/L using native water from a monitoring well showing no detectable PCP. No preservatives were added, and the solutions were not sterilized; accordingly, microbial degradation was suspected.

The screen of the well being continuously pumped, MW 9, was periodically clogged by a greaselike substance, which upon testing proved to be slime-forming bacteria [16]. Examination of the anaerobic and aerobic effluents at the sewage treatment plant indicated that components of creosote were being degraded, especially in the aerobic treatment, but the PCP was not being degraded in either treatment process. Concentrations of PCP in two samples of the sewage effluent taken in 1977 were 2.1 and 3.7 μg/L.

Observation of the bacterial slime in the pumping well and the apparent degradation of PCP in the laboratory were the first indications that biodegradation was operating at the site. In order to investigate this possibility more fully, field and laboratory microbiological determinations were made in August 1977. The evaluations indicated that the creosote and oily materials in contact with the groundwater body were undergoing aerobic microbial breakdown. Dissolved oxygen concentrations in the groundwater supply away from the contamination were generally found to be between 2 and 6 ppm (parts per million). The PAH solutes were depleted in less than 10 days in aerobic shaking-flask experiments using contaminated water samples. PCP was not degraded in the aerobic shaking flasks for the duration of the experiment. No anaerobic breakdown of the PAH was detected, and PCP did not appear to be affected in the aquifer. During laboratory column experiments, there was evidence that between 15% and 40% of PCP was degraded within 7 days on a column packed with native aquifer sand maintained in an anaerobic environment. The degradation occurred in the absence of any additional nutrients or carbon source. Interestingly, no degradation of PCP was observed in the same time period for a companion experiment when glucose was added.

Near the end of the drought period, the water table had fallen below the intake screens of most of the shallow observation wells, and the wells went dry. Until the wells went dry, water samples continued to have consistent amounts of both PAH and PCP. Following the drought, as the water table returned to near the predrought level of 1976, PCP amounts also returned to the previous concentrations. The PAH concentrations, however, lagged behind relative to the return of PCP, and throughout 1980 they remained 1–3 orders of magnitude lower than before the water table normalized. It is thought, based on the observations discussed above, that as the water table fell, PAH compounds remaining in the unsaturated zone were catabolized by microbes in conjunction with the formation of an aerobic environment. Upon return of the water table to predrought levels, concentrations of PAH remained low, suggesting that the PAHs were depleted. Correspondingly, PCP, remaining relatively unaltered, returned almost immediately to near predrought concentrations. Replenishment of PAHs from the source by transport would take longer, with the PAHs moving at a rate of about 100 m/yr.

Because PCP exhibited conservative behavior at the site, it was chosen as a tracer compound in the transport study. During test hole drilling and well installation, water samples and aquifer sediments were analyzed on-site for PCP and several polynuclear aromatic hydrocarbons for the purpose of optimizing the location of the wells and screens. These analyses were by HPLC in a field laboratory near the site. Each analysis was done within an hour of sample collection, and results were available before the next sampling. Using this approach, the USGS successfully placed multilevel wells within and near the edge of the groundwater contamination for the aquifer testing and transport modeling effort. Owing to the drought conditions and the construction of the slurry wall, however, the aquifer testing and modeling effort was no longer feasible and was discontinued.

F. Conclusions

Although the study was interrupted, the information gained from this limited examination opened possible areas for research in future investigations.

The nonaqueous wood-treatment fluids migrated unexpectedly large distances from the point of entry, downward and laterally in the aquifer. The observed migration behavior was found to be consistent with findings from other studies as reported in a review by Villaume [17].

Microbial degradation of the PAHs and possibly the associated NH and other heterocycles was strongly implicated. Bacterial slime observed in a pumped well and the attenuation of PAHs during normal domestic sewage treatment was further evidence that aerobic degradation was at work. PCP did not appear to be affected by the aerobic processes. The possibility of anaerobic degradation of PCP was indicated in laboratory experiments but was not found in the aquifer.

Pentachlorophenol was found to migrate with the groundwater in a conservative manner and was useful in tracing contaminant migration from the source. Since its introduction in the early 1950s, PCP replaced creosote as the primary wood-treatment chemical. Owing to its conservative behavior and the known time of introduction, it was suggested that PCP may be useful as a time and spatial tracer at other sites having similar contamination.

HPLC was found amenable to on-site analysis. It was used to readily detect PCP and naphthalene in water and sediments and was useful in locating the presence of these contaminants during test well installation [18].

V. STUDY OF CONTAMINATED GROUNDWATER AT HAWTHORNE, NEVADA

A. Introduction

Since before World War II, waste fluids resulting from the processing of explosives and demilitarization of munitions were flushed to surface drainage systems consisting of several unlined beds or pits within the ammunition depot at Hawthorne, Nevada. The approximate location of the disposal area is shown in Fig. 13. The site is near Walker Lake, the southern shore of which was approximately 3 km northwest during the investigation, which began in 1975. Walker Lake, a remnant of Lake Lahontan, is a saline, closed-basin lake fed mostly by the Walker River from the north. Much of the water in Walker River is diverted for irrigation. Because of this diversion and low amounts of precipitation, the lake level is generally declining.

The study area is underlain by alluvial and lacustrine deposits, predominantly clay and silt 800–1000 m thick [19]. Well drillers' logs indicate that the uppermost 65 m consists of coarser sand and gravel deposits. Generally, the groundwater moves northwestward toward Walker Lake; in the vicinity of the site, the linear velocity at the water table is estimated on the order of 100 m/yr.

Following the publication of *Water Quality Criteria* [20], which recommended that drinking water contain less than 10 mg/L of nitrate nitrogen, the USGS was asked to identify possible sources of elevated nitrate concentrations found in a water supply well near the Hawthorne facility. As a result

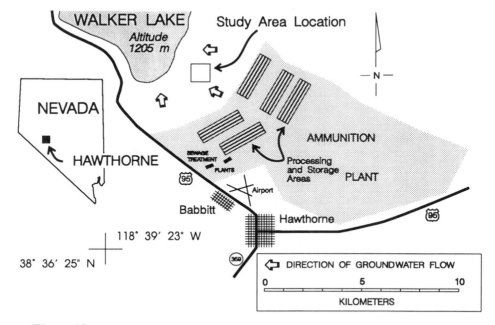

Figure 13   Location of Hawthorne ammunition depot and study site, Hawthorne, Nevada.

of a comprehensive evaluation of available information and testing of existing wells and exploratory wells, contamination from the explosives handling area was ruled out. It was determined that the hydraulic gradient diverged from the water supply wells and that the distances involved precluded significant effects from pumpage. The sources of the nitrate were traced to two nearby sewage treatment ponds. Nitrate levels in the groundwater adjacent to the affected supply well ranged from 0.2 to 19 mg/L nitrate nitrogen [21].

Comprehensive test sampling downgradient from an ammunitions processing facility, however, identified significant groundwater contamination in that area. The shallow groundwater near the collection pits, located within the study area as shown in Fig. 13, was found to contain 2200–2800 mg/L of dissolved solids; this included as much as 17 mg/L organic nitrogen (N), 44 mg/L nitrate N, and 75 mg/L ammonium N. By contrast, chemical analysis of groundwater collected away from the contamination was found to characteristically contain 1100–1700 mg/L dissolved solids and small concentrations, <2 mg/L, of nitrate N. These data are summarized in Table 1.

Table 1 Chemical Character of Contaminated Groundwater and Adjacent Groundwater, Hawthorne, Nevada

Component or property	Range of measured values	
	Contaminated groundwater	Adjacent groundwater
Dissolved solids (mg/L)	2200–2800	1100–1700
Dominant ions	Na, SO$_4$	Na, SO$_4$
Ratio, Cl/(Na + SO$_4$)	0.065–0.085	0.101–0.102
Nitrogen species (mg/L)		
Organic N	0.3–17	0.1–0.4
NO$_3$ as N	2.7–44	<0.1
NO$_2$ as N	0.1–3	<0.1
NH$_4$ as N	0.0–75	0.0–0.1
Total N	3.1–130	0.1–0.6
Carbon (mg/L)		
Organic C	2–5	1–2
Inorganic C	60–160	<80

B. Study Plan

The USGS was asked to make a detailed investigation of the impact of the processing facility on the groundwater downgradient from the disposal sites. Very little quantitative documentation was available concerning the operation of the facility prior to and just after World War II. Records indicate that ammonium picrate wastes were discharged from 1952 to 1958. During 1964–1968, most wastes were composed of TNT and ammonium nitrate. These periods of activity coincided with the Korean War and the Vietnam Conflict.

In a previous study, at the Bangor Annex in the state of Washington in 1975, degradation or alteration of TNT in groundwater samples was observed over time, especially when the contaminated water was exposed to air. Evidence suggested that significant degradation was occurring in the sample bottle and possibly in the well itself. In recognition of these uncertainties, and the particular remoteness of the Hawthorne site from the analytical laboratory, specific analysis for TNT using HPLC was done at the site; subsequently, HPLC and GC analyses were done for comparison at Menlo Park, California.

C. Sampling and Analysis

To find the extent of the contamination, test wells were augered adjacent to and downgradient from the disposal sites; these well sites are shown in Fig. 14.

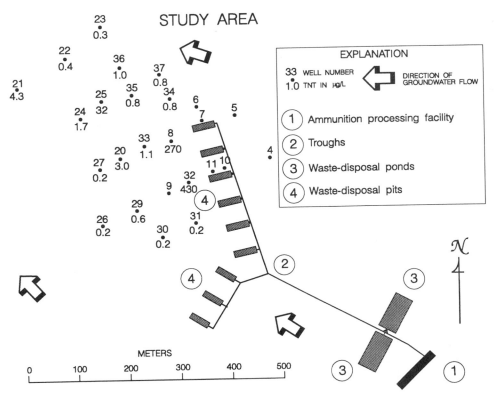

Figure 14 Test well array in area of TNT-contaminated groundwater.

Prior to drilling, each site was leveled, and clean fill was brought in to establish a relatively uncontaminated working surface. The finished wells were cased with 3-in. (7.6-cm nominal) galvanized pipe. Generally, stainless-steel screens, 60 cm long, were set at or just below the water table. The wells were constructed with blank casing extending 3 m below the screen to provide space for sampling. Sample bailers were fabricated from galvanized metallic tubing, nominally 2.5 cm diameter and 3 m in length, and were fitted with a steel check valve at the bottom. A bailer was prepared for each well and remained suspended in the casing between samplings. Groundwater samples were collected by lowering the bailer from a block attached to a portable tripod. Water samples were transferred from the bottom of the bailer by defeating the check valve with a steel wire and allowing the water to drain into a glass sample bottle.

Field analyses for the extended study were done several weeks after the test wells were prepared. The field HPLC was done in a trailer van within a few minutes drive from the sampling area. Samples were collected from the wells in bailer-volume increments of approximately 1 L, and each was analyzed separately within an hour after collection.

Generally, sampling of a particular well was continued until the concentrations of TNT stabilized or until 3–5 casing volumes had been removed and no explosives or associated organic components were detected. Only a few milliliters from each sample was needed to do HPLC. Subsequently, the samples were prepared, iced, packaged, and sent to the laboratory.

D. Results and Discussion

On-site and laboratory analyses for explosives in groundwater samples revealed the presence of only TNT and ADNT in the groundwater. Available records, although sparse, indicate that TNT and ammonium picrate were the only organic explosives processed in major amounts at the facility. Ammonium picrate and picric acid were found in bottom soil from the disposal ponds and drainage ditches but not in groundwater in wells downgradient from the area. Concentrations of TNT in the groundwater ranged from 430 μg/L near the disposal site to less than 1 μg/L within 500 m of the pits, as shown in Fig. 14. Downward migration was limited by a lens of impermeable clay underlying the study site, 7.5–11 m below the land surface. Table 1 lists some other chemical characteristics of contaminated water compared to that of adjacent water.

Comparisons of field and laboratory analyses indicate that TNT alterations were taking place both in the well casing and in unpreserved samples. The degradation was rapid, and 2-ADNT and 4-ADNT were the major alteration products found. Shaking-flask laboratory experiments verified that the microbial alterations were accelerated in the aerobic environment, and within 20 days, untreated well water samples containing 180 μg/L TNT lost 50–75% of the original contamination. Won et al. [10] found total catabolization of TNT in enriched media within 24 hr. Concurrently, the ADNT concentrations increased from 0 to 34 μg/L during the experiment. The use of mercuric chloride as a preservative was effective as a biocide for these particular samples.

E. Conclusions

Losses of TNT in unpreserved water samples were confirmed. These losses were shown by field and laboratory experiments to result from aerobic microbial degradation. Two major alteration products are 2-ADNT and 4-ADNT, which were identified as sampling artifacts and were not detected in fresh or preserved samples of the groundwater. Mercuric chloride, at a concentration of 60 mg/L, was effective as a sample preservative and significantly reduced the rate of alteration of TNT.

Samples collected in the close vicinity of, or "skimmed" from, the water table are apt to be less representative of the groundwater than samples taken from below this transition zone. Aeration, which destabilizes anaerobic processes and promotes potential aerobic processes, may be responsible for sample alterations.

On-site analysis of specific compounds in the contaminated groundwater is useful for assessing stability and evaluating sampling procedures. HPLC provided adequate specificity and sensitivity for the determination of TNT and ADNT on-site. The resulting data compared favorably with laboratory data from the analysis of the same samples [18].

VI. GROUNDWATER CONTAMINATION BY WOOD-PRESERVING CHEMICALS AT ST. LOUIS PARK, MINNESOTA

A. Introduction

Coal tar derivatives from the operation of a coal tar distillation and wood-treatment plant at St. Louis Park, Minnesota (Fig. 15) contaminated the groundwater in the area. The plant operated from 1918 to 1972, and during this period, products and wastes from the distillation and treatment operations were routinely allowed to drain to wetlands near the site. The wetland ponds were in hydraulic contact with the shallow aquifer and provided a main entry point for the contamination. As early as 1932, taste and odor problems caused the shutdown and abandonment of a municipal well about 1000 m from the plant site. The aquifer tapped by this well is the Prairie du Chien–Jordan aquifer, the principal source of groundwater for the Minneapolis–St. Paul metropolitan area. Results of the 1932 investigation by a well company concluded that the source for the taste- and odor-causing contaminants in the groundwater was the creosote plant. It was alleged that several of the plant's wells were actually being used to drain away creosote fluid wastes.

Since about 1938, most of the surface-water inflow to the wetlands was recharged to the underlying peat and the water table aquifer. The inflow included 100–200 L/min of wastewater, increasing to more than 500 L/min during periods of peak activity. This added inflow from the plant raised the water table, creating ponds that significantly increased vertical leakage and intrusion of waste into the aquifer. Much of the contaminated wastewater that was discharged to the wetlands resulted from the first step in the refining process. Approximately 2% of the coal tar received for processing was

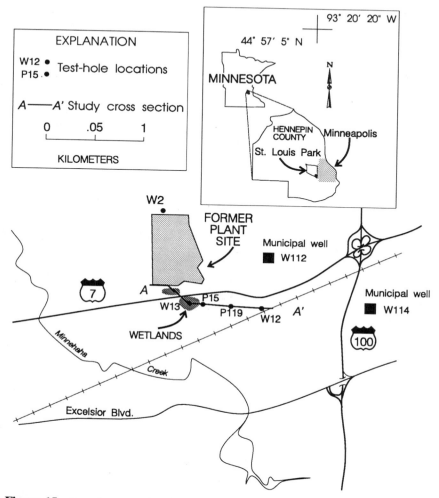

Figure 15 Location map showing site of former coal tar distillation plant, St. Louis Park, Minnesota.

associated with water that was sent directly to waste before the beginning of the distillation. This initial fraction was enriched with the water-soluble and low-boiling components of coal tar. Additionally, sodium hydroxide and sulfuric acid were used at various times in the refining processes. On record, about 3×10^5 L of 70% sodium hydroxide was used and possibly discharged to the ponds from 1940 to 1943. The water table was at or slightly below land surface during the period June 1978–June 1979. The reduction of the pond level is attributed to the plant closure in 1972 and the cessation of waste discharge.

B. Hydrogeologic Setting

The area beneath the study site is underlain by an aquifer system [22]. The principal water supply source for the local municipalities is the Prairie du Chien–Jordan aquifer that lies approximately 115 m below the land surface. The scope of the work described in this section is confined to the water table middle drift aquifer and the uppermost bedrock unit, the Platteville limestone (Fig. 16). The middle drift aquifer consists of glacial till, outwash sand and gravel, lake deposits, and alluvium. The Platteville aquifer is a flat-lying dolomitic limestone interrupted by fracture channels that contain water. The Platteville aquifer is underlain by a confining unit of shale.

The potentiometric surface of the Platteville aquifer was similar to that of the middle drift aquifer, and the generalized flow of groundwater was toward the east. Low vertical hydraulic conductivity near the surface was attributed to the presence of peat and lake deposits.

Several of the aquifers in the aquifer system in the vicinity of the site are contaminated by coal tar. In 1978, coal tar derivatives were found to the north of the site in five municipal wells completed in the Prairie du Chien–

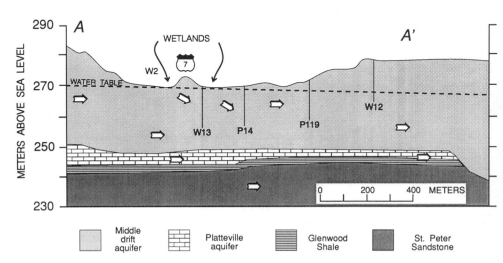

Figure 16 Geologic cross section A–A′ of water table, middle drift aquifer, and uppermost bedrock unit.

Jordan aquifer. One of these wells is approximately 2 miles from the probable source of contamination [22].

C. Objectives and Scope

In July 1978, the USGS began a study of the St. Louis Park problem in cooperation with the Minnesota Department of Health. The objectives were to develop a detailed understanding of the groundwater flow system and to study the behavior and transport of the organic contaminants in the vicinity of the former plant. Part of the study was limited to the water-saturated drift and uppermost bedrock unit and is the subject of this section.

In order to evaluate the transport and fate of the organic solutes in the surficial aquifer system, a number of wells were installed around, and mostly downgradient from, the source of contamination. Additional wells were installed from time to time, adjusting to findings from chemical analysis of the samples. Preliminary chemical investigation of water and sediment cores by the Menlo Park organics project were started in the fall of 1978. To assess the nature and extent of the contamination of the water table aquifer, water samples were collected from wells in the spring and summer of 1978. Subsequently, a comprehensive, multidiscipline study was begun in mid-1980.

D. Observation and Discussion

Chemical analyses of groundwater samples indicated that the direction of contaminant transport in the surficial aquifer system was in good agreement with the direction of groundwater flow estimated from hydraulic information. As at Visalia, California, nonaqueous liquid was found a considerable distance (approximately 250 m) from the distillation plant site. The nonaqueous liquid in the well sample was observed in two layers, approximately 10% of the fluid on the water surface and the remainder sinking to the bottom of the container. In this instance, overland drainage of waste fluids to the wetlands was evident and was possibly the major element in the transport of the source fluid the 250 m distance from the site.

Nearly all the discrete chemical solutes in the water samples were identified, and naphthalene accounted for about 70% of the PAH fraction. Unlike the creosote contamination investigated in Visalia groundwater, phenols were found in many of these samples. Very low concentrations of nitrogen compounds, two orders of magnitude below PAH and phenols, were detected in well W13. Table 2 lists the PAHs and phenolic compounds identified in water from well W13. Only 20–30% of the dissolved organic carbon was separated and identified as individual organic compounds. The remaining dissolved organic carbon behaved as highly polar, oxidized compounds when analyzed by HPLC. Lower boiling components such as benzene and alkyl-

Table 2 Concentrations of Organic Solutes from Creosote Contamination in Groundwater from Well W13, St. Louis Park, Minnesota

Solute	mg/L
PAH fraction	
1,2-Dihydroacenaphthalene	0.16
1-Methylnaphthalene	0.65
2-Methylnaphthalene	1.02
Acenaphthalene	0.17
Benzothiophene	0.40
Biphenyl	0.07
Fluorene	0.08
Indene	1.08
Naphthalene	9.69
Total	13.32
Phenol fraction	
2,3,5,6-Tetramethylphenol	0.75
2,3,5-Trimethylphenol	0.32
2,3,6-Trimethylphenol	0.67
2,3-Dimethylphenol	0.81
2,4,6-Trimethylphenol	0.45
2,5-Dimethylphenol	6.24
2,6-Dimethylphenol	1.02
2-Ethylphenol	0.44
2-Methylphenol	2.25
3,4-Dimethylphenol	1.05
3,5-Dimethylphenol	6.21
3-Methylphenol	3.93
Ethylmethylphenol	1.42
Phenol	0.50
Total	26.06

benzenes, also operationally known as "purgeables," were not included in the analytical scheme at the time.

The first samples collected were maintained on ice and were analyzed within 24 hr of receipt by the Menlo Park laboratory. Analysis of duplicate samples done a few days later indicated that phenols were being lost over time relative to naphthalene, even though the samples were refrigerated (4°C). The groundwater temperature was 11°C, suggesting that any microbes present might be acclimated to cold and that therefore refrigeration was a poor choice of preservation method.

At the study site, it was noticed that as wells were being dewatered in preparation for sampling, the cold water was effervescing and the emanating gas was readily ignited when lit with a match. Upon analysis, the gas was found to consist of methane and carbon dioxide. Because the samples were degassing, the concentrations were only estimated, but the concentrations of methane were above saturation (>20 mg/L) in wells near the source. The concentrations remained generally in a near-saturated condition for a distance of more than 180 m downgradient. Background concentrations for methane from water table wells away from the contamination were below 0.1 mg/L. Methane-producing bacteria were found in samples taken from wells showing the presence of organic solutes and having elevated concentrations of methane [23]. Concentrations of methane and other selected constituents are listed in Table 3.

Chemical analysis for organic solutes occurring in coal tar showed the concentration of contaminants decreasing in the downgradient direction from the source. However, the ratio of phenols and PAHs to sodium did not remain constant as it should if only nonselective processes such as dispersion and dilution were operating. As shown in Fig. 17, the selective attenuation of the organic solutes relative to the sodium concentration is apparent. Sodium was used as a tracer because of its known presence and high concentrations in the waste comparative to the uncontaminated groundwater.

E. Conclusions

Chemical analysis and microbial evaluation gave compelling evidence that anaerobic microbial degradation of the organic solutes, especially the phenols, from the coal tar contamination was occurring in the aquifer at St. Louis

Table 3 Methane Content and Other Properties of Water Samples from Test Wells, St. Louis Park, Minnesota

Quality/property	Well designation				
	W2[a]	W13	P14	P119	W12
Methane (mg/L)	<0.01	15.8	21.0	2.8	3.2
Dissolved oxygen (mg/L)	<0.1	<0.1	<0.1	<0.1	<0.1
Sodium (mg/L)	19	430	230	120	49
Organic carbon (mg/L)	3.4	150	55	14	8.6
Dissolved solids (mg/L)	620	1400	1000	550	610
Temperature (°C)	10.5	9.5	11.0	11.0	11.0
pH	7.01	6.95	7.12	7.00	7.00

[a]W2 is located 1500 m upgradient from W13.

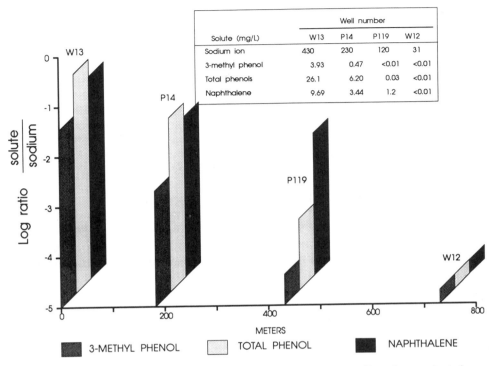

Figure 17 Ratio of concentrations of organic solutes to sodium from selected wells in middle drift aquifer, St. Louis Park, Minnesota.

Park, Minnesota. The field data backed by laboratory experiments show that certain organic compounds are being converted to methane and carbon dioxide. Most of the identifiable organic solutes were depleted within 1000 m of the contamination source. Although dilution and dispersion were appraised, neither of these processes accounted for the reduction of organics compounds relative to sodium (ion).

Methane was present at high concentrations in the contaminated aquifer but not elsewhere, and methanogenic bacteria were found only in the contaminated aquifer. Methane was produced from contaminated well water in the laboratory. Phenol solutions were inoculated with bacterial cultures obtained from specially collected samples from St. Louis Park. Using anaerobic incubation methods, and following a 6-week acclimation period, a steady production of methane was observed, and phenol was completely depleted from the inoculum at the end of the experiment. Naphthalene did not degrade under similar conditions during the period of the experiment. Field data from the analysis of groundwater samples, however, implied some differential attenuation of naphthalene during groundwater transport relative to sodium.

The presence of a valid tracer such as sodium greatly simplified the sampling and interpretation of the behavior of individual components during transport. Although differential attenuation of phenols relative to naphthalene was observed, it is less convincing as an argument. PCP appeared to be a conservative tracer at Visalia, California, but was not found in the groundwater or its use documented at St. Louis Park.

Laboratory experiments were conducted to test the sorption of selected phenols and PAHs on the aquifer sediments using pH 7.0, naturally buffered aquifer water [24]. The PAHs were slightly sorbed, but the phenols were not, lending further evidence to the argument that the attenuation of the organics in the aquifer was due mainly to microbial processes [23]. Although naphthalene was somewhat attenuated in transport and was found to be slightly sorbed by the aquifer sediments, biodegradation was not eliminated from consideration as part of this attenuation process (see Fig. 17).

This study was one of the first to gather convincing evidence for in situ biodegradation by chemical analysis of specific organic constituents and the detection of methane gas, an end product of anaerobic metabolism [25]. The realization of the potency of biodegradation brought about new appreciation for more careful and detailed observations. The importance of establishing clear objectives and consideration of all aspects of water quality determinations was made clear. The need for a multidisciplinary approach and strong peer review for such studies is evident.

VII. GROUNDWATER CONTAMINATION BY WOOD-PRESERVING CHEMICALS AT PENSACOLA, FLORIDA

A. Introduction

From early in 1981 and continuing into 1990, the USGS has been investigating the magnitude, extent, and fate of groundwater contamination by wood-preserving chemicals in the sand aquifer at Pensacola, Florida. The site is in Escambia County on the western tip of the Florida panhandle shown in Fig. 18. The underlying sand aquifer is the principal groundwater supply for the area and consists of fine-to-coarse quartz sand deposits interrupted by discontinuous silt and clay layers [26].

Contamination of the aquifer resulted from treatment practices at the wood-preserving plant that operated on the site for over 80 years. The wastes were reported to consist of creosote, PCP, diesel fuel, and moisture from

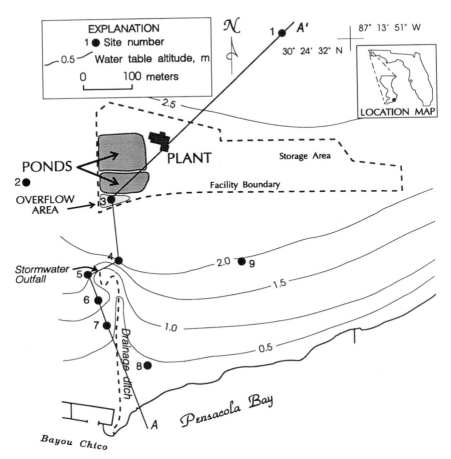

Figure 18   Abandoned creosote works, Pensacola, Escambia County, Florida. A–A' is the placement line for 1983 sampling sites.

dewatering the wood. Diesel fuel was employed as a solvent for the PCP and as a viscosity-controlling agent for the creosote. The wastes were routed to unlined "recirculation-holding" ponds, shown in Fig. 18, and were in direct hydraulic contact with the aquifer. The altitude of the water table is also shown. Until 1950, creosote was used exclusively for the treatment of the wood products, but from about that time the use of PCP steadily increased to equal that of creosote. Approximately 10^5 L per month of each of the treatment fluids, creosote and PCP, was being used just prior to the plant closure in 1981. Approximately 10^4 L of residual waste fluids was discharged to the ponds each week.

B. Plans and Objectives

An investigation of the hydrology and chemistry of the aquifer was started in March 1981 to evaluate the extent and nature of the contamination in the aquifer. This investigation was done in cooperation with the Florida Department of Environmental Regulation. Nine sites were selected for test drilling and water sampling. Locations were established from existing information and from an earlier reconnaissance sampling. One multilevel well cluster was located upgradient of the impoundments and was used to establish background data. Eight additional sites were established downgradient. Other samples were collected in the vicinity of the impoundments at six plant-owned monitoring wells, six private wells, and seven surface-water sampling points. The purpose of this initial investigation was to define the scope and extent of groundwater contamination and to point to any processes affecting the transport of contaminants in the aquifer. The results of this investigation [27] were used by the USGS's Office of Hazardous Waste Hydrology for establishing a national multidisciplinary research demonstration area to apply the latest techniques for characterizing hazardous waste problems.

In 1983, a multidisciplinary research effort was begun to study the processes that affect the occurrence, transport, and fate of the contaminants in the aquifer. The participants included biologists, hydrologists, two organic geochemical teams, inorganic geochemists, geophysicists, and two microbiological teams, one specializing in aerobic systems and the other in anaerobic systems.

The first comprehensive sampling for the study was done in July 1983 and supplemented in October of the same year. Test holes were drilled upgradient from the impoundments (site 1) and in a line downgradient to the bay, line A–A' in Fig. 18. Multilevel well clusters at sites 3, 4, 5, 6, and 7 were drilled near the suspected centroid of contamination and at sites 2, 8, and 9 to sample the anticipated lateral extent of the contamination. In October 1983, the unlined wastewater impoundments were dewatered and capped under the supervision and requirements of the U.S. Environmental Protection Agency Superfund Program.

A year after the pond remediation effort (October 1984), two wells, at sites 16 and 17, shown in Fig. 19, were drilled at a location thought to be a safe distance from the affected groundwater zone for the purpose of obtaining hydrologic measurements. Unexpectedly, the groundwater was found to be contaminated at these locations, showing 2 and 6 mg/L PAH in water from sites 16 and 17, respectively, down to a depth of 30 m. Phenols also were detected at both sites. Additional analyses of samples from sites 18 and

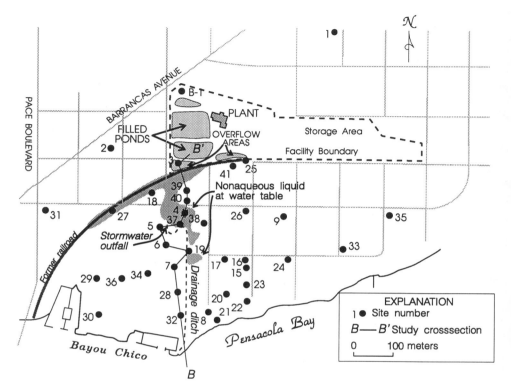

Figure 19 Location of test sites and certain drainage features at abandoned creosote works.

19 confirmed that the contaminant plume was more extensive than originally calculated. These findings brought about a reconsideration of the supposed limits of contaminant migration.

To define the extent of the involvement more adequately, an extensive well drilling and sampling operation was done in March 1985. Water samples from these test wells were analyzed first on-site, then later in the laboratory. Field analyses were performed to immediately detect the presence of specific contaminants, to permit timely information for the test drilling, to help locate the contamination in the groundwater, and to identify the presence of readily degradable solutes that might otherwise undergo alteration before laboratory analysis. HPLC was used at the site for the determination of PAHs, NH, and phenols [18]. After processing, samples were sent to the laboratory in Menlo Park for more comprehensive HPLC and GC analysis and confirmatory identification of the organic solutes by GC/MS and HPLC/MS. Analysis for purgeables, GC/MS, was done by a private laboratory in Tallahassee, Florida. Gas chromatography was also used at the site for the measurement of methane in the water samples [28].

C. Sample Collection and Analysis

To reduce the possibility of cross-contamination, test well drilling proceeded from the least affected parts of the aquifer toward the source. The drilling equipment, such as auger flights and sampling apparatus, was thoroughly steam-cleaned after each use. Apparatus directly in contact with samples was also rinsed with HPLC grade acetone. Test holes were drilled using 3¾-in. (9.5-cm) steel hollow-stem augers. The hollow stem was sealed by O-rings at each auger joint and by a bottom plate in the bit to keep out material during drilling. One-time test holes, drilled where construction of a well was not possible or secure, were sampled through the hollow-stem auger at the selected depth. Generally, the auger stem was opened at the sampling depth, using a screened wall point to push the sealing plate from the bottom. Well construction was completed by inserting the casing into the hollow auger and allowing the sands to collapse around the casing as the auger flights were removed. Galvanized pipe, 2-in. (5-cm) nominal size, screwed to a 1-m-long stainless-steel screen, was used to case wells scheduled for repetitive sampling. The metal casings and screens were steam-cleaned and then rinsed with methanol and liberal amounts of distilled water prior to being placed in the test hole. Two to five wells, set at selected depths, were installed at each site.

Samples were collected immediately after the drilling of the test hole. One-time test holes were sampled through the hollow-stem auger at the selected depth from a screened well point by pumping or bailing water from the hollow auger. For sampling established test wells, a volume of water equivalent to at least five times the calculated amount in the well casing was removed before samples were taken. The pump intake was set near the surface of the water to ensure removal of the standing water from the casing. Water samples for organic solutes were then collected by bailer or by means of a peristaltic pump with a 1-L glass bottle connected between the pump and a length of 6-mm i.d. Teflon tubing. Only the Teflon tubing was inserted down the hole to the screen, and once in place it remained for any subsequent use. The water was drawn from the well through a Teflon stopper into the bottle without contacting the silicone rubber pump tubing. Samples not analyzed at the site were packed on ice and shipped by overnight delivery service to the laboratory. Samples taken after July 1983 were collected by bailer and, in addition to packing in ice, were preserved by adding 65 mg/L of mercuric chloride to each bottle as a biocide.

D. Results and Discussion

Data collected in the 1981 investigation indicated the presence of heavy contamination of the aquifer in the vicinity of the unlined surface impoundments of the creosote plant. The highest concentrations were found immediately downgradient from the overflow-holding pond, and the extent of the contamination was detected in a privately owned well 340 m southeast of the plant. During the reconnaissance phase of the investigation, the total phenol determination [2] was used to trace the contamination in the groundwater. Total phenol measurement was chosen to detect the contamination because phenols are highly soluble in water. Also, there is evidence that phenols are not significantly sorbed during transport in aquifers, such as this one, that have low natural organic content [24].

Generally, the groundwater flow is to the south toward the bay, and mean velocities were found to vary between 0.03 and 0.3 m/day, slower at increasing depths. Previous industrial pumpage and continuing water withdrawals for domestic irrigation influenced the lateral dispersion of the contaminated groundwater to some extent. Low levels of contamination were found in several shallow private irrigation wells southeast of the plant, and it was suspected that this migration was induced by irrigation pumpage.

The behavior of the phenols in the deeper aquifer appeared to be similar to processes at St. Louis Park, Minnesota [23]. The migrating phenol was undergoing attenuation in transport, and the presence of high levels of methane suggested active microbial degradation. The microbial activity was suspected to be a major factor in this downgradient attenuation, and laboratory experiments using bacteria isolated from the groundwater produced methane from contaminated water collected at the Pensacola site.

The data obtained from analysis of samples taken during the 1983 sampling sequence presented a somewhat different condition than was originally contemplated. A rather consistent silt-clay layer was found a few meters below the surface, extending to a depth of 15 m. This layer was less permeable than the sand structure and sectioned the local aquifer into two layers.

In general, the polynuclear aromatic hydrocarbons and phenols migrated with the groundwater as expected, and the concentrations decreased with distance. A significant anomaly, however, was observed approximately 300 m downgradient, suggesting that other means of transport were operating. Concentrations of both PAHs and phenols were greater by nearly an order of magnitude at the 18 m depth, near and below the emergence of drainage water from a buried drain tile discharging to the drainage ditch (Fig. 19).

Subsequent analyses of samples taken from additional wells installed nearby confirmed the original finding. No credible explanation was offered at the time, but it was generally accepted that this finding was somehow influenced by the drainage feature. This anomaly complicated the interpretation of the behavior of the solutes from this point on downgradient. Movement of the groundwater contaminants was unaffected by the drainage in the reach between the ponds and the emergence of the drainage flow, and study of the transport phenomena within this undisturbed reach continued.

Later, analytical data obtained from samples collected in this reach were evaluated, and a number of solutes were selected as amenable to transport modeling. The selected compounds and applicable data are listed in Table 4. Except for the concentrations column, these data, particularly K_{ow} and R, are taken from the accumulated chemical literature and are not site-specific. After attempts to relate these values to distribution coefficients, Franks [29] concluded that these data are not representative of natural material and should be treated only as order-of-magnitude estimates of the properties of these organic compounds. Evaluation of chemical data from groundwater samples collected in 1983 indicated that differential attenuation, or retardation of the more soluble compounds, was occurring. Evidence from laboratory experiments, however, confirmed that sorption was not a major factor of influence on the apparent attenuation [30].

Anaerobic microbial alteration processes were strongly implicated in the attenuation of many of the solutes studied [31]. These findings were consistent with the previous laboratory studies and a similar observation reported at St. Louis Park [19]. The detection of methane was common to both studies. The possibility that the methane originated prior to the entry of the waste into the groundwater was ruled out, as before, by the methane

Table 4 Properties and Measurements of Organic Solutes Found Contaminating the Groundwater and Selected for Transport Modeling at Pensacola, Florida

Compound	Solubility (mg/L)	Maximum measured concentration (mg/L)	Log K_{ow}	Specific gravity	R
Naphthalene	30	15.3	3.0–3.4	1.15	2–6
Phenol	80,000	26.0	1.46	1.07	1–1.3
2,4-Dimethylphenol	4,000	8.4	2.50	1.04	1.5–3.2
3,4-Dimethylphenol	4,000	19.3	2.35	1.04	1.2–2.3
Quinoline	60,000	14.8	2.03	1.09	1–1.6

Source: Franks [30].

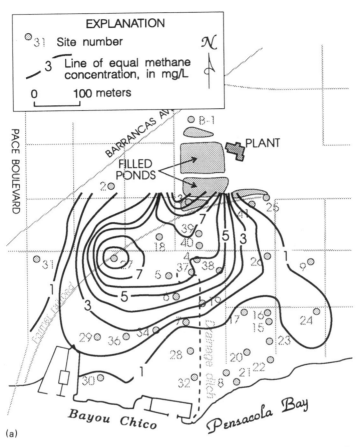

Figure 20 Methane concentrations in groundwater (a) in the 5–15-m depth interval. (b) in the 15–25-m depth interval.

concentration profile in the subsurface. The methane concentration increased with increasing waste-migration distance from the point of entry. If the methane had been in the waste prior to entry into the groundwater, the concentration would have decreased by dispersion in the downgradient direction. The extent of methane in the groundwater of the two layers of the aquifer is drawn in Figs. 20a and 20b [28]. Significantly high concentrations of low molecular weight organic acids were found in the groundwater and are additional indicators of anaerobic microbial activity [30]. A summary of these findings is given in Table 5.

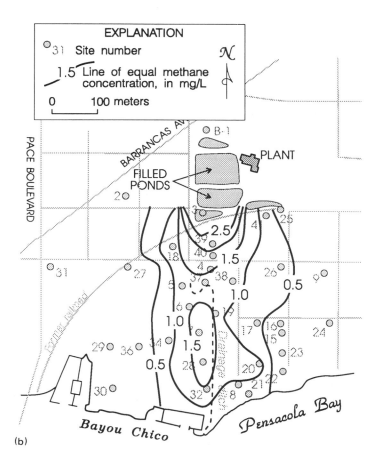

(b)

There was concern that the actual field microbial processes were locally suppressed by the inhibiting action of PCP on the microbes [32]. Although large amounts of PCP were used for wood treatment at the plant, it was detected only close to the source, and usually only when a nonaqueous liquid phase was observed. This was inconsistent with the observations at Visalia, California. The presence of organic acids in the vicinity of the source and the low ambient pH of the groundwater suppressed the ionization, and consequently the solubility, of PCP. The solubility of PCP was experimentally determined to be 4–5 mg/L in native groundwater at pH 5. This was close

Table 5 Concentrations of Organic Solutes in Samples Collected from Several
Depths at Sites 3 and 39, Pensacola, Florida

Site	3	39	39	39
Sample depth (m)	6.1	3.3	5.8	11.0
Compound	Concentration (mg/L)			
PAH compounds (includes OxH, SH)				
Acenaphthalene	10.4	0.00	0.00	0.00
Benzothiophene	0.76	0.47	0.49	0.81
Biphenyl	0.25	0.08	0.10	0.11
Dibenzofuran	0.62	0.31	0.19	0.27
Fluorene	0.58	0.26	0.16	0.22
Indene	0.14	0.76	0.72	0.60
Naphthalene	10.4	8.30	8.30	13.20
Phenanthrene	2.00	0.26	0.06	0.13
1,2-Dihydroacenaphthalene	0.84	0.38	0.33	0.45
1-Methylnaphthalene	0.60	0.31	0.42	0.59
2-Methylnaphthalene	1.20	0.54	0.78	1.09
2-Phenylnaphthalene	0.21			
Total	28.0	11.7	11.6	17.5
Phenolic compounds				
PCP	3.53	0.00	0.77	0.25
Phenol	35.2	9.78	9.53	0.06
1-Naphthol	0.17	0.00	0.39	0.05
2,3,5,6-Tetramethylphenol	0.06	0.00	0.03	0.03
2,3,5-Trimethylphenol	0.63	0.63	0.53	0.16
2,3,6-Trimethylphenol	0.56	0.27	0.34	0.18
2,3-Dimethylphenol	1.62	0.82	0.82	0.19
2,4,6-Trimethylphenol	0.41	0.36	0.12	0.05
2,4-Dimethylphenol	9.68	4.34	4.92	1.33
2,5-Dimethylphenol	0.00	0.56	0.00	0.00
2,6-Dimethylphenol	1.34	0.50	0.60	0.17
2-Ethylphenol	0.33	0.21	0.15	0.07
2-Methylphenol	14.6	6.02	5.59	0.46
2-Naphthol	1.66	0.62	1.07	0.33
3,4-Dimethylphenol	3.23	1.80	1.67	0.95
3,5-Dimethylphenol	16.8	5.90	7.58	1.24
3-Methylphenol	32.1	13.9	11.6	0.61
4-Methylphenol	16.0	7.00	5.81	0.30
Total	138.0	52.6	51.5	6.43
Nitrogen Compounds				
Acridine	0.11	0.00	0.01	0.00
Carbazole	0.80	0.64	0.07	0.98
Isoquinoline	2.24	0.21	0.11	0.00

Table 5 Continued

Site	3	39	39	39
Sample depth (m)	6.1	3.3	5.8	11.0
Compound	Concentration (mg/L)			
Pyridine	0.37	0.21	0.09	0.00
Quinoline	10.50	0.45	0.01	0.00
1-Isoquinolinone	29.7	22.0	5.80	0.00
2,3,6-Trimethylpyridine	0.42	0.29	0.20	0.05
2,3-Dimethylpyridine	0.69	0.21	0.10	0.24
2,4,6-Trimethylpyridine	2.00	0.63	0.27	0.05
2,4-Dimethylpyridine	3.42	1.10	0.60	0.11
2,4-Dimethylquinoline	0.08	0.16	0.05	0.00
2,5-Dimethylpyridine	1.03	0.31	0.14	0.04
2,6-Dimethylpyridine	0.63	0.30	0.20	0.03
2,6-Dimethylquinoline	0.17	0.04	0.04	0.00
2-Methylpyridine	0.91	0.34	0.26	0.00
2-Methylquinoline	2.82	0.81	0.31	0.35
2-Quinolinone	27.6	66.0	19.00	0.00
3-Methylpyridine	1.23	0.30	0.20	0.01
4-Methylpyridine	1.34	0.43	0.25	0.00
4-Methylquinoline	0.21	0.02	0.01	0.00
6-Methylquinoline	0.50	0.02	0.02	0.01
Total	86.8	94.5	27.7	1.87
Acids				
Acetic	67.1	63.9	30.3	3.52
Benzeneacetic	3.94	1.30	1.45	0.00
Benzenepropionic	1.52	1.20	0.49	0.00
Benzoic	22.3	2.32	6.44	0.00
Benzoic, 4-methyl	1.89	0.00	1.41	0.00
Butyric	20.6	2.20	1.50	0.00
Butyric, 2-ethyl	0.00	0.29	0.00	0.00
Butyric, 3-methyl	2.01	0.78	0.59	0.00
Formic	3.88	1.66	1.30	1.00
Hexanoic	2.99	0.00	0.28	0.00
Hexanoic, 3-methyl	0.00	0.30	0.00	0.00
Hexanoic, 2-ethyl	0.17	5.79	1.44	0.00
Hexanoic, 2-methyl	0.00	0.91	0.35	0.00
Pentanoic, 2-methyl	0.06	0.00	0.10	0.00
Propionic	31.4	9.03	3.39	0.09
Propionic, 2-methyl	3.32	1.63	1.25	0.00
Pentanoic	6.55	1.53	0.51	0.00
Total	167.7	92.9	50.8	4.61

to the acidity of the groundwater within the first 20 m downgradient from the source and perhaps accounted for the limited movement of PCP in the aqueous system. The acidity of the groundwater adjacent to the contamination ranged from pH 6.7 to pH 7.1. Within the contamination, the pH range was 5.0–6.3. PCP migrated more than 500 m from the source in Visalia. by contrast, the ambient pH of the water at the Visalia location is neutral to alkaline, and, experimentally, the solubility of PCP in that groundwater exceeded 60 mg/L.

Concentrations of PAH and phenols in water samples collected in March 1985 from the wells nearest the source were found to be comparable to the 1983 data [33]. Not expected were the high concentrations of NH compounds, nearly an order of magnitude higher than previously reported [34]. Analytical data obtained from samples taken from various depths at site 39, for example, are listed in Table 5. Only small amounts of the operationally defined "purgeable compounds" were found between the source and the drainage ditch. Maximum concentrations found were benzene, 170 μg/L; ethylbenzene, 58 μg/L; toluene, 230 μg/L; and xylene, 350 μg/L.

Figure 21 illustrates the concentration profile of NH compounds along the line B–B' in Fig. 19, both in depth and downgradient directions. It was

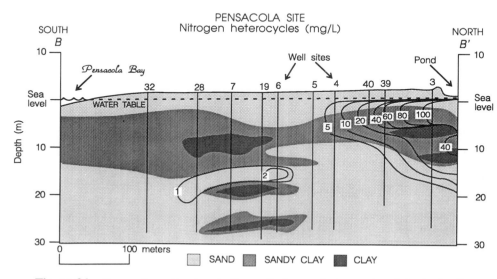

Figure 21 Summation of concentrations of nitrogen heterocycle solutes in depth profile along geologic cross section B–B'.

discovered during on-site analysis that the NH compounds were undergoing very rapid degradation, a half-life of approximately 2 days in the sample bottle held at 4°C. Exposure to the air apparently accelerated the degradation. Immediate analysis or effective sample preservation was found to be necessary for this work if the presence of these transitory compounds was to be detected. Although creosote consists of about 85% PAH, 10% phenols, and 3% NH, OxH, and SH, the bulk of the solutes in water adjacent to the source are phenols and the NH. Figure 22 illustrates the migration of phenols along the B–B′ section, and Fig. 23 shows the movement of PAHs. The PAHs are slightly soluble in water, naphthalene to about 30 mg/L, and the higher molecular weight PAHs even less so. The phenols and NH are highly water-soluble, in the range of grams per liter. Until depleted, the phenols and NH are the predominant solutes leaching from the parent liquid.

Elevated levels of contamination, suggestive of a nearby and relatively fresh liquid source, were again detected in October 1984 in groundwater samples collected just below the drainage discharge at site 19. Samples collected in March 1985 confirmed a greater lateral spread of contamination than was calculated for simple groundwater migration. Inspection of aerial

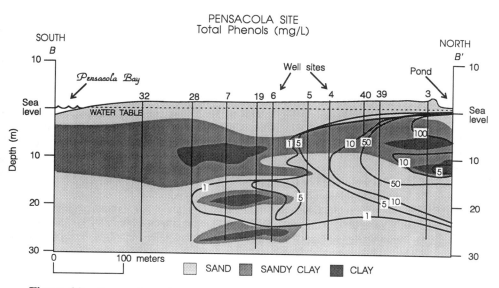

Figure 22 Summation of concentrations of phenol solutes in depth profile along geologic cross section B–B′.

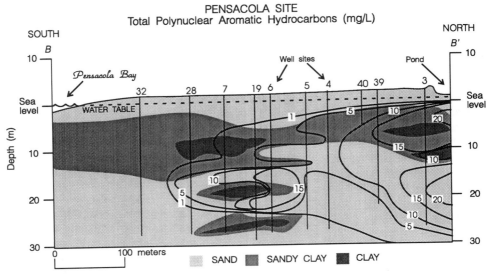

Figure 23 Summation of concentrations of aromatic hydrocarbon solutes in depth profile along geologic cross section B–B′.

photographs and examination of surface soil borings suggested other possible reasons.

Aerial photos showed evidence of a previously existing railroad passing along the south property line of the creosote works and turning to the south across Bayou Chico. The suspected path of the former railroad is drawn in Fig. 19. Water samples from the 6–8-m depth taken at sites 18 and 27, along the former railroad, contained levels of contamination, including degradable compounds, at concentrations as high as those found at the source. Examination of soil borings and further inspections of the aerial photos revealed that the land between the plant and the bay was earlier a wetland area. Prior to the legislation of the Clean Water Act in 1970, the contents of the holding ponds were occasionally allowed to overflow when water levels became high and also during and following heavy rains [27]. The treatment fluid discharged directly into Pensacola Bay, and some collected in the wetland area. Evidence from the aerial photos and soil borings showed that the wetland area was filled and covered over with soil. It is presumed that there was significant drainage of treatment fluid along the drainage features associated with the railroad bed. The rails have since been removed and the roadbed covered over and leveled.

Overland flow was suspected to be responsible for the movement of large amounts of parent material and helps to explain the findings of otherwise inordinately high concentrations of both the recalcitrant PAHs and readily degradable phenols and NH at much greater distances downgradient than would result from ordinary solute transport. As late as 1986, during and immediately following heavy rainstorms, nonaqueous oily fluid was readily observed flowing along the bottom of the drainage ditch and appeared to emerge from a visible oily layer on top of the water table surface intersecting the bank of the ditch. The darkened area south of the ponds and along the abandoned railroad drawn in Fig. 19 is the conceptualized location of the areas affected by overland flow, as reconstructed from visual and chemical examination of selected soil borings. Concentrations of methane were higher in the near-surface layer of the aquifer than in the deeper layer at these locations. (See Fig. 20a.) This suggests that a significant reservoir of relatively unleached source material remains buried and continues to add solutes to the groundwater.

E. Conclusions

The site at Pensacola was selected as a multidisciplinary research demonstration area primarily because in the initial evaluation it was determined to be a relatively idealized field laboratory site for groundwater research. As it turned out, the aquifer was not simple, and the history of overland flow combined with the existence of drainage ditches and roadways worked to complicate the study, calling for several resamplings and reexamination and revision of concepts.

A less permeable silt-clay layer, 5–15 m thick, separates the aquifer into two layers: a shallow 2–6 m layer and a deeper 15–30 m layer. Although the lithology directly beneath the location of the former ponds is not well documented, the nearest test holes at site 3 showed a heavier clay lens at 12 m that appears to extend some distance to the north, possibly causing a layering of groundwater flow at the source and slowing the downward rate of movement of denser components.

The impact of overland flow was of more consequence than was originally thought. Overland flow predominates in much of the upper layer where it is accumulated. The contaminants occur and are transported in the lower permeable layer in a more expected and predictable manner, although suspected intrusion of parent treatment fluid from overland flow did appear in the deeper layer below the storm drain outfall.

When the influences of the physical situation were taken into account, the behavior of the solutes in the aquifer system more closely followed that seen in earlier studies. Detection of methane is solid indication of microbial activity, as is the presence of volatile fatty acids. The decrease in concentrations of phenols and NH relative to PAH further implicates microbial degradation or alteration of these solutes, especially because sorption was shown experimentally not to be a factor. Earlier, the dimethylphenols appeared to be unaffected in transport [30], but neither naphthalene, the most prevalent PAH, nor any dimethylphenols were detected at the farthest distance sampled. Elevated levels of dissolved organic carbon and methane were found in samples taken from the wells farthest downgradient. This suggests that all the major solutes from creosote are being catabolized in this environment.

Vertical profiles of concentrations of organic solutes indicate that in the vicinity of the water table, microbial activity is greatest, rapidly altering the composition of the materials in a transition zone. Furthermore, any nonaqueous liquid situated at the water table provides a poorly defined boundary with respect to quantitative sampling in this zone.

The advantages of on-site chemical analysis for discrete contaminants in addition to pH, alkalinity, and dissolved oxygen have been demonstrated. Furthermore, the high concentrations of NH may not have been detected in this instance without on-site HPLC. Methane determinations and the analysis of water samples for individual solutes during the test drilling improved reliable placement of sampling points, both horizontally and vertically, and helped to correctly define the extent of contamination.

VIII. SUMMARY

The histories of studies of groundwater contamination at six sites by the Organics Project illustrate some of the approaches and problems faced in conducting such investigations. Although much of the information given is now generally known and improved technology has helped eliminate some of the many uncertainties common to groundwater studies, a much broader understanding is still needed. The scientific assessment of the fate and transport of organic solutes in groundwater, in most instances, is complicated and calls for a multidisciplinary approach by knowledgeable and experienced investigators. This was one of the most important needs drawn from the studies reviewed in this section.

The importance of a multidisciplinary approach was recognized during the first study, the chlorinated hydrocarbon waste dump in Hardeman County, Tennessee. The aquifer cores taken in the first sampling were compromised, down-hole, by contaminated surface soil. Additionally, the compounds targeted for analysis to be representative of the contamination were found not to be. Direct communication between the site project chief and the analytical chemists followed by a second sampling trip helped to salvage this investigation. The chemists participated in the planning and collection of the second round of samples.

The problems associated with the isolated approach lessened as more experts from other scientific disciplines were included in the planning and analytical stages of the later investigations. These disciplines include hydrology, chemistry, physics, and biology. Training and utilization of experienced field workers, well drillers, and sampling personnel significantly improved sample reliability. Finally, the importance of on-site chemical analysis for specific contaminants and alteration products, both for analysis and quality assurance of samples, is discussed. HPLC provided adequate specificity and sensitivity for these purposes, and field laboratory data comparisons are given.

REFERENCES

1. Shuter, E., and Teasdale, W.E. Applications of drilling, coring, and sampling techniques to test holes and wells. *U.S. Geological Survey Techniques of Water-Resources Investigations*, book 2, 1989, Chapter F1.
2. Goerlitz, D.F., and Brown, E. Methods for the analysis of organic substances in water. *U.S. Geological Survey Techniques of Water-Resources Investigations*, book 5, 1972, Chapter A3.
3. Burke, J. Programmed temperature gas chromatography (PTGC) and microcoulometric detection of chlorinated insecticides. *Assoc. Offic. Agric. Chem. J., 46*:198, 1963.
4. Peech, M. Hydrogen-ion activity. In *Methods for Soil Analysis,* Part 2, *Chemical and Microbial Properties,* Vol. 9 (C.A. Black, ed.), American Society of Agronomy, Inc., Madison, WI, 1965, p. 171.
5. Rima, D.R., Brown, E., Goerlitz, D.F., and Law, L.M. Potential contamination of the hydrologic environment from the pesticide waste dump in Hardeman County, Tennessee. USGS Open-File Rep., Nashville, TN, 1972.
6. Dion, N.P. A proposal for the investigation of possible groundwater contamination in the Bangor Area, Kitsap County, Washington. USGA Open-File Rep., Tacoma, WA, 1974.
7. Molenaar, D. Geology and ground-water resources. Water Resources and Geology of the Kitsap Peninsula and Certain Adjacent Islands. Washington Div. Water Resour. Water Supply Bull. *18*:24, 1965.
8. Tracy, J.V., and Dion, N.P. Evaluation of ground-water contamination from cleaning explosive-projectile casings at the Bangor Annex, Kitsap County, Washington Phase II. USGA Water Resour. Invest. Rep., 62-75, Tacoma, WA, 1976.
9. Goerlitz, D.F., and Law, L.M. Gas chromatographic method for the analysis of TNT and RDX explosives contaminating water and soil-core material. USGS Open-File Rep. 75-182, 1974.
10. Won, W.D., Heckley, R.J., Glover, D.J., and Hoffsommer, J.C. Metabolic disposition of 2,4,6-trinitrotoluene. *Appl. Microbiol. 27*(3):513, 1974.
11. McCormick, N.G., Cornell, J.H., and Kaplan, A.M. Biodegradation of hexahydro-1,3,5-trinitro-1,3,5-triazine. *Appl. Environ. Microbiol. 42*(5):817, 1981.

12. von Rumker, R., Lawless, E. W., and Meiners, A. F. Production, distribution, use and environmental impact potential of selected pesticides. U.S. Environmental Protection Agency, EPA 540/1-74-001, 1975. (American Chemical Society. *Chem. Eng. News, 56*:15, 1978.)

13. Croft, M.G. Subsurface geology of the late tertiary and quaternary water-bearing deposits of the southern part of the San Joaquin Valley, California. USGS Water Supply Paper 1999-H, 1972.

14. Kirsch, E.J., and Etzel, J.E. Microbial decomposition of pentachlorophenol. *J. Water Pollution Contam. Fed., 45*(2):359, 1973.

15. Goerlitz, D.F. Determination of pentachlorophenol in water and aquifer sediments by high-performance liquid chromatography. USGS Open-File Rep. 82-124, 1982.

16. Ehrlich, G.G. Personal communication, 1977.

17. Villaume, J.F. Investigations at sites contaminated with dense, non-aqueous phase liquids (NAPLs). *Ground Water Monit. Rev., 5*(2):60, 1985.

18. Goerlitz, D.F., and Franks, B.J. Use of on-site high performance liquid chromatography to evaluate the magnitude and extent of organic contaminants in aquifers. *Ground Water Monit. Rev., 9*(2):122, 1989.

19. Van Denburgh, A.S., and Goerlitz, D.F. Mobility of nitrogen-bearing wastes and related compounds in shallow ground water near Hawthorne, Nevada. Geological Society of America, Abstracts with Programs, Phoenix, AZ, Vol. 9, No. 7, 1987, p. 875.

20. Committee on Water Quality Criteria. *Water Quality Criteria*, Environmental Protection Agency Pub. R3-73-033, 1972, p. 73.

21. Van Denburgh, A.S. Personal communication, 1990.

22. Hult, M.F., and Schoenberg, M.E. Preliminary evaluation of ground-water contamination by coal-tar derivatives, St. Louis Park Area, Minnesota. USGS Open-File Rep. 81-72, St. Paul, MN, 1981.

23. Ehrlich, G.G., Goerlitz, D.F., Godsy, E.M., and Hult, M.F. Degradation of phenolic contaminants in ground water by anaerobic bacteria: St. Louis Park, Minnesota. *Ground Water, 20*(6):703, 1982.

24. Goerlitz, D.F. A column technique for determining sorption of organic solutes on the lithological structure of aquifer. *Bull. Environ. Contam. Toxicol., 32*(3): 261, 1984.

25. Healy, J.B., Jr., and Daughton, C.G. Issues relevant to biodegradation of energy-related compounds in ground water—a literature review. Univ. California Bull. UCB/SEEHRL86-10, Berkeley, CA, 1986, p. 6.

26. Franks, B.J. Principal aquifers in Florida. USGA Water-Resour. Invest. Open-File Rep. 82-255, 1982.

27. Troutman, D.E., Godsy, E.M., Goerlitz, D.F., and Ehrlich, G.G. Phenolic contamination in the sand-and-gravel aquifer from a surface impoundment of wood treatment wastes, Pensacola, Florida. USGS Water Invest. Rep. 84-4230, 1984.

28. Baedecker, M.J., Franks, B.J., Goerlitz, D.F., and Hopple, J.A. Geochemistry of a shallow aquifer contaminated with creosote products. In *U.S. Geological*

Survey Toxic Waste–Groundwater Contamination Program (S.E. Ragone, ed.), Proc. Second Tech. Meeting. Cape Cod, MA, Oct. 21–25, 1985, USGA Open-File Rep. 86-481, 1988, Chapter A, p. 27.

29. Franks, B.J. Transport of organic contaminants in a surficial sand aquifer—hydrogeologic and geochemical processes affecting transport movement. Dissertation, Florida State Univ., Tallahassee, FL, 1988, p. 37.

30. Goerlitz, D.F., Troutman, D.E., Godsy, E.M., and Franks, B.J. Migration of wood preserving chemicals in contaminated ground water in a sand aquifer at Pensacola, Florida. *Environ. Sci. Technol., 19*(10):955, 1985.

31. Godsy, E.M., and Goerlitz, D.F. Anaerobic microbial transformations of phenolic and other selected compounds in contaminated ground water at a creosote works, Pensacola, Florida. In Movement and Fate of Creosote Waste In Ground Water, Pensacola, Florida: U.S. Geological Survey Toxic Waste–Ground-Water Contamination Program (H.C. Matraw and B.J. Franks, eds.), USGS Water Supply Paper 2285, 1984, Chapter H, p. 55.

32. Godsy, E.M., Goerlitz, D.F., and Ehrlich, G.G. The effect of pentachlorophenol on methanogenic fermentation of phenol. *Bull. Inviron. Contam. Toxicol., 36*: 271, 1987.

33. Goerlitz, D.F. Re-examination of the occurrence and distribution of creosote compounds in ground water. In U.S. Geological Survey Toxic Waste–Ground-Water Contamination Program (S.E. Ragone, ed.), Proc. Second Techn. Meeting, Cape Cod, MA, Oct. 21–25, 1985, USGA Open-File Rep. 86-481, 1988, Chapter A, p. 21.

34. Pereira, W.E., and Rostad, C.E. Investigations of organic contaminants derived from wood-treatment processes in a sand and gravel aquifer near Pensacola, Florida. In *Selected Papers in the Hydrologic Sciences 1986* (S. Subitzky, ed.), USGS Water-Supply Paper 2290, 1986, p. 65.

12

Pollution of Groundwater by Organic Compounds Leached from Domestic Solid Wastes: A Case Study from Morley, Western Australia

Chris Barber, David John Briegel, Terrence R. Power, and Janis Kaye Hosking

Commonwealth Scientific and Industrial Research Organization, Perth, Western Australia, Australia

Integrated sampling and analytical procedures have been developed for the determination of volatile and nonvolatile organic compounds in groundwater. This formed part of a study of pollution at a domestic waste site at Morley, a suburb of Perth in Western Australia. The sampling procedures, which involve purge-sampling and the taking of fresh groundwater samples in situ using syringe samplers, could be easily adapted for routine monitoring of groundwater pollution.

Volatile organic compounds (benzene, toluene, TCE, and PCE) were determined by purge-and-trap GC techniques. Sample degassing and sample transfer were avoided by purging within the removable syringe used for sampling. Reproducibilities on replicate samples obtained over a 2-year period were generally good.

Concentrations of benzene, toluene, TCE, and PCE in groundwater were all low. Peak concentrations, particularly of benzene, were observed before breakthrough of inorganic leachate constituents (chloride and ammonia) and nonvolatile carbon (TOC). This was tentatively attributed to delayed microbial utilization of these compounds in the groundwater system, which reduced concentrations within the heavily polluted part of the plume.

Nonvolatile organic compounds were characterized by a fractionation technique that uses exchange resins to separate dissolved carbon into hydrophobic and hydrophilic acids, bases, and neutral fractions. This technique was used to assess the attenuation of nonvolatile organics within the aquifer at Morley. Hydrophobic and hydrophilic acids were dominant in all samples. Attenuation was found to be limited. Some removal of the hydrophobic acid fraction was observed. This was attributed to sorption of humic acid rather than the more hydrophilic fulvic acid.

I. INTRODUCTION

The Swan coastal plain aquifer in Western Australia provides the state capital, Perth, with 35% of its water for domestic use [1]. Sixty percent of this groundwater is abstracted from the unconfined aquifer within coastal sands and limestones lying between the scarp bordering the Darling ranges and the Indian Ocean (Fig. 1). There are also 60,000 private wells operating within the Perth metropolitan area, mainly to provide irrigation water. Because of a rapid expansion of the urban area across the coastal plain, groundwater in the unconfined aquifer is being stressed, both from overpumping [1] and as a result of pollution [2].

Pollution of groundwater by nitrate (from septic tanks and the use of garden fertilizers) has been recognized in a well field within an urbanized area close to Perth [2,3]. Other sources of pollution, particularly of organic compounds leaking from underground storage tanks, were identified as potentially more damaging to the groundwater resource, following experience in North America and Europe [2]. Consequently, interest was focused on investigation of organic pollution within the unconfined aquifer.

Sanitary landfills of domestic and commercial solid wastes have also produced localized pollution of groundwater on the Swan coastal plain [2]. The extent of leaching of specific organic compounds from these wastes, however, has not yet been investigated. A major project is currently being carried out to develop more efficient techniques for the assessment of pollution of groundwater from waste sites. This paper reports on part of this research on leaching of organic compounds.

The study had two main objectives:

1. To develop an integrated sampling/analysis system, specifically for the determination of volatile and nonvolatile organic contaminants in groundwater
2. To assess the extent of leaching of these compounds from domestic solid wastes and determine their mobility within the saturated zone of the unconfined aquifer

The study was carried out around a working landfill in the suburb of Morley, within the Perth metropolitan area.

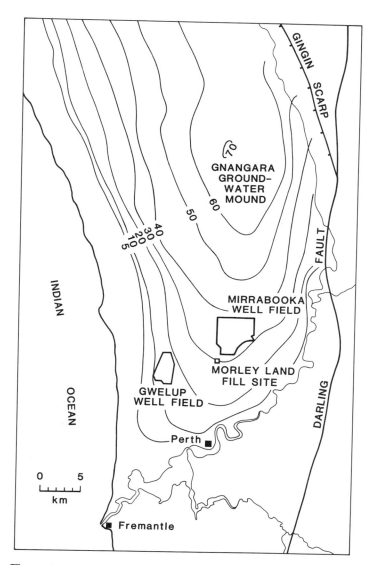

Figure 1   Map showing the northern part of the Swan Coastal Plain aquifer in Western Australia, the Gnangara groundwater mound with water table contours (m), well fields, and the Morley landfill site.

II. HYDROGEOLOGICAL SETTING

The Swan coastal plain lies on the eastern edge of the Perth Basin, which contains late Tertiary and Quaternary sediments. Beneath the plain, superficial deposits varying in depth from 10 m to 100 m unconformably overlie the older basin sediments. These deposits contain discontinuous units varying from quartz sand to coastal limestone (aeolianite). In the Perth area, these form groundwater basins, locally referred to as mounds. The Gnangara mound lies to the north of the city of Perth and the Swan River (Fig. 1).

The Morley waste site is situated on the southern limb of the Gnangara mound (Fig. 1). The site is underlain by fine to medium quartz sands approximately 70 m thick, above the Cretaceous clays of the Osborne Formation. Hydraulic conductivities at the site have been estimated to vary between 19 and 70 m/day, giving groundwater velocities in the range of 60-100 m/yr. The regional groundwater gradient is approximately 0.002 to the southwest. Depth to groundwater near the site varies from 10 to 40 m. Groundwater beneath the site within the unconfined sand aquifer is of good quality, with total dissolved solids of 150 mg/L, pH usually around 5, and dissolved oxygen 5-6 mg/L. Further details of the hydrogeological setting of the site are given in Salama et al. [4].

III. MORLEY LANDFILL SITE

The sanitary landfill has been in operation since 1980, infilling part of a working sandpit where excavations are taking place within part of a dune ridge. The sandpit is up to 40 m deep, with excavation to within a few meters of the water table, occasionally below it. The site license for disposal of domestic solid waste at the site issued in 1980 required a buffer zone of at least 1 m of sand to remain (or be emplaced) between wastes and groundwater. The site thus operates on the basis of dilution and dispersion of leachate in groundwater, as opposed to the common North American practice of containment, collection, and treatment of leachate. At the Morley site, dispersion of leachate contamination in groundwater downgradient of the site was expected to take place beneath an extensive area of bushland not scheduled for development.

The site received baled domestic solid waste between 1980 and 1987. After this time, crude domestic wastes were deposited. Sand was used as intermediate cover. Some liquid waste (septage) was treated and spray-irrigated in evaporation ponds on lined areas of the site. Some liquids and sludges from road drains have been deposited directly onto wastes since wastewater treatment ceased in 1988. Generally, the amount of wastewater that was allowed to infiltrate into the wastes was minimal. No hazardous liquid (solvents, oils) or solid wastes have been deposited at the site. Leachate is thus likely to reflect the leaching and decay of domestic and commercial wastes.

Twenty-two observation boreholes were drilled to monitor the development of pollution around the site (Fig. 2). These include boreholes emplaced specifically for monitoring organic contamination. In addition to regular monitoring of these boreholes, several soil-gas surveys have been carried out to delineate the groundwater plume of pollution [5]. Geoelectric soundings [vertical electrical sounding (VES) and early-time transient electromagnetic sounding (TEM)] have been carried out at 6-month intervals to assess the extent of the plume and to help determine the rate of spread of contamination [6]. Semianalytical and numerical modeling is also being used to provide indications of the long-term impact of the site on groundwater quality.

Borehole monitoring and geoelectric sounding have shown that the pollution plume extends from the site to the southwest, in the local and regional direction of groundwater flow. The extent of the plume in late 1989 and borehole positions are shown in Fig. 2. The pollution plume showed considerable expansion between 1986 and 1990. Monitoring and modeling also indicate

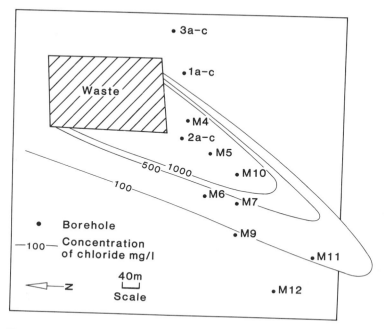

Figure 2 Extent of pollution of groundwater at the Morley site, given by the maximum concentrations of chloride observed in boreholes in November 1989.

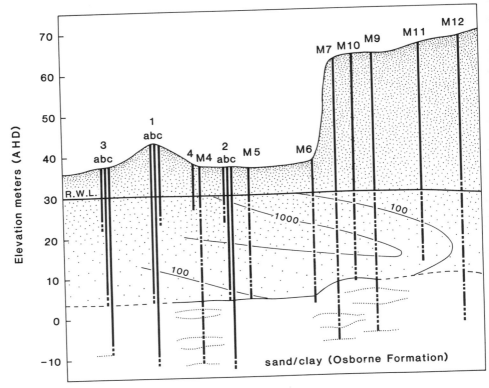

Figure 3 Section through the sand aquifer at Morley showing borehole screen locations and distribution of chloride with depth as determined by boreholes 2a-c, M5,
M10, and M11 only.

some vertical movement of the plume within the aquifer. This is indicated in
the section in Fig. 3.

Boreholes were drilled specifically for sampling and analysis of organic
pollutants in eight locations identified as boreholes M4-M7 and M9-M12 in
Fig. 3. Seven of these boreholes were located within the identified plume, although one had to be abandoned (M4 in Fig. 2). Borehole M12 was located
outside the plume for background measurements. Each of these boreholes
was cased with thick-walled PVC. They were screened with 2-m lengths of
slotting every 4 m along the casing to as close to the base of the aquifer as
drilling permitted. Lengths of casing were machine-slotted to specification
by local contractors, giving multiscreened boreholes. The locations of screens
for each borehole within the aquifer are shown in Fig. 3. The reason for this

borehole design and details of the design of the sampling equipment are discussed below.

IV. DESIGN OF BOREHOLES AND SAMPLING EQUIPMENT

Design criteria and borehole/equipment construction are summarized in Table 1.

The achievement of all design criteria would have involved astronomical cost. The overall design concept was thus a compromise to achieve maximum benefit within a limited budget. The main compromise was with the borehole construction. Ideally, this should have consisted of all metal parts rather than the thick-walled PVC that was chosen.

Table 1 Design Criteria for Borehole Construction and Sampling Equipment, and Design Features Specifically for Investigation of Volatile Organic Compounds in Groundwater

Design criteria for recovery of samples for organic analysis	Design chosen
Multilevel sampling of groundwater.	(a) Large diameter (75-mm) multiscreen borehole.
	(b) Straddle-packer, pump, and in situ syringe samplers for sample recovery from different depths.
Purging of borehole fluids (casing storage water) required by pumping (site requirement for pumping against a head of 40 m).	(a) Straddle-packer, pump, and in situ syringe samplers for sample recovery from different depths.
	(b) 60-mm-diameter diaphragm (bladder) pump incorporated between packers.
No plastics or organic polymers and solvents to contact groundwater before sample is taken for organic analysis (alternatively, contact to be minimized).	Samples to be taken in situ in removable syringe samplers. Casing to be PVC without solvent-welded joints. Diaphragm pump casing and access lines to be metal; syringe samplers to be glass with metal housings.
Sampling and analysis to be coordinated, sample storage time to be minimized, sample transfer (from container to container) to be avoided. Degassing to be minimized. Multiple samples to be taken for quality control checks.	Samples taken in situ in four removable syringe samplers. Equipment adapted for purge-and-trap concentration of volatile organics within each syringe prior to analysis.

Although initially we were against the use of PVC for borehole casing, tests carried out in Canada [7,8] suggested that this material had minimal effect on absorption of organic compounds, at least with contact times up to 24 h. No PVC solvents were used during borehole installation. Instead, collar joints held by stainless-steel screws were used for joining lengths of casing.

Each screen was isolated within the boreholes by using small flange packers when not sampling and straddle packers when sampling (Fig. 4).

A single-stroke diaphragm pump was constructed as an integral part of the straddle packer assembly for purge pumping. In situ syringe samplers similar to those described by Pankow et al. [9] were also used for recovery of samples in situ between the inflatable packers. These allowed multiple (replicate) samples to be taken at any one time.

Apart from the slotted PVC casing and a small section of inflated Viton packer material, no organic materials came into contact with groundwater within the part of the borehole isolated by the inflated packer. The pump housing was constructed of stainless steel. The syringe samplers were constructed from glass and were easily installed and removed from the assembly. Connectors and vacuum/gas lines that operated the piston of the syringes were constructed of metal.

V. OPERATIONAL PROCEDURE

An electrically operated winch was used to raise and lower the packer assembly within the boreholes. Prior to lowering the assembly, syringe samplers, which were solvent-washed and cleaned, were emplaced in each sampler housing between the packers. Each syringe was primed (barrel fully up), and a small positive gas pressure was applied to keep the syringe barrel in this position during purge pumping.

Each borehole was usually sampled from the topmost screen down. Packers were emplaced so as to straddle the screened interval and were inflated to isolate the screen from the rest of the borehole. The purge pump was activated from the surface by using gas pressure; pumpstroke timing was electronically controlled. Each purge generally took 1 hr to achieve a constant electrical conductivity (EC) in discharge water, but purging often was carried out for up to 2 hr.

Samples of discharge water were taken immediately before activation of the syringe samplers for analysis of inorganic constituents and nonvolatile compounds. The pH value of pump discharge was also taken at this time. Syringe samplers were activated by releasing the gas pressure and applying a vacuum to lower the barrels of the syringes and take a sample of fresh groundwater. After this, the packers were deflated, and the whole assembly was raised to the surface for recovery of the syringes and water samples.

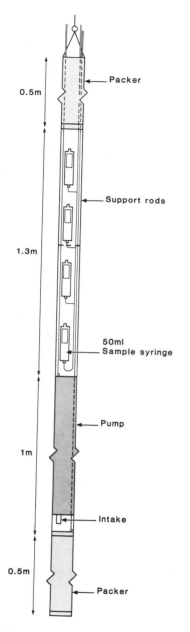

Figure 4 Straddle packer assembly used for recovery of samples for organic and inorganic analysis, showing positioning of syringe samples and pump.

Syringes were sealed using stainless-steel Luer caps. These were stored under refrigeration before transportation to the laboratory for processing and analysis, which was carried out generally within 24 hr.

VI. PRETREATMENT AND ANALYSIS

A. Volatile Organic Compounds

Concentrations of benzene, toluene, trichloroethylene (TCE), and perchloroethylene (PCE) were determined by a modified purge-and-trap/gas chromatography technique (Fig. 5). A modified Luer fitting (inset in Fig. 5) was used for purging within the syringe sampler. This avoided the transfer of sample from one container to another; losses by adsorption and volatilization were thus avoided. Purged volatiles were collected on glass-lined steel concentrator traps containing Tenax GC adsorbent.

The concentrator traps were desorbed thermally using an SGE Unijector connected in-line to a Hewlett-Packard Model 5890A gas chromatograph fitted with a 25-m long, 0.22-mm diameter Hewlett-Packard BP5 bonded open tubular fused silica capillary column and flame ionization detector. During desorption, trapped organics were held at the head of the capillary column by cooling to $-10\,°C$. Ultrahigh-purity helium was used as carrier gas. Generally, because of low concentrations found in samples, benzene and toluene could not be confirmed by GC/MSD. The presence of chlorinated compounds was confirmed using an electron capture detector.

Efficient recoveries ($<95\%$) of the four compounds could be achieved by purging aqueous samples for 20 min at $75\,°C$.

B. Nonvolatile Organic Compounds

The nonvolatile organic carbon fraction of landfill leachate is likely to consist predominantly of humic and fulvic acids and, in high-strength leachates, also of short-chain carboxylic acids [10]. In this study, short-chain carboxylic acids (C_1-C_6) were determined by both GC and HPLC techniques. Also, organic compounds in leachate-polluted groundwater were characterized using fractionation techniques developed by Leenheer and Huffman [11]. These techniques separate fractions using a range of exchange resins under various conditions. This is analogous to natural processes of exchange that may occur within an aquifer. Fractionation of dissolved carbon in groundwater samples taken along a flow line thus provides a means of assessing potential sorption/desorption of organic compounds in groundwater.

Organic constituents were separated into broad categories on the basis of hydrophobicity/hydrophilicity and acid/base/neutral character. The separation of hydrophobic and hydrophilic fractions was achieved on a nonionic

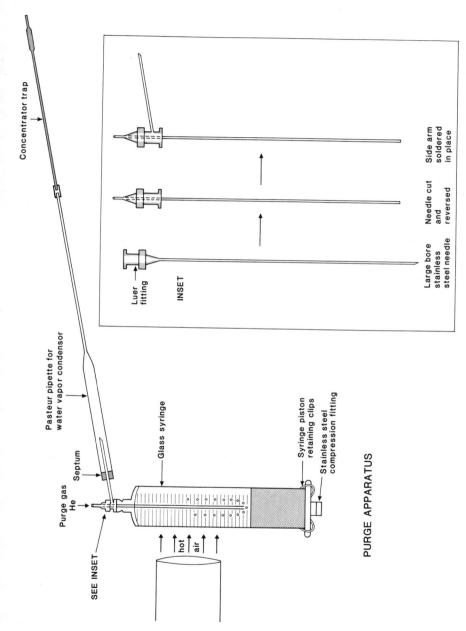

Figure 5 Modified purge-and-trap system for use with syringes for sample collection, and Tenax concentrator traps.

Amberlite XAD-8 resin. Hydrophobic bases were desorbed from XAD-8 resin by acid washing, and hydrophobic acids by washing with alkaline solution. Neutral constituents were determined by difference. Hydrophilic bases were adsorbed on a strong-acid ion-exchange resin, and hydrophilic acid by strong-base exchange resins. Hydrophilic neutral compounds were not retained on either [11].

Generally 180-mL samples of pump discharge from each screen of a borehole were fractionated at any one time. Dissolved carbon was determined using a Beckman Tocamaster carbon analyzer. Recoveries of carbon (calculated from summation of fractions as a percentage of determined DOC) were generally above 80% for samples containing carbon in excess of 20 mg/L.

VII. RESULTS AND DISCUSSION

A. Equipment Performance

Experience with installation of boreholes and the use of the equipment has led us to a number of conclusions on suitable systems for monitoring volatile organic contaminants. It became clear that the system shown in Fig. 4 could not be used for routine monitoring because of time constraints (1 day required to obtain samples giving the full depth profile in each borehole) and portability problems.

The reasons for the choice of the multiscreen bore/straddle packer system in Fig. 4 were somewhat site-specific. For example, depth to groundwater was up to 40 m below the surface. This precluded the use of standard 40-mm nested piezometers, the preferred borehole construction for multilevel sampling. Most (available) purge pumps suitable for use in 40-mm diameter borehole casing are unable to pump against a 30-40-m head of water.

The use of nested piezometers would have greatly simplified operations at the Morley site. If a suitable pump for purging of casing storage had been available, the equipment could have been easily adapted for more routine monitoring applications. There is wide scope for making the techniques suitable for routine use. Such a system would retain the benefits of obtaining several replicates of fresh groundwater (up to four separate samples) in situ using the syringes. This avoids problems from degassing, absorption on discharge lines, and sample transfer.

B. Volatile Organic Compounds (Benzene, Toluene, TCE, PCE)

Concentrations of the four volatile organic compounds in groundwater near the waste site were generally low, all being less than 30 ppb (Table 2). Benzene showed the highest concentrations in midplume boreholes where groundwater

Table 2 Concentrations as Ranges and Arithmetic Means[a] for Benzene, Toluene, TCE, and PCE in Groundwater from All Sampled Depths in Multiscreen Boreholes, Morley Landfill, Western Australia, 1988-1989[a]

Location	Borehole	Benzene (μg/L)	Toluene (μg/L)	TCE (μg/L)	PCE (μg/L)
Midplume	M5	0.1-8 (1.4)	0.1-4 (1.2)	<0.1-4 (0.5)	<0.1-0.5 (0.2)
	M10	0.1-27 (3.8)	0.1-7 (1.8)	<0.1-6 (0.5)	<0.1-0.5 (0.2)
Midplume downgradient	M11	<0.1-0.6 (0.5)	0.1-4 (1.1)	<0.1-0.2 (<0.2)	<0.1-0.3 (<0.1)
Off-plume (background)	M12	<0.1-0.2 (<0.1)	<0.1-1.5 (0.3)	<0.1-0.2 (<0.1)	<0.1-0.3 (<0.1)

[a]Means given in parentheses.

was polluted by leachate (boreholes M10 and M5). Benzene, toluene, TCE, and PCE were consistently higher in groundwater within the plume, compared with background groundwater (Table 2).

Reproducibility determined by analysis of replicate syringe samples taken in the field using the integrated techniques was good. Ranges of concentrations for benzene from replicate determinations carried out during development of the equipment in 1988 and 1989 are compared with mean values in Fig. 6. This comparison shows that over time, sampling error was generally greater than the error introduced by processing and analysis as given in Table 3. In all cases, coefficients of variation (for benzene) were better than ±100%, most were better than ±50%, and approximately one-third (for samples taken most recently when procedures were fully developed) were within ±10% of mean values (similar to processing/analytical error).

Variability introduced during sampling affects the volatile organic constituents more than inorganic nonvolatiles such as chloride, as can be seen from a comparison of Figs. 6 and 7. In these figures, coefficients of variation for chloride in replicate syringe samples and pump discharge after purging (Fig. 7) are lower than those for benzene (just based on replicate syringe sample analyses in Fig. 6). This indicates that groundwater within the screened sections of the tested boreholes was reasonably uniform in composition following purging (at least for inorganic constituents that are present at high concentrations). We conclude that disturbance during sampling and possibly during transportation to the laboratory introduces significant variability into the data for volatile organics. The effects of pumping on loss of volatile organics (by determination of these in pump discharge) were not investigated in this study.

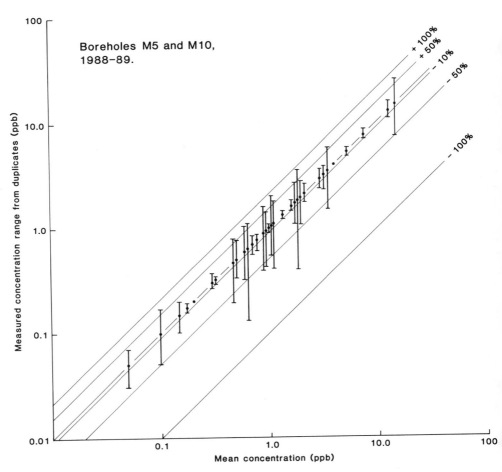

Figure 6 Assessment of reproducibility of sampling/analytical procedures for benzene, comparing mean concentrations with full ranges determined by analysis of between two and three replicate samples. Coefficients of variation are shown for comparison.

Despite some error introduced by the sampling procedures adopted for volatile organics, the techniques were more than adequate to detect the sub-ppb concentrations and distinguish between polluted and background groundwater. The technique currently gives reproducibilities similar to those obtained in previous studies (e.g., Pankow et al. [9]).

All the volatile organic compounds show significant trends in concentration with time in boreholes close to the waste site. Data for benzene, toluene,

Table 3 Summary Statistics for Estimation of Analytical Reproducibility[a]

	Benzene	Toluene	TCE	PCE
Amount (ng) in 1 mL	46.7	45.6	76.6	85.4
Coefficient of variation, $[(2s/\bar{x}) \times 100]\%$	6.7	14	11	7.8

[a]Data based on two replicate analyses of four standard gas samples taken from the same standard gas mixture. All samples preconcentrated on Tenax traps, which were then thermally desorbed into the GC column.

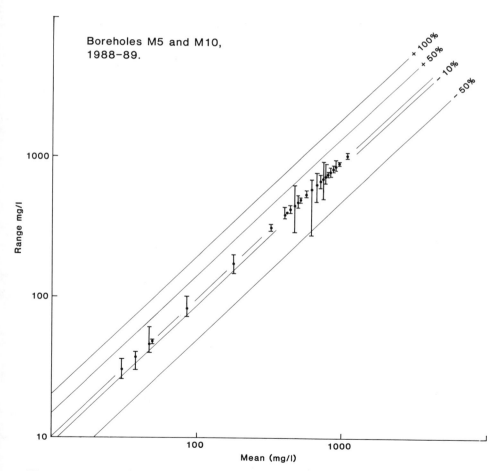

Figure 7 Assessment of reproducibility of sampling/analytical procedures for chloride, as in Fig. 6.

TCE, and PCE in all samples (including replicates) from boreholes M5 and M10 are shown in Figs. 8 and 9. These are compared with variations with time in concentrations of chloride, ammonia, and TOC in the same samples in Figs. 10 and 11.

At borehole M5, 120 m downgradient from the edge of the waste site, concentrations of chloride, TOC, and ammonia in groundwater increased over time until late 1988. After this, they remained at approximately the same concentration through 1989 (Fig. 10). The highest concentrations were found in the middle and lower parts of the aquifer, as shown in Fig. 3. The results suggest that breakthrough of leachate occurred in at least a part of the aquifer in late 1988.

Borehole M10, 200 m from the edge of the site, showed breakthrough of chloride, TOC, and ammonia in late 1989 (Fig. 11). Again, groundwater was most polluted in the middle to lower parts of the sand aquifer. The results from boreholes M5 and M10 are consistent with the estimated average rate of groundwater flow at the site of 70–80 m/yr using Darcy's law [6].

In contrast, concentrations of benzene and other volatile organic compounds peaked prior to the breakthrough of inorganic contaminants, first

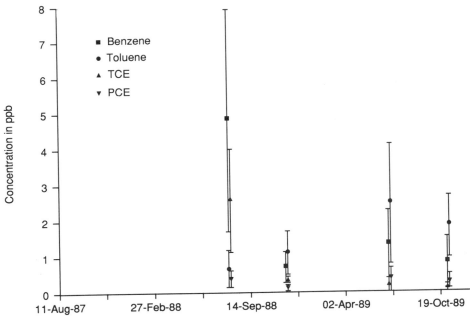

Figure 8 Variation with time in the concentration of volatile organic compounds in groundwater in borehole M5.

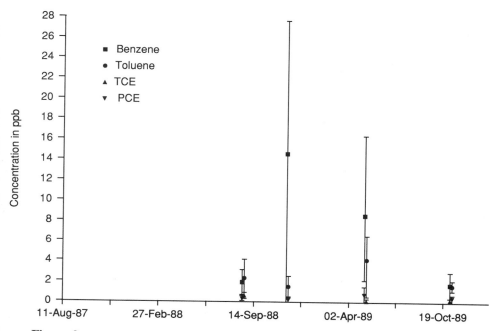

Figure 9 Variation with time in the concentration of volatile organic compounds in groundwater in borehole M10.

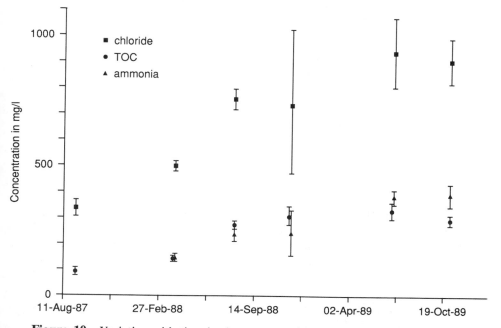

Figure 10 Variation with time in the concentration of chloride, TOC, and ammonia in borehole M5.

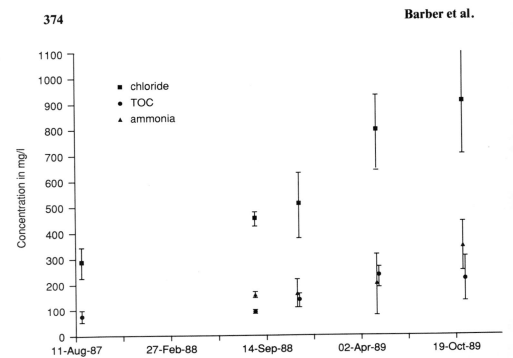

Figure 11 Variation with time in the concentration of chloride, TOC, and ammonia in borehole M10.

in borehole M5 and then in borehole M10. These decreased to lower levels when breakthrough of chloride, TOC, and ammonia occurred (Figs. 8 and 9). The highest concentrations of organic volatiles always occurred in the middle to lower parts of the aquifer, where the highest concentrations of chloride, ammonia, and TOC were found.

The appearance of early peak concentrations of volatile organics is unusual, and the reason is difficult to determine from the limited time-series data. We consider that changes in waste composition are unlikely to account for the observed variation in concentration of organic volatiles in groundwater. Enhanced concentrations of volatiles would be expected later in the site's history, following commencement of the disposal of crude (as opposed to baled) domestic solid wastes and increased disposal of liquid wastes and sludges. There is no evidence of nonaqueous-phase liquid wastes being disposed of at the site.

Transport of organic compounds laterally away from the landfill by advection in landfill gas does occur at the Morley site [5]. This would increase the mobility of these compounds. Concentrations observed in landfill gas close to borehole 4 are too low (Table 4) to account for the higher concentrations

Table 4 Concentrations of Volatile Organics in Landfill Gas and Soil Gas Around the Morley Waste Site, and Equivalent Concentrations of Organics in Leachate in Equilibrium with Landfill Gas at 20°C, Using Henry's Law

	Benzene	Toluene	TCE	PCE
	Landfill gas (concentrations in ppb)			
Borehole M4	0.5-1.0	0.4-0.7	<0.01	0.15-1.13
	Soil gas (concentrations in ppb)			
Borehole M5	<0.01-0.06	0.03-0.14	<0.01-0.24	<0.01-0.66
Borehole M10	<0.01-0.07	0.02-0.06	<0.01-0.04	<0.01-0.7
Borehole M11	0.01-0.11	0.01-0.3	<0.01-0.13	<0.01-0.6
Borehole M12	<0.01-0.01	0.07-0.08	<0.1	<0.01-0.08
	Leachate			
Henry's constants [14] (kPa·m^3/mol)	0.55	0.66	0.9	2.3
Equivalent concentration in leachate at 20°C (μg/L)	2-4	1.3-2.3	<0.03	0.2-1.4

found in polluted groundwater (Table 2), assuming that the gas is in equilibrium with leachate and using Henry's coefficients [12]. However, with the exception of TCE, concentrations are similar to mean values in Table 2. The distribution of higher concentrations of organic volatiles in groundwater also coincides with the inorganic plume of contamination in the lower part of the aquifer. Thus, although it is probable that advection of organic volatiles in landfill gas has a significant effect on leaching of these compounds, it is difficult to see how this could cause peaking of volatile organics before breakthrough of inorganics.

Transformation of benzene and toluene [13] and of PCE and TCE [14] has been reported under anaerobic conditions in groundwater. It seems most likely that at the Morley site, microbial transformation of trace concentrations of benzene, toluene, PCE, and TCE has increased as groundwater has become more polluted. Presumably the rate of degradation is linked to the availability of groundwater of other organic substrates that would support bacteria capable of metabolizing trace concentrations of compounds like benzene and TCE. The coincidence between increased concentrations of TOC and decreased trace levels of volatile organics provides support to this thesis. Volatile organic compounds within the dispersed front of pollution at the site would be more persistent owing to delayed breakdown of these compounds by bacteria. We tentatively conclude that this gives rise to an apparent early arrival of volatile organics at boreholes M5 and M10.

C. Nonvolatile Organic Compounds

The sand aquifer at Morley is generally thought to have a limited attenuation (sorption) capacity for inorganic and organic compounds. Salama et al. [4] noted that the sands contained negligible amounts of clay minerals. The sands are also low in organic matter (0.3–0.7% by weight) and sesquioxides (metal oxide adsorbents). Chloride, TOC, and ammonia show similar increases in concentration with time in boreholes M5 and M10 (Figs. 10 and 11). Thus TOC and ammonia show little retardation relative to chloride within the aquifer, which is consistent with the limited attenuation capacity of the sand formation.

The (small) extent of removal of TOC from groundwater relative to chloride can be assessed by comparing the TOC/chloride ratio with chloride concentration. This has been done using data from boreholes M5 and M10 in

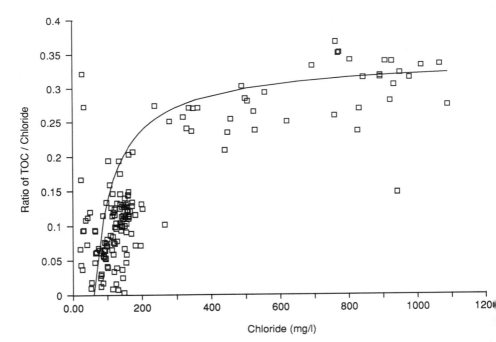

Figure 12 Comparison of the concentration of chloride with the TOC/chloride ratio for samples recovered from boreholes M5, M10, and M11 in 1988–1989. An empirical dilution curve for leachate (TOC/chloride ratio of 0.32) in groundwater (background concentrations of 60 mg/L chloride and 2 mg/L TOC is also shown for comparison).

Fig. 12. An empirical dilution curve for leachate in groundwater has also been superimposed in Fig. 12. This was calculated by assuming an average TOC/chloride ratio in leachate of 0.32, estimated from site data. It was also assumed that unpolluted groundwater contained 60 mg/L of chloride and 2 mg/L of TOC as measured in borehole M12.

Although there is some scatter in Fig. 12, the data generally follow the dilution curve, with the majority of points lying slightly below the theoretical curve, which suggests a limited retardation of TOC relative to chloride during development of the pollution plume.

Analyses of leachate and polluted groundwater from the site by HPLC and GC all indicated negligible concentrations of short-chain carboxylic acids. Thus it would be expected that the nonvolatile organic carbon consists principally of fulvic acids (MW approximately within the range 500–1000) and humic acids (MW >10,000), similar to the stabilized leachates described by Chian and deWalle [10].

Fractionation analysis of TOC from all boreholes (Fig. 13) indicates that hydrophobic and hydrophilic acids predominate. Hydrophilic bases (e.g.,

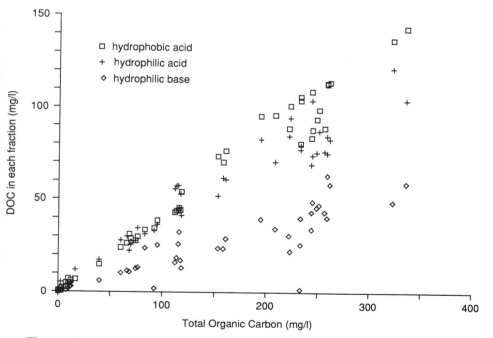

Figure 13 Fractionation analysis of nonvolatile organic carbon in groundwater for samples from boreholes M5, M10, and M11.

amino acids) generally make up 10–15% of total nonvolatile carbon in the polluted groundwater. Other fractions are not present in significant amounts.

There are small changes in the relative proportions of each fraction as a function of TOC in Fig. 13. At high TOC concentrations (i.e., midplume near the waste site), hydrophobic acids predominate. At low concentrations (some distance downgradient from the site), hydrophobic acids and hydrophilic acids are present in roughly equal amounts. Thus with increasing distance from the site, the hydrophobic acid fraction is depleted relative to the hydrophilic fractions at lower concentrations of TOC (i.e., within the leading edge of the pollution plume). Hydrophobic acids therefore show a limited retardation in movement away from the waste site compared with other organic species under near-neutral pH conditions within the pollution plume.

Steinberg and Muenster [15] report adsorption of high molecular weight organic compounds (with molecular weights greater than 1000) from lake water by colloidal alumina and clays. They also noted that the more hydrophilic, lower molecular weight (500–1000) fractions were only weakly adsorbed. It thus seems likely that at the Morley site, humic acids (which are hydrophobic) are removed by sorption within the aquifer, whereas the more abundant fulvic acids (which are hydrophilic) are not.

VIII. SUMMARY

Integrated sampling and analysis procedures have been developed for the determination of concentrations of volatile and nonvolatile organic compounds in groundwater. The borehole constructions and equipment used in our study would be inappropriate for use in routine monitoring. However, it is anticipated that these techniques could be easily adapted for use in routine studies, providing improvements in sampling/analytical quality control. Coefficients of variation of better than 10% should be achievable for volatile organics.

Concentrations of benzene, toluene, TCE, and PCE were low within groundwater around the waste site. Results suggest that microbial utilization of trace concentrations of these compounds commences only when significant levels of other nonvolatile substrates are present (e.g., measured as TOC). The study indicates that microbial activity under anaerobic conditions is likely to reduce contamination of the sand aquifer by volatile organic compounds.

Fractionation analysis of dissolved nonvolatile carbon has been successfully applied to samples of groundwater polluted by leachate from the waste site. Attenuation of organic compounds was limited within the sand aquifer. Some removal of nonvolatile hydrophobic acids relative to hydrophilic acids and bases was observed. This was attributed to sorption of high molecular weight humic acids.

ACKNOWLEDGMENTS

We thank Tuyen Thi Pham for carrying out carbon fractionations and TOC analyses, and Michael Lambert for helping with the development, testing, and using down-hole sampling equipment. Mr. Andy Giacommel of AGE Developments, Perth, designed and constructed the original packer/pump/sampler assembly. The manuscript was reviewed and greatly improved by Drs. Robert Gerritse and Greg Davis and Mr. Bradley Patterson.

REFERENCES

1. Cargeeg, G.C., Boughton, G.N., Townley, L.R., Smith, G.R., Appleyard, S.J., and Smith, R.A. *Perth Urban Water Balance Study*, Vol. 1, *Findings*, Water Authority of Western Australia, Leederville, W.A., 1987.
2. Atwood, D.F., and Barber, C. The effects of Perth's urbanisation on groundwater quality—a comparison with case-histories in the USA. Proc. Swan Coastal Plain Groundwater Management Conference, Perth, W. Australia, W. Australia Water Resources Council, Perth, 1989, pp. 177–190.
3. Appleyard, S.J., and Bawden, J. The effects of urbanisation on nutrient levels in the unconfined aquifer underlying Perth, Western Australia. Proc. Intl. Conf. on Groundwater Systems Under Stress, Brisbane, May 1986, Australian Water Resources Council, 1986, pp. 587–594.
4. Salama, R.B., Davis, G.B., and Barber, C. Characterising the hydrogeological variability of a sand aquifer in the region of a domestic waste disposal site. Proc. IAHS Symposium on Groundwater Management, Quantity and Quality, October 1989, IAHS Publ. 188, 1989, pp. 215–226.
5. Barber, C., Davis, G.B., Briegel, D., and Ward, J.K. Factors controlling the concentration of methane and other volatiles in groundwater and soil-gas around a waste site. *J. Contam. Hydrol., 5*:155–169, 1990.
6. Davis, G.B., Barber, C., and Buselli, G. Borehole and surface geophysical monitoring and simple modelling of groundwater polluted by waste leachates, Proc. 3rd Intl. Minewater Congress, Melbourne, Australia, October 1988, pp. 261–270.
7. Gillham, R.W., and O'Hannesin, S.F. Sorption of aromatic hydrocarbons by materials used in construction of groundwater sampling wells. ASTM Symposium: Standard Development for Groundwater and Vadose Zone Monitoring Investigations, Albuquerque, NM, Jan. 27–29, 1988.
8. Reynolds, G.W., Hoff, J.T., and Gillham, R.W. Sampling bias caused by materials used to monitor halocarbons in groundwater. *Environ. Sci. Technol., 24*:135–142, 1990.
9. Pankow, J.F., Isabelle, L.M., Hewetson, J.P., and Cherry, J.A. A syringe and cartridge method for downhole sampling for trace organics in groundwater. *Ground Water, 22*(3):330–339, 1984.
10. Chian, E.S.K., and DeWalle, F.B. Characterisation of soluble organic matter in leachate. *Environ. Sci. Technol., 11*:158–163, 1977.

11. Leenheer, J.A., and Huffman, W.D., Jr. Analytical method for dissolved organic carbon fractionation. USGS Water Resource Investigations Paper 79-4, 1979.

12. Mackay, D., and Shiu, W.Y. A critical review of Henry's law constants for chemicals of environmental interest. *J. Phys. Chem. Ref. Data, 10*(4):1175–1199, 1981.

13. Major, D.W., Mayfield, C.L., and Barker, J.F. Biotransformation of benzene by denitrification in aquifer sand. *Groundwater, 26*(1):8–14, 1988.

14. Vogel, T.M., Criddle, C.S., and McCarty, P.L. Transformations of halogenated aliphatic compounds. *Environ. Sci. Technol., 21*:722–736, 1987.

15. Steinberg, C., and Muenster, U. Geochemistry and ecological role of humic substances in lakewater. In *Humic Substances in Soil, Sediment and Water: Geochemistry, Isolation and Characterisation* (G.R. Aiken, D.M. McKnight, R.L. Wershaw, and P. MacCarthy, eds.), Wiley, New York, 1985, pp. 105–145.

13

Developing Methods for the Analysis of Toxic Chemicals in Soil and Groundwater: The Case of Ville Mercier, Quebec, Canada

Hooshang Pakdel, Geneviève Couture, Christian Roy, Anne Masson, Jacques Locat, and P. Gélinas

Université Laval, Sainte-Foy, Quebec, Canada

Suzanne Lesage

National Water Research Institute, Environment Canada, Burlington, Ontario, Canada

I. INTRODUCTION

Pollution has become a serious worldwide problem not only from an environmental point of view but also because many pollutants are dangerous to human health. Since large U.S. and Canadian populations draw their drinking water from subsurface water supplies, the issue of organic contamination of groundwater has initiated considerable concern in regulatory agencies in both the public and private sectors. Consequently, the role of analytical chemistry in public health protection has become more important, with an emphasis on trace analysis. Pollutants such as chlorinated and nonchlorinated volatile hydrocarbons, phenols, polycyclic aromatic hydrocarbons (PAHs), polychlorinated dibenzodioxins (PCDDs), and polychlorinated dibenzofurans (PCDFs) are present in complex mixtures. In addition, some of them are potentially dangerous even at levels of parts per thousand. Analysis of environmental samples is a very difficult task: a single technique must have high resolution capacity, detection power, and specificity. Coupling the mass spectrometer with the gas chromatograph (GC/MS) or liquid chromatograph (LC/MS) has increased the specificity of the mass spectrometer as a detector, particularly in the single-ion-monitoring mode, and has allowed the analysis of one or more pollutants in complex mixtures. The use of the purge-and-trap (PT) concentrator with automatic sampler in combination with GC/MS has enabled chemists to analyze a large number of water or soil samples within a short period. The rapid development of advanced software has resulted in the widespread use of GC/MS or LC/MS systems to quantify pollutants in environmental samples. GC/MS, LC/MS, and PT/GC/MS have the advantage of simplifying the cleanup procedures necessary to eliminate any possible interferences. Commonly analyzed matrices are air, water, and soil. In this chapter we deal particularly with oil, water, and soil.

Primary sources of organic chemical pollution of the subsurface include seepage from unlined lagoons and other surface impoundments; improper landfilling; improper surface and subsurface disposal practices; leaks in pipes, storage tanks, and other processing or transport equipment; and accidental spills. Many of the organic liquids emanating from such sources have low water solubilities and may migrate through the subsurface environment as nonaqueous-phase liquids or organic vapors.

The Ville Mercier site, located in southern Quebec on the south shore of the St. Lawrence River, is an excellent example of heavy organic contamination. The geological setting of the study area is discussed briefly in Section II. From 1968 to 1972, approximately 40,000 m^3 of liquid wastes, including refinery oil residues, solvents (chlorinated hydrocarbons), phenolic compounds, acids, pesticides, insecticides, polymers, paint residues, and mercaptans, was dumped in two abandoned sand and gravel pits near Ville Mercier [1]. The dumping of organic wastes resulted in the contamination of the groundwater, with the contamination plume now estimated to cover an area of about 10–15 km^2 [2]. The reports so far indicate that decontamination operations (extraction well and treatment plant) have removed only about 5% of the total contaminants [3].

The analytical significance of contaminants arises from their known or suspected mutagenic and carcinogenic character. Mutagenic and carcinogenic activity is connected with chemical structure, which varies with substituting groups and also with isomeric differences. 1,2-Dichloroethane and phenolics, for example, are good indicators of change in the quality of groundwater. They have high aqueous solubility and stability (low biodegradability) with weak adsorbability. Many of the halogenated hydrocarbons are characterized as dense nonaqueous-phase liquids (DNAPLs). Their densities exceed that of water, but they have lower viscosities. Their solubility in water is in the range of 100–5000 mg/L. Being very soluble in oils, DNAPLs sink to the bottom of the aquifers and fill fractures in bedrocks. Oil-based DNAPLs are more viscous than water, so significant quantities of organic matter are trapped in the pore spaces and rock fractures. In that position, DNAPLs still dissolve slowly in moving groundwater. Their high vapor pressure and molecular weight result in density-induced sinking of chemical vapors downwards. Diffusion results in lateral migration of vapor through the vadose zone and ultimately in significant groundwater contamination.

In contrast, petroleum chemicals (mainly benzene, toluene, xylene, and benzene derivatives) categorized as light nonaqueous-phase liquids (LNAPLs) tend to form pools and spread laterally [2] because of their low densities. Consequently, groundwater contamination cannot be eliminated in the long term without removal of the LNAPL and DNAPL sources. Chlorinated volatile hydrocarbons have low to moderate toxicity and carcinogenic and mutagenic properties. Vinyl chloride is among the most carcinogenic volatile compounds [4]. Benzene, toluene, and xylene have proven to systemically affect the blood, central nervous system, skin, and bone marrow, and through vapor exposure they can affect the eyes and respiratory system.

Polycyclic aromatic hydrocarbons (PAHs), in general, are ubiquitous environmental pollutants and are formed from both natural and anthropogenic sources. The latter are by far the major contributors. Natural sources include forest fires [5], volcanic eruptions [6], and degradation of biological materials, which has led to the formation of these compounds in various sediments and fossil fuels [7]. Major anthropogenic sources include the burning of coal refuse banks, coke production, automobiles, commercial incinerators, and wood gasifiers. With rapid development of the industry throughout the world, the natural balance between the production and natural degradation of PAHs has been disturbed. During the last thirty years, many studies have been undertaken to characterize these compounds. A bibliography of over 400 references on studies involving advanced analytical techniques has been published [8].

Phenols have a broad spectrum of toxicity depending on the nature of the substituents. Some phenolic compounds are of concern only because they cause taste and odor problems in drinking water; others, such as the pentachlorophenols, are highly toxic [9]. The occurrence and biological effects of phenolic compounds in aquatic ecosystems were reviewed by Buikema et al. [10]. Chlorination of drinking water and sewage treatment may also lead to the production of chlorinated phenols and other substituted phenolic compounds. Phenols enter and are transported in the environment via adsorption, diffusion, volatilization, leaching, surface movement, and atmospheric movement [11]. The worldwide use of phenols has resulted in extensive and wide distribution of these substances throughout the environment.

Polychlorinated biphenyls (PCBs) are ubiquitous environmental contaminants. They are present as complex mixtures, and Aroclors are their commercial sources. They have been used extensively as dielectric fluids, plasticizers, and heat exchange fluids. PCBs exhibit a high degree of biological and chemical stability and lipid solubility, and they tend to accumulate in food chains and human adipose tissues [12]. Commercial PCBs, in common with other halogenated aromatics such as the polychlorinated dibenzo-*p*-dioxins (PCDDs), dibenzofurans (PCDFs), and polybrominated biphenyls (PBBs) elicit a number of common toxic and biological effects. Their structure and mechanism of action have been reviewed elsewhere [13–15]. Whether or not pesticides represent a threat to groundwater is dependent upon factors related to the pesticides themselves, the soil, and application factors. Organophosphorus compounds can rapidly degrade, either biologically or chemically. In contrast, organochlorinated compounds resist biodegradation, and thus they can be recycled through food chains [16].

The objective of the present study was to investigate the fate of organic chemicals in the soil and groundwater of Ville Mercier and to assess the extent of the pollution plume. Various analytical techniques were employed to identify and quantify single components. However, owing to their relative chemical similarity and complexity of the matrices and numerous samples collected during this investigation, identification and quantification of the pollutants was rather difficult. The determination of trace level components in complex matrices usually is accomplished by performing a variety of manual cleanup and reconcentration steps before analysis by any of the available chromatographic methods. GC/MS and PT/GC/MS methods, which were used to reduce sample preparation steps and analysis times, will be discussed in detail. Three oil samples were analyzed in detail for different hydrocarbon types from highly contaminated wells, and methods were developed to concentrate, identify, and quantify the volatile chlorinated hydrocarbons as well as PCBs. Qualitative and quantitative results of the priority pollutants from the oil phase, sand, and gravel aquifer and bedrock aquifer are discussed.

II. GEOLOGICAL SETTING

The geological setting of the study area, in a radius of about 20 km, has been established from previous work by many researchers [17–19]. Information related to the geology and hydrogeology of the area can also be found in previous reports [20–22]. Some aspects of the bedrock features were adapted from Denis and Rouleau [23]. Finally, new data are reported in this paper.

A. Bedrock Geology

The bedrock geology of the region consists mostly of sedimentary rocks of the Cambrian and Ordovician eras. The structural setting of the rock formations is quite simple, as their strata are subhorizontal and their deformations are limited to large-amplitude folds (synclinal and anticlinal).

The Cambrian rocks are represented by both the Covey Hill and Cairnside formations of the Postdam group. Both formations are made of hard indurated sandstone, the former arkosic and reddish and the latter quartzitic and whitish. They form outcrops in the western part of the area near the St. Lawrence River.

The Ordovician rocks of the area are composed of the Theresa and Beauharnois formations of the Beekmantown group. The Theresa formation is composed of sandstone and dolomite, and the Beauharnois formation consists of dolomitic sandstone and shale. According to the Globensky's [17] map, the Mercier site would be within the cartographic boundaries of the Beauharnois formation. However, all cores collected at the site consist only of sandstones similar to that of the Theresa formation. The underlying bedrock unit composing the subsurface bedrock aquifer in the area is only within the Theresa formation.

Rock cores collected in the seven drillholes are generally homogeneous. This gray-white quartzitic sandstone has fine, rounded to subrounded grains with a brownish-beige alteration. More coarsely grained zones are visible within one core sample. Calcite veins and veinlets can also be seen in some cases. Small clusters of pyrite are occasionally found in cavities. Visual examination of the rock reveals that it is composed of more than 96% quartz; the remaining 4% is feldspars, carbonates (dolomite), alteration minerals, and iron oxides. Sedimentary structures have been identified in many cores such as the cross-lamination in the R6 site (see Fig. 2). Many load casts, characterized by a high concentration of iron oxides, were found in all cores.

B. Structural Geology

The bedrock of the area has been affected by some tectonic activity that resulted in the development of a few folds and faults. Near the Mercier site we can find two major folds: the Ste-Martine synclinal and the Aubrey anticlinal. Trends of these folds are N 43° and N 32°, respectively. As expected, most bedding planes dip toward their respective fold axes, usually at an angle of 2–4°, with values as high as 7° near fold axes, and are oriented either parallel or perpendicular to fold axes. The closest fault to the Mercier site is the Havelock fault, a few kilometers to the east, trending more or less to the north. As the bedding is more or less horizontal, the joints are subvertical (dipping at more than 80°) and also trend normal or parallel to the fold axes. There seems to be little influence of the Havelock fault on the attitude of the joint system in the area. Detailed presentation of the fracture system (mostly the bedding) obtained from observations in the field and from the cores collected at the Mercier site (23) is as follows.

Two major subvertical fracture families oriented at 90° to each other are present. The average orientations of the families are N 120° and N 30° in decreasing order of density. In the first family (N 120°), average fracture spacing varies from 0.1 to 1.0 m. In the second family (N 30°), it varies from 0.1 to 0.45 m. The aperture of fractures has not been measured because of the weathering of the rock, which often produces an apparent aperture that is not representative.

In cores, most of the fractures are not visibly open. They are filled with iron oxides, altered minerals, or calcite in veins and veinlets a few millimeters thick. The average spacing of filled fractures varies between 0.12 and 0.24 m. The great variation in fracture spacing is reflected by extremely variable rock quality designation (RQD) values (0–100%), although the recuperation was often near 100% (generally between 60 and 75%). Open fractures are much less frequent, and it is not easy to determine their aperture from visual examination. Spacing of these fractures ranges from 0.15 to 4.5 m depending on the drill site. Minerals, iron oxides, and traces of rust brown oxidation cover their flat, sometimes irregular, surfaces. The degree of roughness of the surfaces is low; they are generally smooth, sometimes rough.

C. Erosion and Bedrock Topography

The drainage system pattern, when developed in or on the bedrock, is clearly controlled by the joint system associated with the local structural geology. This is well shown at Châteauguay, where river bends follow the same direction as the bedding or the fracture system. This is also true for a part of the Esturgeon River just upstream from Ste. Martine.

Bedrock surface topography shows the presence of two depressions in the study area: one just south of the Mercier site (hereafter called the Mercier depression) and a second about 2 km to the northwest of Ste. Martine (called the Ste. Martine depression). The lowest elevation of the Mercier depression is located at its center and is just 3 m above sea level, so that maximum thickness of drift reaches 40 m. As for the surface drainage system, the orientation of these depressions is also trending in the same direction as the vertical joints system. Bedrock surface rises to an elevation of 45.7 m to the north of the site and 30.5 m to the south [1].

Regional groundwater flow in the bedrock is oriented toward Ste. Martine and appears to follow the direction of the joint system. Our investigations have indicated that the bedrock aquifer at the Mercier site and downstream toward Ste. Martine can be restricted to the Theresa formation of the Beekmantown group. However, if deeper groundwater flow in the bedrock aquifer has to be taken into consideration, then one could include the Cairnside formation of the Postdam group that underlies the Theresa formation. Still, this could only be confirmed by a deep borehole at the Mercier site.

D. Quaternary Geology

Following a long period of erosion, Quaternary deposits were laid down on an irregular bedrock surface. The actual study site is above a bedrock depression that is deeper than the rock floor of the Esturgeon River and close to the mean sea level. Quaternary deposits are about 30–40 m thick under the contaminated area, and the thickness of these deposits decreases to about 10 m near the Esturgeon River to the south. The study area lies on top of a ridge that has long been considered an esker extending about 11 km in a NNE–SSW direction and overlooking the local marine clay plain by 10–15 m [1]. We will see below that the actual interpretation of the ridge differs slightly. The distribution of Quaternary deposits in this area is taken from a report by Dion et al. [19,24], which has adapted the work of Lasalle [18].

The various sedimentary units have been amalgamated into three hydrostratigraphic units that apply to the Mercier site and are presented in more detail hereafter.

1. Basal Till

The base of the stratigraphic section is composed of a discontinuous basal till (St. Jacques till), very compact and dense. The high density of the till is probably the result of erosion and mixing of rock debris from fine-grained (shales) and coarse-grained (limestones or sandstones) rocks. The unit was put in place under the glacier, and its high density is due to the weight of the ice combined with a very wide grain-size distribution, which enables a compaction with low porosity. This unit was encountered in most boreholes that were drilled in the fall of 1988. In one instance dry fragments were observed in this compact layer. A major question is the extent of that layer because its observed thickness ranges from less than a meter to about 3 m. Chances are that windows (till-free areas) exist and that coarser sand and gravel may be in direct contact with the bedrock aquifer.

2. Glacio-Marine Sediments Complex (Fan Deposit)

The glacio-marine sediments are made up of a complex succession of fine-grained and coarse-grained sediments. Borehole data and available sections clearly indicate a great variability in both grain size and extent of this deposit. They compose most of the hydrostratigraphic unit found near the study area. As one goes away from the side of the glacio-marine deposit, the degree of heterogeneity decreases and the sediments are mostly made of sand occasionally interbedded with marine clays. Poulin [1] considered this deposit to be an esker. The actuality is closer to that envisioned by Sharpe [25], who interpreted such deposits as glacio-marine rather than ice-contact features. Actually, they are about the same except that the subglacial river was entering the Champlain Sea rather than a lake. This is evidenced by the various fossiliferous sandy sediments found interbedded with the glacio-fluvial sediments. Although the sediments composing the deposit at the Mercier site are coarser than those found at the Gloucester test site [26], the sedimentary environments are very similar.

This sedimentary complex contains various subunits indicating the transition from very coarse (cobbles and pebbles) ice-contact, poorly stratified sediments to marine sandy sediments. Between these extreme conditions, typical glacio-fluvial sediments are often found at very steep angles. Laterally, these sediments are interbedded with other well-sorted sands. The former liquid waste disposal pits, as well as the purge wells and the waste treatment plant, were located in that unit. The bulk of the contaminants are dispersed in that sand and gravel unit.

3. Marine Sediments

Marine deposits are mostly composed of fossiliferous sandy and clayey sediments. They are found up to the surface in most of the area. As indicated by the presence of the fossil *Hiatella arctica*, they were deposited in a brackish (low-salinity) water body. From the borehole and surface data, these sediments are interdigited with the glacio-marine sediments complex. In that zone, they are particularly coarse but rapidly become finer and finer away from the glacio-marine complex. These marine sediments evolve from a gravely to a clayey sediment.

III. EXPERIMENTAL

Soil, water, and oil samples were investigated for volatile hydrocarbons, PAHs, phenols, and PCBs. The procedures applied are discussed separately. The separation scheme is shown in Fig. 1.

Before the experiment, all glassware was thoroughly washed, oven-dried, and rinsed with good-quality solvents (distilled or pesticide grade). Anhydrous Na_2SO_4 was dried at 500°C for at least 6 hr before use. Alumina (Aldrich, activated, neutral, Brockmann I) was used for fractionation. All the extracts were stored in glass vials with Teflon or silicone stoppers.

A. Extraction and Purification

1. Water Samples

About 1 L of sample was filtered to remove the solids, then transferred to a 2-L separation funnel. Its pH was adjusted to 11 with 10 N NaOH. The solution was then extracted three times with 60 mL of dichloromethane.

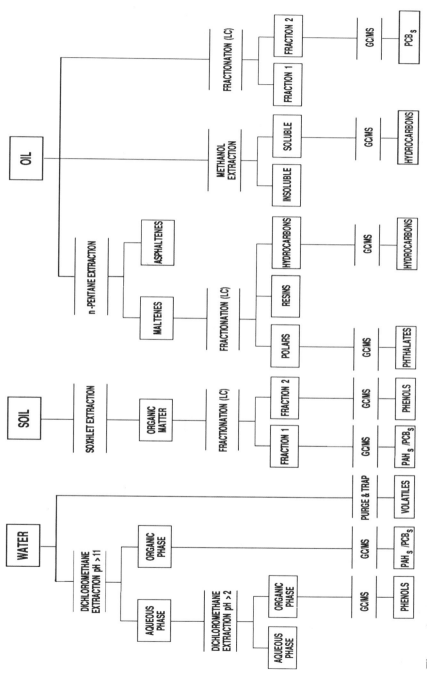

Figure 1 Summary of chemical separation scheme and methods of analyses.

The funnel was left standing for about 10 min for separation of the aqueous phase from the organic phase. The organic phase was decanted and then evaporated to one-third of its volume on a rotary evaporator under moderate vacuum at room temperature. The solution was then dried on a 6.5-cm glass column packed with 17 g of anhydrous Na_2SO_4. The sample volume was reduced to 1 mL, and the sample was then stored at about 5 °C for analysis of neutral and basic compounds. The aqueous phase was then acidified to pH $\leqslant 2$ with 4.5 N H_2SO_4 and extracted three times with 60 mL of dichloromethane. The organic phases were mixed and dried as mentioned above and stored for the analysis of acidic compounds. A similar blank extraction was made without water to monitor any possible contamination from solvents and glassware.

2. Soil Samples

a. Soxhlet Extraction

Typically about 200 g of the as-received soil sample was extracted in a Soxhlet apparatus for 24-28 hr, depending on the sample, with a mixture of dichloromethane and methanol (2:1 volume ratio). The extract was filtered and dried on anhydrous Na_2SO_4 (see Section III.A.1), evaporated to dryness, and weighed.

b. Liquid Chromatographic Purification

The organic matter extracted from the soil was fractionated on an alumina packed column into two fractions. A 80 mm × 5.5 mm i.d. glass column was first packed with approximately 1.5 g of alumina for 25 mg of sample. The organic matter was first dissolved in a small quantity of toluene, and the soluble part was transferred to the column. The first fraction (F1) was then collected with 30 mL of hexane/toluene (3:2). Subsequently the toluene-insoluble part of the organic matter was dissolved in methanol (all soluble) and transferred to the same column and eluted with 30 mL of methanol (F2). Both F1 and F2 were evaporated to 1 mL and were then analyzed for PAHs/PCBs and phenols, respectively. A similar blank extraction was made without soil, and the residue was analyzed.

3. Liquid Oil Samples

a. Solvent Extraction and Liquid Chromatographic Fractionation for Hydrocarbon Type Analysis

All the samples studied are identified in Table 1 and Fig. 2. About 5 g of T5 sample was dissolved in 150 mL of distilled n-pentane, stirred for a few hours, and then filtered. The n-pentane-soluble fraction (maltene) was dried over anhydrous Na_2SO_4 and filtered. The solvent was then evaporated, and maltene was recovered. The n-pentane-insoluble fraction (asphaltene) was dis-

Table 1 Ville Mercier Oil Samples

Sample	Date of sampling	Depth (ft)	% H$_2$O	Remarks
T4	?	25-30	~90	Contains oil
T5	Oct. 28, 1988	15-20	0.0	Oil sample
R4E1	Oct. 28, 1988	—[a]	39	Mainly oil

[a]Sample represents oil and water that could be pumped from the top 15 m of the bedrock in drillhole R4.

solved in dichloromethane and dried. Asphaltene was recovered after removal of the solvent. The other two samples were similarly fractionated into maltenes and asphaltenes. About 500 mg of maltene fraction was further fractionated into three subfractions (F1, F2, and F3) on 15 g of dichloromethane prewashed silica gel (60-220 mesh, activated at 120 °C for 2 hr). Fraction F1 (aliphatic, alicyclic, and aromatic hydrocarbons) was eluted with 120 mL of 15% dichloromethane in n-pentane. F2 (polar organics) was eluted with 120 mL of dichloromethane. F3 (resins) was eluted with 50 mL of 30% methanol in dichloromethane. F1 was further fractionated on a 20 cm × 20 cm silica gel TLC plate (1 mm thick, activated at 110 °C for 4 hr) into three fractions with n-pentane: F1-1 ($0.0 \leqslant R_f \leqslant 0.14$, polars); F1-2 ($0.14 \leqslant R_f \leqslant 0.55$, aromatics), and F1-3 ($R_f \geqslant 0.55$, aliphatics and alicyclics). The fractionation scheme is shown in Fig. 1.

b. Solvent Extraction for Analysis of Volatile Chlorinated
   and Nonchlorinated Hydrocarbons

About 13 mg of the oil sample (T5) was dissolved in 10 mL of cold methanol. The volatiles and part of the other hydrocarbons were extracted with methanol, but about 55% of the oil remained insoluble. Depending on the volatile concentration, 2 μ L of the methanol solution in this experiment was further diluted in 10 mL of water and analyzed by the purge-and-trap method.

c. Liquid Chromatographic Fractionation and
   Solvent Extraction for PCB Analysis

To a 25 cm × 1.65 cm i.d. glass column filled with pesticide grade n-hexane, 15 g of activated alumina (at 115 °C for 2 hr), 3 cm of anhydrous Na$_2$SO$_4$, and 2 mm of freshly activated copper powder were added. The column was initially washed several times with n-hexane, and then 1 mL of the standard Aroclor solution of 1242, 1254, and 1260 (1.6 μg each in 1 mL of methanol) was transferred to the column. The column was eluted with n-hexane, and eight fractions (25 mL each) were collected. GC/MS analyses revealed that only fractions 2, 3, and 4 contained PCBs. A similar separation was made

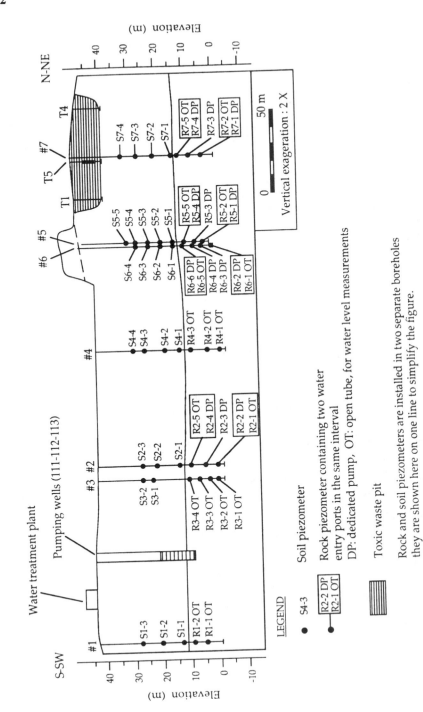

Figure 2   Identification of Mercier site monitoring wells.

on about 28 mg of the oil sample (T5). To the column was added 1 mL of methanol prior to elution with *n*-hexane. Fractions 2-4 were collected, mixed, and evaporated to dryness. To the extracted organic matter was added 1 mL of methanol, and the mixture was analyzed by GC/MS.

B. Analysis

1. Purge-and-Trap (PT) Analysis

a. Instrumentation

All the samples were analyzed using a Tekmar ALS2016 autosampler (16 samples loaded for analysis) and a Teckmar LSC2000 concentrator interfaced to a HP5890 gas chromatograph with a HP5970 mass-selective detector. The helium carrier gas transported the organic vapors from ALS2016 to a sorbent-packed trap in LSC2000, where they were collected. The trap was then heated to desorb the volatile compounds. A Tekmar Model 2000 capillary interface cooled with liquid nitrogen was used to focus the organic vapors cryogenically on the GC column. A 120-cm-long piece of uncoated and deactivated fused silica tubing was used as the transfer line to connect the trap to the column inlet using a brass union. The column inlet was passed through a liquid nitrogen cooled trap at $-150\,°C$. During desorption, the sample was cryogenically trapped in the column inlet. When desorption was complete, the cooled trap was rapidly heated at a rate of 800 °C/min to inject the sample into the column.

Purge-and-trap and GC/MS conditions were as follows:

Sample size	5 mL
Standby temperature	35 °C
Purge flow	40 mL/min
Purge time	12 min
Dry purge time	6.0 min
Desorb preheat temperature	175 °C
Desorb time	4 min at 180 °C
Injection temperature	175 °C
Bake time	40 min at 200 °C

- The trap was a 30 cm × 2.5 mm i.d. stainless-steel column packed with 15 cm of Tenax and 8 cm of charcoal.
- The GC columns were two 30 m × 0.32 mm i.d. DB624 (J & W) fused silica with 1.8 μm film thickness, joined with a glass seal connector.

- The GC oven temperature was held at 35 °C for 2.5 min, then programmed to 175 °C at 4 °C/min and then to 250 °C at 10 °C/min.
- The mass spectrometer interface temperature was 170 °C, and the source temperature was 250 °C. The spectrometer was scanned from 40 to 250 amu at 2 scans/sec.

b. Standard Solutions for PT Analysis

- *Purgeable A*. Carbon tetrachloride; chlorobenzene; chloroform; dibromochloromethane; 1,1-dichloroethane; 1,1-dichloroethylene; 1,2-dichloropropane; dichloromethane; tetrachloroethylene; 1,1,2-trichloroethane; trichloroethylene; and trichlorofluoromethane—200 µg each in 1 mL of methanol
- *Purgeable B*. Benzene; bromodichloromethane; bromoform; 1,2-dichloroethane; *trans*-1,2-dichloroethylene; 1,3-dichloropropene; ethylbenzene; 1,1,2,2-tetrachloroethane; toluene; and 1,1,1-trichloroethane—200 µg each in 1 mL of methanol
- *Purgeable C*. Bromomethane, chloroethane, chloromethane, and vinyl chloride—200 µg each in 1 mL of methanol
- *Surrogate solution*. Bromochloromethane—1000 µg in 1 mL of methanol

Solutions A and B were prepared by measuring 16 µL of each solution in 1 mL of methanol. Unprocessed spring water (5 mL) was spiked with 5 µL of solutions A and B and analyzed. Spring water (5 mL) was spiked with 2.5 µg of purgeable C and analyzed. The surrogate solution (80 µL) was diluted in 1 mL of methanol. Spring water (5 mL) was spiked with 0.5 µL of the diluted surrogate solution and analyzed.

2. Gas Chromatographic/Mass Spectrometric (GC/MS) Analyses

a. Standard Solutions for GC/MS Analyses

Three sets of PAH, phenol, and PCB standard solutions were prepared as follows:

Pure single PAHs were purchased (Aldrich, BDH, Eastman Kodak, and Supelco), and a solution with concentrations in the range of 0.5-2.4 µg of each in 1 mL of toluene was prepared (see Table 2).

A mixture of 13 phenolic compounds (see Table 3) was purchased from Supelco and diluted to 100-2500 µg in 1 mL of methanol.

Commercial Aroclor standards (1242, 1254, and 1260) were purchased from Supelco, and a diluted mixture with 1.6 µg of each in 1 mL of methanol was prepared.

b. Volatiles Analysis

Target compounds were identified by retention time of the total ions and by comparison of background subtracted spectra with a computerized NBS

Table 2 Polycyclic Aromatic Hydrocarbons

Compound name	Monitoring ion (m/e)
Naphthalene	128
Acenaphthylene	152
Acenaphthene	154
Fluorene	166
Phenanthrene	178
Anthracene	178
Fluoranthene	202
Pyrene	202
Benz[a]anthracene	228
Chrysene	228
Benz[a]anthracene-d_{12}[a]	240
Benz[a]pyrene	252
Benzo[k]fluoranthene	252
Benzo[b]fluoranthene	252
Indeno[1,2,3-cd]pyrene	276
Dibenz[α,h]anthracene	278
Benzo[ghi]perylene	276

[a]Internal standard.

Table 3 Phenolic Compounds

Compound name	Monitoring ion (m/e)
Phenol	94
2-Chlorophenol	128
2-Nitrophenol	139
2,4-Dimethylphenol	107
2,4-Dichlorophenol	162
4-Chloro-3-methylphenol	107
2,4,6-Trichlorophenol	196
2,4-Dinitrophenol	184
4-Nitrophenol	65
2-Methyl-4,6-dinitrophenol	198
Pentachlorophenol	266
o-Cresol	108
p-Cresol	107

library of 42,000 mass spectra. All the compounds were separated with minimum interference and acceptable resolution. Quantification was done by the external standard method using three sets of purgeable standards—purgeables A, B, and C (see Section IV.B).

The mass spectrometer's average responses for the A, B, C, and surrogate solutions were measured from the total ion current chromatograms and compared with the samples. All samples were spiked with 0.5 μL of the diluted surrogate solution, and its concentration was reported for quality control purposes. The acceptable recovery range was 80–120%. The instrument was calibrated daily. Other compounds, not included in the standard solutions, were tentatively identified by comparing their background subtracted spectra with the NBS library.

c. PAH Analysis

Analysis of PAHs was made on an HP5890 gas chromatograph coupled to an HP5970 mass-selective detector. Analytical GC/MS conditions were:

- Injection mode: split (1:40) at 290 °C.
- Column: 30 m × 0.25 mm i.d. DB5 capillary (J & W) with 0.25-μm film thickness.
- Oven temperature: maintained at 50 °C for 2 min, then programmed to 100 °C and 290 °C at 30 and 5 °C/min, respectively. The oven temperature was maintained isothermally at 290 °C for 10 min.
- Mass-selective detector source temperature: 270 °C.

Data acquisition was done in the selected ion monitoring (SIM) mode using one ion for each target compound (see Table 2 for the ions). Quantification was done on those selected ions by the internal standard method using benz[a]anthracene-d_{12} as the internal standard (purchased from Supelco). A solution of 20 μg of benz[a]anthracene-d_{12} in 1 mL of dichloromethane was prepared. Ten microliters of the solution was added to the samples, which were then analyzed.

d. Phenol Analysis

Phenols were analyzed on the same GC/MS system and column as were used for PAHs. The injection mode was splitless at 250 °C, and the column temperature was held at 50 °C for 2 min, then programmed to 100, 180, and 290 °C at rates of 30, 5, and 30 °C/min, respectively. Similarly, data acquisition was done in the SIM mode using one ion for each target compound (see Table 3 for the ions). Quantification was done on those selected ions by the external standard method.

e. PCB Analysis

Analysis for PCBs was carried out on the same GC/MS system and column as were used for the PAHs and phenols. The injection mode was splitless at

Table 4 Polychlorinated Biphenyls

Aroclor	Monitoring ion range (m/e) (amu)
1221	188-190
1232	222-224
1242	256-258
1248	326-328
1254	324-326
1260	394-396

290 °C, and the oven temperature was held at 50 °C for 2 min, then programmed to 150 and 290 °C at rates of 30 and 5 °C/min, respectively. Similarly, data acquisition was done in the SIM mode using two ions for each target Aroclor (see Table 4 for the ions). Quantification was done on those selected ions by the external standard method. Since each Aroclor contains a number of isomers with a characteristic distribution pattern, the quantification was made by comparing the distribution patterns of standard and samples.

IV. RESULTS AND DISCUSSION

A. Hydrocarbon Types in the Main Source of Contamination

Tables 5-7 reveal the following global results regarding petroleum hydrocarbon distributions in three samples collected from the contaminated sites of Ville Mercier (Table 1). All three samples contain relatively high amounts of asphaltenes (n-pentane-insoluble and high molecular weight organic compounds; Table 5). Unlike conventional liquid petroleum oil, waste oils, heavy oils, and solid fossil fuels usually contain a high percentage of asphaltenes. In addition, T5 oil contained a significant quantity of very volatile compounds including chlorinated and nonchlorinated hydrocarbons (Section III.B), which could be related to a particular source other than petroleum oil or are produced by in situ biodegradation of the waste oil. Maltene fractions (n-

Table 5 Maltene and Asphaltene Content of Oil Samples (wt % Moisture, Ash-Free Basis)

Sample	Maltene	Asphaltene
T4	90.9	9.1
T5	91.4	8.6
R4E1	87.6	12.4

Pakdel et al.

Table 6 Maltene Subfractions (wt %)

Maltene sample	Aliphatics, alicyclics, and aromatics	Polar cmpds	Resins
T4	82.1	7.7	10.2
T5	73.9	14.6	11.5
R4E1	77.7	14.4	7.9

pentane-soluble organic compounds) of T4 and R4E1 oils contain respectively lower percentages of polar and resin oils organic compounds than T5 oil (Table 6). Since the T4 sample was mainly water, polar compounds (e.g., phenols) with high water solubility could have been separated from hydrocarbons by water. Both T4 and R4E1 oils contained a high percentage of aliphatic/alicyclic hydrocarbons (Table 7) but were lower in PAHs than the T5 sample. Either this was due to a diverse origin of R4E1 oil or the PAHs were washed away by water.

Preliminary GC/MS analyses of the total aliphatic/alicyclic fractions (Table 7) of T5, T4, and R4E1 oils showed a Gaussian distribution of n-alkanes (the most abundant compounds) in the range of C_{10}-C_{33} with maxima at C_{14}, C_{20}, and C_{16}, respectively. The T5 sample contained a slightly higher proportion of C_{10}-C_{15} than the other two samples, typical light crude petroleum hydrocarbon distribution. For the same reason, T5 oil had a lower asphaltene content than the other two oil samples (Table 5).

A preliminary GC/MS analysis of the polar fractions (Table 6) revealed the occurrence of carboxylic acid methyl esters including phthalic acid esters (PAEs). Two major PAEs were positively identified: dioctyl phthalate and dibutyl phthalate. Various fatty acid methyl esters were also found.

Phthalate esters or phthalic acid esters are the esters of orthophthalic acid that are synthesized from phthalic anhydride and are used as plasticizers. There are some indications that certain organisms can also synthesize PAEs during their normal metabolic activities [27]. The majority of PAEs have

Table 7 Hydrocarbon Types in Total Aliphatic, Alicyclic, and Aromatic Fraction (F1) (wt %)

Sample	Polars (F-1)	Aromatics (F1-2)	Aliphatics and alicyclics (F1-3)
T4	3.2	40.9	55.9
T5	4	51.1	44.9
R4E1	3.3	41.8	54.9

been produced for a long time in increasing quantities, and some have reached the environment as pure compounds or mixtures. PAEs are readily adsorbed by organic residues and solid surfaces in aqueous environments [28]. Environmental concern for PAEs arises because of the relatively high and increasing rates of PAE release into the environment. Although some PAEs, for example, dioctyl phthalate, are suspected carcinogens, their toxicological consequences have not been fully explored [29]. Kohli et al. [28] have published a comprehensive and critical review on their occurrence and on analytical methods for them. PAEs have a strong characteristic peak at m/e 149 in their mass spectra, except for dimethyl phthalate, for which the characteristic peak is at m/e 163. Several picograms of PAEs could easily be detected using the SIM-GC/MS method. PAEs usually coelute with fatty acid esters in cleanup procedures. However, they are unambiguously identified by their characteristic mass spectral fragmentation pattern.

B. Chlorinated and Nonchlorinated Volatile Hydrocarbons

Purge-and-trap analysis is the most common method of analysis for quantification of volatile organics in water. It was first developed by Bellar and Lichtenberg [30]. The trap normally employed is 2,6-diphenylene oxide polymer-Tenax (60–80 mesh) or a combination of Tenax (2/3) and silica gel, grade 15 (1/3) in a tube of 0.27 cm i.d. and 25 cm long [4]. However, it was found that Tenax with activated carbon (see Section III.B.1) was an effective trapping material for volatile analyses by PT in combination with GC/MS. The Tenax trap was unable to retain vinyl chloride, and Tenax with silica gel produced unacceptable water background for the mass spectrometer. The column flow was controlled externally by a pressure regulator installed on the helium flow stream. The column showed satisfactory resolution with only a few coeluting compounds. Vinyl chloride, the most volatile compound, was detected at about 3 min.

Sixteen samples were loaded to an ALS2016 autosampler, which enabled a high degree of automation. All the GC/MS analyses were performed automatically in sequence.

The efficiency with which a volatile organic compound is removed from water during PT concentration is dependent upon vapor pressure, solubility, and temperature. The system performance quality control was accomplished by spiking a surrogate sample to all analyzed water samples. An acceptable recovery range of 80–120% was obtained for all the samples reported. 1,2-Dichloroethane was found to be one of the most abundant volatile compounds in the samples analyzed. Its concentration reached about 4.2% in one oil sample and 46 mg/L in a water sample. Because of their high concentration, carryover of the volatile compounds was observed during the consecutive runs. Consequently most of the samples were diluted 100-fold in this investigation, and the PT bake time was maintained at 45 min.

The widespread occurrence of volatile hydrocarbons in groundwater is generally a result of lateral diffusion of LNAPLs and downward penetration of DNAPLs from the main pollution source. Figure 2 illustrates a cross section of the waste site where the samplings were made. The quantitative analytical results are shown in Table 8. Those compounds that were identified but not quantified are listed in Tables 9A and 9B. The separation and concentration techniques employed for the T5 oil sample (Fig. 1) enabled us to separate volatile compounds from the oil because the introduction of oil to the PT sampling tube is practically impossible and would result in contamination of the system and make the GC/MS analysis extremely complex.

The S1-2 to S6-4 samples were from sand and gravel, and those from R2-3 to R7-1 were from bedrock (Fig. 2). WTP was an effluent water sample from the water treatment plant that initially contained a significant quantity of pollutants. A sample (not shown in Table 8) taken from a site located 2 km away and at a higher elevation from the main pollution source showed only traces of volatiles amounting to about 10 μg/L. The R7-1 sample had a significantly lower concentration of organic compounds than R2-3 and R6-2, which may indicate, for example, their faster migration in bedrock than in sand and gravel. The S3-1 sample from a location close to R2-3, but in a sand and gravel formation, contained at least 50-fold more organics. A similar conclusion was reached earlier by Martel and Ayotte [2]. As Table 8 demonstrates, T5 is the main source of contamination. All the chlorinated and nonchlorinated solvents were most likely dumped together with waste oils at the T5 site.

The concentration methodology applied in this investigation is a very simple, efficient, and fast method for oil analysis. The PT analysis itself is a simple method but requires a dedicated system free of any leakage. PT/GC/MS also requires a high-resolution capillary column and fully automated MS software to handle all separations, concentrations, analyses, and output of results.

C. Polycyclic Aromatic Hydrocarbons from Water and Soil

The detection of trace quantities of PAHs in air, water, sediments, biota, and multiple-phase mixtures is of interest because it helps to trace their origins and assess the biological risks of their presence.

The procedure for separating organic compounds from water and soil and concentrating them for PAH analysis is discussed in Section III and illustrated in Fig. 1. A wide variation of PAH concentrations (0–200 mg/L) on the one hand and possible interference from many ions on the other hand

Table 8 Volatile Hydrocarbons in Water and Oil Samples (in μg/L)

Sample identification[a] Elevation (m)	T5-1[b] 43.2	R7-1 2.6	S6-1 14.2	R6-2 0.3	S6-4 27.1	S4-1 12.7	S4-4 28.6	S3-1 22.2	R2-3 3.1	S1-2 18.9	WTP[c] 44
Vinyl chloride	—	—	1415	—	3956	8773	—	288	—	26	44
Chloroethane	—	—	—	—	—	—	—	—	—	17	—
1,1-Dichloroethene	52×10^5	—	1182	1032	412	7019	—	9	1798	1	—
Dichloromethane	20×10^5	—	30	—	10	372	19	1	—	13	337
trans-1,2-Dichloroethene		—	—	—	—	—	—	16	72	2	—
1,1-Dichloroethane		—	105	356	—	1200	372	106	1056	2	—
cis-1,2-Dichloroethene		—	d	106	697	18335	—	282	232	1	47
Chloroform	40×10^5	12	87	63	61	916	—	64	468	—	29
1,1,1-Trichloroethane	32×10^5	tr	—	—	311	d	101	d	200	d	—
Benzene	418×10^5	27	—	—	—	d	—	d	—	d	6
1,2-Dichloroethane	181×10^5	—	32587	30920	2737	46515	41	599	29156	41	1
Trichloroethene		—	1311	3602	140	12950	—	102	6996	—	—
1,2-Dichloropropane		—	—	12	—	—	—	2	—	—	—
Toluene	137×10^5	2	321	789	—	2746	918	16	1996	—	58
1,1,2-Trichloroethane	191×10^5	9	8109	17815	985	17122	18	127	20784	1	1
Tetrachloroethene	489×10^5	—	2966	5784	—	9859	27	1	7848	—	—
1,3-Dichloropropane		—	—	det	—	—	—	—	640	—	—
Chlorobenzene	6.5×10^5	—	224	1787	—	1382	—	—	445	—	—
Ethylbenzene	33×10^5	—	105	358	—	1093	11	—	704	—	1
m,p-Xylene	158×10^5	tr	238	839	11	2611	199	—	1503	—	4
o-Xylene		—	—	380	—	1832	—	—	699	—	1
1,1,2,2-Tetrachloroethane	det	—	100	485	—	1600	—	—	1096	—	—
Total	1757.5×10^5	50	48780	64328	9320	134325	1706	1613	75663	104	485

a See Fig. 2 for sample location.
b T5 oil sample.
c WTP, after water treatment plant.
d, detected; tr, trace.

Table 9A   Other Compounds Identified

Compound	T5-1	S1-2	S3-1	S4-1	S4-4	S6-1	S6-4	WTP
2-Methylpropene			d		d	d	d	
2-Butene								
1,3-Butadiene		d						
Diethyl ether		d	d					
Carbon disulfide							d	
Thiobis methane			d					
Pentane	d							
Ethyl acetate		d						d
Tetrahydrofuran		d						
Heptane	d							
2-Methylheptane	d							
1-Bromo-2-chloroethane					d			
2-Hexanone				d	d			d
1,2- or 1,3-Chloro-1-propene		d	d	d	d		d	
1,2-Dichlorobutane			d	d		d	d	
3-Chloro-2-(chloromethyl)-1-propene						d		
trans-1,4-Dichloro-(1 or 2)-butene	d	d	d	d	d	d	d	d
2-Heptanone					d			
1,3-Dichlorocyclobutane		d	d	d	d		d	
1,1-, 1,2-, or 2,3-Dichloropropene				d	d			
2-, 3-, or 4-Ethyltoluene	d		d					
Decane	d							
2,6,7-Trimethyldecane	d							
Trichlorobutane			d	d		d		
3-Methyldecane	d							
2-, 3-, or 4-Propyltoluene	d							
1-Propynylbenzene								d
Octahydro-4,7-methano-1H-indene	d							
sym-Dichloroethyl ether								
(2-Chloroethyl)benzene			d	d		d		
Tetramethylbenzene	d			d		d		

d, detected; WTP, water treatment plant.

Table 9B   Other Compounds Found in the Bedrock Water Samples

Compound	R2-3	R6-3	R6-6
Sulfur dioxide			d
2-Chloro-1,3-butadiene	d	d	d
2,3-Dichloropropene	d		
4-Methyl-2-pentanone		d	d
1,3,5,7-Cyclooctatetraene	d	d	d
2-Hexanone	d		
Ethylaminoethanol	d		
1,1,2,2-Tetrachloroethane	d		
1,2-Dichlorobutane	d	d	d
1,3-Dichlorobutane	d	d	d
1,2,4-Trimethylbenzene	d		
Ethylmethylbenzene	d	d	
Bicyclooctatriene [4.2.0]			d
Trichlorobutane	d	d	d
1,2,3-Trimethylbenzene	d	d	d
Limonene	d		
Indene	d		
Methyl-3-methylethylbenzene	d		
2-Chloro-6-methylphenol	d	d	
2-Chloroethylbenzene	d	d	d
2,3-Dihydro-5-methylindene	d		
Methylindene	d		
Trichlorotrifluoropropane	d		
1-(4-chlorophenyl)ethanone	d		
Methyl naphthalene	d		
Biphenyl	d		
Ethyl naphthalene	d		
1,3-Dimethyl naphthalene	d		
1,2-Dimethyl naphthalene	d		
1-(2-Propenyl)naphthalene	d		

Source: From Lesage, S. Characterization of groundwater contaminants using dynamic thermal stripping and adsorption/thermal desorption-GC-MS. *Fresenius J. Anal. Chem.*, 339: 516, 1991.

will interfere with many of the analytical procedures. Separation and subfractionation of the organic compounds, especially from water samples, removes interfering substances and concentrates PAHs to allow their detection. The procedure explained in this investigation enabled us to separate and fractionate the organic substances into two fractions in which PAHs/PCBs and phenols were concentrated (PCBs and phenol will be discussed in the following sections).

Relatively abundant molecular ions are produced in the electron impact ionization mode with little fragmentation. There are several PAH isomers among the priority pollutants with identical mass spectra. Good chromatographic separation on a nonpolar capillary column (~30 m) enables their unambiguous characterization, particularly in the single ion monitoring mode.

Solvent extraction of water samples at two different pH yielded two distinct fractions, one concentrated with PAHs/PCBs and one with phenols. Chromatographic fractionation of the Soxhlet extracts separated the PAHs/PCBs and phenols in soil samples. The simplest approach for PAH/PCB quantification was found to be GC/MS analysis of the first fraction.

Tables 10 and 11 present PAH analysis results of soil and water samples, respectively, collected from various locations (Fig. 2).

It is generally accepted that contaminant transport in large river systems is predominantly a function of the sediment characteristics [31]. PAHs can be adsorbed onto both organic and inorganic particulate matter. From Table 10, soil PAH contamination decreased laterally from the main source T5 and reached a minimum value at P1, about 350 m from T5. The P4 monitoring well surprisingly had a high level of contamination compared to the other sites. Similar to the volatile hydrocarbons, PAH concentrations were relatively high in water samples collected from the S4-1 and S5-5 sites (see Tables 8 and 11). Owing to their adsorption characteristics, the soil samples (Table 10) contained relatively higher PAH concentrations than the water samples (Table 11). All the samples had high naphthalene content because of its relatively high solubility compared to that of other PAHs. Water after treatment (WTP sample) showed only 0.4 μg/L of naphthalene.

The methods of sample separation and concentration were adapted from the methods used earlier by other researchers in this investigation [8,32]. Because oil was the main source of the contamination, most of the samples also contained various other types of hydrocarbons. Only PAHs, which are considered priority pollutants, were characterized in this work. High-resolution capillary columns with reproducible chromatograms and stable retention times (± 0.04 min) enabled unambiguous assignments of all the chromatographic peaks without further cleanup.

D. Phenols

The priority pollutant list issued by the U.S. Environmental Protection Agency includes 11 phenols, including phenol, cresol, chlorinated phenols, and nitrophenols. Considerable work has been reported on the recovery and removal of phenolic compounds from industrial wastes and water. Adsorption, ion exchange, solvent extraction, and chemical derivatization methods have been

Table 10 Polycyclic Aromatic Hydrocarbons in Soil Samples (in µg/kg)

Sample identification:	P1-SS1	P1-SS3	P2-SS3	P2-SS7	P4-SS1	P4-SS4	P4-SS7	P5-SS1	P5-SS8	P7-SS4	P7-SS12	T5[a]
Elevation (m):	26.5	14.5	29	19	26.5	18.5	13.5	31.5	16	32.5	17.5	39.5
Naphthalene	11.3	3.5	3.1	10.3	60.7	2.8	10.7	4.3	4.9	12.0	2.0	28,800
Acenaphthylene	nd	nd	nd	nd	0.4	nd	67.0	0.4	nd	16.1	1.6	15,700
Acenaphthene	nd	nd	52.1	nd	nd	nd	70.7	nd	nd	10.2	1.8	13,500
Fluorene	nd	nd	14.9	nd	nd	nd	257.2	0.9	0.6	49.9	10.1	23,100
Phenanthrene	1.3	nd	17.3	2.8	3.7	2.6	855.6	23.8	11.1	416.3	130.9	65,200
Anthracene	nd	nd	nd	nd	0.4	nd	169.8	5.2	2.5	238.3	34.4	45,700
Fluoranthene	1.5	nd	2.7	0.9	5.8	3.3	146.6	11.8	6.2	120.5	60.4	17,800
Pyrene	2.3	nd	5.5	2.0	5.5	4.6	217.4	14.2	8.4	139.5	54.3	20,100
Benzo[a]anthracene	0.4	nd	nd	nd	0.5	nd	50.0	2.6	1.4	61.1	13.4	9,300
Chrysene	1.8	nd	4.7	1.1	2.4	3.0	57.8	4.3	3.7	72.4	15.2	9,000
Benzo[a]pyrene	1.4	nd	10.8	1.1	0.6	2.5	4.0	3.0	2.8	78.41	1.2	8,400
Benzo[k]fluoranthene	nd	nd	nd	nd	nd	nd	nd	nd	nd	nd	nd	nd
Benzo[b]fluoranthene	nd	nd	15.1	1.1	nd	nd	16.9	0.9	0.7	42.4	5.5	6,500
Indeno[1,2,3-cd]pyrene	nd	nd	21.2	nd	nd	nd	5.9	nd	nd	21.6	3.6	2,000
Benzo[ghi]perylene	0.8	nd	nd	1.3	nd	1.9	7.2	1.5	1.7	30.4	3.3	2,200
Dibenzo[ah]anthracene	nd	nd	nd	nd	nd	nd	nd	nd	nd	nd	0.8	nd
Total	20.8	3.5	147.4	20.6	80.0	20.7	1936.8	72.9	44.0	1309.1	338.5	267,300

P samples are from the same sites as S samples.
nd: not detected
[a] Soil under T5 oil.

Table 11 Polycyclic Aromatic Hydrocarbons in Water Samples (in μg/L)

Sample:	WTP	R2-3	R6-2	R6-6	R7-1	S1-2	S3-1	S4-1	S4-4	S5-5	S5-5[a]	S6-1	S6-4	T5
Elevation (m):	44	3.1	0.3	10.6	2.6	18.9	22.2	12.7	28.6	30.9	30.9	14.2	27.1	43.2
Naphthalene	0.4	16.0	155.7	58.2	0.4	nd	0.2	105.2	0.9	455.6	21.2	29.9	nd	117.2
Acenaphthylene	nd	3.9	7.0	5.3	nd	nd	nd	6.1	nd	24.8	0.5	1.8	nd	11.9
Acenaphthene	nd	4.6	3.2	2.3	nd	nd	nd	9.6	nd	38.0	0.6	1.8	nd	4.4
Fluorene	nd	4.8	3.5	2.4	nd	nd	nd	10.2	nd	23.9	0.3	1.8	nd	5.5
Phenanthrene	nd	6.8	4.0	1.7	nd	nd	nd	11.6	nd	13.7	0.3	2.1	nd	11.7
Anthracene	nd	1.3	1.0	0.4	nd	nd	nd	1.4	nd	nd	nd	nd	nd	4.7
Fluoranthene	nd	0.4	nd	nd	nd	nd	nd	0.4	nd	nd	nd	nd	nd	3.4
Pyrene	nd	0.5	nd	nd	nd	nd	nd	0.4	nd	nd	nd	nd	nd	3.6
Benzo[a]anthracene	nd	nd	nd	nd	nd	nd	nd	nd	nd	nd	nd	nd	nd	1.1
Chrysene	nd	nd	nd	nd	nd	nd	nd	nd	nd	nd	nd	nd	nd	1.1
Benzo[a]pyrene	nd	nd	nd	nd	nd	nd	nd	nd	nd	nd	nd	nd	nd	0.0
Benzo[k]fluoranthene	nd	nd	nd	nd	nd	nd	nd	nd	nd	nd	nd	nd	nd	nd
Benzo[b]fluoranthene	nd	nd	nd	nd	nd	nd	nd	nd	nd	nd	nd	nd	nd	0.3
Indeno[123-cd]pyrene	nd	nd	nd	nd	nd	nd	nd	nd	nd	nd	nd	nd	nd	nd
Benzo[ghi]perylene	nd	nd	nd	nd	nd	nd	nd	nd	nd	nd	nd	nd	nd	nd
Dibenzo[ah]anthracene	nd	nd	nd	nd	nd	nd	nd	nd	nd	nd	nd	nd	nd	nd
Total	0.4	38.3	174.4	70.3	0.4	0	0.2	144.9	0.9	556.0	22.9	37.4	0	164.9

R samples are from bedrock formation in the same location as S samples; S samples are from sand and gravel.
T5, water in contact with T5 oil; WTP, after water treatment plant.
[a]Well is less purged.

extensively utilized. Organic solvents are used very often to extract phenols from environmental samples. In general, all techniques suffer from a lack of a common solvent for quantitative extraction of all compounds from aqueous media. The efficiency of various solvent extraction techniques was evaluated by Afghan et al. [33]. In order to obtain a similar degree of sensitivity for phenols of environmental interest, several derivatization procedures were developed to introduce electrophoric groups into the phenol molecule. For trace level analysis, most of the procedures are nonuniversal and suffer from the instability of the derivatives, high background levels from residues of reagents, the high cost of the reagents, and incomplete reactions [34]. It seems that the anion-exchange technique has some advantages and is partially successful [35]. In this investigation, however, solvent extraction for water samples and Soxhlet extraction for soil samples were applied for the separation of organic matter, which was then followed by elution chromatography.

The gas chromatogram of our reference mixture, which contained 11 phenols, clearly demonstrated that all these compounds can be determined as free phenols on a 30-m DB5 capillary column. The chromatogram showed Gaussian peaks with little tailing, a prerequisite for quantitative determination. The neutrality or optimum inertness of the column and the injection port liner made direct analysis for phenols possible. A quartz liner was used in this investigation because an untreated glass liner failed to separate all the compounds and produced strong band broadening. GC/MS was used for identification and quantification. The detection limit was increased by using the SIM mode.

Various soil and water samples were collected from different sites, and their phenolic compounds were separated and analyzed. 4-Methylmercaptophenol was also identified in significant abundance and is included in Tables 12 and 13 (see Fig. 2 for the sample sites). T5 soil (Table 12) contained more phenol and 2-chlorophenol than T5 water. It also had a surprisingly large amount of 2,4-dimethylphenol and 4-nitrophenol, which could not be found in the T5 water (see Table 13). 2,4-Dimethylphenol and 4-nitrophenol were observed in water samples only at sites 4 and 5. In contrast, a large amount of o- and p-cresol and some 2,4-dichlorophenol and 4-chloro-3-methylphenol were found in T5 water but not in the T5 soil. These compounds were found in soil at sites 4 and 5. Water at sites 4 and 5 and soil at site 4 contained significant levels of phenolics compared to the other locations in their immediate vicinities. This may indicate that either dimethylphenol and 4-nitrophenol were dumped together on the soil surface before the oil dump or that they were dumped into sites 4 and 5. Due to its high solubility, phenol was transported 250 m laterally from site T5 along with the water flow and deep into the bedrock. Cresol was found in a large proportion in T5 water

Table 12 Phenolic Compounds in Soil Samples (in $\mu g/kg$)

Sample identification:	P1-SS1	P1-SS3	P2-SS3	P2-SS7	P4-SS1	P4-SS4	P4-SS7	P5-SS1	P5-SS8	P7-SS4	P7-SS12	T5[a]
Elevation (m):	26.5	14.5	29	19	26.5	18.5	13.5	31.5	16	32.5	17.5	39.5
Phenol	1.9	9.2	1.3	1.6	1.1	6.2	9.0	5.5	1.6	1.4	2.1	12350
2-Chlorophenol	nd	1.0	0.3	nd	nd	nd	nd	nd	nd	0.6	nd	160
o-Cresol	0.1	nd	nd	2.1	nd	nd	8.8	1.0	nd	nd	nd	nd
4-(Methylmercapto)phenol	nd	0.5	nd	nd	nd	nd	nd	nd	nd	nd	0.4	nd
p-Cresol	5.2	nd	nd	35.6	nd	77.1	60.7	2.7	6.1	nd	2.4	nd
2-Nitrophenol	nd	nd	0.6	nd	nd	nd	nd	nd	nd	nd	nd	nd
2,4-Dimethylphenol	0.2	nd	0.2	7.3	2.5	36.0	28.2	0.6	nd	nd	0.5	4280
2,4-Dichlorophenol	nd	nd	nd	nd	nd	nd	3.3	nd	nd	nd	nd	nd
4-Chloro-3-methylphenol	0.5	nd	0.6	0.9	nd	nd	0.9	nd	nd	nd	nd	nd
2,4,6-Trichlorophenol	nd	nd	nd	nd	nd	nd	nd	nd	nd	nd	nd	nd
2,4-Dinitrophenol	nd	nd	nd	nd	nd	nd	11.7	nd	nd	nd	nd	nd
4-Nitrophenol	nd	nd	22.9	7.1	33.8	nd	2.6	nd	nd	nd	nd	70400
2-Methyl-4,6-dinitrophenol	nd	nd	nd	nd	nd	nd	nd	nd	nd	nd	nd	nd
Pentachlorophenol	nd	nd	nd	1.8	nd	nd	nd	7.0	nd	nd	nd	nd
Total	7.9	10.7	25.9	56.4	37.4	119.3	125.2	16.8	7.7	2.0	5.4	87190

P samples are from the same sites as S samples.
nd, not detected.
[a]Soil Under T5 oil.

Table 13 Phenolic Compounds in Water Samples (in $\mu g/L$)

Sample identification: Elevation (m):	WTP 44	R2-3 3.1	R6-2 0.3	R6-6 10.6	R7-1 2.6	S1-2 18.9	S3-1 22.2	S4-2 16.7	S4-4 28.6	S5-5 30.9	S6-1 14.2	S6-4 27.1	T5[a] 43.2
Phenol	nd	72.6	0.6	64.9	0.1	nd	0.1	1.0	4.5	51.7	0.4	0.5	11078
2-Chlorophenol	nd	6.5	nd	5.4	0.5	nd	nd	nd	0.4	2.1	nd	0.1	15.9
o-Cresol	nd	nd	nd	nd	nd	nd	nd	nd	29.9	nd	nd	0.3	4057
4-(Methylmercapto)phenol	nd	nd	nd	nd	nd	nd	nd	nd	5.1	nd	nd	0.3	12.6
p-Cresol	nd	nd	nd	nd	nd	nd	nd	nd	150.7	nd	nd	0.6	9833
2-Nitrophenol	nd	0.7	nd	nd	nd	nd	nd	nd	1.3	0.2	nd	nd	nd
2,4-Dimethylphenol	nd	113.3	3.8	48.0	nd	nd	nd	73.5	101.7	136.9	1.6	1.2	nd
2,4-Dichlorophenol	nd	1.2	nd	0.4	nd	nd	nd	0.2	nd	1.1	nd	0.3	14
4-Chloro-3-methylphenol	nd	1.3	nd	nd	nd	nd	nd	nd	nd	nd	nd	nd	44.6
2,4,6-Trichlorophenol	nd	0.1	nd	nd	nd	nd	nd	nd	nd	0.1	nd	nd	nd
2,4-Dinitrophenol	nd	11.3	nd	nd	nd	nd	nd	nd	nd	1.5	nd	nd	nd
4-Nitrophenol	nd	6.5	nd	nd	nd	nd	0.3	nd	1.0	14.0	nd	nd	nd
2-Methyl-4,6-dinitrophenol	nd	nd	nd	nd	nd	nd	nd	nd	nd	nd	nd	nd	nd
Pentachlorophenol	nd	3.3	0.3	0.6	nd	nd	nd	nd	nd	0.2	nd	nd	nd
Total	0	216.8	4.7	119.3	0.6	0.0	0.4	74.7	294.6	207.8	2.0	3.3	25055.1

nd, not detected; WTP, after water treatment plant.
[a]Water in contact with T5 oil (main oil phase).

but surprisingly not in significant quantities at the other locations except for site 4.

Tables 12 and 13 mainly demonstrate that the T5 site is an extremely contaminated area. Phenols have a longer retention period in bedrock than in sand and gravel and are distributed over a vast area. High water solubility of the phenolic compounds, various adsorption affinities of the geological formations, prolonged water pumping, and the natural water flow resulted in a wide distribution of phenolic compounds without a distinct trend. The contamination level at site 4 is relatively high, which may be related to the composition of the site sediments. The organic compounds seemed to have low mobility at this site.

E. Polychlorinated Biphenyls

The analytical problems with respect to PCBs are complicated by the fact that there are 209 chlorinated biphenyl isomers spanning 10 homologous series (1–10 chlorine atoms per biphenyl), particularly if the PCB pattern found in the samples deviates from that of defined technical mixtures or a combination of such mixtures [36]. A commercial mixture (Aroclor) may contain as many as 60 chlorobiphenyl isomers. Once an Aroclor enters the environment, it can be altered in terms of internal composition by admixture with other PCB isomers produced, for example, by chemical processes or incineration that are not the same as the chemical reactions used for the manufacture of Aroclor. Those PCBs do not necessarily produce any fixed patterns of chlorobiphenyl isomers. Changes in Aroclors can also result from differential partitioning of the PCB isomers between various environmental compartments (i.e., water, sediment, and biota). The PCB patterns in natural samples are often different from those of PCB standards or standard mixtures. An explanation for this is the differences in metabolism and water solubility of the isomers [37]. The characterization and identification of PCBs become even more complicated when they are found in complex matrices such as sediments. The contaminants must be separated from the matrices, and this is a most time-consuming and difficult task. Furthermore, other interfering materials, such as sulfur and hydrocarbons, must be removed or separated from PCB residues to enable their accurate qualitative and quantitative analyses.

Quantitative analyses are reported using EC and MS detectors. Electron capture detection has high sensitivity, but the GC/MS method using selected ion monitoring (SIM) is preferred over ECD because of its selectivity and low sensitivity to interferences from other compounds. There are no standards or automated methods for quantification of PCBs by GC/MS, but several approaches have been used as outlined below. The high cost of the instrument and difficulties in automating the analyses by GC/MS make the adaptation of this method for routine PCB analysis unlikely. However, GC/MS is an excellent method for confirmatory analyses. Because ECD response to different PCB isomers is so highly variable, this approach for complete quantification of PCBs involves considerable uncertainty. Mass spectrometry using selected ion monitoring has recently been proposed to improve the sensitivity of the mass spectrometric approach and to give quantitative information on the homologous composition of the PCBs [38].

Commercially available surrogate PCB standards or Aroclors have been used to measure total PCBs [39]. Accordingly, a set of 31 available congeners, each representing a group of congeners, has been developed. One PCB has also been used to quantify a group of isomers with similar levels of chlorination, from which the total PCB concentration was calculated. The major PCB congeners were also analyzed on the basis of the weight composition of congeners in Aroclors [40]. In all of these methods the concentration of the available PCB congeners and the estimated concentration of the unavailable congeners were summed to obtain the total PCBs. The ideal but most time-consuming analytical procedure is one that identifies and measures each individual chlorobiphenyl isomer [41], which requires resolution of all isomers from each other and from other substances. To fulfill these requirements, analyses must be performed on different columns with very high efficiency and selectivity. However, PCB quantification would be valid if the components could be individually measured and then summed to obtain a measure of total PCB content. The conventional quantification method reports the PCB content in environmental and biological samples by referring it to a particular Aroclor mixture, for example, Aroclors 1242, 1254, and 1260 (12 carbon atoms and 42, 54, and 60% chlorine atoms, respectively). The GC detector is first calibrated using a commercial Aroclor mixture, and then the appropriate commercial Aroclor profile is matched to the sample profile. This approach, however, is subject to error if the GC profiles do not resemble the Aroclor patterns.

Studies reported here use a capillary GC/MS and SIM method of detection and quantification. Since peak pattern fingerprints, in both environmental samples and the standard mixture of Aroclors 1254, 1242, and 1260, were apparent, with no significant environmental alteration or contamination from sources other than commercial Aroclors, the profile-matching technique was used in this investigation. All the samples, except those from the T5 site (mainly oil), were directly analyzed for PCBs after their fractionation on an alumina column into PAH/PCB and phenolic fractions.

Various soil, water, and oil samples were analyzed for the PCB content. The PCB compositions of the contaminated samples are listed in Table 14. PCBs were not detected in the majority of water samples collected. As observed for PAHs and phenols, the oil from T5 was found to be the main

Table 14 Polychlorinated Biphenyls (PCBs) (in μg/kg soil or μg/L water)

Sample identification	Location (elevation in m)	Aroclors 1242	1254	1260	Total
Soil samples					
P2-SS1	32.5	nd	0.4	nd	0.46
P5-SS1	31.5	nd	4.3	1.8	6.8
P5-SS5	22.5	nd	1.4	0.5	1.9
P5-SS6	21	nd	2.0	0.6	3.0
P5-SS8	16	nd	2.0	1.0	3.0
P6-SS1	29.5	nd	0.42	nd	0.42
P6-SS7	19	0.3	nd	0.2	0.5
P6-SS8	17	nd	4.7	4.4	9.1
P6-SS10	14.5	nd	2.1	1.4	3.5
P6-SS11	13	nd	1.0	0.7	1.7
P7-SS1	38.5	nd	136.4	132.5	268.9
P7-SS4	32.5	nd	20.1	12.9	33.0
P7-SS6	27.5	32.0	17.1	4.9	54.0
P7-SS8	23.5	nd	1.7	0.4	2.1
P7-SS12	17.5	111.2	29.7	6.5	147.4
P7-SS13	16	nd	8.2	0.9	9.1
Soil T1	40	nd	98.0	42.6	140.6
Soil T5	39.5				
Oil T5	43.2	2300×10^3	845×10^3	116×10^3	3261×10^3
Water samples					
WTP[a]	44	nd	0.1	0.1	0.2
R4	−3.0	nd	1.8	2.2	4.0
R6-4	6.4	nd	0.6	0.6	1.2
Water T5	43.2	7.7	7.8	1.7	17.2

P2–P7 samples are from the same sites as S samples in Fig. 2.
nd, not detected.
[a]Water treatment plant, collected before treatment.

source of PCBs. This investigation reports finding only three Aroclors—1242 (pentachlorobiphenyls), 1254 (hexachlorobiphenyls), and 1260 (heptachlorobiphenyls). This method of quantification is simpler and faster than the other methods. It also revealed some information regarding the possible in situ biodegradation of PCBs and particularly any PCB contributions from sources other than commercial Aroclors. In addition, the results showed an approximately similar weight ratio of each Aroclor deposited on land.

The method of extraction used is very similar to those frequently appearing in the literature. Some modifications were made, particularly in cleanup and PCB concentration in oil samples. The validity of the total PCB recovery after elution chromatography was evaluated by elution of a few micrograms of Aroclor mixture in methanol. Quantitative recovery was obtained by analysis of the mixture before and after elution.

Since various samples contained PCB concentrations far above the acceptable limits, further detailed study of the total PCB congeners will have limited interest for the regulatory agencies at this stage.

The majority of the soil samples analyzed contained significant quantities of elemental sulfur (S_8) identified by GC/MS. In many cases, sample solutions were saturated with sulfur, which was not removed in this investigation. However, the oil sample (T5) could not be analyzed without a preliminary cleanup as discussed in Section III.A.3. Accordingly, (see Section III.A.3) a concentrated PCB fraction was separated. Its total ion scanning GC/MS analysis did not show any compounds other than PCBs (above 90% purity). Aliphatic hydrocarbons were found to be the main source of interference for GC/MS analyses of PCBs in oil.

It was found earlier that the contamination level in sediments is a function of their characteristics [31]. It has also been reported that PCB levels in sediments correlate with their organic carbon content [42]. In close agreement, the highest level of Aroclor was found in P7-SS1 (Table 14), which also had the highest organic carbon content (3900 mg/kg soil). The PCB levels declined from the main source toward the water treatment plan installed about 300 m from the T5 site. The water samples contained relatively little PCBs, as expected from their low aqueous solubility.

Figure 3a shows the total ion chromatogram of the standard Aroclor mixture of 1242, 1254, and 1260 in a 1:1:1 weight ratio. These three were the major Aroclors at the Ville Mercier contaminated site. Figures 3b–3d show respectively the reconstructed m/e 256–258, 324–326, and 394–396 ion chromatograms. Figures 4a and 4b are the total ion chromatograms of the PCB fraction of T5 oil and P7-SS12 soil samples, respectively. A separate analysis did not reveal any octachlorobiphenyls in these samples (Aroclor 1262 or 1266). Selected ion chromatograms of T5 and P7-SS12 samples are shown in Figs. 5 and 6, respectively. Comparison of Figs. 3b–3d with Figs. 5 and 6 revealed distribution patterns similar to those of commercial Aroclors. It also allowed the calculation of individual Aroclor abundances at contaminated sites. Detailed quantification of the total PCBs has not been made at this stage.

In this method of quantification three or four characteristic chromatographic peaks were selected for each Aroclor. Their areas were matched with corresponding peaks in the chromatograms of the samples, and their

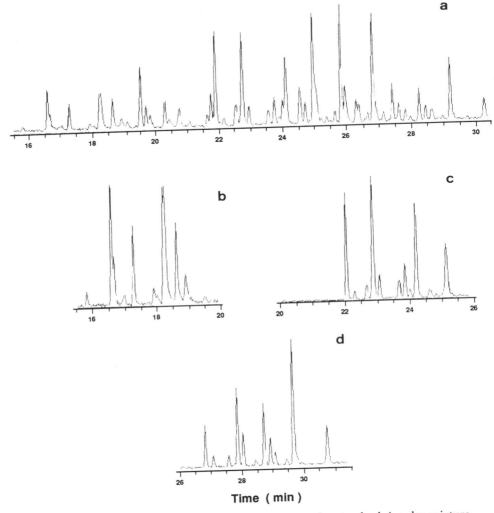

Figure 3 Total ion and selected chromatograms of a standard Aroclor mixture. (a) Total ion; (b) *m/e* 256–258; (c) *m/e* 324–326; (d) *m/e* 394–396.

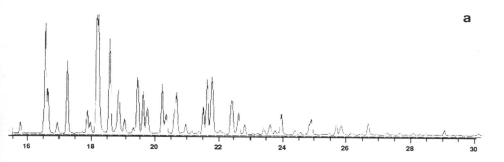

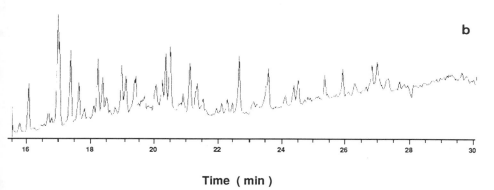

Time (min)

Figure 4   Total ion chromatograms of PCBs extracts. (a) An oil sample; (b) a soil
sample.

abundance were calculated. Since all the Aroclor characteristic peaks showed
a similar abundance, no alteration in PCB congeners distribution has occurred
since the deposition of the commercial Aroclors on the land almost 20 years
ago. Both qualitative and quantitative similarities were observed, however,
in Aroclor distribution patterns of the standard mixture and those of the
samples, which indicated the absence of significant PCB biodegradation.

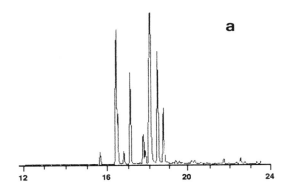

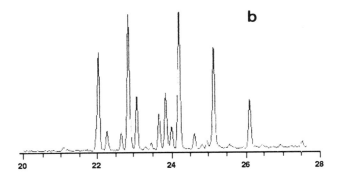

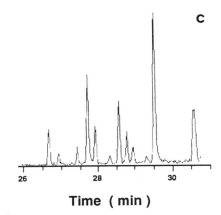

Time (min)

Figure 5 Ion chromatograms of PCBs extracts from an oil sample. (a) *m/e* 256–258; (b) *m/e* 324–326; (c) *m/e* 394–396.

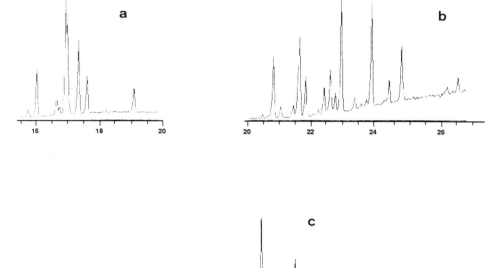

Figure 6 Ion chromatograms of PCB extracts from a soil sample. (a) *m/e* 256–258; (b) *m/e* 324–326; (c) *m/e* 394–396.

V. CONCLUDING REMARKS

A major aim of the work described in this paper has been to make some contribution to our knowledge of the Ville Mercier contaminated site and the extent to which the contamination has occurred and spread. Particular attention has been paid in the present work to the separation and concentration of hydrocarbons—volatile hydrocarbons, polycyclic aromatic hydrocarbons (PAHs), phenols, and polychlorinated biphenyls (PCBs)—and methods of analyses for them. Several soil, water, and oil samples were taken from different locations between the main dump site and the water treatment plant.

A fractionation method prior to the PT/GC/MS analysis was developed for volatile hydrocarbons in the oil samples. Twenty-two volatile compounds were identified and quantified. Their total concentration varied from about 1 μg/kg to 100 mg/kg in the sand and gravel formation and from 6 to 200 mg/kg in the bedrock formation, and was more than 10% in the oil. Tetrachloroethene and 1,2-dichloroethane were the most abundant contaminants in the water samples. Hydrocarbon type analysis of the oil samples revealed high concentrations of asphaltenes (high molecular weight compounds), ranging from 8.6% to 12.4%. Tetrachloroethene and 1,2-dichloroethane represented 4.9% and 4.2% of the oil, respectively. The distribution of n-alkanes, total PAHs, polar compounds, and aliphatic/alicyclic compounds in the oil samples showed significant differences from one sample to another, presumably due to their diverse origins.

GC/MS analyses of the polar fractions of the oil samples showed the presence of two main phthalate esters as well as a series of fatty acid esters.

The organic matter separated by solvent extraction (water samples) and Soxhlet extraction (soil samples) was fractionated by elution chromatography into two fractions concentrated with PAHs/PCBs and phenols. GC/MS analyses using the selected ion monitoring mode (GC/MS-SIM) was used to identify priority pollutant PAHs, phenols, and PCBs in the Ville Mercier soil and groundwater without further cleanup. The PAH content ranged from 0 to 260 mg/kg in soil samples and from 0 to 0.5 mg/L (mainly naphthalene) in water samples. Phenols were found in the range of 0–90 mg/kg (mainly 4-nitrophenol) in soil samples and 0–30 mg/L (mainly phenol) in water samples.

An Aroclor (commercial PCB) profile-matching method was applied to measure the Aroclor content of oil, soil, and water samples. A method of fractionation and concentration was developed to characterize PCBs in oil samples. The PCB fraction was mainly composed of Aroclors 1242, 1254, and 1260 with purity greater than 90%. Their concentration was 0–140 μg/kg in soil, 0–17 μg/L in water, and 0.3% in the oil phase.

The T5 site (oil phase) was found to be the main source of contamination and merits extensive efforts to obtain more information about the origin of the contamination.

Owing to the complexity of the contaminating oils and their flow properties, surfacial excavation, the use of in situ biotechnology, displacement of the DNAPLs by water vapor, and the use of surfactants may all be considered for decontamination. In addition, the technique of vacuum pyrolysis that was recently tested to decontaminate the Ultramar refinery site in St-Romuald, Québec, which is covered with approximately 60,000 barrels of oil residues in a watertight lagoon, could be ultimately applied to decontaminate the Ville Mercier main oil dump site.

ACKNOWLEDGMENTS

This work was supported by Environment Canada and the Natural Science and Engineering Research Council of Canada.

REFERENCES

1. Poulin, M. Groundwater contamination near a liquid waste lagoon, Ville Mercier, Québec. M.Sc. Thesis, University of Waterloo, Ontario, 1977.
2. Martel, R., and Ayotte, P. Etat de la situation sur la contamination de la nappe souterraine dans la région de Ville Mercier. Ministère de l'Environnement du Québec, 1989.
3. Locat, J. Gélinas, P., Isabel, D., Rouleau, A., and Roy, C. Assessment of the organics contaminant plume at Ville Mercier, Québec, Canada. 42ᵉ Conférence Canadienne de Géotechnique, Winnipeg, 1989.
4. Oliver, B.G. Analysis of volatile halogenated and purgeable organics. In *Analysis of Trace Organics in the Aquatic Environment* (B.K. Afghan and S. S.Y. Chau, eds.), CRC Press, Boca Raton, FL, 1989, p. 1.
5. Blumer, M., and Youngblood, W.W. Polycyclic aromatic hydrocarbons in soils and recent sediments. *Science, 188*:53, 1975.
6. Ilnitsky, A.P., Mischenko, V.S., and Shabad, L.M. New data on volcanoes as natural sources of carcinogenic substances. *Cancer Lett., 3*:227, 1977.
7. White, C.M., and Lee, M.L. Identification and geochemical significance of some aromatic components of coal. *Geochim. Cosmochim. Acta, 44*:1825, 1980.
8. Bartle, K.D., Lee, M.L., and Wise, S.A. Modern analytical methods for environmental polycyclic aromatic compounds. *Chem. Soc. Rev., 10*(1):113, 1981.
9. Environment Canada. Substances in the List of Priority Chemicals (1979)—Chlorophenols. CCB-In-4-80 Infonotes. Hull, Québec, 1979.
10. Buikema, A.L., Jr., McGinniss, M.J., and Cairns, J., Jr. Phenolics in aquatic ecosystems: a selected review of recent literature. *Marine Environ. Res., 2*:87, 1979.
11. Pavlov, B., and Terentyev, T. *Organic Chemistry*, Gordon and Breach, New York, 1965.
12. Kreiss, K. Studies on populations exposed to polychlorinated biphenyls. *Environ. Health Perspect., 60*:193, 1985.
13. Safe, S., Bandiera, S., Sawyer, T., Robertson, L., Safe, L., Parkinson, A., Thomas, P.E., Ryan, D.E., Reik, L.M., Levin, W., Denomme, M.A., and Fujita, T. PCBs: structure–function relationships and mechanism of action. *Environ. Health Prespect., 60*:47, 1985.
14. Oliver, B.G., Baxter, R.M., and Lee, H.-B. Polychlorinated biphenyls. In *Analysis of Organics in the Aquatic Environment* (B.K. Afghan and A.S.Y. Chau, eds.), CRC Press, Boca Raton, FL, 1989, p. 31.
15. Clement, R.E., and Tosine, H.M. Analysis of chlorinated dibenzo-*p*-dioxins and dibenzofurans in the aquatic environment. In *Analysis of Trace Organics in the Aquatic Environment* (B.K. Afghan and A.S.Y. Chau, eds.), CRC Press, Boca Raton, FL, 1989, p. 151.

16. Marble, L.K., and Delfino, J.J. Extraction and solid phase cleanup methods for pesticides in sediment and fish. *Am. Lab., November*:23, 1988.

17. Globensky, Y. Géologie de la région de Saint-Chrysostôme et de Lachine (sud). Ministère de l'Energie et des Ressources du Québec, MM 84-02, 1986.

18. Lasalle, P. Géologie des dépôts meubles de la région de Saint-Jean-Lachine. Ministère des l'Énergie et des Ressources du Québec, DPV-780, 1981.

19. Dion, D.-J., Cockburn, D., and Caron, P. Levé géotechnique de la région de Beauharnois-Candiac. Ministère de l'Énergie et des Ressources du Québec, MB 86-56, 1986.

20. Freeze, R.A. Hydrogéologie de la région de Lachine-St-Jean, Québec (au sud du St-Laurent), parties des cartes 31H/5 et 31H/6, moitié ouest. Geological Survey of Canada, Bull. 112, 1964.

21. Foratek International Inc. Etude hydrogéologique de faisabilité du captage des eaux contaminées extraites de la nappe aquifère de Ville Mercier. Pour le Ministère de l'Environnement du Québec par M. Poulin, rapport No. 154, 1982.

22. SNC Inc. Etude de faisabilité pour le traitement des eaux souterraines contaminées de Mercier. Pour le Ministère de l'Environnement du Québec, 1982.

23. Denis, C., and Rouleau, A. Pollution des eaux souterraines à Ville-Mercier: caractérisation hydrogéologique du substratum fracturé. Rapport soumis au GREGI, Université Laval, CERM 90-03, 1990.

24. Dion, D.-J., Cockburn, D., and Caron, P. Levé géotechnique de la région de Beauharnois-Candiac. Ministère de l'Énergie et des Ressources du Québec, Cartes, DV 85-05, 1985.

25. Sharpe, D. Glaciomarine fan deposition in the Champlain Sea. In *The Late Quaternary Development of the Champlain Sea Basin* (N.R. Gadd, ed.), Geological Association of Canada, Spec. Paper 35, 1989, p. 63.

26. Jackson, R.E., Patterson, R.J., Graham, B.W., Bahr, J.M., Bélanger, D.W., Lockwood, J., and Priddle, M. Hydrogéologie des contaminants organiques toxiques d'un site d'enfouissement, Gloucester (Ontario). Institut National de Recherches en Hydrologie, Direction Générale des Eaux Intérieures, Ottawa, 1985.

27. Pierce, R.C., Mathur, S.P., Williams, T.D., and Boddington, M.J. Phthalate esters in the aquatic environment. Nat. Res. Council Canada Publ. NRCC No. 17583, Associate Committee on Scientific Criteria for Environmental Quality, 1980.

28. Kohli, J., Ryan, J.F., and Afghan, B.K. Phthalate esters in the aquatic environment. In *Analysis of Trace Organics in the Aquatic Environment* (B.K. Afghan and A.S.Y. Chan, eds.), CRC Press, Boca Raton, FL, 1989, p. 243.

29. Autian, J. Toxicity and health threats of phthalate esters: review of the literature. Report ORNL-TIRC-72-2, Oak Ridge Natl. Lab., Toxicology Information Response Center, 1972.

30. Bellar, T.A., and Lichtenberg, J.J. Determination of volatile organics at microgram-per litre levels by gas chromatography. *J. Am. Water Works Assoc., 66*: 739, 1974.

31. Holdrinet, F.R., Braun, H.E., Thomas, R.L., Kemp, A.L.W., and Jaquet, J.M. Organochlorine insecticides and PCBs in sediments of Lake St. Clair (1970 and 1979) and Lake Erie (1971). *Sci. Tot. Environ., 8*:205, 1977.

32. Bravo, L.G., and Rejthar, L. Quantitative determination of trace concentration of organics in water by solvent extraction and fused silica capillary gas chromatography: aliphatic and polynuclear hydrocarbons. *Int. J. Environ. Anal. Chem., 24*:305, 1986.

33. Afghan, B.K., Belliveau, P.E., Larose, R.H., and Ryan, J.F. An improved method for determination of trace quantities of phenols in natural waters. *Anal. Chim. Acta, 71*:355, 1974.

34. Renberg, L. Gas chromatographic determination of phenolic compounds in water, as their pentafluorobenzoyl derivatives. *Chemosphere, 10*:767, 1981.

35. Richard, J.J., and Fritz, J.S. The concentration, isolation, and determination of acidic material from aqueous solution. *J. Chromatogr. Sci., 18*:35, 1980.

36. Wittlinger, R., and Ballschmiter, K. Global baseline pollution studies XI: congener specific determination of polychlorinated biphenyls (PCB) and occurrence of alpha- and gamma-hexachlorocyclohexene (HCH), 4,4'-DDE and 4,4'-DDT in continental air. *Chemosphere, 16*:2497, 1987.

37. Weigelt, V. Capillary gas chromatographic PCB sample analysis in marine species—comparison between concentration and chlorine substitution in the PCB components in marine organisms in a German bay. *Chemosphere, 15*:289, 1986.

38. Gebhart, J.E. Hayes, T.L., Alford-Stevens, A.L., and Budde, W.L. Mass spectrometric determination of polychlorinated biphenyls as isomer groups. *Anal. Chem., 57*:2458, 1985.

39. Cooper, S.D., Moseley, M.A., and Pellizzari, E.D. Surrogate standards for the determination of individual polychlorinated biphenyls using high-resolution gas chromatography with electron-capture detection. *Anal. Chem., 57*:2469, 1985.

40. Capel, P.D., Rapaport, R.A., Eisenreich, S.J., and Laoney, B.B. PCBQ: computerized quantification of total PCB and congeners in environmental samples. *Chemosphere, 14*:439, 1985.

41. Mullin, M.D., Pochini, C.M., McCrindle, S., Romkes, M., Safe, S.H., and Safe, L.M. High resolution PCB analysis: synthesis and chromatographic properties of all 209 PCB congeners. *Environ. Sci. Technol., 18*:468, 1984.

42. Kaiser, K.L.E., Comba, M.E., Hunter, H., Maguire, R.J., Tkacz, R.J., and Platford, R.F. Trace organic contaminants in the Detroit River. *J. Great Lakes. Res., 11*:386, 1985.

IV
GEOCHEMICAL INVESTIGATIONS

14

The Determination and Fate of Unstable Constituents of Contaminated Groundwater

Mary Jo Baedecker and Isabelle M. Cozzarelli

Geological Survey, U.S. Department of the Interior, Reston, Virginia

Geochemical investigations of contaminated groundwater are essential to an understanding of the fate and transport of organic compounds. The degradation of organic compounds by oxidation-reduction reactions results in large changes in the concentrations of unstable chemical constituents in groundwater. Methods of analysis used in field investigations are presented for alkalinity, dissolved oxygen, methane, hydrogen sulfide, ferrous iron, ammonia, organic carbon, nitrate, and sulfate. Methods for collecting samples and determining concentrations of methane in groundwater are presented in detail because such information is lacking in the literature. Microbial degradative processes that consume or generate these chemical constituents alter the Eh of the uncontaminated groundwater and may produce organic acids that interfere with the determination of bicarbonate. These processes also fractionate the stable isotopes of carbon and nitrogen. The geochemistries of two aquifers, one contaminated with creosote waste and the other contaminated with industrial and municipal waste, are significantly different from those of the natural environments as a result of biogeochemical reactions.

I. INTRODUCTION

An understanding of the groundwater flow system and the chemical reactions that control the movement and fate of contaminants is needed to manage our water resources properly. Much has been published since the 1960s on the geochemistry of groundwater and on the equilibrium approach to describing and calculating the speciation of chemical constituents (see Back and Freeze [1], for a historical review). Equilibrium models and mass balance calculations have been applied to describe geochemical processes in water that has high concentrations of organic compounds [2,3]. Quasi-equilibrium is generally assumed in geochemical modeling of reactions that involve labile organic compounds, even though these reactions are kinetically controlled and irreversible. More information is needed to understand organic processes so that reactions can be better incorporated into geochemical models.

Organic compounds are unstable at the Earth's surface. Evaluation of thermodynamic data indicates that these unstable compounds are out of equilibrium with their present environment and will react to form CO_2 and CH_4. Oxidation and reduction reactions (chemical reactions in which the participating elements lose or gain electrons) are important in degradative processes in environments contaminated with organic compounds. Degradative reactions, which are largely microbially mediated, affect the oxidation-reduction potential and the fate and transport of many elements in sediment and water. As a result of these reactions, changes occur in the mineral equilibria of the dissolved species and the mineral composition of the aquifer. Aerobic degradation is the most significant process that oxidizes organic compounds to carbon dioxide and water in shallow subsurface environments. Also of importance are processes that transform organic compounds in the absence of oxygen. Some processes, such as methanogenesis via the acetate pathway, result in the oxidation and reduction of organic compounds, but other processes couple the oxidation of organic compounds to the reduction of iron, manganese, sulfate, nitrate, bicarbonate, or other inorganic electron acceptors. The biological and physicochemical reactivity of these electron acceptors makes them unstable in contaminated groundwater environments. The products of the oxidation-reduction reactions include gaseous and ionic species, some of which also are unstable in groundwater environments. Geochemical studies of unstable constituents in groundwater require the use of special collection and analytical procedures in the field and careful analytical techniques in the laboratory.

The first objective of this chapter is to present methods for determining unstable constituents in contaminated groundwater environments. The second objective is to describe the geochemical processes and some problems associated with interpreting geochemical data at sites where high concentrations of organic compounds are found. Controls on concentrations of unstable constituents in contaminated groundwater are discussed, and the use of stable isotope data in organic geochemical problems is described. Two important problems associated with interpreting geochemical data at contaminated sites are the definition of oxidation-reduction potentials and the effect of organic acids on alkalinity. The third objective of the chapter is to present results of investigations of the geochemical processes occurring at two sites contaminated with organic compounds.

II. MEASUREMENT OF UNSTABLE COMPOUNDS IN GROUNDWATER

Methods are presented in this section for collecting water samples and determining alkalinity and the concentrations of selected unstable constituents, including dissolved oxygen, methane, hydrogen sulfide, ferrous iron, ammonia, organic carbon, nitrate, and sulfate. These are methods that we have used in our geochemical investigations of contaminated groundwater sites, but other methods may be equally valid. Methods for the analysis of methane in groundwater are not well documented in the literature and are therefore given here in detail.

A. Alkalinity

Alkalinity, defined as the capacity of a solution to neutralize acid, is one of the most commonly reported analyses in geochemical studies of groundwater. The determination of alkalinity as a property of natural water dates back to the mid-nineteenth century when Thomas Clark [4] reported on his methods for examining wastewater. In uncontaminated groundwater, the principal source of carbon species that produce alkalinity is the solution of CO_2 from soil gases and the dissolution of carbonate minerals. An additional source of CO_2 in aquifers that contain natural organic material or contaminants is the microbial degradation of organic compounds.

Alkalinity is most accurately determined in the field at the time a water sample is collected rather than in the laboratory. The theory and method of determining alkalinity by potentiometric titration are discussed by Weber and Stumm [5]. GRAN plots are commonly used to determine alkalinity in seawater [6,7] and acidity in water with low specific conductance such as rainwater [8]. These plots use functions that linearize titration curves and thus make it possible to determine alkalinity with a few points rather than from the inflection point on an entire titration curve. The use of this method to determine alkalinity in contaminated water that contains organic acids needs additional investigation. The method described here is the standard potentiometric method.

a. Procedure

Total alkalinity is determined in the field on an aqueous sample that has been filtered through a 0.4-μm filter placed in line with the sampling pump. A 20-mL sample is pipetted into a small beaker and titrated, while stirring, with equal increments of a strong acid using a microburet such as a 2.0-mL Gilmont.* The titration is taken to a pH of 3.0. The endpoint pH occurs at the maximum change of pH per added volume of titrant. Alkalinity is calculated from the volume of acid required to reach the endpoint pH, which is determined graphically by plotting (1) the titration curve (pH as a function of the titrant volume) and locating the inflection point or (2) the first derivative of the titration curve as a function of the titrant volume and locating the maximum [9].

Although alkalinity represents the equivalent sum of all the species present in solution that are titratable with a strong acid, it is commonly reported only as bicarbonate and carbonate. Titration to a fixed endpoint pH of 4.5 will yield incorrect values for bicarbonate at low concentrations of bicarbonate or in solutions with other titratable anions. Other inorganic species that contribute to alkalinity include hydroxide, silicate, and borate complexes [10]. Also, low molecular weight organic acids such as acetic acid contribute to alkalinity. Groundwater with high concentrations of organic compounds that are degraded in anoxic environments may contain organic acids that persist. Organic acids can complicate the determination of the bicarbonate titration endpoint. This is discussed in more detail in Section III.C.

B. Dissolved Oxygen

Dissolved oxygen is an important constituent to determine in groundwater because its presence or absence affects geochemical and microbiological processes. Depending on the amount of oxygen in the aquifer, organic compounds degrade by different pathways and thus at different rates. Concentrations of oxygen in groundwater range from saturation (about 11.3 mg/L at 10 °C) to zero. Aquifers contaminated with organic compounds usually become anoxic when the demand for oxygen from microbial degradation or mineral precipitation exceeds the influx of oxygen from recharge water.

a. Procedure

Dissolved oxygen is determined in the field either by colorimetric techniques or by an oxygen-sensitive probe. Because probes are difficult to calibrate at

---

*The use of trade names is for identification purposes only and does not constitute endorsement by the U.S. Geological Survey.

0.00 mg/L, measurements of low levels of dissolved oxygen (less than 1.0 mg/L) may be inaccurate. Thus, in contaminated groundwater with low concentrations of dissolved oxygen, results from the Winkler colorimetric method [11] are more accurate and reproducible than measurements with probes. In performing the Winkler titration, reducing the sample volume from the standard 300 mL [11] to 60 mL is better suited for field studies. The entire sample is titrated directly in the bottle using a 2-mL Gilmont buret. At dissolved oxygen concentrations above 6.5 mg/L, an aliquot is removed and titrated. The limit of detection is 0.01 mg/L.

In the field, 60-mL BOD bottles that were previously calibrated (volumetrically) are filled and allowed to overflow with at least three volumes of water. The bottles are kept stoppered until the addition of 0.5 mL each of the following reagents in order: potassium fluoride, manganous sulfate, and alkaline-iodide sodium azide. The reagents used are available from Hach Company (Loveland, CO). The stopper is replaced carefully to preclude air bubbles, and the bottle is shaken twice, each time allowing the floc to settle near the bottom of the bottle. About 0.5 mL of concentrated sulfuric acid is added by dripping the acid down the neck of the bottle (to minimize mixing of atmospheric oxygen in the sample). The bottle is placed on a magnetic stirrer, a small stir bar is added, and after the floc is completely in solution, the entire sample is titrated to a straw color with 0.025 N sodium thiosulfate solution (0.005 N for samples with O_2 concentrations less than 0.5 mg/L). A small amount of blue indicating Thyodene powder (Fisher Scientific, Pittsburgh, PA) is added to the solution, and the titration is continued until the first disappearance of color. Calculations are made using the following equation:

$$DO\ (mg/L) = \frac{1}{mL_{sm}} \times \frac{mL_{bt} + mL_{rg}}{mL_{bt}} \times mL_{tr} \times N_{tr} \times 8000 \qquad (1)$$

Where N_{tr} is the normality of the titrant (eq/L); mL_{tr}, the volume of the titrant (mL); mL_{bt}, the volume of the sample bottle (mL); mL_{rg}, the volume of the reagents (mL); mL_{sm}, the volume of sample titrated (mL); and 8000, the unit conversions (atomic weight/equivalent for O_2 × mg/g).

C. Methane

Data on the distribution of methane in hydrogeologic environments are scarce because the concentrations are low in natural groundwater [12]. Methane is found in aquifers with a source of organic material, such as coal, peat deposits, and organic waste. Methods for the determination of methane in hydrologic studies are not well documented. Field determinations are difficult because methane analyses require a gas chromatograph, which generally is unavailable in the field. In the method described below, water samples are collected and preserved in the field, then analyzed in the laboratory by gas chromatography (GC).

a. Procedure

Samples are collected by syringe from the stream of water from a well pump and transferred to a serum bottle. The serum bottles are prepared in the laboratory in the following manner. The bottle (24 mL) is flushed with N_2, a few grains of $HgCl_2$ are added (to produce a concentration of approximately 55 mg/L, which is sufficient to inhibit microbial activity), and a rubber stopper is inserted. The stopper is secured with an aluminum crimp seal. The bottle is weighed (± 0.01 g), and about 15 cm³ of N_2 is removed with a syringe so that the bottle is under a partial vacuum. In the field, 10-mL Glaspak syringes with luer tips and four-way stopcocks (Propper Company, Long Island City, NY) are rinsed several times while being held in the flowing stream of well water to ensure that all air bubbles are expelled and only water unexposed to air is drawn into the syringe (Fig. 1A). A needle is attached quickly to the four-way stopcock, and some water is expelled to flush the needle. The water sample (2–10 mL depending on the methane concentrations) is injected through the rubber stopper of the serum bottle; then the stopcock is closed to the bottle (Fig. 1B) and the needle and syringe assembly is withdrawn. The bottles are stored at 4°C to slow bacterial activity in the water sample.

In the laboratory, the sample bottle is weighed again to measure the volume of water collected. After analysis of the headspace, the bottle is filled with water and weighed a third time to obtain the total volume of the bottle. The headspace volume is calculated by difference.

Methane is analyzed by gas chromatography at room temperature using a column (1.5 m length, 0.3 cm o.d.) packed with Porapak Q. The gas chromatograph is equipped with a flame ionization detector heated to 100°C. The serum bottles, at room temperature, are shaken vigorously for a few minutes so that the methane in the headspace is in equilibrium with the methane in the water. Most of the methane will be in the headspace. An accurately measured 0.5-2-mL (± 0.01 mL) sample of the headspace is injected into the gas chromatograph. Methane standards mixed in air, nitrogen, or helium are commercially available in concentrations of 10-10,000 ppm by volume, which is an adequate range for groundwater samples. With this analytical procedure, a detection limit of 10 μg/L dissolved methane is obtained with a 10-mL water sample and 14-mL headspace. A detection limit of 1 μg/L can be obtained with a 20-mL water sample and by using a standard with lower methane concentrations made by diluting a standard with nitrogen.

Equation (2) given below assumes that the same volumes for the standard and sample were injected into the column and that the response of the standards is linear. If the conditions are different, a correction factor is calculated for the volumes injected, and a standard curve is then plotted to calculate the concentration of methane in the gas sample. Equations (2) and (3)

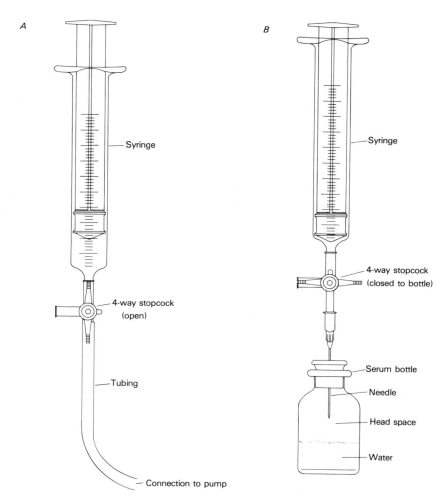

Figure 1 Syringe, four-way stopcock, and serum bottle used for sampling dissolved methane from a well. A, position of the assembly to fill the syringe with water; B, assembly after the syringe is emptied to the serum bottle (drawing not to scale).

(the ideal gas law) are used in Eq. (4), which is used to calculate the amount of methane in the original water sample in millimoles per liter.

$$C_{gas} = \frac{PA_{gas} \times C_{std}}{PA_{std}} \tag{2}$$

$$PV = nRT \tag{3}$$

$$c_{wat} = C_{gas} \times \frac{V_{gas}}{V_{wat}} \times \frac{10^{-3}}{0.0821 \times T} \tag{4}$$

where C is the concentration of methane (ppm by volume); PA, the integrated peak area of methane from the chromatogram; P, the pressure (atm); V, the volume of gas or water (L); n, the number of moles of gas; R, the ideal gas constant (0.0821 L·atm·K^{-1}·mol^{-1}); T, the temperature of the sample at time of analysis (K); and c, the concentration of methane (mmol/L); and the subscripts gas, std, and wat refer to headspace in the sample bottle, standard, and water in sample bottle, respectively.

Equation (4) is used to calculate the amount of methane in the original water sample based on the amount of methane measured in the headspace. A more rigorous calculation includes the amount of methane that remains in the water after equilibration of the water and headspace by shaking. The amount of methane remaining is calculated from Henry's law and the constant for the solubility of methane in water [13]. The error in aqueous methane concentrations that results from using only Eq. (4) is small for most groundwater environments. For example, the amount of methane remaining in water was 2.3% of the total methane in a 10-mL water sample with a methane concentration of 0.1 mmol/L.

D. Hydrogen Sulfide

Hydrogen sulfide is volatile and unstable in the presence of oxygen. Therefore, methods of analysis of aqueous sulfide involve either fixing sulfide species as metallic sulfides, measurement with a sulfide electrode, or formation of a color complex. At low sulfide concentrations, colorimetric methods are preferred because standards for use with the sulfide electrode are unstable in the field, and large volumes of sample water are required to form metallic sulfides for analysis. Also, some metallic sulfides are unstable in the presence of oxygen. A methylene blue colorimetric method for determining the concentration of aqueous sulfide species [14] uses small sample volumes, is easy to use in the field, and gives reproducible results.

a. Procedure

In this method, H_2S, HS^-, and S^{2-} react with diamine (N,N-dimethyl-p-phenylenediamine sulfate) and FeCl$_3$ (ferric chloride) in an acidic reagent solution to form a methylene blue color complex that is measured spectrophotometrically. Details of the method for five concentration ranges of aqueous sulfide are reported by Lindsay and Baedecker [14]. Two concentration ranges of standards, 0.312-31.2 μmol/L and 31.2-312 μmol/L, are usually sufficient to measure concentrations for sulfides (0.01-10.0 mg/L) in groundwater. Diamine and $FeCl_3$ reagents and a standardized sulfide solution are prepared in the laboratory. A standard curve is plotted for mixtures of the two reagents and sulfide solutions of known concentrations.

In the field, the diamine and $FeCl_3$ reagents are mixed in equal amounts, and 2 mL of the solution is taken up in a 10-mL Glaspak syringe fitted with a four-way stopcock to exclude air bubbles. The syringe and stopcock are connected in-line with the water pumped from the well. After being flushed with water, the stopcock is opened to introduce water into the syringe, where the water mixes with the reagents. A splitter may be necessary to divert some of the water from the pump to allow the syringe to fill slowly with water. The syringe is filled to 10 mL, the stopcock is closed to the syringe, and the sample is kept in the dark on ice for at least 30 min. If sulfides are present, a blue complex forms. The samples can be transferred to vials and refrigerated for as long as 1 week, if necessary. The absorbance of the samples is measured on a spectrophotometer at a wavelength of 670 nm. The reagents remain stable for at least a year if they are kept in the dark under refrigeration.

E. Ferrous Iron

The concentration of dissolved iron in oxygenated shallow groundwater is low. The activity of total ferric iron species in oxidized water that is in equilibrium with ferric hydroxide at a pH greater than 4.8 is less than 10 μg/L [10]. However, the solubility of iron increases significantly in groundwater where the pH is lower or where oxygen is depleted and the Eh of a system is low. Microbial degradation of organic compounds in groundwater can affect the concentrations of ferrous iron in solution directly by reducing ferric iron or indirectly by causing changes in the pH and Eh of the system. The ability of microorganisms to reduce iron has long been recognized. Recently, the coupling of anaerobic oxidation of hydrocarbons to dissimilatory iron reduction was demonstrated [15].

a. Procedure

Ferrous iron concentrations are determined by the colorimetric bipyridine method [11]. Water samples are filtered through a 0.1-μm filter placed in-line with the sampling pump to remove particulate iron hydroxides or carbonates. Ferrous iron and 2,2'-bipyridine form a red complex that is measured spectrophotometrically at 520 nm. Solutions and standards are prepared as described in Brown et al. [11]. Standard curves are prepared for two concentration ranges of 0-10 mg/L and 10-100 mg/L dissolved iron. Samples are prepared as follows. For the low concentration range, pipet 0.5 mL of a 2.0 g/L bipyridine solution and 1 mL of double-deionized water into a small plastic bottle and add a 10-mL water sample. After the solution is allowed to stand for 30 min, add 1 mL of a 350 g/L sodium acetate solution. For the high concentration range, a 1-mL sample and 9 mL of deionized water are substituted for the 10-mL sample. The colored complex is stable for at least 4 weeks when samples are refrigerated.

Total iron concentrations can also be determined by the bipyridine technique. The technique is the same as for ferrous iron except that 1 mL of a hydroxylamine hydrochloride solution is substituted for the 1 mL of deionized water [11]. The hydroxylamine hydrochloride reduces ferric iron to ferrous iron. Iron standards are prepared by the total iron method because iron in stock solutions is in the ferric oxidation state. To compare ferrous and ferric iron data, the same method, such as the colorimetric method described above, should be used to determine the concentrations of ferrous and total iron (ferric iron is calculated by difference). However, to determine only the concentration of total iron, instrumental methods such as atomic absorption or direct current plasma techniques rather than colorimetric methods are generally used.

F. Ammonia

Sources of ammonia in groundwater are the biologically mediated reduction of nitrate, the degradation of nitrogen-containing organic compounds, and fertilizers. At the pH of most groundwater (less than pH 9), ammonia is present predominantly as the ammonium ion (NH_4^+).

a. Procedure

Ammonia concentrations are determined with an ion-sensitive electrode such as the Orion Model 95-12 electrode. The electrode uses a gas-permeable membrane that allows ammonia from the sample to diffuse to the internal solution of the electrode [16]. The partial pressure of ammonia in a given sample is proportional to its concentration, according to Henry's law. The electrode response is described by the Nernst equation and is linear to 5×10^{-6} M (0.09 mg/L).

A water sample is filtered through a 0.4-μm filter and collected in a glass vial, acidified with concentrated ultrapure nitric acid (0.1 mL per 20 mL of sample), and stored on ice. In the laboratory, the electrode is standardized over a range of 10^{-2}-10^{-6} M, and a standard curve is prepared. A 5-mL sample is added to 500 μL of a pH buffering solution containing NaOH and EDTA (ethylenediaminetetraacetic acid) (Orion ISA solution 951211). The addition of EDTA prevents clogging of the membrane of the ammonia electrode by metal complexes such as iron hydroxides [17]. The sample is stirred in a closed system, and the ammonia is allowed to diffuse into the membrane solution until a steady reading is obtained, usually less than 5 min. Ammonia can also be determined by colorimetric techniques. Colorimetric techniques, such as the indophenol method [18], may yield erroneous results due to interferences from dissolved iron, sulfides, amines, or organic acids that may be present in the sample [19].

G. Organic Carbon

The mean concentration of dissolved organic carbon (DOC) in natural groundwater is 1.2 mg/L [20], whereas concentrations for contaminated groundwater can reach several hundred milligrams per liter. DOC is operationally defined as that fraction of the organic carbon that passes through a 0.4-μm filter. Because the amount of colloidal organic material in contaminated water may be appreciable, filtration of samples through filters with a pore size smaller than 0.4 μm might be a better measure of the fraction of carbon that is dissolved. The pore size of the filters used to collect DOC samples should be reported with the data.

a. Procedure

Bottles are cleaned by washing and then baking at 400 °C. Teflon-lined caps are used for the bottles. The water samples are filtered through a 0.4-μm filter, preserved with $HgCl_2$ (resulting in a concentration of 55 mg/L), and chilled to slow bacterial activity. Samples are analyzed instrumentally by wet oxidation or combustion techniques. A standard curve is obtained for quantitation using potassium biphthalate as a standard.

Wet oxidation and combustion methods of analysis require acidification and sparging of the sample to remove inorganic carbon. This process also removes volatile organic carbon (VOC), such as monoaromatic hydrocarbons. In some environments, the VOC may be a large part of the organic carbon pool. The concentrations of VOC are determined by passing the sparge gas through a Tenax trap and then removing and combusting this fraction [21]. This technique is possible on some organic carbon analyzers. Separate standard curves for the dissolved and volatile fractions are determined for quantitation.

H. Nitrate and Sulfate

The inorganic anions nitrate and sulfate commonly are major chemical constituents in groundwater. Nitrate in water generally signifies contamination, whereas sulfate has natural sources in addition to possible contaminant sources. The amount of natural sulfate in groundwater depends on the solubilities of sulfur-bearing minerals. Nitrate and sulfate are mobile and stable in oxygenated water. In anoxic groundwater, these anions serve as terminal electron acceptors in microbial processes that involve the oxidation of organic material. Thus, determination of the concentrations of these constituents in groundwater provides an increased understanding of the progress of biogeochemical reactions in hydrogeologic environments. Techniques for measuring nitrate and sulfate concentrations are well documented [18].

a. Procedure

Water samples are filtered through 0.4-μm filters, stored in plastic bottles, and refrigerated until analyzed. Samples are analyzed within 6 weeks of collection to limit further degradation of sulfate and nitrate in the sample bottle. The anions are determined by ion-exchange chromatography. Other techniques of analysis, such as colorimetric and turbidimetric techniques, are more time-consuming and yield less reproducible results than analysis by ion chromatography.

III. INTERPRETATION OF UNSTABLE CONSTITUENTS IN GROUNDWATER

A. Concentrations of Unstable Constituents at Contaminated Sites

Concentrations of unstable constituents are given for seven contaminated groundwater sites in Table 1. The sources of contamination at these sites are creosote, crude oil, gasoline, and domestic and industrial wastes in landfills. The hydrogeologic environments are shallow unconfined aquifers in the coastal plain or glacial outwash sediments in the United States and Canada. At each of these sites, the contaminant plume is less than 30 m below the land surface. Chemical data for the Borden site are maximum values found at the landfill. For the other sites, the data are for the most contaminated groundwater at one sampling location where the water was not directly in contact with a separate contaminant phase, such as creosote or crude oil. Concentrations of chemical constituents for uncontaminated groundwater are given where data were available.

The measurement of specific conductance in organic-rich water can have little relation to the amount of contamination because most organic compounds are nonionic. For example, at Pensacola (a creosote-contaminated site) the specific conductance was 32% higher in contaminated groundwater than in the uncontaminated groundwater, but at Galloway (a gasoline-contaminated site) the specific conductance was 63% higher than in the uncontaminated groundwater. These readings do not reflect the fact that the water

Table 1 Selected Geochemical Parameters in Groundwater Contaminated with Organic Compounds

| | (1) Conroe, TX Creosote waste Fluvial | | (2) Pensacola, FL Creosote waste Coastal plain | | (3) Bemidji, MN Crude oil Glacio-fluvial | | (4) Galloway, NJ Gasoline Coastal plain | | (5) Borden, Ont. Landfill Glacio-fluvial | | (6) Army Creek, DE Landfill Coastal plain | | (7) North Bay, Ont. Landfill Glacio-fluvial | |
Site: Organic source: Aquifer:	contam	bkgrnd	contam	bkgrnd	contam	bkgrnd	contam	bkgrnd	contam[a]	bkgrnd	contam	bkgrnd	contam	bkgrnd
Specific conductance (μS/cm)	125	88	370	280	NR	NR	285	175	NR	NR	850	NR	NR	NR
pH	5.60	4.00	5.65	6.40	6.94	7.68	5.99	4.69	6.5	7.4	6.55	5.10	6.87	NR
Dissolved oxygen (mg/L)	0.15	1.38	0.0	0.70	0.0	9.0	0.0	6.4	0.0	NR	0.0	7.6	NR	NR
Dissolved organic carbon (mg/L)	NR[11]	NR	192	8.5	32.0	2.7	29.4	1.2	38.4	9.6	23	<0.1	290	3.3
Volatile organic carbon (mg/L)	NR	NR	8.4	<1.0	21	<0.5	NR	NR	NR	NR	NR	NR	NR	NR
Alkalinity (as HCO_3^-) (mg/L)	19	0	136	112	642	221	148	2.2	1500	275	368	12	1520	NR
Sulfate (mg/L)	5.0	7.0	0.76	23	<0.02	1.8	13.2	18.4	1642	19	3.2	7.8	23.0	NR
Nitrate (mg/L)	0.7	1.3	<0.2	0.3	<0.01	1.3	0.0	13.7	53	<1	0.4	17	1.0	NR
Iron (mg/L)	0.65	0.42	20.6	0.1	46.3	0.03	19.9	<0.03	30.8	1.4	65	0.02	43	NR
Ammonium (mg/L)	0.66	0.21	6.5	<0.6	0.12	<0.02	2.59	<0.05	28.8	<1	55	<0.01	276	NR
Hydrogen sulfide (mg/L)	0.14	0.02	3.95	0.01	<0.01	<0.01	1.20	<0.02	0.12	0.06	NR	NR	NR	NR
Methane (mg/L)	NR	NR	10	0.02	9.6	<0.01	0.01	<0.01	12.8	0.24	10.2	0.0	NR	NR

bkgrnd, uncontaminated native groundwater; contam, contaminated groundwater; NR, not reported.

[a] Data are maximum concentrations in plume.

Sources: (1) Bedient et al. [22]; (2) Cozzarelli et al. [23]; (3) Baedecker and Cozzarelli, unpub. data; Baedecker et al. [24]; (4) Baedecker and Cozzarelli, unpub. data; (5) Nicholson et al. [3]; (6) Baedecker and Back [2]; Barker et al. [25].

at the Pensacola site is more contaminated and has 6.5 times as much DOC as contaminated water at the Galloway site.

In general, the distribution of DOC is useful to delineate a plume of contaminated groundwater at landfills or sites contaminated with organic compounds [3,26]. DOC data in Table 1 are for nonvolatile organic carbon. DOC data in the literature generally include only the nonvolatile organic compounds because the volatile organic carbon (VOC) in groundwater is not routinely measured. At sites where the contamination contains organic compounds such as gasoline or organic solvents, the VOC is a large component of the total DOC. For example, at a crude oil spill site (Bemidji) the VOC was 40% of the total DOC (Table 1). In contrast, at the Pensacola site the volatile fraction was only 4% of the total DOC, reflecting the low volatility of the components in creosote.

The pH of the contaminated water at these sites is 5.60-6.94, a narrow range compared to the pH of the uncontaminated groundwater, which was 4.00-7.68. With the exception of the Conroe site, these sites are anoxic, and anaerobic degradation of organic compounds is an active process. The major controls on the pH in the contaminated water are (1) microbial activity that generates CO_2 (or HCO_3^-) and organic acids and (2) the existing H_2CO_3/HCO_3^- buffering system. In these studies, the pH of contaminated water increased where the pH of background water was less than 6.0 and decreased where the pH of background water was greater than 6.0 (Table 1).

In the contaminated water the anions NO_3^- and SO_4^{2-} were depleted relative to the native water at all sites except at the Borden site, where these constituents are components of the waste in the landfill [3]. The major process controlling the concentrations of these anions is microbial oxidation of dissolved organic carbon coupled to reduction of nitrate and sulfate. The species Fe^{2+}, NH_4^+, H_2S, and CH_4 are end products in a series of oxidation-reduction reactions involving anoxic degradation of organic compounds in groundwater. The concentrations of these species are higher in the contaminated water than in the native water. At the Conroe site, concentrations of these constituents are low. However, the groundwater contained small amounts of oxygen, which suggests that anaerobic degradation of organic carbon is not a major process.

At all of the sites except Conroe, the concentrations of Fe^{2+} in the contaminated part of the aquifer were equal to or greater than 20 mg/L. Iron is present as oxide coatings in most glacial and coastal plain sediments, and some of the iron can be reduced by microbial and chemical processes. Much of the iron in solution in organic-rich aquifers may result from bacterial processes that couple the oxidation of organic compounds to the reduction of iron. The high concentrations of NH_4^+ (up to 276 mg/L) in anoxic groundwater at these sites may result from degradation of nitrogenous organic compounds and reduction of nitrate. The maximum concentration of sulfide found at these sites was 4.0 mg/L at Pensacola. Sulfide phases can precipitate as amorphous iron sulfides and pyrite from anoxic groundwater in the presence of high iron concentrations. The principal sources of methane in these aquifers are reduction by methanogens of CO_2 or HCO_3^- and degradation of simple organic compounds such as low molecular weight organic acids. Methane concentrations in the contaminated water are 9.6-12.8 mg/L except at the Galloway site, where concentrations are less than 0.01 mg/L. The values for methane at the Galloway site are low, probably because methanogenesis is inhibited by sulfate and nitrate reduction. The concentrations of SO_4^{2-} and NO_3^- are high in the uncontaminated groundwater, and in the anaerobic part of the aquifer these aqueous species are reduced. Data for the contaminated water at the Borden site are maximum values; water containing CH_4 is from a different part of the plume than where the high SO_4^{2-} and NO_3^- concentrations were found [3].

B. Oxidation-Reduction Potentials

In most natural aqueous geochemical environments the redox potential of the system is a qualitative determination that may not be quantitatively significant [27-29]. This potential, represented by Eh, is a measure of the activities of species at equilibrium that participate in oxidation-reduction reactions. The significance of Eh in contaminated groundwater is even more uncertain because a controlling redox couple may not exist in organic-rich aquifers; several competing oxidation-reduction reactions may occur simultaneously. Also, these reactions may occur in microenvironments that are difficult to sample. The spatial zonation of redox reactions was recognized in natural aquifers [30-34] and in contaminated aquifers [2,3,35]. Problems associated with evaluation of results of field Eh measurements have led to the calculation of Eh from redox couples of aqueous species with the Nernst equation. However, these calculated values are questionable in organic-contaminated environments because the Nernst equation is derived from thermodynamic principles that assume chemical equilibrium, whereas organic material degrades by irreversible reactions that are kinetically controlled.

A comparison of Eh values calculated for several redox couples in a large data set from WATSTORE (Water Storage and Retrieval System of the U.S. Geological Survey) by Lindberg and Runnells [36] showed that the range of calculated values was more than 1.0 V for a single sample. The authors concluded that none of the waters had reached thermodynamic equilibrium and that certain redox-sensitive species may be useful only as a qualitative indication of the redox state of the water. These results support the earlier conclusion by Stumm [27] that when multicomponent redox reactions occur in natural water or treated wastewater, Eh calculations may be meaningless.

Where oxygen is measured in groundwater, Eh can be calculated using the relation of Sato [37] for the H_2O_2/O_2 couple in chemical equilibrium programs such as WATEQF [38]. Barcelona et al. [39] calculated Eh with the H_2O_2/O_2 couple using measured concentrations of both species as input data in the computerized chemical equilibrium model WATEQ4F [40]. The determination of H_2O_2 is difficult because the concentrations are low and peroxide is unstable. The concentrations of H_2O_2 in groundwater for 111 field determinations were about 20 nmol/L, and values for blanks were about 7.8 nmol/L [41]. In reducing water, several redox couples, including SO_4^{2-}/S^{2-}, NO_3^-/NH_4^+, NO_3^-/NO_2^-, Fe^{3+}/Fe^{2+}, $Fe(OH)_3(s)/Fe^{2+}$, As^{5+}/As^{3+}, and $HCO_3^-/CH_4(aq)$ have been used to determine Eh. Calculations are made with activities of measured values of aqueous chemical species and the Nernst equation (generally using geochemical equilibrium models).

Redox potentials for the redox couples SO_4^{2-}/S^{2-}, NO_3^-/NH_4^+, Fe^{3+}/Fe^{2+}, $Fe(OH)_3(s)/Fe^{2+}$, and $HCO_3^-/CH_4(aq)$ were calculated for contaminated groundwater from four sites (Fig. 2). The activities of aqueous species used in the determinations of Eh were calculated from measured values using

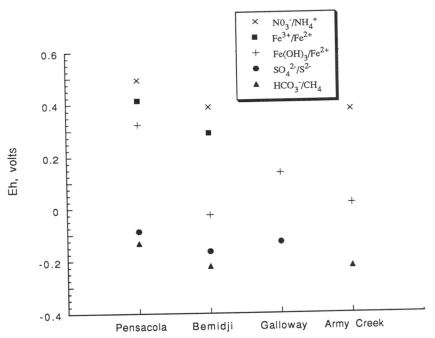

Figure 2 Calculated redox potentials of groundwater from aquifers at four sites contaminated with organic compounds.

WATEQF or WATEQ4F. For the SO_4^{2-}/S^{2-}, NO_3^-/NH_4^+, and Fe^{3+}/Fe^{2+} couples, the Eh was calculated in the program WATEQ4F. For the $Fe(OH)_3$ (s)/Fe^{2+} and HCO_3^-/CH_4(aq) couples, Eh was calculated using the Nernst equation and published thermydynamic data [42]. For these Eh calculations, chemical data were from selected sites where the groundwater was highly reducing at Pensacola, Bemidji, Army Creek, and Galloway (see Table 1 for water chemistry). The water was characterized by high concentrations of Fe^{2+}, CH_4, and H_2S and was from sections of the aquifer containing gray sediments. An exception was at Army Creek, where H_2S was not measured. Values of Eh calculated for the SO_4^{2-}/S^{2-} and HCO_3^-/CH_4(aq) couples were negative and in close agreement (within 0.046 V for Pensacola and within 0.056 V for Bemidji) (Fig. 2). The Eh values calculated for the $Fe(OH)_3$(s)/Fe^{2+}, NO_3^-/NH_4^+, and Fe^{3+}/Fe^{2+} couples are not in agreement with the other couples at the sites and are not reasonable values for reducing environments. Two exceptions are at Bemidji and Army Creek, where the $Fe(OH)_3$(s)/Fe^{2+} values are close to zero. However, groundwaters at these sites are highly reducing, and values of -0.1 to -0.2 V calculated from the SO_4^{2-}/S^{2-} and HCO_3^-/CH_4(aq) couples are more reasonable than the values from the iron couple.

The nitrogen couple is invalid for describing the Eh of the groundwater at these sites, probably because several reactions are capable of reducing and oxidizing nitrogen species. Thus, no single reaction is controlling the nitrogen speciation. The Eh values calculated from the Fe^{3+}/Fe^{2+} couple cannot be valid because the measured Fe^{3+} concentrations used in the calculations cause the calculated Eh values to be positive even in water with high concentrations of Fe^{2+}. Measured values of Fe^{3+} in anoxic water at pH 5-8 reflect the presence of colloidal iron rather than dissolved iron [43]. In these examples of highly reducing groundwater, only the SO_4^{2-}/S^{2-} and HCO_3^-/CH_4(aq) couples yield Eh values that are consistently reasonable.

A test of the sensitivity of Eh calculations shows that an increase by a factor of 1000 in the activity of CH_4 for the HCO_3^-/CH_4(aq) couple changes the Eh by a negligible amount of 0.010 V. Likewise, an increase in the activity of S^{2-} by a factor of 1000 changes the Eh for the SO_4^{2-}/S^{2-} couple by 0.020 V. This small effect of activities on Eh calculations was recognized by others [36,39]. The calculated values of these two redox potentials depend almost entirely on the standard potential ($E°$) and the activity of H^+. The redox potential calculated for the Fe^{3+}/Fe^{2+} couple is independent of pH and is sensitive to concentrations of Fe^{2+} in solution. An increase in the activity of Fe^{2+} by a factor of 1000 changed the calculated redox potential by 0.170 V.

These results indicate that the calculated redox potential depends on the selected redox couple and that no one couple can be recommended for calculating Eh in contaminated environments. Eh values calculated for contaminated groundwater are useful as qualitative tools and are necessary to include in equilibrium models for speciation of redox-sensitive species. In oxidizing groundwater, Eh calculations from O_2 measurements are the most reliable. In reducing water, where sulfate reduction or methanogenesis is a major process, Eh values calculated from the SO_4^{2-}/S^{2-} or $HCO_3^-/CH_4(aq)$ couples are the most reliable indicators of the redox potential.

C. Effects of Organic Acids on Alkalinity

Alkalinity titrations in water containing high concentrations of organic compounds are difficult to interpret and may yield erroneous results if the alkalinity is attributed entirely to carbonate and bicarbonate species. Low molecular weight carboxylic acids (C_6 or less) that are produced as intermediates by microbial activity may accumulate in groundwater if the production rate of the acids is greater than the utilization rate. The geochemical significance of carboxylic acids in determining alkalinity of oilfield brines was recognized earlier [44]. In that study, Carothers and Kharaka [44] found that at some temperatures aliphatic acid anions dominate over carbonate species. Dissociation constants, expressed as pK_a, are 3.91-4.90 for the monoprotic carboxylic acids—acetic, propanoic, butanoic, pentanoic, hexanoic, benzoic, o-toluic, m-toluic, and cyclohexanoic. The presence of these acids will affect the determination of alkalinity because their pK_a values are lower than the pK_a of bicarbonate (6.48). These acids and other low molecular weight acids were found in contaminated groundwater in concentrations of 0.35-130 mg/L [45,46]. The source of these acids in these contaminated aquifers is the biodegradation of organic compounds under anoxic conditions. Water that has measurable concentrations of organic acids is characterized by an absence of dissolved oxygen and high concentrations of carbon dioxide, ferrous iron, and methane.

The effect of the organic acid anion acetate ($C_2H_3O_2^-$) on titrations to determine bicarbonate alkalinity is demonstrated in Fig. 3A. Experiments were run at two amounts of bicarbonate and specified amounts of acetate. In the first set of titration curves (Fig. 3A), 0, 5, 10, or 20 μeq (microequivalents) of acetate was added to solutions containing 20 μeq of bicarbonate, and the solutions were titrated with 0.02 N sulfuric acid (H_2SO_4). In the second set of titration curves (Fig. 3B), 0, 25, 50, or 100 μeq of acetate was added to solutions containing 100 μeq of bicarbonate, and the solutions were titrated with 0.1 N H_2SO_4.

As the ratio of acetate to bicarbonate increased, the slope of the titration curves near the inflection point decreased; consequently, identification of the inflection points became difficult. For the curves with 20 μeq of HCO_3^-, the inflection point changes from 4.90 to 4.85 as the amount of $C_2H_3O_2^-$

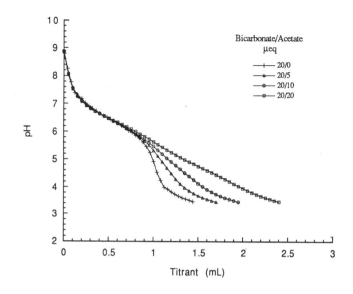

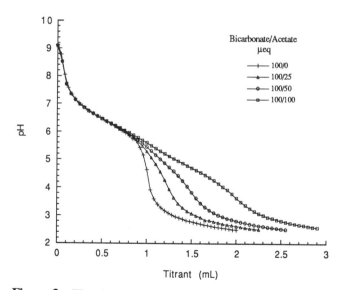

Figure 3 Titration curves of acetate-bicarbonate mixtures showing the changes in the titration curve with addition of acetate to solutions of 20 μeq of bicarbonate and 100 μeq of bicarbonate. Solutions titrated with 0.02 N and 0.1 N sulfuric acid for the 20 μeq and 100 μeq bicarbonate solutions, respectively.

increased from 0 to 10 μeq. In the final mixture with 20 μeq of both anions, the inflection point could not be determined from the curve or from a plot of the first derivative. The inflection point for the curves with 100 μeq of HCO_3^- changed from 4.64 to 3.63 with increasing amounts of $C_2H_3O_2^-$ (Fig. 3B). The alkalinities calculated from these experiments are compared to the known amounts of HCO_3^- and $C_2H_3O_2^-$ in solution (Table 2). The amounts of the two anions were additive at the higher concentrations (100 μeq of HCO_3^-). That is, the amount of titrant depended only on the total HCO_3^- and $C_2H_3O_2^-$ in solution. This was not true at lower concentrations; the calculated alkalinity was less than the total amounts of HCO_3^- and $C_2H_3O_2^-$ in solution. These results indicate that corrections made to the determination of alkalinity in water to account for the presence of organic acid anions depend on the concentrations of the dissolved bicarbonate and organic anions. Even if the amounts of organic acids are determined in a water sample, the alkalinity correction is nonlinear.

Alkalinity curves were compared for two contaminated environments that have high concentrations of organic compounds (Fig. 4). At both sites, a major process was biodegradation under reducing conditions. The sources of the organic material were creosote waste at the Pensacola site (Fig. 4A) and industrial and municipal waste at the Army Creek site (Fig. 4B). The inflection points of both titration curves, pH 3.9 at the Pensacola site and pH 2.9 at the Army Creek site, are lower than the inflection points would be for titrated bicarbonate solutions. Organic acids were identified in these waters, but not all of the compounds that contribute to the total alkalinity were identified. The inflection point at the Army Creek site is lower than that obtained in other laboratory experiments where mixtures of bicarbonate and anions

Table 2 Titration Data for Bicarbonate and Acetate Solutions

Solution titrated[a] (μeq)		Total μeq in solution	Total μeq measured	pH of inflection point
HCO_3^-	$C_2H_3O_2^-$			
20	0	20	20.0	4.90
20	5	25	22.6	4.85
20	10	30	24.8	4.85
20	20	40		—
100	0	100	100	4.64
100	25	125	125	3.94
100	50	150	150	3.82
100	100	200	200	3.63

[a]Solutions of bicarbonate (HCO_3^-) and acetate ($C_2H_3O_2^-$) in the amounts listed were titrated with sulfuric acid.

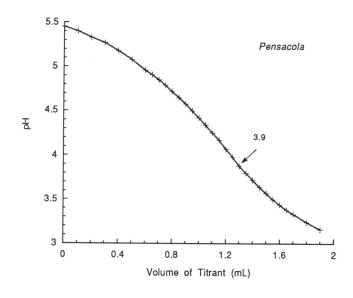

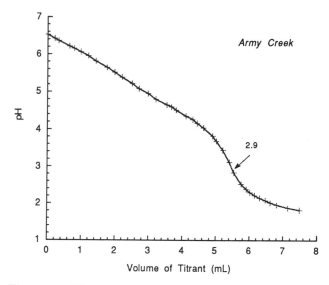

Figure 4 Alkalinity titration curves of contaminated groundwater containing high concentrations of dissolved organic carbon from a creosote waste site at Pensacola, Florida and an industrial and municipal landfill at Army Creek, Delaware. pH of equivalence point is marked.

of several monoprotic carboxylic acids (acetate, proprionate, and benzoate) were titrated with acid. Oxalic acid was identified in an analysis of water for volatile organic compounds at Army Creek. Oxalic acid, a diprotic acid, has pK_a's of 4.3 and 1.42. Possibly, the presence of this compound contributed to the low inflection point observed during the titration of Army Creek groundwater samples.

Calculations of speciation of solutes and mineral solubilities may be in error in cases where the measured alkalinity data include organic anions but are attributed to only carbonate species. The activity of bicarbonate used in carbonate equilibria equations will be higher than the true value, and the calculated abundances of dissolved carbonate species and the saturation states of carbonate minerals will be in error. Although some geochemical models contain thermodynamic data for organic acids, only a few species are included. For example, the geochemical model SOLMINEQ.88 [47] contains thermodynamic data for three organic acid anions. In contaminated groundwater from Pensacola, eight organic acids were identified, and other acids may be present that contribute to the alkalinity. It is difficult to determine which compounds contribute to total alkalinity because of the large number of organic compounds that are found in contaminated environments.

In the geochemical investigations at Pensacola and Army Creek, we used measured alkalinity values for the calculation of bicarbonate but recognized that the results could be incorrect. Experimentation using total inorganic carbon (TIC) measurements to calculate bicarbonate gave errors in the balance of cations and anions. In these experiments, the total anions were depleted relative to the total cations. This was expected in contaminated water because organic acid anions balance a part of the cations in solution. However, in natural water that did not contain low molecular weight organic acids, the balances between cations and anions also were not in agreement. Because TIC is a measurement of all inorganic carbon species, the error probably reflects a loss of CO_2 during sample collection or storage prior to analysis. It is difficult to collect, store, and analyze samples for TIC without exposing them to air. Additional studies are needed to determine the most accurate method to determine bicarbonate concentrations in contaminated water.

D. Fractionation of Isotopes

The stable isotope ratios most commonly measured in groundwater are $^{18}O/^{16}O$, $^{2}H/H$, $^{15}N/^{14}N$, $^{13}C/^{12}C$, and $^{34}S/^{32}S$. The variation in abundances of isotopes is expressed as δ values in permil (‰).

$$\delta_x = \left| \frac{R_x}{R_{std}} - 1 \right| \times 1000$$

where R_x and R_{std} are $^2H/H$, $^{18}O/^{16}O$, $^{15}N/^{14}N$, $^{13}C/^{12}C$, or $^{34}S/^{32}S$ of the sample and standard, respectively.

Isotopes of carbon and nitrogen are used more often than those of the other elements to understand geochemical and hydrological processes in contaminated environments. The fractionation of isotopes between two substances occurs as a result of equilibrium isotopic exchange reactions or kinetic isotope reactions; fractionation results in the enrichment or depletion of the heavy isotope in a chemical species.

The equilibrium fractionation factor between two phases is expressed as α_{A-B}, where

$$\alpha_{A-B} = \frac{1000 + \delta_A}{1000 + \delta_B}$$

where A and B are the chemical species in the oxidized and reduced forms, respectively. The difference in isotopic content of two species may be expressed as Δ_{A-B}, where $\Delta_{A-B} = \delta_A - \delta_B$.

In environments contaminated with organic compounds, kinetically controlled reactions are more important in determining the distribution of isotopes than are equilibrium reactions. The use of isotopic data in groundwater investigations of contaminated environments is increasing our understanding of processes that involve unstable constituents.

Because hydrogen and oxygen are components of the water molecule, investigations of the stable isotopic ratios of these elements are generally related to sources of water, geochemical evolution of a water type, and age of water [48]. In shallow contaminated environments, the amount of deuterium (D) and ^{18}O enrichment or depletion is small compared to the amount of water that is transported through the aquifer, and isotopic shifts are difficult to ascertain. In our investigations, little enrichment or depletion of these isotopes was found in groundwater at two sites contaminated with creosote and crude oil. However, at the Army Creek landfill site, $\delta^{18}O$ was unchanged whereas δD was enriched $8.3\%_0$ compared to the uncontaminated groundwater. In a landfill in Germany the deuterium and ^{18}O were enriched in groundwater downgradient from a landfill compared to the surrounding water [49]. However, differences in the composition of recharge water as a possible explanation for the enrichment could not be ruled out.

The fractionation of stable carbon isotopes in pristine and contaminated hydrogeologic environments is well documented in the literature [12,50,51]. The large isotopic shifts in inorganic carbon are from the formation of methane that results in isotopically depleted methane and enriched carbon dioxide. The $\delta^{13}C$ values for total inorganic carbon and methane for six aquifers contaminated with organic material are shown in Table 3. The carbon isotopic values for ΣCO_2 (total inorganic carbon) at these sites were -11.6 to $+18.4\%_0$.

Table 3 $\delta^{13}C$ $(^0/_{00})^a$ Values of ΣCO_2 and CH_4 in Contaminated Groundwater

Site	ΣCO_2	CH_4	$\Delta_{\Sigma CO_2\text{-}CH_4}$	Ref.
Army Creek	+ 18.4	—	—	2
Pensacola	− 7.5	− 60.2	52.7	23
Bemidji	+ 2.35	− 56.1	58.5	24
Landfill A	+ 16.6	− 52.1	68.7	50
Landfill B	+ 16.1	− 48.5	64.6	50
Borden	− 11.6	− 76.9	65.3	12

aRelative to PDB standard.

A range of − 25 to + 23‰ is reported for carbon isotopic values in shallow aquifers contaminated with organic compounds other than natural gas [51; Baedecker and Cozzarelli, unpublished data]. The data depend on the isotopic composition of the source organic material, the isotopic composition of natural ΣCO_2 from soil gas or dissolution of carbonates, and the degree of bacterial methane formation. The data in Table 3 are the heaviest ΣCO_2 and lightest CH_4 values that were measured in groundwater at a site.

The $\delta^{13}C$ values for CH_4 at these sites were − 76.9 to − 48.5‰. Values reported for biogenic methane in contaminated aquifers generally are about − 77 to − 38‰ [12,23; Baedecker and Cozzarelli, unpublished data]. Reported values as isotopically heavy as − 2.9‰ for CH_4 in a landfill may result from oxidation in an anaerobic environment by methane-oxidizing bacteria that preferentially use isotopically light carbon [51]. Laboratory experiments by Coleman and Risatti [52] indicated that bacterial oxidation of methane may result in enrichment of ^{13}C and D in the methane. In shallow unconfined aquifers, oxidation of CH_4 may occur so rapidly that the isotopically enriched CH_4 cannot be sampled. In a shallow aquifer at Pensacola, we found isotopically depleted methane (− 50.4‰) at a concentration of 0.056 mmol/L in groundwater with a trace of oxygen. At 0.5 m closer to the land surface, the oxygen concentrations were higher and the concentration of methane was 0.003 mmol/L, which suggests that the methane was oxidized over a short vertical distance. At the site closer to the land surface, the concentration of dissolved CH_4 was too low for us to measure its isotopic composition.

The differences in isotopic values between CO_2 and CH_4 at these sites were 52.7-68.7 (Table 3). The fractionation factor $(\alpha_{CO_2\text{-}CH_4})$ for the Borden site calculated by Barker and Fritz [12] was 1.058 using the isotopic composition of $CO_2(g)$ in equilibrium with groundwater. This value was less than the equilibrium fractionation factor of 1.076 calculated for $CO_2(g)$ and CH_4 using the equilibrium data of Richet et al. [53]. Barker and Fritz [12] found that carbon isotopic equilibrium in the CO_2–CH_4 system was not attained in any of the groundwaters they investigated, indicating that kinetic, not equilibrium, processes control the isotopic composition of the CO_2 and CH_4.

Studies of the changes in the isotopes of nitrogen in groundwater have focused primarily on identifying sources of nitrogen from fertilizers or animal waste [54–57], determining the extent of denitrification [58], and assessing of groundwater flow paths [59]. The use of $^{15}N/^{14}N$ ratios in hydrogeochemical studies of contaminated environments was reviewed by Heaton [60]. An example of the use of isotopes to determine sources of nitrogen is the investigation of nitrogen isotopes at a munitions waste disposal site [61]. Analysis of the $\delta^{15}N$ values of NO_3^- in groundwater downgradient from the disposal site allowed Spalding and Fulton [61] to trace the source of NO_3^- contamination to animal waste upgradient from the site rather than to the munitions waste.

Data are scarce on the use of nitrogen isotopes in interpreting geochemical processes in point-source contaminated environments. In anoxic groundwater associated with waste disposal sites and chemical spills, nitrate concentrations are low because nitrate is reduced to NH_4^+, N_2O, NO_2^-, or N_2. The isotopic values for nitrogen in reduced species, such as NH_4^+, are more difficult to interpret than values obtained for NO_3^-. The isotopic values depend on the source of NH_4^+, which may be from the degradation of organic nitrogen compounds or denitrification. Physical processes such as exchange of NH_4^+ on clays or volatilization [54] can also affect the isotopic composition of the NH_4^+ pool.

IV. GEOCHEMICAL PROCESSES AT TWO SITES

A. Army Creek Site

The Army Creek, Delaware landfill was filled in the 1960s with municipal and industrial waste. The site is in the unsaturated and saturated zones of the surficial sands in the Atlantic coastal plain. Beneath the sands is a semiconfining clay unit of variable thickness. Contaminated groundwater from the waste infiltrated through the sands and clay to the underlying Potomac aquifer and moved downgradient. A system of recovery wells was installed in the 1970s in the Potomac aquifer between the landfill and downgradient supply wells to prevent the leachate from reaching supply wells [62]. The wells in the landfill are 7–9 m below the land surface, and the downgradient recovery and supply wells are 30–60 m below the land surface. Pumping the recovery wells caused mixing of contaminated and uncontaminated groundwater; this mixing led to steep gradients in the distributions of dissolved chemical species in the groundwater.

The important geochemical reactions in the landfill are the degradation of organic compounds by aerobic and anaerobic processes and exchange of cations from mineral surfaces [2]. Organic compounds are degraded in the landfill primarily by microbially mediated reactions as shown by the fractionation of carbon isotopes and the formation of large amounts of HCO_3^-, NH_4^+, Fe^{2+}, and CH_4. The concentrations of dissolved oxygen were 0.0–8.4 mg/L (Fig. 5). Oxygen is depleted in the landfill because it is used as an electron acceptor at a faster rate than it is replenished in the aquifer. The pH shifted from about 5.1 in the background water to 6.6 in the leachate, primarily from the generation of HCO_3^- by microbial degradation. The highest HCO_3^- concentration in landfill water was 380 times background values. The concentrations of the major cations, Ca^{2+}, Mg^{2+}, Na^+, and K^+ in the landfill water were 50–190 times higher than those in uncontaminated water. Exchange of cations from clay minerals is the most likely source for the cations in the landfill water. A large amount of NH_4^+ was formed by nitrate reduction, and NH_4^+ readily exchanges for cations on clays. The refuse deposited in the landfill also may be a source of cations.

The distribution of dissolved chemical constituents downgradient from the landfill is affected by the local lithology and pumping of wells in the area [62]. The mixing of leachate and oxygenated groundwater is accelerated by pumping of the recovery wells. An area of less contaminated water is downgradient from the midsection (most narrow part) of the landfill (Fig. 5). Groundwater levels are higher in this localized area than in the surrounding area, which suggests that this small area may be a recharge zone and that the leachate is diluted. Alternatively, the underlying clay layer may be less permeable beneath the narrow part of the landfill, and therefore less leachate moves downgradient.

The $\delta^{13}C$ values were measured for total inorganic carbon dissolved in water from the landfill, downgradient from the landfill, and in contaminated groundwater (Fig. 5). The $\delta^{13}C$ values for ΣCO_2 in groundwater from the landfill were $+10.3$ to $+18.4\%_0$. The uncontaminated groundwater in this area (supply wells, Fig. 5) had carbon isotopic values of -24.3 to $-26.6\%_0$, which are typical of soil-gas CO_2 solubilized by recharge water. This large shift in isotopes to enriched values in the landfill results from the formation of isotopically light methane, which leaves the remaining CO_2 pool enriched in ^{13}C. Methane was in the landfill water in concentrations up to 22 mg/L.

The δD values for groundwater at the site had significant variations, whereas the $\delta^{18}O$ values were in a narrow range. The δD values at three locations in the landfill were -37.05 to $-34.45\%_0$ compared to background values of -44.95 to $-44.30\%_0$ for six locations in the uncontaminated part of the aquifer (Fig. 5). The δD values in contaminated groundwater downgradient from the landfill were intermediate in these two ranges, except for the area

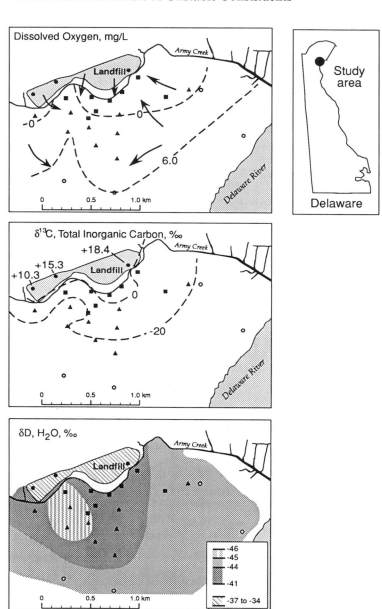

Figure 5 The distribution of dissolved oxygen (mg/L), $\delta^{13}C$ of total dissolved inorganic carbon ($^0/_{00}$), and δD of water ($^0/_{00}$) in groundwater at Army Creek landfill, Delaware. Arrows in top figure indicate direction of groundwater flow.

directly downgradient from the midsection of the landfill, where the values were depleted in deuterium compared to background values. The mechanism that resulted in the enriched δD values in water in the landfill cannot be explained by evaporation because the ^{18}O values were not affected. The $\delta^{18}O$ values for the landfill were -7.9 to $-7.35\%_{00}$, and for the uncontaminated water, -7.95 to $-7.60\%_{00}$. Although the process that results in deuterium enrichment is not known, the enrichment may result from the decomposition of a type of organic material discarded in the landfill. The distributions of the δD of water and $\delta^{13}C$ values of inorganic carbon in groundwater at the site are similar to the distribution of dissolved oxygen (Fig. 5). All of these measurements indicate the presence of a zone where little contamination has occurred downgradient from the midsection or narrowest part of the landfill.

B. Pensacola Site

The Pensacola, Florida site is a shallow coastal plain aquifer contaminated with creosote compounds. Wastewater from a wood-treatment process, which contains cellular components of wood, creosote, small amounts of diesel fuel, and pentachlorophenol, was discharged to unlined surface-water impoundments (ponds). Groundwater was contaminated by leakage of waste through the ponds and by surface runoff when the ponds overflowed [63]. The surficial aquifer, which consists of quartz sand and gravel interbedded locally with silt and clay, is recharged north of the site and discharges to the south along the coast and bay (Fig. 6) [63]. The hydrogeology and geochemistry of the contaminated zone is complex. The contaminant plume is separated into a shallow plume (about 2–8 m below the land surface) and a deeper plume (about 17–35 m below the land surface) by a discontinuous clay lens in the aquifer. The shallow plume discharges to the south along the bay and locally to a drainage ditch about 200 m south of the ponds.

The shallow contaminant plume at the site was delineated by the concentration of dissolved CH_4, which was 0.001–11.1 mg/L (Fig. 6). The geometry of the plume was also defined by on-site analyses of naphthalene and total phenols, components of creosote [64]. Of these highly soluble constituents, CH_4 was the most widespread, which suggests that the other organic compounds were removed from solution by sorption or attenuated by microbial processes. Alternatively, near the source of creosote, where the concentrations of dissolved gases are high, CH_4 may be transported as a gas phase. In the upper 10 m of the aquifer, concentrations of some chemical constituents changed by as much as a factor of 100 within distances of a few meters vertically and horizontally [23,65].

An anoxic zone downgradient of the ponds had high concentrations of organic compounds found in creosote (A–A', Fig. 6) (see Goerlitz, Chapter

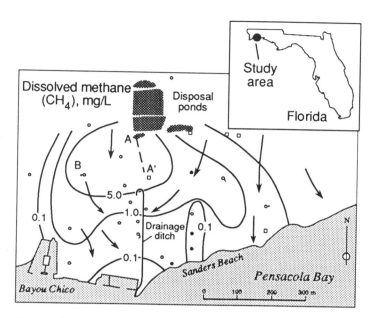

Figure 6 The distribution of dissolved methane (mg/L) in upper 6 m of the surficial aquifer downgradient of creosote waste impoundments at Pensacola, Florida. Arrows indicate the direction of groundwater flow. Transect A–A' and site B indicate locations of detailed geochemical investigations.

11, this volume). At a depth of 5 m below the land surface, concentrations of most unstable constituents decreased along a flow path in this zone (Fig. 7). Exceptions were HCO_3^- and CH_4, which remained at about the same concentrations or increased slightly. The increase in HCO_3^- from 90 and 120 m is due to a change in pH, from 5.65 to 6.00, rather than to an increase in total inorganic carbon. The pronounced decrease in SO_4^{2-} is due to sulfate reduction. The distributions of Fe^{2+} and S^{2-}, which were similar along the flow path, are controlled in part by precipitation of amorphous FeS or pyrite [66].

The DOC decreased from 19 to 1.7 mmol/L, and the TIC decreased from 19 to 7.0 mmol/L along the 145-m-long flow path. Assuming a dispersion factor of 4 based on the loss of chloride along the flow path, the loss or gain of chemical species can be calculated. The calculated concentrations of DOC, TIC, and CH_4 at the end of the 145-m-long flow path are 4.8, 4.8, and 0.21 mmol/L, respectively. The observed concentrations of these chemical constituents were 1.7, 7.0, and 0.81 mmol/L, respectively. The decrease in DOC

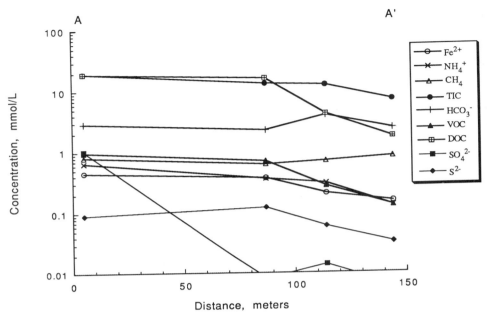

Figure 7 Concentrations (mmol/L) of dissolved iron, hydrogen sulfide, sulfate, ammonium, methane, bicarbonate, total inorganic carbon (TIC), organic carbon (DOC), and volatile organic carbon (VOC), along transect A–A′ of Fig. 6, at the Pensacola, Florida site.

and increase in TIC and CH_4, in observed values relative to calculated values, indicate that biodegradative processes, including methanogenesis, are occurring in this area. This is supported by data that showed loss of specific phenolic and N-heterocyclic compounds in this zone and by microcosm experiments that demonstrated anaerobic biodegradation of these compounds [67]. The decrease in DOC along the flow path exceeds the increase in CH_4 and TIC by 0.3 mmol/L. The lack of a carbon balance suggests that CO_2 is lost by outgassing from the system or that inorganic carbon is precipitated in the aquifer. The data indicate that both processes can occur. The groundwater is supersaturated with respect to carbonate minerals (calculated with WATEQF), and CO_2 and CH_4 may outgas because the partial pressure of dissolved gases was greater than 1. Also, part of the DOC may be sorbed on the aquifer solids along the flow path.

Similar changes in concentrations of unstable constituents were found in a vertical section of the aquifer (location B, Fig. 6). Chemical analyses of groundwater from 11 sampling sites demonstrated that the geochemical processes are significantly different in a 9-m-thick vertical section of the aquifer. Oxygen concentrations decreased from 0.10 mmol/L at the water table (2 m below the land surface) to less than a detectable amount (0.3 μmol/L) within 2 m below the water table and remained at that level to the bottom of the aquifer section. In the top 3 m of the aquifer, the concentrations of organic and inorganic carbon, iron, and sulfide increased by at least a factor of 10 (Fig. 8). The DOC increased to about 10 mmol/L, and the CH_4 to about 1.0 mmol/L, and the concentrations of both constituents were constant at depths greater than 5 m below the land surface. The SO_4^{2-} concentrations increased with depth initially and then decreased sharply where SO_4^{2-} reduction became an active process. As SO_4^{2-} decreased, S^{2-} increased and then remained at a constant concentration of 0.1 mmol/L. The concentration of Fe^{2+} is variable in this section; maximum Fe^{2+} concentrations are reached at 3 m and 5 m where concentration levels are higher than S^{2-}, but concentrations decrease below 5 m. The nearly constant concentrations of CH_4 and S^{2-} from 6 to 11 m below the land surface may indicate that diffusive as well as advective processes are important for these gaseous species.

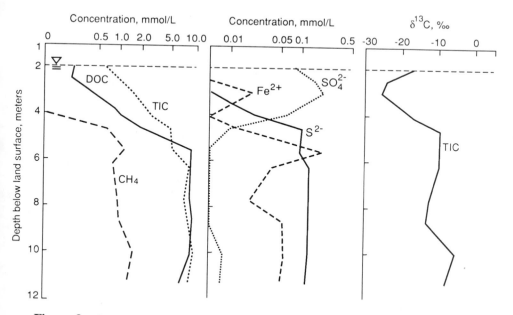

Figure 8 Concentrations (mmol/L) of dissolved iron, hydrogen sulfide, sulfate, methane, total inorganic carbon (TIC), and organic carbon (DOC), and $\delta^{13}C$ of total CO_2 (‰) with depth at location B of Fig. 6, at the Pensacola, Florida site.

The $\delta^{13}C$ values of TIC in the upper 2–3 m (Fig. 8) were isotopically lighter than the values deeper in the aquifer. The lightest value of $-25.5\%_{00}$ is depleted by $5\%_{00}$ compared to background $\delta^{13}C$ values, which were $-20.5\%_{00}$. This depletion is most likely the result of oxidation of isotopically light methane generated a few meters deeper in the aquifer. The $\delta^{13}C$ values of TIC from the 5–11-m depth interval were -4.45 to $-12.15\%_{00}$. These values are 8–16‰ enriched in ^{13}C compared to background values. This enrichment is most likely a result of methanogenesis.

C. Summary of Site Investigations

The nature of the organic material at the Army Creek landfill is considerably different from that of the Pensacola impoundments. The landfill contains mixed municipal and industrial material, whereas the impoundments are filled with wastewater containing creosote. Yet many of the reactions that occur are similar. At both sites a zone of active methanogenesis is near the source of organic material, and iron reduction is an important process. Although iron is reduced directly by microbial processes [15], abiotic iron reduction cannot be eliminated. Because these sites have existed for long periods of time (20–40 years), processes that occur more slowly than microbial processes, such as abiotic iron reduction, may contribute to the total concentration of dissolved iron. Some nitrate and sulfate reduction occurs at these sites. More nitrate reduction and less sulfate reduction is occurring at the landfill site than at the creosote site. Both sites are hydrologically complex, and geochemical interpretations are difficult. Although the source material was not well defined at the landfill, the location of the refuse was well known. The opposite was true at the creosote site, where the source was known to be predominantly creosote with pentachlorophenol, cellular material from wood, and small amounts of diesel fuel. This material was deposited primarily in the ponds, but other buried pockets of creosote were found during the investigation. Creosote in the subsurface was from underground transport of the organic phase, overland flow from the ponds followed by sinking of the organic phase, and dumping of creosote material outside the pond area. Thus, following a flow path to describe geochemical reactions is difficult. At the landfill site, the flow path was easy to delineate. However, the distribution of chemical constituents downgradient from the landfill was affected largely by pumping of the recovery and supply wells.

Oxidation–reduction processes result in the fractionation of isotopes; in contaminated environments, methanogenesis is the major process that fractionates carbon isotopes. The $\delta^{13}C$ values for dissolved inorganic carbon in the zones of active methanogenesis were isotopically lighter at the creosote site (-9.2 to $-3.5\%_{00}$) than at the landfill site ($+10.3$ to $+18.4\%_{00}$). Most of this difference is probably due to the isotopic composition of the organic material and the amount of methanogenesis that has occurred in the aquifers. Creosote at the Pensacola site had a $\delta^{13}C$ value of $-24.83\%_o$; however, the isotopic composition of organic material in the Army Creek landfill was unknown and, more than likely, variable because many types of refuse were deposited. The $\delta^{13}C$ values of ΣCO_2 have shifts of $30\%_o$ or more in zones of active methanogenesis near sources of organic material. These isotopic shifts may be useful in providing information on the progress of degradation of organic compounds in contaminated environments.

REFERENCES

1. Back, W., and Freeze, R.A. *Chemical Hydrogeology*, Benchmark Papers in Geology / 73, Hutchinson Ross, Stroudsburg, PA, 1983.
2. Baedecker, M.J., and Back, W. Hydrogeological processes and chemical reactions at a landfill. *Ground Water*, 17:429, 1979.
3. Nicholson, R.V., Cherry, J.A., and Reardon, E.J. Migration of contaminants in groundwater at a landfill: a case study. 6: Hydrogeochemistry. *J. Hydrol.*, 63:131, 1983.
4. Clark, T. On the examination of water for towns, for its hardness, and for incrustation it deposits on boiling. *Chem. Gaz.*, 5:100, 1847.
5. Weber, W.J., and Stumm, W. Mechanism of hydrogen ion buffering in natural waters. *Am. Water Works Assoc. J.*, 55:1553, 1963.
6. Gran, G. Determination of the equivalent point in potentiometric titrations. *Acta Chem. Scand.*, 4:559, 1950.
7. Edmond, J.M. High precision determination of titration alkalinity and total carbon dioxide content of sea water by potentiometric titration. *Deep-Sea Res.*, 17:737, 1970.
8. Galloway, J.N., Cosby, B., Jr., and Likens, G.E. Acid precipitation: measurement of pH and acidity. *Limnol. Oceanogr.*, 24:1161, 1979.
9. Wood, W.W. Guidelines for collection and field analysis of ground-water samples for selected unstable constituents. *Techniques of Water-Resources Investigations of the U.S. Geol. Survey*, Chapter D2, U.S. Govt. Printing Office, Washington, DC, 1976.
10. Hem, J.D. *Study and Interpretation of the Chemical Characteristics of Natural Water*, 3rd edition, U.S. Geol. Survey Water-Supply Paper 2254, U.S. Govt. Printing Office, Washington, DC, 1989.
11. Brown, E., Skougstad, M.W., and Fishman, M.J. Methods for collection and analysis of water samples for dissolved minerals and gases. *Techniques of Water-Resources Investigations of the U.S. Geol. Survey*, U.S. Govt. Printing Office, Washington, DC, 1970.
12. Barker, J.F., and Fritz, P. The occurrence and origin of methane in some groundwater flow systems. *Can. J. Earth Sci.*, 18:1802, 1981.
13. Rettich, T.R., Handa, Y.P., Battino, R., and Wilhelm, E. Solubility of gases in liquids. 13. High-precision determination of Henry's constants for methane and ethane in liquid water at 275 to 328 K. *J. Phys. Chem.*, 85:3230, 1981.

14. Lindsay, S.S., and Baedecker, M.J. Determination of aqueous sulfide in contaminated and natural water using the methylene blue method. In *Ground-Water Contamination: Field Methods* (A.G. Collins and A.I. Johnson, eds.),, ASTM Spec. Tech. Pub. 963, Philadelphia, PA, 1988, pp. 349–357.

15. Lovley, D.R., Baedecker, M.J., Lonergan, D.J., Cozzarelli, I.M., Phillips, E.J.P., and Siegel, D.I. Oxidation of aromatic contaminants coupled to microbial iron reduction. *Nature, 339*:297, 1989.

16. Orion Research. Orion instruction manual for Model 95-12 ammonia electrode. Orion Research, Inc., Cambridge, MA, 1986.

17. Simon, N.S., and Kennedy, M.M. The distribution of nitrogen species and adsorption of ammonium in sediments from the tidal Potomac River and estuary. *Estuar., Coastal Shelf Sci., 25*:26, 1987.

18. Fishman, M.J., and Friedman, L.C. Methods for determination of inorganic substances in water and fluvial sediments. *Techniques of Water-Resources Investigations of the U.S. Geol. Survey*, 3rd ed., U.S. Govt. Printing Office, Washington, DC, 1989.

19. Ngo, T.T., Phan, A.P.H., Yam, C.F., and Lenhoff, H.M. Interference in determination of ammonia with the hypochlorite–alkaline phenol method of Berthelot. *Anal. Chem., 54*:49, 1982.

20. Leenheer, J.A., Malcolm, R.L., McKinley, P.W., and Eccles, L.A. Occurrence of dissolved organic carbon in selected ground-water samples in the United States. *J. Res. U.S. Geol. Survey, 2*:361, 1974.

21. Barcelona, M.J. TOC determinations in ground water. *Ground Water, 22*:18, 1984.

22. Bedient, P.B., Rodgers, A.C., Bouvette, T.C., Tomson, M.B., and Wang, T.H. Ground-water quality at a creosote waste site. *Ground Water, 22*:318, 1984.

23. Cozzarelli, I.M., Baedecker, M.J., and Hopple, J.A. Effects of creosote products on the aqueous geochemistry of unstable constituents in a surficial aquifer. In *U.S. Geological Survey Program on Toxic Waste—Ground-Water Contamination*, Proc. Third Tech. Meeting. Pensacola, FL, U.S. Geol Survey Open-File Rep. 87-109, 1987, pp. A15–A16.

24. Baedecker, M.J., Siegel, D.I., Bennett, P., and Cozzarelli, I.M. The fate and effects of crude oil in a shallow aquifer, I. The distribution of chemical species and geochemical facies. In *U.S. Geological Survey Toxic Substances Hydrology Program*, Proc. Tech. Meeting, Phoenix, AZ, U.S. Geol. Survey Water-Resour. Invest. Rep. 88-4220, 1989, pp. 13–20.

25. Barker, J.F., Tessmann, J.S., Plotz, P.E., and Reinhard, M. The organic geochemistry of a sanitary landfill leachate plume. *J. Contam. Hydrol., 1*:171, 1986.

26. Siegel, D.I., Baedecker, M.J., and Bennett, P. The effects of petroleum degradation on inorganic water–rock reactions. *Fifth International Symposium on Water-Rock Interaction*, Reykjavik, Iceland, 1986, pp. 524–526.

27. Stumm, W. Redox potential as an environmental parameter; conceptual significance and operational limitation. *Adv. Water Pollut. Res., 1*:283, 1967.

28. Thorstenson, D.C. The concept of electron activity and its relation to redox potentials in aqueous geochemical systems. U.S. Geol. Survey Open-File Rep. 84-072, 1984.

29. Hostettler, J.D. Electrode electrons, aqueous electrons, and redox potentials in natural waters. *Am. J. Sci., 284*:734, 1984.

30. Back, W., and Barnes, I. Relation of electrochemical potentials and iron content to ground water flow patterns. U.S. Geol. Survey Prof. Paper 498-C, U.S. Govt. Printing Office, Washington, DC, 1965.

31. Champ, D.R., Gulens, J., and Jackson, R.E. Oxidation–reduction sequences in ground water flow systems. *Can. J. Earth Sci., 16*:12, 1979.

32. Jackson, R.E., and Patterson, R.J. Interpretation of pH and Eh trends in a fluvial-sand aquifer system. *Water Resour. Res., 18*:1255, 1982.

33. Thorstenson, D.C., Fisher, D.W., and Croft, M.G. The geochemistry of the Fox Hills–Basal Hell Creek aquifer in southwestern North Dakota and northwestern South Dakota. *Water Resour. Res., 15*:1479, 1979.

34. Edmunds, W.M., Miles, D.L., and Cook, J.M. A comparative study of sequential redox processes in three British aquifers. Proceedings of Symposium, *Hydrochemical Balances of Freshwater Systems*, Intl. Assoc. Hydrol. Sci. Publ. 150, 1984, pp. 55–70.

35. Golwer, A., Matthess, G., and Schneider, W. Effects of waste deposits on groundwater quality. Proceedings of Symposium, *Groundwater Pollution*, Moscow, 1971, Intl. Assoc. Hydrol. Sci. Publ. 103, 1975, pp. 159–166.

36. Lindberg, R., and Runnells, D. Ground water redox reactions: an analysis of equilibrium state applied to Eh measurements and geochemical modeling. *Science, 225*:925, 1984.

37. Sato, M. Oxidation of sulfide ore bodies. *Econ. Geol., 55*:928, 1960.

38. Plummer, L.N., Jones, B.F., and Truesdell, A.H. WATEQF—a FORTRAN IV version of WATEQ, a computer program for calculating chemical equilibrium of natural waters. U.S. Geol. Survey Water Resour. Invest. 76-13, 1978.

39. Barcelona, M.J., Holm, T.R., Schock, M.R., and George, G.K. Spatial and temporal gradients in aquifer oxidation–reduction conditions. *Water Resour. Res., 25*:991, 1989.

40. Ball, J.W., Nordstrom, D.K., and Zachman, D.W. WATEQ4F—a personal computer FORTRAN translation of the geochemical model WATEQ2 with revised data base. U.S. Geol. Survey Open-File Rep. 87-50, 1987.

41. Holm, T.R., George, G.K., and Barcelona, M.J. Fluorometric determination of hydrogen peroxide in groundwater. *Anal. Chem., 59*:582, 1987.

42. Wagman, D.D., Evans, W.H., Parker, V.B., Schumm, R.H., Halow, I., Bailey, S.M., Churney, K.L., and Nuttall, R.L. The NBS tables of chemical thermodynamic properties—selected values for inorganic and C_1 and C_2 organic substances. *J. Phys. Chem. Ref. Data 11, Suppl. 2*:2-1, 1982.

43. Hem, J.D., and Cropper, W.H. Survey of ferrous–ferric chemical equilibria and redox potentials. U.S. Geol. Survey Water-Supply Paper 1459-A, U.S. Govt. Printing Office, Washington, DC, 1959.

44. Carothers, W.W., and Kharaka, Y.K. Aliphatic acid anions in oil field waters—implications for the origin of natural gas. *Am. Assoc. Petr. Geol. Bull., 62*:2241, 1978.

45. Goerlitz, D.F., Troutman, D.E., Godsy, E.M., and Franks, B.J. Migration of wood-preserving chemicals in contaminated groundwater in a sand aquifer at Pensacola, Florida. *Environ. Sci. Technol., 19*:955, 1985.

46. Cozzarelli, I.M., Eganhouse, R.P., and Baedecker, M.J. Transformation of monoaromatic hydrocarbons to organic acids in anoxic groundwater environment. *Environ. Geol. Water Sci., 16*:135, 1990.

47. Kharaka, Y.K., Gunter, W.D., Aggarwal, P.K., Perkins, E.H., and DeBraal, J.D. SOLMINEQ.88: a computer program for geochemical modeling of water-rock interactions. U.S. Geol. Survey Water Resour. Invest. Rep. 88-4227, 1988.

48. Fontes, J.C. Environmental isotopes in groundwater hydrology. In *Handbook of Environmental Isotope Geochemistry* (P. Fritz and J.C. Fontes, eds.), Elsevier, New York, 1980.

49. Fritz, P., Matthess, G., and Brown, R.M. Deuterium and oxygen-18 as indicators of leachwater movement from a sanitary landfill. In *Interpretation of Environmental Isotope and Hydrochemical Data in Groundwater Hydrology*, Proc. Advisory Group Meeting Intl. Atomic Energy Agency, Vienna, 1976, pp. 131–142.

50. Games, L.M., and Hayes, J.M. On the mechanisms of CO_2 and CH_4 production in natural anaerobic environments. In *Environmental Biogeochemistry*, Vol. 1, *Carbon, Nitrogen, Phosphorus, Sulfur and Selenium Cycles* (J.O. Nriagu, ed.), Ann Arbor Science, Ann Arbor, MI, 1976, p. 51.

51. Games, L.M., and Hayes, J.M. Carbon isotopic study of the fate of landfill leachate in groundwater. *J. Water Pollut. Control Fed., April*:668, 1977.

52. Coleman, D.D., and Risatti, J.B. Fractionation of carbon and hydrogen isotopes by methane-oxidizing bacteria. *Geochim. Cosmochim. Acta, 45*:1033, 1981.

53. Richet, P., Bottinga, Y., and Javoy, M. A review of hydrogen, carbon, nitrogen, oxygen, sulphur, and chlorine stable isotope fractionation among gaseous molecules. *Ann. Rev. Earth Planet. Sci., 5*:65, 1977.

54. Kreitler, C.W. Determining the source of nitrate in ground water by nitrogen isotope studies, Univ. Texas, Austin, TX. Bur. Econ. Geol. Rep. Invest., 83, 1975.

55. Kreitler, C.W. Nitrogen-isotope ratio studies of soils and groundwater nitrate from alluvial fan aquifers in Texas. *J. Hydrol., 42*:147, 1979.

56. Kreitler, C.W., Ragone, S.E., and Katz, B.V.E. Nitrogen-isotope ratios of ground-water nitrate, Long Island, New York. *Ground Water, 10*:404, 1978.

57. Flipse, W.J., and Bonner, F.T. Nitrogen-isotope ratios of nitrate in ground water under fertilized fields, Long Island, New York. *Ground Water, 23*:59, 1985.

58. Bottcher, J., Strebel, O., Voerkelius, S., and Schmidt, H.L. Using isotope fractionation of nitrate-nitrogen and nitrate-oxygen for evaluation of microbial denitrification in a sandy aquifer. *J. Hydrol., 114*:413, 1990.

59. Wells, E.R., and Krothe, N.C. Seasonal fluctuation in $\delta^{15}N$ of groundwater nitrate in a mantled karst aquifer due to macropore transport of fertilizer-derived nitrate. *J. Hydrol., 112*:191, 1989.

60. Heaton, T.H.E. Isotopic studies of nitrogen pollution in the hydrosphere and atmosphere: a review. *Chem. Geol., 59*:87, 1986.

61. Spalding, R.F., and Fulton, J.W. Groundwater munition residues and nitrate near Grand Island, Nebraska, U.S.A. *J. Contam. Hydrol., 2*:139, 1988.

62. Baedecker, M.J., and Apgar, M.A. Hydrogeochemical studies at a landfill in Delaware. In *Groundwater Contamination, Studies in Geophysics*, National Academy Press, Washington, DC, 1984, pp. 127–138.

63. Franks, B.J. Hydrogeology and flow of water in a sand and gravel aquifer contaminated by wood-preserving compounds, Pensacola, Florida. U.S. Geol. Survey Water Resour. Invest. Rep. 87-4260, 1988.

64. Franks, B.J., Goerlitz, D.F., and Baedecker, M.J. Defining extent of contamination using onsite analytical methods. In *Petroleum Hydrocarbons and Organic Chemicals in Ground Water—Prevention, Detection and Restoration*, Proc. NWWA/API Conference, Houston, TX, National Water Well Association, Dublin, OH, 1985, pp. 265–275.

65. Baedecker, M.J., Franks, B.J., Goerlitz, D.F., and Hopple, J.A. Geochemistry of a shallow aquifer contaminated with creosote products. In *U.S. Geological Survey Program on Toxic Waste—Ground-Water Contamination*, Proc. Second Tech. Meeting, Cape Cod, MA, U.S. Geol. Survey Open-File Rep. 86-481, 1988, pp. A17–A20.

66. Hearn, P., Brown, Z., and Dennen, K. Analysis of sand grain coatings and major-oxide composition of samples from a creosote works, Pensacola, Florida. *Movement and Fate of Creosote Waste in Ground Water, Pensacola, Florida: U.S. Geological Survey Toxic Waste—Ground-Water Contamination Program* (H.C. Mattraw, Jr. and B.J. Franks, eds.), U.S. Geol. Survey Water-Supply Paper 2285, U.S. Govt. Printing Office, Washington, DC, 1986, pp. 19–26.

67. Godsy, E.M., and Goerlitz, D.F. Anaerobic microbial transformations of phenolic and other selected compounds in contaminated ground water at a creosote works, Pensacola, Florida. *Movement and Fate of Creosote Waste in Ground Water, Pensacola, Florida: U.S. Geological Survey Toxic Waste—Ground-Water Contamination Program* (H.C. Mattraw, Jr. and B.J. Franks, eds.), U.S. Geol. Survey Water-Supply Paper 2285, U.S. Govt. Printing Office, Washington, DC, 1986, pp. 55–58.

15

The Dissolution of Dense Immiscible Solvents into Groundwater: Implications for Site Characterization and Remediation

Richard L. Johnson

Oregon Graduate Institute, Beaverton, Oregon

Chlorinated solvents are frequently observed in contaminated groundwater at concentrations well below 1 mg/L. In contrast, laboratory dissolution experiments indicate that concentrations should be near saturation values of 100–10,000 mg/L. This chapter examines the roles of (1) the mass-transfer-limited dissolution away from pools of solvent, (2) dispersion from erratically distributed fingers and pools of solvent, and (3) the pumping and groundwater sampling processes on observed sample concentrations. Analytical models suggest that these factors play a major role in reducing concentrations to the low levels commonly observed. The predicted low concentrations suggest that remediation of sites where solvents occur will require a very long time (e.g., decades).

I. INTRODUCTION

Dense nonaqueous-phase liquids (DNAPLs), such as the chlorinated hydrocarbon solvents (CHCs), are frequently observed in contaminated groundwater at concentrations in the range 10^{-3}–1 mg/L. In contrast, the solubilities of the important CHCs are 100–10,000 mg/L (Table 1). The low concentrations are often observed even at sites where the presence of pure-phase solvents in the groundwater zone is known or strongly suspected. The disparity between observed and theoretical concentrations is thought to result from mass transfer limitations to the dissolution process and dilution of the aqueous plume due to the irregular distribution of the residual DNAPLs within the groundwater zone. In addition, the groundwater pumping and sampling processes may substantially dilute groundwater samples. These factors complicate efforts to characterize the contaminant distribution at solvent-contaminated sites. Another important consequence of mass-transfer-limited dissolution is that spills of chlorinated solvents may require very long times to remediate.

Laboratory and theoretical studies suggest that even under nearly ideal conditions, the infiltration of a DNAPL into the saturated zone is very erratic [3,4]. The immiscible fluid penetrates the water table as a number of scattered "fingers" or ganglia rather than in one coherent plug. In addition, laboratory experiments indicate that heterogeneous layers (both small- and large-scale) play a dominant role in controlling the movement of the DNAPLs. Therefore, the size, shape, and distribution of immiscible fingers in the saturated zone resulting from a large-scale immiscible solvent spill is thought to be highly complex. If the infiltrating solvent encounters a low-permeability stratum, it can spread to form a thin pool. This process is controlled to a high degree by the characteristics of the low-permeability zone (e.g., bedding angle, continuity of the stratum, solvent entry pressure).

Table 1 Drinking Water Concentration Limits and Solubilities for a Number of Important Chlorinated Solvents

Compound	MCL[a] (mg/L)	Solubility[b] (mg/L)
Benzene	0.005	1780 (25°C)
Carbon tetrachloride	0.005	785 (20°C)
p-Dichlorobenzene	0.075	79 (25°C)
1,2-Dichloroethane	0.005	8690 (25°C)
1,1-Dichloroethylene	0.007	400 (20°C)
Tetrachloroethylene	0.005	200 (20°C)
1,1,1-Trichloroethane	0.2	720 (25°C)
Trichloroethylene	0.005	1100 (20°C)
Vinyl chloride	0.002	2700 (25°C)

[a]Code of Federal Regulations [1].
[b]Mabey et al. [2].

Unfortunately, there are few case histories of accidental spills in which DNAPLs were located in the groundwater zone. Indeed, direct observation of any DNAPLs is the exception rather than the rule for most sites. As a consequence, the presence of pure solvent is typically inferred from aqueous-phase concentration information. Concentrations far below solubility limits are most common and can lead to the incorrect assumption that no liquid product is present. Unfortunately, it is more often correct to assume that immiscible-phase product is present in the subsurface whenever aqueous solvent plumes are observed.

Because of the importance of the DNAPL problem, a number of experimental and mathematical modeling studies have been undertaken in recent years. Laboratory experiments have demonstrated that dissolution from fingers results in high concentrations. In early laboratory experiments using a one-dimensional column filled with porous medium and DNAPL, van der Waarden et al. [5] found that concentrations rose to saturation values after only a few centimeters of contact. More recently, Anderson et al. [6] have reported that dissolution from fingers in a three-dimensional aquifer model produces near-saturation concentrations.

On the other hand, laboratory experiments (Schwille [4]) and modeling by Johnson and Pankow [7] indicate that dissolution from thin pools of product is mass-transfer-limited. Subsequent modeling by Anderson et al. [8] suggests that the low concentrations frequently observed in the field are due to a combination of mass transfer limitations from pools and dilution due to dispersion from erratically distributed fingers and pools. These will be discussed in greater detail in the following sections.

II. BACKGROUND

Releases of DNAPLs often occur at or near ground surface. The liquids initially move through the vadose zone as a separate liquid phase and begin to move downward under the influence of gravity. Because the medium tends to retain some of the infiltrating solvent (residual saturation, S_r), the further the fluid penetrates, the smaller the fraction of the spill that remains mobile. The factors that affect the trapping can be quite complex, but in general there is a tendency for the residual saturation of an immiscible liquid to increase as the permeability of the porous medium decreases [9]. For media with a porosity of 0.40, Schwille [9] found that S_r varied from approximately 2% (of the pore volume) in highly permeable media to 15% in media with low permeability ($> 10^{-11}$ m^2). Hoag and Marley [10] measured vadose zone retention capacities ranging from 14 to 55% for gasoline in initially dry coarse and fine sands. When the sands were initially water wet, the retention of gasoline decreased.

Depending on the size of the spill and the depth to the water table, the immiscible solvent may continue to flow downward until it reaches the fully water-saturated pores at the capillary fringe. If it is more dense than water (such as a CHC), it will have the potential to penetrate the water table and make its way down through the aquifer. For this to happen, however, the solvent must first be able to displace water from the pores. This requires sufficient pressure head to overcome the capillary pressure barrier between the water and the immiscible solvent. Villaume et al. [11] showed how to calculate the critical height of an infiltrating immiscible fluid necessary to generate sufficient pressure to overcome pore capillary pressures in an ideal medium. Critical height estimates for PCE range from centimeters for coarse sands to tens of meters for fine silts and clays [12]. Saturated silts and clays can therefore present a very significant barrier to the infiltration of immiscible liquids. The coarse sands and gravels found in many aquifers, however, may be easily infiltrated by these fluids.

When a fluid saturating a porous medium is displaced by the pressure of another fluid, the interface between them may become unstable. This instability is manifested by "fingers" of the driving fluid penetrating the displaced fluid [13]. This phenomenon, well known in the petroleum industry, is generally referred to as "viscous fingering" because the viscosity difference between the two fluids is one of the key factors contributing to the instability [14]. Chuoke et al. [15] showed that when a more dense, less viscous fluid moves downward to displace a less dense, more viscous liquid, the fluid interface will always be unstable. This is the case for most dense immiscible solvents as they displace water from a saturated porous medium.

As a result of viscous finger formation, upon reaching the water table a DNAPL spill may split into a number of small fingers that push their way into the aquifer. This behavior has been observed for PCE spills penetrating the saturated zone in sandbox models [16,17]. Such behavior could have two important effects on the subsequent contamination of the aquifer. First, if the infiltrating fingers now occupy a smaller total cross-sectional area, the solvent will be able to penetrate to a greater depth than one might estimate from residual saturation considerations. This will increase both the difficulty and the cost of possible remediation schemes. Second, dissolved contaminants may then emanate from a number of narrow sources separated by regions of clean water. The resulting contaminant plume may therefore be more dilute than expected. Both of these factors can greatly complicate attempts to model immiscible fluid transport and contamination in the saturated zone.

A. Dissolution Experiments

To examine the process of dissolution from fingers of residual solvent under conditions where the groundwater was not forced to flow through the residual, Anderson et al. [6] used a three-dimensional physical model in which they could construct "fingers" of known geometry. In most of their experiments they used a finger 0.15 m in diameter produced by flooding a portion of the aquifer contained within a metal cylinder with tetrachloroethylene (PCE) and allowing it to drain. The resultant source had a residual PCE saturation of 13%, distributed in a relatively uniform manner. Groundwater was allowed to flow both through and around the finger. Thus it was expected that any decrease in concentrations due to the reduced permeability within the finger would be taken into account. Groundwater velocities in the range of 0.1–1 m/day were examined. In all cases the PCE concentration exiting the finger was at saturation.

In a subsequent experiment in the same physical model, PCE was applied in an "uncontained" manner into the homogeneous sand. The resulting distribution of PCE was highly irregular (Fig. 1). The bulk of the product was retained in very thin pools that occurred on the small-scale bedding planes

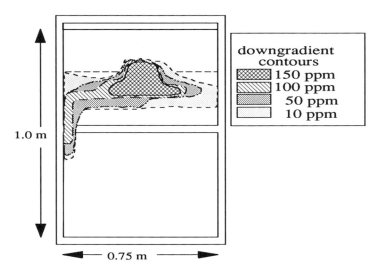

Figure 1 Vertical cross section (perpendicular to flow) of the aqueous plume from a spill of PCE into a laboratory physical model filled with a homogeneous medium sand.

present in the sand (as the result of the packing process). The short fingers connecting these pools were on the order of 0.01–0.03 m in diameter. Groundwater velocities were varied from 0.1 to 1.6 m/day. Groundwater samples collected immediately downgradient from the source indicated that concentrations remained at saturation values in all cases.

To examine the process of dissolution from pools of solvent, Schwille [4] conducted a series of physical model experiments in which pools of trichloroethylene (TCE) or 1,1,1-trichloroethane (TCA) were located at the bottom. The physical model had dimensions of 1.5 m long by 0.5 m wide by 0.28 m deep. Groundwater velocity was varied over the range 0.45–2.7 m/day. The concentrations of TCE or TCA in the tank effluent were monitored and used to calculate mass removal rates for the experiments. The boundary conditions approximated by the model are seen in Fig. 2.

B. Modeling

Johnson and Pankow [7] used these data to examine the mass transfer process and to calculate mass transfer coefficients (M_a) for a range of subsurface conditions. Because the time required for the complete dissolution of a pool is generally long in comparison with the contact time between the pool and the flowing groundwater, the dissolution process can be approximated using a steady-state form of the advection–dispersion equation. Also, because the pools are generally wide in comparison to the transverse mixing process, a two-dimensional form of the equation can be used:

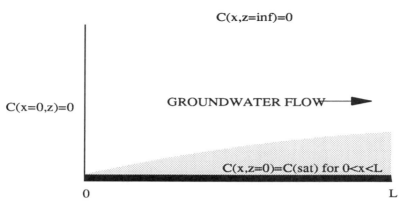

Figure 2 Boundary conditions for Eq. (2) and for Schwille's [4] experiments.

$$\bar{v}\frac{\partial C}{\partial x} = \frac{\partial^2 C}{\partial z^2} \tag{1}$$

where $\bar{v}$ is the average groundwater velocity and D_z is the vertical dispersion coefficient (i.e., transverse to groundwater flow). D_z is frequently expressed as the sum of molecular and mechanical dispersion terms [18]:

$$D_z = D_e + \alpha_z \bar{v} \tag{2}$$

where D_e is the effective aqueous diffusion coefficient (m^2/sec) in the medium, and α_z is the vertical transverse dispersivity (m).

Effective aqueous-phase diffusion coefficients (D_e) for porous media are typically on the order of 10^{-10} m^2/sec. This can be substantially reduced if the solute of interest partitions strongly to the soil matrix. However, sorption does not play a role at steady state because there is no net change in the distribution of mass with time. D_e values for TCE in porous media are in the range of $(2.5-3.0) \times 10^{-10}$ m^2/sec [19]. Because sorption is not important at steady state, and because D_e is not strongly dependent upon the molecular size or the properties of the porous medium, values of D_e in the range $(1-4) \times 10^{-10}$ m^2/sec are applicable for a wide range of contaminants and porous media. This makes it possible to estimate DNAPL dissolution behavior over a wide range of conditions.

Vertical dispersion in groundwater systems is generally a weak process. At low groundwater velocities (e.g., 0.1 m/day), vertical dispersion is on the order of molecular diffusion, even in heterogeneous aquifers [20,21]. In laboratory experiments, transverse dispersivities are frequently observed to be on the order of millimeters. In laboratory physical model experiments, Anderson [16] observed transverse dispersivities of 3×10^{-4} m. Thus, the mechanical dispersion component is expected to dominate the dissolution process only when velocities are greater than 0.1 m/day.

Hunt et al. [19] presented an analytical solution of Eq. (1) for the boundary conditions depicted in Fig. 2:

$$C(L_p, z) = C_{SAT} \operatorname{erfc} \frac{z}{2\left(\dfrac{D_z L_p}{\bar{v}}\right)^{1/2}} \tag{3}$$

where C_{SAT} is the saturation concentration of the compound of interest, and erfc signifies the complimentary error function. Inspection of Eq. (3) indicates that for a given set of porous-media conditions, the mass transfer is proportional to the solubility of the contaminant.

To estimate dissolution times of pools, we first need an estimate of the mass transfer coefficient M_a. M_a is a function of the length and thickness of the pool, the solubility of the contaminant of interest, and the groundwater velocity. Using PCE as an example, with $D_e = 2.7 \times 10^{-10}$ m²/sec, $\alpha_z = 0.0003$ m, and a solubility of 200 mg/L, Eq. (3) can be integrated to obtain values for M_a as a function of pool length for a variety of velocities (Fig. 3). Johnson and Pankow [7] used calculated mass transfer coefficients to estimate the time required for dissolution from pools of various lengths and volumes as a function of groundwater velocity. They concluded that, at typical groundwater velocities (e.g., 0.1–0.3 m/day), spills of only a few hundred liters will persist for a time ranging from decades to hundreds of years.

To examine how the distribution of CHCs as fingers and pools affect observed aqueous concentrations and dissolution times, Anderson et al. [8] examined several hypothetical CHC distributions. They used analytical solutions for dissolution from individual fingers and pools and superpositioned a number of conditions to arrive at complex spill geometries.

To represent dissolution from a single finger of residual solvent, an analytical solution for a vertical parallelepiped source was used. Water emerging

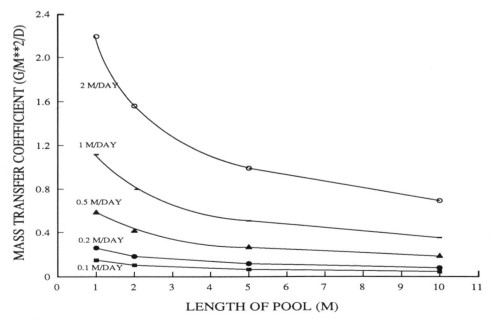

Figure 3 Mathematical model results showing mass transfer coefficients M_a as a function of pool length for a several groundwater velocities.

from the source was assumed to be saturated with solvent (e.g., 200 mg/L for PCE). The analytical solution for this case is given by [22]:

$$
\begin{aligned}
C(x,y,z,t) = \frac{M_v}{4n} \int_0^t & \left[\text{erfc}\ \frac{x - x_2 - vt'}{2\sqrt{D_x t'}} - \text{erfc}\ \frac{x - x_1 - vt'}{2\sqrt{D_x t'}} \right] \\
\times & \left[\text{erfc}\ \frac{y - y_0}{2\sqrt{D_y t'}} - \text{erfc}\ \frac{y + y_0}{2\sqrt{D_y t'}} \right] \\
\times & \left[\frac{z_2 - z_1}{L} + \frac{2}{\pi} \sum_{k-1}^{\infty} \frac{1}{k}\ \cos \frac{k\pi z}{L} \left[\sin \frac{k\pi z_2}{L} - \sin \frac{k\pi z_1}{L} \right] \right. \\
\times & \left. \exp\ \frac{-k^2\pi^2 D_z t'}{L^2}\ dt' \right]
\end{aligned}
\tag{4}
$$

where v is the mean groundwater velocity; n, porosity; L, thickness of the aquifer; M_v, volumetric mass transfer coefficient $(M/L^3 T)$; x_1, upstream x coordinate of the source; x_2, downstream x coordinate of the source; y_0, half-width of the source; z_1, upper z coordinate of the source (from the top of the aquifer); z_2, lower z coordinate of the source (from the top of the aquifer); D_x, coefficient of longitudinal dispersion; D_y, coefficient of horizontal transverse dispersion; and D_z, coefficient of vertical transverse dispersion. The integral was solved using Gaussian quadrature. Mass-balance calculations were also employed to provide information about the lifetimes of the fingers in the aquifer.

To stimulate saturated concentrations coming from the source, the volumetric mass transfer coefficient M_v in Eq. (4) was calculated from the equilibrium solubility C_s using the equation

$$
M_v = \frac{C_{SAT} n \bar{v}}{(x_2 - x_1)}
\tag{5}
$$

where $x_2 - x_1$ is the dimension of the source of the direction of flow. The value of C_{SAT} was set equal to 200 mg/L, the aqueous solubility of PCE at 20°C.

To stimulate dissolution from thin pools, the following analytical solution was utilized [22]:

$$
\begin{aligned}
C(x,y,z,t) = \frac{M_a}{4nL} \int_0^t & \left[\text{erfc}\ \frac{x - x_2 - vt'}{2\sqrt{D_x t'}} - \text{erfc}\ \frac{x - x_1 - vt'}{2\sqrt{D_x t'}} \right] \\
\times & \left[\text{erfc}\ \frac{y - y_0}{2\sqrt{D_y t'}} - \text{erfc}\ \frac{y + y_0}{2\sqrt{D_y t'}} \right] \\
\times & \left[1 + 2 \sum_{k-1}^{\infty} \cos \frac{k\pi z}{L}\ \cos \frac{k\pi z_0}{L}\ \exp\ \frac{-k^2\pi^2 D_z t'}{L^2} \right] dt'
\end{aligned}
\tag{6}
$$

where M_a is the surface mass transfer coefficient (M/L^2T) and z_0 is the depth of the horizontal pool from the top of the aquifer.

In Eq. (6), the thickness of the pool was assumed to be insignificant relative to the thickness of the aquifer. Therefore, only one z coordinate was necessary to define the depth of the pool. Gaussian quadrature was again employed for the solution of the integral expression. Unlike the first model, however, it was not assumed that contaminant concentrations would necessarily reach solubility levels in the water flowing over a pool. This was because dissolution in the pool model was restricted to the water–immiscible solvent interface, where it is less efficient than dissolution from small residual droplets that have a large surface area/volume ratio. Under these circumstances, mass transfer is strongly affected by the velocity of the water and the size of the pool.

For the model studies of Anderson et al. [8] it was assumed that the volume of PCE lost was sufficient to penetrate an aquifer that was 15 m thick. The porosity of the aquifer was 0.35, and the mean water velocity was 30 cm/day. The immobile PCE in the aquifer had an S_r of 15%. The source of all groundwater contamination was restricted to the solvent in the aquifer. Transport to or from the vadose zone was not considered. The length of a pool was set at 3 m. Based on Eq. (3) and a dispersivity of 3×10^{-4} m, the M_a for this case would be 0.24 g m^{-2} day^{-1}.

The simulation shown in Fig. 4 represents a hypothetical solvent spill that penetrated the water table in one location and then split into smaller fingers and spread out into a number of pools as it made its way to the bottom of the aquifer. The vertical lines represent the locations of the residual fingers in this simulation. The horizontal lines represent pools where the penetrating immiscible fluid encountered a zone of slightly lower permeability and

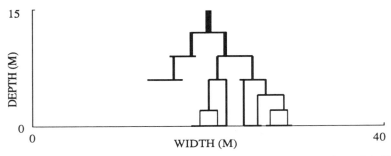

Figure 4 Schematic drawing of the system of fingers and pools modeled by Anderson et al. [8].

spread out until it was able to again move downward. The initial finger was 0.1 m in diameter, and the diameters decreased with depth down to 0.02 m near the bottom of the aquifer.

Figure 5 shows the PCE concentrations from the steady-state contaminant plume 100 m downgradient of the Fig. 4 source. The relative contributions to the plume from the fingers and pools are illustrated in Figs. 5a and 5b, respectively. Because most of the DNAPL mass is in the pools, the fingers will become depleted much more quickly. As a consequence, the pools will produce a low-concentration plume that is very long lived. Since experience has shown that real-world contaminant plumes are often at low concentrations and not rapidly removed by pump-and-treat remediation schemes, it

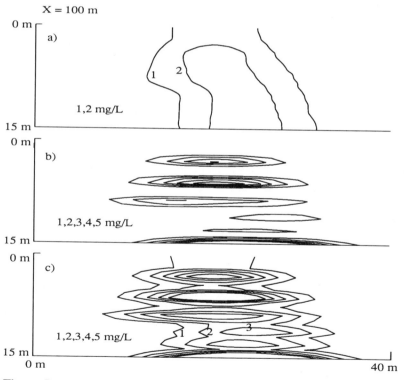

Figure 5 Vertical cross section (perpendicular to flow) of PCE concentrations 100 m downgradient from the source of contaminants pictured in Fig. 4. (a) From the fingers; (b) from the pools; (c) from the fingers and pools.

would appear that many solvent spills in saturated porous soils consist of a number of scattered pools from which mass transfer is not very efficient.

C. Groundwater Pumping and Sampling

Aqueous-phase concentrations observed during groundwater pumping and sampling may not accurately reflect concentrations actually present in the subsurface. As seen above, vertical dispersion is a relatively weak process; thus aqueous plumes can have a very limited vertical extent. During pump-and-treat operations, it is common to draw water from the entire vertical thickness of the aquifer (either intentionally or as the result of upconing). As a consequence, large quantities of clean water can be drawn into the pumping and observation wells. This can lead to an underestimation of maximum groundwater concentrations. Because laboratory dissolution experiments have demonstrated that saturation concentrations are often at saturation values, a common misinterpretation of low observed aqueous concentrations is that no immiscible phase is present in the subsurface. It is probably better to assume that if aqueous plumes of solvents are detected, then immiscible solvents are present.

Similarly, monitoring well screens are typically 2–10 m in length; thus if the contaminant plume is substantially thinner, measured groundwater concentrations will be underestimated. Furthermore, if the aquifer is heterogeneous, then the bulk of the water drawn into the well may come from a higher permeability zone in which concentrations are low. Another difficulty associated with observation wells is that they are frequently spaced widely apart (e.g., 50–100-m intervals). As Anderson et al. [8] demonstrated, movement of only a short distance (e.g., 5–15 m) off-center from a plume can result in substantial reduction in observed concentrations. Thus, it is likely that most monitoring wells will miss the zone of maximum concentration. Again, this can be misinterpreted as indicating that no immiscible phase is present. An important implication of the spatial variability of solvent plumes is that more detailed, three-dimensional sampling will be required to adequately characterize the distribution of solvent in the subsurface.

If the bulk of a spilled solvent occurs as pools in the subsurface, then remediation by pumping groundwater will be limited by dissolution from the pools. Increasing the velocity of the groundwater will increase the rate at which dissolution occurs, but in many cases decades will be required to completely remove the pools. It is important to consider this very long time frame in the design of pump-and-treat systems. In many cases it will be most efficient to design systems that maintain control of the hydraulic gradient and that can be focused on more highly contaminated zones while minimizing the volume of water pumped. This will help to reduce annual operation and maintenance costs, which will become the major expense of the remediation activity.

III. SUMMARY

Dense immiscible-phase liquids tend to distribute themselves irregularly within the unsaturated and groundwater zones. As a consequence, it is very difficult to predict their behavior in porous media. Another consequence of their irregular distribution is that their penetration into the groundwater zone can be substantially deeper than would be predicted on the basis of residual saturation data. As a result, it is very difficult to characterize the distribution of DNAPLs in the subsurface once a spill has occurred, and the distribution and extent of the contaminant source(s) often must be inferred from aqueous concentration data.

If the aquifer is heterogeneous, a substantial portion of the DNAPL mass may exist as thin pools in the saturated zone. If the bulk of the mass is in pools, then the aqueous plumes that are formed by the pools will be vertically quite thin. As a consequence, groundwater samples from pumping or observation wells may have aqueous concentrations that are substantially lower than the maximum concentrations that occur in the groundwater. This complicates the characterization of the extent and distribution of DNAPLs in the subsurface.

Laboratory experiments suggest that dissolution from fingers of residual solvent will result in concentrations that are near saturation. However, if the fingers are narrow and widely distributed, then dispersion will substantially reduce the concentrations from the fingers. If most of the DNAPLs exist in pools, rather than fingers, then the fingers may be dissolved away in a few years, leaving only the pools.

Experimental and mathematical models suggest that dissolution from pools of immiscible solvents is slow. At low groundwater velocities, diffusion is the dominant mass transport mechanism. At higher velocities (e.g., >10 cm/day), vertical dispersion controls dissolution. However, vertical dispersion is a relatively weak process. As a result, dissolution is relatively slow, observed groundwater concentrations are frequently very low, and pools of solvents will persist for decades or longer.

The combination of chemical and physical factors discussed above plus the widespread use of dense chlorinated solvents has resulted in a groundwater contamination problem that is proving to be both time-consuming and expensive to remediate. The presence of DNAPLs as pools in the subsurface means that ''quick fix'' solutions are not likely to be effective and that longer term strategies should be developed. The challenge of the next decade will be to better understand the subsurface behavior of these chemicals in order to develop appropriate, cost-effective techniques for dealing with this difficult problem.

REFERENCES

1. Code of Federal Regulation, 40CFR141.61, 1980; 40CFR141.61, 1991.
2. Mabey, W.R., Smith, J.H., Podoll, R.T., Johnson, H.L., Mill, T., Chou, T.-W., Gates, J., Waight Partridge, I., Jaber, H., and Vandenberg, D. *Aquatic Fate Process Data for Organic Priority Pollutants*, Rep. No. 440/4-81-014, U.S. Environmental Protection Agency, Washington, DC, 1982.
3. Kueper, B.H., and Frind, E.O. An overview of immiscible fingering in porous media. *J. Contam. Hydrol.* 2(2):95–110, 1988.
4. Schwille, F. *Dense Chlorinated Solvents in Porous and Fractured Media* (J.F. Pankow, transl.), Lewis, Chelsea, MI, 1988.
5. van der Waarden, M., Bridie, A.L.A.M., and Groenewoud, W.M. Transport of mineral oil components to groundwater. I. Model experiments on the transfer of hydrocarbons from a residual oil zone to trickling water. *Water Res., 5*: 213–226, 1971.
6. Anderson, M.R., Johnson, R.L., and Pankow, J.F. The dissolution of dense immiscible solvents into groundwater: laboratory experiments involving a well-defined source geometry. *Ground Water, 30*:250–256, 1992.
7. Johnson, R.L., and Pankow, J.F. Dissolution of dense immiscible solvents into groundwater. 2. Dissolution from pools of solvent and implications for the remediation of solvent-contaminated sites. *Environ. Sci. Technol., 26*:1992, in press.
8. Anderson, M.R., Johnson, R.L., and Pankow, J.F. The dissolution of dense immiscible solvents into groundwater. 3. Modeling contaminant plumes from fingers and pools of solvent. *Environ. Sci. Technol., 26*:1992, in press.
9. Schwille, F. Migration of organic fluids immiscible with water in the unsaturated zone. In *Pollutants in Porous Media: The Unsaturated Zone Between Soil Surface and Groundwater* (B. Yaron, G. Dagon, and J. Goldshmid, eds.), Springer-Verlag, New York, 1984, pp. 27–48.
10. Hoag, G.E., and Marley, M.C. Gasoline residual saturation in unsaturated uniform aquifer materials. *J. Environ. Eng. Div., ASCE, 112*(3):586–604, 1986.
11. Villaume, J.F., Lowe, P.C., and Unites, D.F. Recovery of coal gasification wastes: an innovative approach. *Proc. Third Natl. Symp. on Aquifer Restoration and Ground-Water Monitoring*, 1983, pp. 434–445.
12. Anderson, M.R. The dissolution of residual dense non-aqueous phase liquid (DNAPL) from a saturated porous medium. *Proc. NWWA/API Conf. on Petroleum Hydrocarbons and Organic Chemicals in Ground Water*, 1987, pp. 409–428.
13. Saffman, P.G., and Taylor, G. The penetration of a fluid into a porous medium or Hele-Shaw cell containing a more viscous liquid. *Proc. Roy. Soc. Lond. A, 245*:312–329, 1958.
14. Homsy, G.M. Viscous fingering in porous media. *Annu. Rev. Fluid Mech., 19*:271–311, 1987.

15. Chuoke, R.L., van Meurs, P., and van der Poel, C. The instability of slow, immiscible, viscous liquid–liquid displacements in permeable media. *Trans. AIME, 216*:188–194, 1959.
16. Anderson, M.R. The dissolution and transport of dense non-aqueous phase liquids in saturated porous media. Ph.D. Dissertation, Oregon Graduate Center, Beaverton, OR, August 1988.
17. Kueper, B.H. Abbot, W., and Farquhar, G. Experimental observations of multiphase flow in heterogeneous porous media. *J. Contam. Hydrol., 5*:83–95, 1989.
18. Freze, R.A., and Cherry, J.A. *Groundwater*, Prentice-Hall, Englewood Cliffs, NJ, 1979.
19. Hunt, J.R., Sitar, M., and Udell, K.S. Nonaqueous phase liquid transport and cleanup. 1. Analysis of mechanisms. *Water Resour. Res. 24*(8):1247–1259, 1988.
20. Sudicky, E.A., Cherry, J.A., and Frind, E.O. Migration of contaminants in ground-water at a landfill. 4. A natural gradient dispersion test. *J. Hydrol.,* 81–107, 1983.
21. Sudicky, E.A. A natural gradient experiment on solute transport in a sand aquifer: spatial variability of hydraulic conductivity and its role in the dispersion process. *Water Resour. Res., 22*(13):2069–2082, 1986.
22. Sudicky, E.A. University of Waterloo, Waterloo, Ontario, Canada, personal communication to Michael R. Anderson of the Oregon Graduate Institute, 1988.

16

Geochemical Evaluation of Polychlorinated Biphenyls (PCBs) in Groundwater

Stanley Feenstra

Waterloo Centre for Groundwater Research, University of Waterloo, Waterloo, Ontario and Applied Groundwater Research Ltd., Mississauga, Ontario, Canada

Groundwater contamination by PCBs has not been reported frequently in the literature compared to that of many other types of organic chemicals. The potential for groundwater contamination by PCBs is typically considered to be low owing to the reported low solubility of most Aroclors in water and the strong tendency of most PCB compounds to be sorbed on soil materials. General conclusions regarding the immobility of PCBs in groundwater are based frequently on the gross behavior of the PCB Aroclor mixtures only and do not take into account the fact that the PCB Aroclors are complex mixtures of a large number of PCB compounds that vary widely, but systemically, in properties such as solubility in water, potential for sorption, and potential for biodegradation.

The evaluation of the hydrogeochemical processes affecting PCBs in groundwater can be aided substantially through examination of the relative concentration of the different PCB isomers in addition to the total PCB concentration. PCB Aroclors have characteristic isomer concentrations, and processes such as dissolution of the Aroclors, sorption on geologic media, aerobic biodegradation, and anaerobic biodegradation can result in recognizable changes in the isomer patterns of the PCBs in groundwater. Dissolution of Aroclors will result in higher relative concentrations of the lower chlorine isomers of the PCB in the groundwater and lower relative concentrations of the lower chlorine isomers in the residual Aroclor. Sorption processes will further increase the relative concentration of the lower chlorine isomers in the groundwater due to the preferential sorption of the higher chlorine isomers. Aerobic and anaerobic biodegradation processes will affect the isomer patterns in different ways. Aerobic degradation will result in lower relative concentrations of the lower chlorine isomers owing to preferential degradation of these compounds. Anaerobic degradation will result in higher relative concentrations of the lower chlorine isomers owing to dechlorination of higher chlorine isomers. The examination and evaluation of such changes in isomer concentrations allows interpretation of hydrogeochemical processes, which is not possible from examination of total PCB concentrations alone.

I. INTRODUCTION

Polychlorinated biphenyls (PCBs) were produced commercially in North America under the trade name Aroclor from 1930 to 1977. The principal use of Aroclors was as dielectric fluids in electrical transformers and capacitors. A total of 209 different PCB compounds or "congeners" are possible involving from one to ten chlorine substitutions. PCB congeners having the same number of chlorine substitutions are referred to as isomers. The chemical structure of PCB and nomenclature for the location of chlorine substitution is shown in Fig. 1. Approximately 70-125 PCB congeners are found in common Aroclors (1). Aroclor products are identified by a four-digit code number. The first two digits denote the type of molecule, with 12 indicating a PCB Aroclor. Other Aroclors can be chlorinated terphenyls or blends of biphenyls and terphenyls. The last two digits denote the weight percent of chlorine in the Aroclor. The most common Aroclors are 1242, 1254, and 1260, which made up approximately 80% of the North American production of the PCB Aroclors. Aroclor 1016 was developed after these Aroclors and did not follow this designation. Aroclor 1016 contains approximately 42% chlorine as does Aroclor 1242 but a lower proportion of Cl_4 and Cl_5 isomers than Aroclor 1242.

Compared to other types of organic chemicals such as chlorinated solvents, groundwater contamination by PCBs has not been reported frequently in the literature. The potential for groundwater contamination by PCBs is typically considered to be low because of the reported low solubility of most Aroclors in water (< 300 μg/L) and the strong tendency for most PCB compounds to be sorbed onto soil materials. However, there are situations where

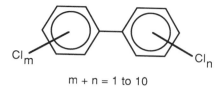

m + n = 1 to 10

Figure 1 Chemical structure of PCBs.

PCBs can be mobile in the subsurface. When PCBs are dissolved in lighter-than-water nonaqueous-phase liquids (LNAPLs) such as petroleum hydrocarbons or denser-than-water nonaqueous-phase liquids (DNAPLs) such as chlorinated solvents, the PCBs will not be sorbed significantly on geologic materials and their migration will be controlled by the migration of the LNAPL or DNAPL. Where PCBs are released into fractured rock environments, the potential for removal of dissolved PCBs from the groundwater by sorption may be much less than in most granular geologic materials.

In addition, general conclusions regarding the immobility of PCBs in groundwater are based frequently on the gross behavior of the PCB Aroclor mixtures only, and do not take into account the fact that the Aroclor mixtures comprise a large number of PCB congeners that vary widely, but systematically, in properties such as solubility, potential for sorption, and potential for biodegradation. In the subsurface, processes such as dissolution of immiscible PCB-bearing oils into the groundwater, sorption on geologic materials, and biodegradation will fractionate various PCB congeners between the immiscible phase, sorbed phase, and dissolved phase. The mixture of PCB congeners in the groundwater that results from the action of these processes will usually be substantially different from that in the original Aroclor. Therefore, it is important to recognize that the behavior of these PCB congeners in groundwater cannot be evaluated using solubility, sorption, and biodegradation data for Aroclors.

This chapter describes a conceptual framework for the geochemical evaluation of PCBs in groundwater that takes into account the widely varying properties of the PCB congeners that occur in the most common PCB Aroclors. This framework involves the examination and consideration of isomer concentrations as well as total PCB concentrations in groundwater. It evaluates the influence of dissolution, sorption, and biodegradation processes on PCB mobility in groundwater. The application of this conceptual framework to the evaluation of groundwater contamination at a PCB storage facility in Smithville, Ontario, Canada is described.

II. COMPOSITION AND CHARACTER OF PCB AROCLORS

Most individual PCB congeners in their pure form are crystalline solids at environmental temperatures, whereas Aroclor 1242 is an oil, Aroclor 1254 is a viscous oil, and Aroclor 1260 is a sticky resin. This is due to the mutual depression of the melting points of the components in the mixtures. Analysis of Aroclor mixtures indicates that each Aroclor contains a characteristic distribution of PCB congeners. It is possible to determine the concentrations of most of the individual congeners present using high-resoltuion gas chromatography with an electron capture detector (GC/ECD) or gas chromatography with a mass spectrometer detector (GC/MS), but such analyses are usually performed only for research studies. The general compositions of Aroclors can be expressed as relative isomer concentrations determined, or estimated, from GC analyses [2]. These isomer concentrations are the weight per weight concentrations of each isomer group relative to the total PCB concentration. This representation of relative isomer concentrations will be used throughout this chapter. Typical isomer concentrations of Aroclors 1242, 1254, and 1260 are shown in Fig. 2.

Liquid Aroclors and many other chlorinated organic compounds make up a group of chemicals that, in their pure form, can be categorized as dense nonaqueous-phase liquids (DNAPLs). "Nonaqueous" refers to the fact that, like petroleum hydrocarbons, they are immiscible with water. "Dense" refers to the fact that, unlike petroleum hydrocarbons, they are more dense than water. Liquid Aroclors have densities ranging from 1.2 to 1.5 g/cm^3. Mixtures of mineral oils and PCB Aroclors containing more than approximately 30% Aroclor 1242 can also have densities significantly greater than that of water. Because a DNAPL is more dense than water, a DNAPL that is released at the ground surface can penetrate readily into the subsurface through both the unsaturated zone and the saturated zone [3]. The tendency for DNAPL chemicals to sink through the saturated zone differs from that of petroleum hydrocarbons, which will float on the groundwater because of their lower density.

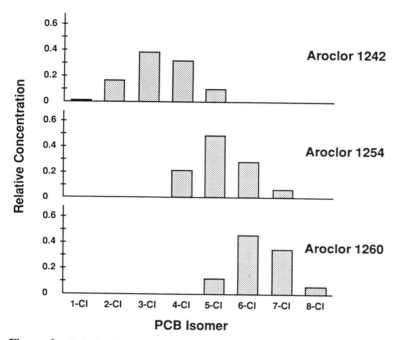

Figure 2 Relative isomer concentrations for common PCB Aroclors. (Data from Webb and McCall [2].)

III. SOLUBILITY OF PCBs IN GROUNDWATER

Compared to many other organic chemicals, such as chlorinated solvents, which are frequently found to cause groundwater contamination, PCB compounds have low solubilities. The water solubility of individual PCB congeners reported in the literature typically represents solubility measured in water in equilibrium with pure solid-phase compounds. Solid-phase solubility values range from several thousand micrograms per liter for Cl_1 isomers to less than 1 μg/L for Cl_8-Cl_{10} isomers. PCB contamination in the subsurface is associated frequently with the release of pure Aroclor liquids or liquid mixtures of Aroclors and chlorinated benzenes, which comprise transformer or capacitor fluids and liquid hydrocarbons contaminated by PCB compounds. In considering dissolution of PCBs from PCB-bearing oils, solid-phase solubility values are not applicable because the PCB congeners are present in a liquid phase rather than a solid phase [4], [5]. Liquid-phase solubility values for PCB congeners can be calculated from their melting points and the system temperature or measured in the laboratory. Liquid-phase solubility values [$C(l)$] for PCB congeners are higher than solid-phase solubility values [$C(s)$], and the difference between $C(l)$ and $C(s)$ values is greater for high chlorine congeners, which have higher melting points. The solubility of PCB congeners decrease with increasing chlorine substitution (see Fig. 3).

For PCB-bearing LNAPLs or DNAPLs in contact with water, the dissolved concentrations of the component compounds can be estimated as

$$C_i = \chi_i S_i \tag{1}$$

where C_i is the dissolved phase concentration of component i, χ_i is the mole fraction of component i in the immiscible organic phase, and S_i is the liquid-phase solubility of component i in water. This relationship is based on the behavior of ideal liquids according to Raoult's law. Laboratory experimental studies suggest that this relationship is a reasonable approximation for mixtures of structurally similar hydrophobic organic compounds [6]. For more complex hydrocarbon mixtures this approximation may be in error, but the error in the estimated dissolved concentration is unlikely to be greater than a factor of 2 [7] unless cosolvents such as surfactants or miscible alcohols are present in the water.

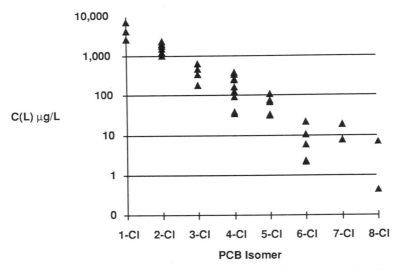

Figure 3 Relationship between liquid-phase solubility [$C(l)$] of PCB congeners and the number of chlorine substitutions. (Data from Shiu and Mackay [4].)

The overall solubilities of PCB Aroclors are reported to be 288 μg/L for Aroclor 1242, 42 μg/L for Aroclor 1254, and 2.7 μg/L for Aroclor 1260 [8]. These are the total dissolved PCB concentrations that would be expected in groundwater in contact with pure Aroclor mixtures. Lower concentrations would be observed in the groundwater if the source of PCB contamination were LNAPLs containing low concentrations of PCB Aroclors. For example, if an LNAPL contained 1% PCB Aroclor 1242, 1254, or 1260, the dissolved concentrations would be approximately 3 μg/L for Aroclor 1242, 0.4 μg/L for Aroclor 1254, or 0.03 μg/L for Aroclor 1260.

The relative isomer composition of the dissolved-phase PCB resulting from dissolution of a PCB-bearing oil will depend on the solubility of the PCB compounds present and the concentrations of the compounds in the oil. For the case of Aroclor 1242, the most soluble PCB congeners are the

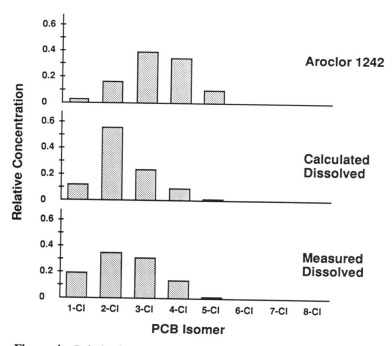

Figure 4 Relative isomer concentrations of the calculated dissolved-phase PCB in equilibrium with Aroclor 1242. Measured dissolved-phase concentrations from Chou and Griffin [8].

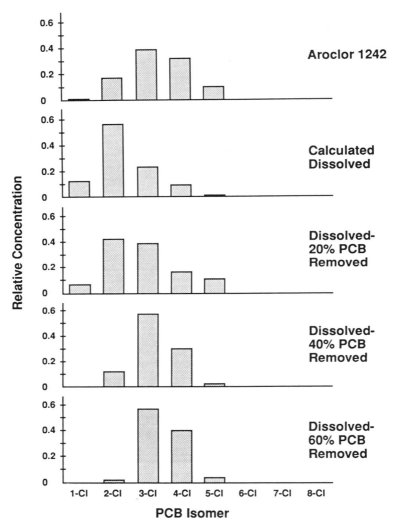

Figure 5   Changes in relative isomer concentrations of dissolved-phase PCB resulting from the progressive dissolution of Aroclor 1242.

Cl_1, Cl_2, and Cl_3 isomers. In the case of Aroclor 1254, the most soluble congeners are the Cl_4 and Cl_5 isomers. The resultant relative isomer concentrations of the dissolved-phase PCB resulting from the dissolution of an Aroclor can be calculated based on the isomer composition of the Aroclor, an average water solubility for each PCB isomer group, and Eq. (1). Although the magnitude of the dissolved PCB concentrations will vary with the total concentration of the Aroclor in the oil phase, the relative concentrations of the PCB isomers will be maintained. Calculations for the dissolution of Aro-chlor 1242 are shown in Fig. 4. The calculated dissolved-phase PCB concentrations are shown with the observed dissolved-phase concentrations from laboratory testing [8]. The calculated and observed dissolved-phase concentrations are generally comparable and show the expected higher concentrations of the lower chlorine isomers relative to the original Aroclor composition. The slight differences between the calculated and observed dissolved-phase concentrations are likely a result of the use of an average solubility for each PCB isomer group in the calculations and possible differences in the composition of the Aroclors used in the laboratory testing.

The dissolved-phase PCB in groundwater resulting from the dissolution of PCB-bearing oils would be expected to exhibit isomer compositions similar to those calculated or observed above. These calculations represent only the initial dissolved-phase concentrations. Indeed, because the more soluble lower chlorine isomers are dissolved preferentially with time, the composition of the residual oil will change as a significant portion of the mass of the PCB in the oil is dissolved. As the composition of the residual oil changes, the composition of the dissolved phase will change correspondingly. The change in oil composition and the composition of the dissolved phase can be estimated using Eq. (1) with the calculation of an incremental dissolution of the oil mass [5]. Figure 5 shows the resultant dissolved-phase composition as Aroclor 1242–bearing oil is dissolved for 20%, 40%, and 60% depletion of the Aroclor in the oil. With increasing depletion of the Aroclor mass, the more soluble Cl_1 and Cl_2 isomers are removed preferentially from the oil, and their concentrations in the oil and in the water are reduced. Correspondingly, the relative concentrations of the less soluble Cl_3, Cl_4, and Cl_5 isomers increase. In most situations in the subsurface, the mass depletion of PCB from an oil phase is expected to be very slow because of the low overall solubility. Many years may be required for significant removal of PCB from the oil by dissolution and the corresponding changes in the oil-phase and dissolved-phase compositions.

IV. ATTENUATION PROCESSES

A. Sorption

Polychlorinated biphenyl compounds are strongly sorbed on geologic media compared to other organic chemical compounds that are frequently encountered in groundwater. The sorption of PCBs from groundwater onto soil and rock is considered to occur primarily by partitioning of the PCBs on solid natural organic matter. Strongly hydrophobic organic compounds such as

PCBs may also be sorbed on clay minerals or other soil particles, but sorption on these phases is not believed to be significant when soil organic matter is present.

The degree of sorption of organic compounds on geologic media has been related to the octanol–water partition coefficient (K_{ow}) of the chemical and the organic carbon content (f_{oc}) of the medium, where the partition coefficient on the bulk geologic medium is defined as

$$K_d = \frac{\text{concentration on solids}}{\text{concentration in water}} = K_{oc} f_{oc} \qquad (2)$$

and the K_{oc} is related to the K_{ow} by an empirical relationship such as [9]

$$\log K_{oc} = \log K_{ow} - 0.21 \qquad (3)$$

The higher the K_{ow} value, the more strongly the compound will be sorbed on geologic media. The log K_{ow} values for selected congeners [4] are shown in Fig. 6. The log K_{ow} values for the PCB congeners increase with increasing chlorine substitution. The log K_{ow} values range from approximately 4.5 for Cl_1 isomers to approximately 7.0 for Cl_6, Cl_7, and Cl_8 isomers. The lower chlorine isomers should be much less strongly sorbed on geologic media than

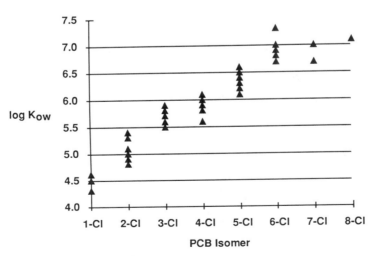

Figure 6 Relationship between the octanol–water partition coefficients (K_{ow}) of PCB congeners and the number of chlorine substitutions. (Data from Shiu and Mackay [4].)

higher chlorine isomers and should be considerably more mobile in groundwater.

Overall log K_{ow} values of 5.29, 6.11, and 6.61 are reported in the literature for Aroclors 1242, 1254, and 1260, respectively [8]. Such values represent a weighted average for the mixture because some of the congeners present in the Aroclors will have higher K_{ow} values and some will have lower values. Consideration must be given to the lower chlorine isomers most likely to be dissolved in the groundwater and their lower K_{ow} values in order to assess properly the mobility of a given PCB mixture in groundwater.

The relative rate of contaminant migration in a porous medium is expressed as the retardation factor,

$$R = \frac{\text{velocity of groundwater}}{\text{velocity of contaminant}} = 1 + \frac{\rho_b}{n} K_d \qquad (4)$$

where ρ_b is the dry bulk density and n is the porosity. Using Eqs. (2)–(3), the retardation factors for the components of Aroclor 1242 (Cl_1–Cl_5 isomers) were calculated for varying values of f_{oc} (Fig. 7). These calculations indicate that all components of Aroclor 1242 should be retarded significantly in porous media having even low f_{oc} values, although Cl_1 and Cl_2 isomers should be 20–100 times more mobile than Cl_4 and Cl_5 isomers. In media having f_{oc} values greater than 0.005, even low-chlorine PCB isomers should be virtually

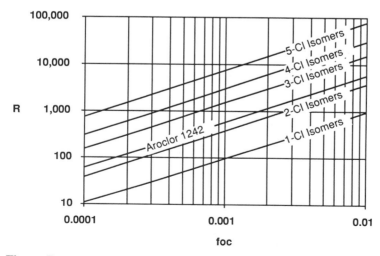

Figure 7 Relationship between calculated retardation factors (R) and the fraction organic carbon (f_{oc}) in soil for various PCB isomers and Aroclor 1242.

immobile. These conclusions are based on the transport of dissolved PCB in the absence of any other compounds such as surfactants or cosolvents, or colloids or emulsions, which may facilitate the transport of PCBs.

The relative isomer composition of the sorbed-phase PCB will be significantly different from the dissolved-phase PCB because the higher chlorine isomers are more strongly sorbed than the lower chlorine isomers. This effect is illustrated in Fig. 8, which was obtained using the dissolved-phase composition resulting from the dissolution of Aroclor 1242 and Eq. (2) to calculate the sorbed-phase composition on a soil. The magnitude of the sorbed-phase concentrations will be determined by the f_{oc} of the soil, but the relative isomer concentrations shown in Fig. 8 will remain the same.

The sorbed-phase PCB composition is similar to the composition of the original Aroclor 1242. This means that PCB sorbed on soil cannot be distinguished readily from PCB dissolved in residual LNAPLs or DNAPLs retained on the soil on the basis of their relative isomer compositions. Similarly, the dissolved-phase PCB composition that can be derived from leaching

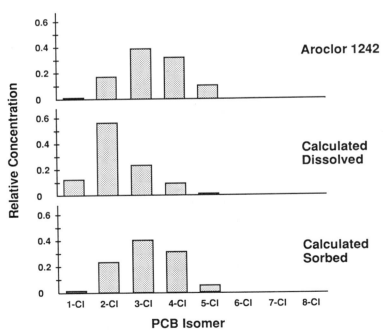

Figure 8   Relative isomer concentrations of calculated sorbed-phase PCB in equilibrium with the dissolved-phase PCB derived from Aroclor 1242.

of sorbed-phase PCB from soil cannot be distinguished from dissolved-phase PCB derived from the dissolution of PCB-bearing LNAPL or DNAPL.

In the same way that differences in solubility can cause higher relative concentrations of the lower chlorine isomers in the water in contact with a PCB-bearing oil, the differences in sorption potential expressed by the K_{ow} can also change the relative isomer concentrations as PCB-containing ground-water flows through geologic media. Because the higher chlorine isomers are more strongly sorbed, the relative concentrations of these isomers will be reduced in the dissolved phase, thereby increasing the relative concentra-tion of lower chlorine isomers in the dissolved phase.

This effect is illustrated using a one-dimensional analytical solution [10] to simulate advective–dispersive flow of PCB isomers through a sandy aquifer. Aquifer parameters adopted for the model are taken from studies of the sand aquifer at Canadian Forces Base Borden, Ontario, Canada conducted by the University of Waterloo [11]. The linear groundwater velocity is as-signed to be 44 m/yr based on a hydraulic conductivity of 10^{-4} m/sec, a hydraulic gradient of 0.005, and a porosity of 0.35. The longitudinal dis-persivity is assigned to be 0.5 m. The f_{oc} of the sand is assigned to be 0.0002 [12]. The K_d values and retardation factors calculated for the PCB isomers are shown in Table 1. Mean velocities of the PCB isomers are calculated to range from 2.2 m/yr for Cl_1 isomers to 0.029 m/yr for Cl_5 isomers and are all considerably lower than the linear groundwater velocity of 44 m/yr.

Longitudinal profiles of calculated concentrations with distance for the PCB isomers are shown in Fig. 9. These longitudinal profiles indicate that the concentrations of the various PCB isomers change relative to one an-other with distance of travel and with time. For example, although Cl_2 and Cl_3 isomers occur at higher concentrations than Cl_1 isomers close to the source, preferential sorption of the more chlorinated isomers removes these isomers so that at a distance of approximately 20 m, Cl_1 isomers occur at the highest

Table 1   Calculated Migration Parameters for Porous Medium Model

PCB isomer	log K_{ow}	log K_{oc}	K_d (mL/g)	Retardation factor	Velocity of contaminant (m/yr)	Distance of travel in 20 yr (m)
Cl_1	4.5	4.29	3.9	20	2.2	44
Cl_2	5.1	4.89	16	77	0.57	11
Cl_3	5.7	5.49	62	300	0.14	2.8
Cl_4	6.0	5.79	120	600	0.073	1.5
Cl_5	6.4	6.19	310	1500	0.029	0.58

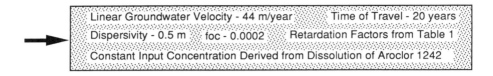

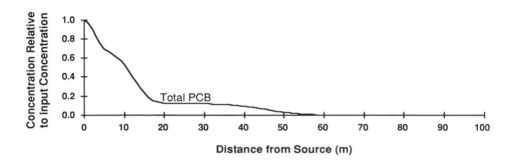

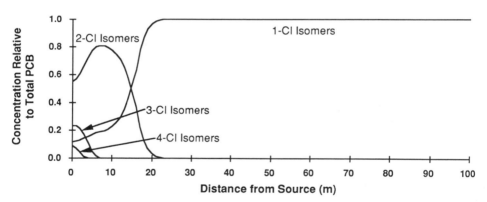

Figure 9   Results of the simulation of one-dimensional transport of PCB isomers through a porous medium.

relative concentration. This trend is clearly shown in Fig. 10, which plots the relative PCB isomer concentrations with increasing distance from the source after a time of 20 years.

B. Biodegradation

Polychlorinated biphenyl compounds are generally regarded as persistent environmental contaminants that accumulate in aquatic sediments, surficial

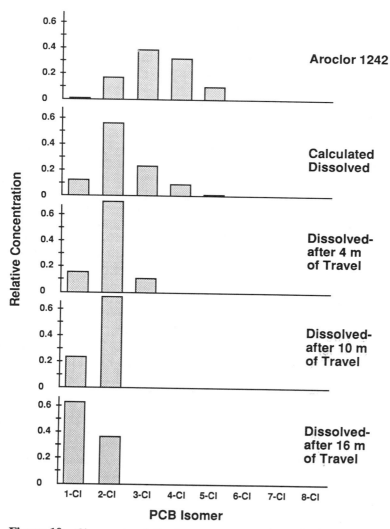

Figure 10 Changes in relative isomer concentrations of calculated dissolved-phase PCB derived from Aroclor 1242 with distance traveled for simulation of transport through a porous medium.

soils, and biota. Despite the fact that lower chlorine isomers (Cl_1, Cl_2, and Cl_3) make up a significant portion of the PCB compounds that have been released to the environment, analysis of PCB-contaminated sediment, soil, and biota typically indicates a predominance of higher chlorine isomers. This general observation suggests that the lower chlorine isomers may be preferentially lost or degraded in the environment. This observation, made in the early 1970s, led to the discovery of PCB biodegradation in soils, aquatic sediment, and sewage sludge, primarily by aerobic biodegradation processes. Then during the early 1980s, it was observed that PCB-contaminated sediment in the Hudson River of New York was depleted in the higher chlorine isomers compared to the levels expected based on the Aroclor released to the river. This observation led to the discovery of anaerobic biodegradation processes. There is currently laboratory and field evidence to demonstrate that naturally occurring bacterial populations in many soils, aquatic sediments, freshwaters, and marine waters are capable of degrading PCBs under aerobic conditions. As found in the Hudson River, PCB degradation under anaerobic conditions has been demonstrated in freshwater sediments. However, there have been no studies reported that demonstrate specifically the aerobic or anaerobic degradation of PCB in groundwater environments.

The aerobic biodegradation of PCB was first demonstrated in the laboratory in 1973. Cl_1 and Cl_2 PCB isomers were degraded by organisms isolated from samples of sewage sludge [13]. Since that pioneering study, a large number of bacteria from freshwater and marine environments and soils have been demonstrated to degrade PCB under aerobic conditions. The rate of degradation of the various PCB congeners is dependent on the number of chlorine substitutions and the position of the substitution [14]. Lower chlorine (Cl_1 and Cl_2) isomers are more readily degraded than higher chlorine (Cl_3, Cl_4, Cl_5) isomers. Degradation is considered to occur by attack of the unsubstituted or less substituted ring by the enzyme 2,3-dioxygenase at the 2,3 positions on the ring. Chlorine substitution at the 2 (*ortho*) or 3 (*meta*) position is considered to reduce the effectiveness of the attack by 2,3-dioxygenase on the PCB molecule. Substitution at these positions is more likely in higher chlorine isomers. Polychlorinated diene biphenyl and polychlorinated dihydroxy biphenyl compounds are unique metabolites of aerobic biodegradation of PCB. These compounds further degrade to polychlorinated benzoic acids [14].

Aerobic biodegradation of the lower chlorine isomers in an Aroclor that contains both lower and higher chlorine isomers will result in a reduction in the total PCB concentration and higher relative concentrations of the higher chlorine isomers in the remaining PCB. This effect is illustrated by the results of laboratory experiments [15] of PCB biodegradation conducted using mixed bacterial cultures isolated from an agricultural soil from Illinois to which PCB-contaminated sewage sludge was applied for 7 years. Aroclor 1242 was added, and the PCB isomer concentrations were monitored at the end of 5-day and 15-day incubation periods. The degradation of the various isomers was extensive and ranged from 100% for the Cl_1 isomer to 6% for the Cl_4 isomers. Overall PCB degradation was 91%. The values for percent degradation of the Cl_1, Cl_2, Cl_3, and Cl_4 isomers can be converted to the relative isomer concentrations shown in Fig. 11. This figure shows that in addition to a decline in the total PCB concentration, relative concentrations of the higher chlorine isomers are increased because of the preferential degradation of the lower chlorine isomers. At many sites, the total PCB concentrations are highly variable, and if the precise source of the contamination and pathways for migration are not known, it is not possible to correlate a decline in concentration in the groundwater with the occurrence of biodegrada-

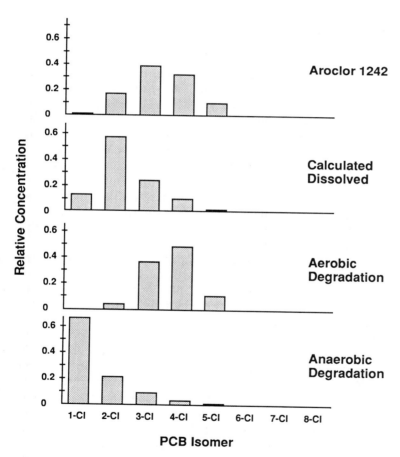

Figure 11 Changes in relative isomer concentrations of PCB by aerobic and anaerobic biodegradation. Isomer concentrations for aerobic degradation are for the dissolved-phase PCB resulting from biodegradation of dissolved-phase PCB derived from Aroclor 1242 [15]. Isomer concentrations for anaerobic degradation are for sorbed-phase PCB resulting from biodegradation of sorbed-phase Aroclor 1242 [18].

tion processes. However, if aerobic biodegradation in the groundwater has occurred, there should be evidence of the loss of the lower chlorine isomers. This effect may not be as clear in the analysis of soils. For PCB sorbed on the soil or dissolved in NAPL retained by the soil, the relative concentrations of lower chlorine isomers may also be reduced by dissolution processes in addition to aerobic biodegradation.

The aerobic biodegradation of PCBs in groundwater has not been reported in the literature. Studies of aerobic degradation in soils and aquatic systems provide an indication of the types of bacteria capable of degrading PCBs. Many of these bacteria are facultative aerobes that may be present in shallow groundwater systems. Little is known about the redox conditions, nutrient levels, or other conditions required by these bacteria, so it cannot be determined if suitable conditions typically occur in groundwater. However, aerobic biodegradation of PCBs, if it occurs in groundwater, will certainly proceed at much slower rates than are exhibited in laboratory studies, owing to lower temperatures, lower nutrient levels, and lower levels of organic substrates.

The occurrence of anaerobic biodegradation of PCBs in the environment was first reported [16] based on the examination of GC patterns in PCB-contaminated sediments in the upper Hudson River of New York. These GC patterns showed lower relative concentrations of higher chlorine isomers and higher relative concentrations of the lower chlorine isomers compared to the GC pattern of the Aroclors originally discharged to the river. These findings were supported by similar findings in the anaerobic sediments of Silver Lake in Massachusetts [17]. The PCBs found in Silver Lake had significantly reduced relative concentrations of Cl_6–Cl_8 isomers and higher relative concentrations of Cl_1–Cl_4 isomers compared to the original Aroclor 1260. Because of the variability in the total PCB concentrations in the sediment, it is not possible to determine whether dechlorination reactions can cause complete dechlorination of the lower chlorine isomers, thereby ultimately eliminating those PCBs from the sediments.

Recent laboratory studies [18] have demonstrated the anaerobic biodegradation of Aroclor 1242. Mixed bacterial cultures were isolated from PCB-contaminated sediment from the Hudson River and incubated in a reduced anaerobic mineral medium spiked with 700 mg/kg of Aroclor 1242. There was little change over a 4-week incubation period, but significant dechlorination of the Cl_3–Cl_5 isomers was observed after 8 weeks and 16 weeks. No decline in the total molar concentration of PCBs was observed, suggesting that complete dechlorination does not occur. The results of these experiments are shown in Fig. 11. The anaerobic biodegradation reactions occurred much more slowly than aerobic biodegradation reactions observed in laboratory studies.

Anaerobic biodegradation of PCBs in groundwater has not been reported in the literature. It is not known if the requisite bacterial populations and environmental conditions for degradation can occur in groundwater systems. If anaerobic biodegradation reactions do occur in groundwater, the loss of the higher chlorine isomers and associated formation of the lower chlorine isomers would be similar to that expected due to attenuation of dissolved PCB concentrations by sorption. This may make identification of anaerobic biodegradation in the subsurface difficult.

C. Matrix Diffusion

In fractured rock environments there is a greater potential for extensive migration of dissolved PCBs in groundwater because groundwater velocities are frequently much higher than in porous media, and the potential for sorption of PCBs on the materials composing the fracture surfaces is generally lower than in porous media. Nevertheless, fractured sedimentary rocks such as sandstones, shales, and carbonates frequently do have a substantial matrix porosity (5–20%), and the diffusion of PCBs into the matrix will reduce concentrations in the groundwater flowing through the fractures [19–21]. The attenuation of PCBs by matrix diffusion will be substantially enhanced as a result of sorption on the matrix solids.

The degree of attenuation of PCB migration through fractured media will depend on the diffusivity of the PCB compounds and the degree of sorption. Lower chlorine PCB isomers will have slightly higher diffusivities than higher chlorine isomers because of their lower molecular weights. Aqueous diffusion coefficients for PCB isomers were estimated from the measured diffusion coefficient of benzene and the relative molecular weight of benzene and the PCB isomers [22,23]. The estimated aqueous diffusion coefficient for Cl_1 isomers is approximately 50% greater than that for Cl_8 isomers (see Table 2). The higher the diffusion coefficient, the higher the potential

Table 2 Estimated Free-Solution Diffusion Coefficients for PCB Isomers

PCB isomer	Free-solution diffusion coefficient at 20°C (m²/sec)
Cl_1	6.4×10^{-10}
Cl_2	6.2×10^{-10}
Cl_3	5.8×10^{-10}
Cl_4	5.5×10^{-10}
Cl_5	5.3×10^{-10}
Cl_6	5.1×10^{-10}
Cl_7	4.8×10^{-10}
Cl_8	4.7×10^{-10}
Chloride	2.0×10^{-9}

degree of attenuation due to matrix diffusion. However, this relatively small difference in diffusion coefficients is unlikely to result in significantly greater attenuation of the lower chlorine isomers compared to the higher chlorine isomers. Nevertheless, in situations where PCBs are sorbed on organic carbon materials in the rock matrix, as described in Section IV.A, the higher chlorine isomers will also be sorbed preferentially to the lower chlorine isomers, and the relative concentrations of the PCB isomers will change with distance traveled.

This effect is illustrated using a one-dimensional analytical solution for solute transport through a single fracture in a porous matrix [24]. Parameters are selected to represent a relatively permeable ($K = 10^{-5}$ m/sec) fractured carbonate aquifer with a fracture spacing of 0.5 m. With these assumptions and assuming a smooth planar fracture, the effective fracture aperture is calculated to be 250 μm based on the hydraulic conductivity and the fracture spacing [25]. For a hydraulic gradient of 0.005, the groundwater velocity in the fractures is 1200 m/year. The rock matrix is assigned a porosity of 5%, an f_{oc} of 0.001, and an effective diffusion coefficient of 2×10^{-11} m²/sec for chloride. This effective diffusion coefficient is similar to the values measured for a cemented sandstone with a porosity of 9% [26]. Effective diffusion coefficients for PCB isomers are calculated in proportion to the value for chloride. For these conditions, it was found that the use of a single-fracture model rather than a multiple interacting fracture model was satisfactory because the migration of PCBs into the matrix was limited to less than 0.01 m in 20 years due to sorption on the matrix. Consequently, the diffusion of PCBs into the matrix is not influenced by conditions in adjacent fractures.

The results of the fracture model simulations are shown in Fig. 12. The trends in the attenuation of PCB isomers are similar to those shown in Fig. 9 for the porous medium model. Close to the source, Cl_2 and Cl_3 isomers occur at the highest concentrations, but at distances greater than 40 m, Cl_1 isomers occur at the highest concentrations due to preferential sorption of the higher chlorine isomers. This trend is clearly shown in Fig. 13 by plots of the relative PCB isomer concentrations.

V. CONCEPTUAL MODEL FOR EVALUATION OF PCBs IN GROUNDWATER

The evaluation of the hydrogeochemical processes affecting PCBs in groundwater can be aided substantially through the examination of relative isomer concentrations. PCB Aroclors have characteristic isomer concentrations, and processes such as dissolution of the Aroclors, sorption on geologic media, aerobic biodegradation, and anaerobic biodegradation can result in recognizable changes in the isomer patterns of the PCBs in groundwater.

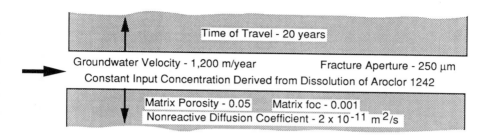

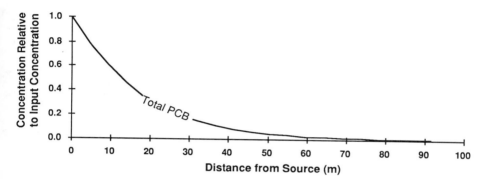

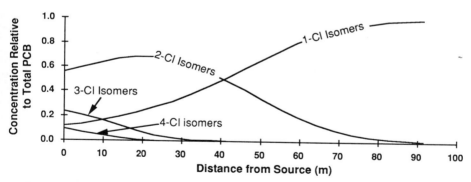

Figure 12   Results of simulation of one-dimensional transport of PCB isomers through a single fracture in a porous matrix. Retardation factors are 1030, 4100, 16,300, 32,600, and 81,800 for Cl_1, Cl_2, Cl_3, Cl_4, and Cl_5 isomers, respectively.

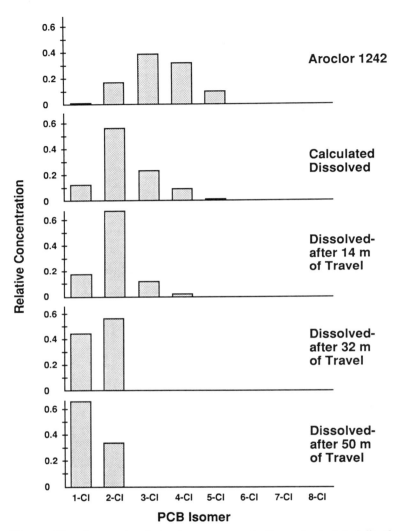

Figure 13   Changes in relative isomer concentrations of calculated dissolved-phase PCBs with distance traveled for simulation of transport through a fractured porous medium.

Dissolution of Aroclors will result in higher relative concentrations of the lower chlorine isomers of the PCB in the groundwater and lower relative concentrations of the lower chlorine isomers in the residual Aroclor. Sorption processes will further increase the relative concentration of the lower chlorine isomers in the groundwater due to the preferential sorption of the higher chlorine isomers. Aerobic and anaerobic biodegradation processes will affect the isomer patterns in different ways. Aerobic degradation will result in lower relative concentrations of the lower chlorine isomers due to preferential degradation of these compounds. Anaerobic degradation will result in higher relative concentrations of the lower chlorine isomers due to dechlorination of higher chlorine isomers. The examination and evaluation of such changes in isomer concentrations should aid in the interpretation of hydrogeochemical processes on PCBs in groundwater.

In almost all groundwater studies, PCB analyses of groundwater, soils, LNAPLs, and DNAPLs are expressed as the total PCB concentration only, and it is not possible to evaluate the geochemical processes that may influence PCB migration in groundwater. However, the analytical methods commonly employed by commercial laboratories for GC/ECD or GC/MS analysis of PCBs would allow calculation or estimation of PCB isomer concentrations if analysts were made aware of the need for this information by hydrogeologists.

VI. PCB STORAGE SITE, SMITHVILLE, ONTARIO, CANADA

A. Setting

A PCB storage and transfer facility operated at Smithville, Ontario from 1978 to 1985. Hydrogeological investigations [27] following closure of the site identified significant contamination by PCB oils of the overburden, bedrock, and groundwater underlying the site (see Fig. 14). The site is underlain by 4–5 m of Late Wisconsin silty clay of glaciolacustrine origin, atop dolostone bedrock of the Lockport Formation, which is of Silurian age. The silty clay is weathered and fissured, with vertical and subvertical fissures extending to the bedrock surface. These fissures provided a pathway for the movement of PCB oils into the bedrock. The upper 10–15 m of the bedrock consists of a horizontally layered sequence of dolostones that dip to the south at approximately 1–5°. The upper portion of the bedrock is relatively permeable, having hydraulic conductivity ranging from 10^{-4} to 10^{-5} m/sec. The horizontal permeability of this zone is considered to be primarily the result of several distinct open bedding plane partings, which can be correlated across the site for distances of up to several hundred meters.

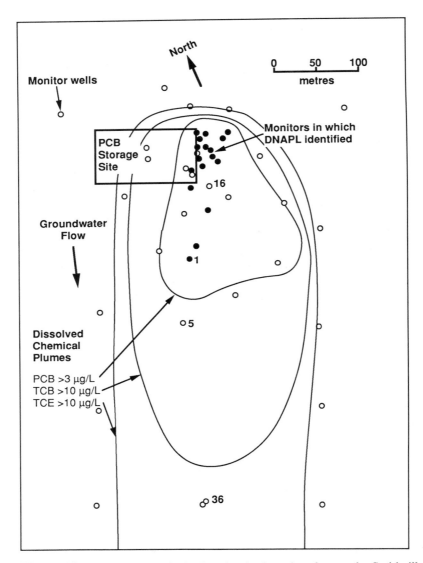

Figure 14 Extent of chemical migration in the subsurface at the Smithville site.

Groundwater velocities in the upper bedrock cannot be reliably estimated because the effective porosity of the fractured dolostone cannot be readily determined. Groundwater velocities in the upper bedrock may be as high as several hundred meters per year. Groundwater flow in the upper bedrock is to the south.

B. DNAPL Zone

Dense nonaqueous-phase liquid is present in the overburden and bedrock underlying the site as determined on the basis of the visual observation of free DNAPL within fissures in the silty clay overburden, recovery of free DNAPL from the bottom of the borings in the bedrock, the visual observation of residual DNAPL in samples of dolostone core, and the presence of high dissolved chemical concentrations exceeding the estimated solubility of the component chemicals in groundwater samples. The density of the DNAPL ranges from 1.2 to 1.3 g/cm^3. The DNAPL contains approximately 45% PCBs (estimated 1:1:1 mixture of Aroclors 1242, 1254, and 1260), 10% 1,2,4-trichlorobenzene (1,2,4-TCB), 3% 1,2,3-triclorobenzene (1,2,3-TCB), 2% trichloroethylene (TCE), and 40% mineral oils. The DNAPL has penetrated downward through the overburden and into the bedrock. The borings in which DNAPL was identified in the bedrock are shown in Fig. 14. Lateral migration of DNAPL away from the site appears to be primarily controlled by the slope of the bedrock strata. DNAPL has been identified as far as monitoring well location 1, approximately 150 m south of the site.

C. Dissolved Chemical Plumes

Groundwater that comes into contact with DNAPL acquires dissolved concentrations of the constituent chemicals in the DNAPL. Based on ideal solution behavior [Eq. (1)] and the concentrations of the principal constituents of the DNAPL, equilibrium dissolved concentrations for total PCB, 1,2,4-TCB, 1,2,3-TCB, and TCE at the Smithville site are calculated to be 100, 3900, 2400, and 39,000 $\mu g/L$, respectively. These concentrations represent the dissolved concentrations that would be expected in groundwater in contact with the source DNAPL. Although TCE occurs at much lower concentrations in the DNAPL than TCB and PCBs, it is the predominant constituent of the dissolved chemical plume because of its much higher solubility.

Chemicals that are dissolved in the groundwater will move with the groundwater according to the direction and velocity of groundwater flow and the influence of attenuation mechanisms such as sorption, matrix diffusion, and biodegradation. TCE has migrated the farthest from the site. At a monitoring well situated 300 m beyond the farthest extent of DNAPL migration (location 36), TCE concentrations average 350 μg/L, or 0.9% of the estimated dissolved-phase concentrations (C/C_0) at the DNAPL zone. At this location, 1,2,4-TCB and 1,2,3-TCB concentrations average 2.6 and 1.1 μg/L, respectively, or C/C_0 values of 0.07% and 0.05%, respectively. The lower relative concentration for TCB indicates a greater degree of attenuation than TCE. Total PCB concentrations at this location are consistently less than detection at 0.02 μg/L or C/C_0 less than 0.02%.

At a monitoring well situated 75 m beyond the farthest extent of DNAPL migration (location 5), TCE concentrations average 1000 μg, or 2.6% of the estimated dissolved-phase concentrations at the DNAPL zone. At this location, 1,2,4-TCB and 1,2,3-TCB concentrations average 49 and 21 μg/L, respectively, or C/C_0 of 1.2% and 0.9%. Total PCB concentrations at this location are only intermittently greater than the detection limit of 0.02 μg/L or C/C_0 of 0.02%.

Sorption of TCE, TCB, and PCB on solid-phase organic carbon in the bedrock formations may attenuate contaminant migration in the groundwater. It is not possible to estimate the degree of attenuation expected for these contaminants based on log K_{ow} values because there have been no analyses of the organic carbon content of the dolostone. However, the K_{ow} values give an indication of the relative mobility of the various contaminants. 1,2,4-TCB and 1,2,3-TCB have log K_{ow} values of 4.02 and 4.11, respectively [28]. Such values are lower than those for most PCB compounds, but they are close to the log K_{ow} values for Cl_1 and Cl_2 isomers, which are in the range of 4.5–5.0. The lower chlorine isomers that are expected to be the predominant PCB species in the groundwater are not expected to be significantly less mobile than TCB. TCE has a log K_{ow} of 2.29 [29] and would not be expected to be attenuated to the same degree as TCB or PCB because of sorption.

The relative mobility of PCB, TCB, and TCE evident at the Smithville site compares generally with that predicted on the basis of log K_{ow} values. TCE is most mobile, and PCB is least mobile. However, the observed mobility of PCB compared to TCB is much less than would be predicted on the basis of log K_{ow} values. This suggests that some other attenuation mechanism may be contributing to reduce the mobility of PCB.

D. Examination of Isomer Concentrations

Some of the PCB analyses of DNAPL and groundwater were performed using GC/ECD methods that allowed estimation of relative isomer concentrations. The method involved comparison of the sample peaks to 24 dominant peaks [2] in a mixed standard of 1:1:1 Aroclors 1242, 1254, and 1260. The 24 peaks represent Cl_2–Cl_8 isomers. Peaks representing Cl_1 isomers were not quantitated. The laboratory reported the total PCB concentrations as the sum of these 24 dominant peaks. For numerous samples, several GC runs were available because of repeated analyses. The GC peak patterns for replicate GC runs were very similar. Isomer concentration plots were estimated from the GC peak data based on published data on the isomer composition of each peak [2]. The GC peak plots for the oil samples from monitoring wells were very similar, and individual peak concentrations were averaged before calculation of isomer concentrations. The GC peak plots for the groundwaters from the DNAPL zone were also very similar and were similarly averaged before calculation of isomer concentrations.

The monitoring well installations at the site include open boreholes in the bedrock with steel casings, stainless-steel piezometers, and PVC piezometers. The results of groundwater samples collected from PVC piezometers were not considered here because of the potential for sorption of organic contaminants on the piezometer materials.

E. Evaluation of Isomer Concentrations

The relative isomer concentrations for all the oil samples were found to be very similar. The isomer concentrations for the oils from monitoring wells were identical to those of the oils stored in tanks and drums on the site. A plot of the mean isomer concentrations for the oils from the monitoring wells is shown in Fig. 15. The fact that the oils from the monitoring wells are identical to the oils stored in the tanks on site indicates that there has not been a substantial amount of dissolution of the DNAPL in the subsurface. A substantial amount of dissolution would have resulted in a preferential loss of the lower chlorine isomers. The PCBs dissolved in water in contact with such an oil phase would consist predominantly of lower chlorine isomers. A plot of the isomer concentrations for the calculated dissolved phase is also shown in Fig. 15.

The relative concentrations of the PCB isomers in the DNAPL zone groundwaters shown in Fig. 15 are virtually identical to those of the oils and are not similar to the calculated dissolved concentrations. Many of the DNAPL zone groundwaters have aqueous-phase PCB concentrations that exceed the calculated solubility (100 μg/L), ranging from 1000 μg/L up to 870,000 μg/L. There was no separate phase oil visible in these samples. Such high PCB concentrations suggest that the PCB in the groundwater samples may be present as an emulsion or sorbed on colloids rather than dissolved in the groundwater. The similarity between the isomer concentrations of the oils and the source area groundwaters also indicates that the PCB present in these samples is present as an emulsion or colloid rather than as dissolved PCB. However, even the DNAPL zone groundwaters with much lower PCB concentrations (<100 μg/L) exhibit relative isomer concentrations identical to

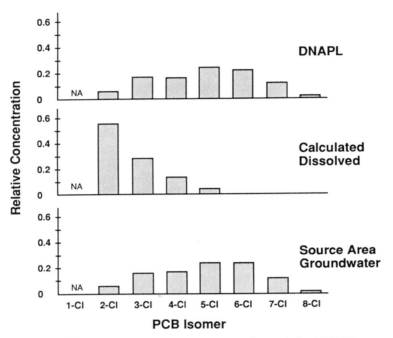

Figure 15 Relative PCB isomer concentrations of the DNAPL recovered from borings at the Smithville site and calculated dissolved-phase concentrations derived from the DNAPL and the groundwater from the area in and around the DNAPL source. NA, not analyzed.

those of the DNAPL. At these lower concentrations, the PCBs present would be expected to be principally in dissolved form, but the isomer concentrations suggest that PCBs are present as an emulsion or sorbed to colloids even in these samples.

Only the groundwater samples collected from well 16 on the perimeter of the DNAPL zone exhibited isomer concentrations that approach those calculated for dissolved-phase PCBs in contact with the DNAPL. These data are shown in Fig. 16. The isomer concentrations for the sample collected on July 19, 1987 is very similar to that of the calculated dissolved phase. The isomer concentrations for the samples collected from the same well on August 9, 1987 appear to be intermediate between those of the calculated dissolved phase and those of the oils (see Fig. 16). The mean total PCB concentration for these samples is 67 μg/L, higher than the 22 μg/L for the July 19 sample.

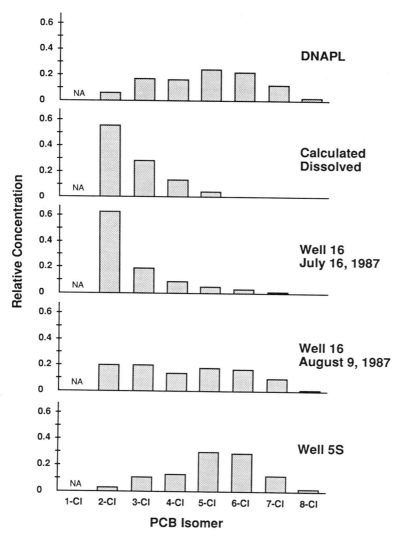

Figure 16   Relative PCB isomer concentrations of selected groundwater samples from periphery of the DNAPL source. NA, not analyzed.

The PCB present in August 9 samples may be a mixture of dissolved PCBs and emulsified PCBs resulting in a blended pattern of isomer concentrations. Substantially different patterns of relative isomer concentrations occur in well 5, the most distant downgradient well in which PCBs have been detected (see Fig. 16). These samples show patterns that are highly depleted in Cl_2 and Cl_3 isomers. Such a pattern would be consistent with aerobic biodegradation of the lower chlorine isomers.

The plume of dissolved TCB has migrated significantly farther than the dissolved PCB despite the general similarity between the log K_{ow} values for TCB and the lower chlorine PCB isomers expected to be predominant in the groundwater. Aerobic biodegradation may be the mechanism responsible for the reduced PCB mobility. The relative isomer concentrations observed in the groundwater furthest downgradient from the DNAPL source may implicate aerobic biodegradation as a mechanism that may contribute to the attenuation of PCB migration in the groundwater at the Smithville site.

VII. CONCLUSIONS

Polychlorinated biphenyl Aroclors are complex mixtures of a large number of PCB congeners that vary widely, but systematically, in properties such as solubility in water, potential for sorption, and potential for biodegradation. Aroclors have characteristic isomer concentrations. In the subsurface, processes such as dissolution of the Aroclors, sorption on geologic media, aerobic biodegradation, and anaerobic biodegradation can result in distinctive changes in the isomer concentrations of the PCBs in groundwater.

Dissolution of Aroclors will result in higher relative concentrations of lower chlorine isomers of the PCB in the groundwater and lower relative concentrations of the lower chlorine isomers in the residual Aroclor. Sorption processes may further increase the relative concentrations of the lower chlorine isomers in the dissolved-phase PCBs due to the preferential sorption of the higher chlorine isomers. Aerobic and anaerobic biodegradation processes will affect the isomer patterns in different ways. Aerobic degradation will result in lower relative concentrations of the lower chlorine isomers due to preferential degradation of these compounds. Anaerobic degradation will result in higher relative concentrations of the lower chlorine isomers owing to dechlorination of higher chlorine isomers. As a result of sorption and biodegradation processes, PCBs should be relatively immobile in many groundwater environments. PCBs may be more mobile when dissolved in LNAPL or DNAPL or in media having low f_{oc} values. Examination of changes in PCB isomer or congener concentrations should aid in the evaluation of hydrogeochemical processes that affect the mobility of PCBs in groundwater.

An examination of PCB isomer concentrations at the PCB storage facility in Smithville, Ontario allowed a variety of interpretations that were not possible from examination of total PCB concentrations alone. Isomer concentrations of the DNAPL from storage tanks on the site and recovered from monitoring wells indicated that there has been no substantial mass depletion of PCB from the DNAPL in the subsurface due to dissolution. Isomer concentrations in the groundwaters from within and surrounding the DNAPL zone indicate that the PCB present in these waters may exist as an emulsion or sorbed on colloids rather than in dissolved form. The isomer concentrations in the groundwater at the downgradient margin of the PCB plume suggests that aerobic biodegradation of PCB may be the process contributing to reduce PCB mobility in groundwater at the site.

REFERENCES

1. Alford-Stevens, A.L. Analyzing PCBs. *Environ. Sci. Technol., 20*:1194, 1986.
2. Webb, R.G., and McCall, A.C. Quantitative PCB standards for electron capture gas chromatography. *J. Chromatogr. Sci., 11*:366, 1973.
3. Feenstra, S., and Cherry, J.A. Subsurface contamination by dense non-aqueous phase liquid (DNAPL) chemicals. Proc. Int. Groundwater Symposium, International Association of Hydrogeologists, May 1–4, 1988, Halifax, Nova Scotia, 1988, pp. 61–69.
4. Shiu, W.Y., and Mackay, D. A critical review of aqueous solubilities, vapor pressures, Henry's law constants, and octanol–water partition coefficients of the polychlorinated biphenyls. *J. Phys. Chem. Ref. Data, 15*:912, 1986.
5. Shiu, W.Y., Maijanen, A., Ng, A.L.Y., and Mackay, D. Preparation of aqueous solutions of sparingly soluble organic substances: II. Multicomponent systems—hydrocarbon mixtures and petroleum products. *Environ. Toxicol. Chem., 7*: 125, 1988.
6. Banerjee, S. Solubility of organic mixtures in water. *Environ. Sci. Technol., 18*: 587, 1984.
7. Leinonen, P.J., and Mackay, D. The multicomponent solubility of hydrocarbons in water. *Can. J. Chem. Eng., 51*:230, 1973.
8. Chou, S.F.J., and Griffin, R.A. Solubility and soil mobility of polychlorinated biphenyls. In *PCBs and the Environment* (J.S. Waid, ed.), CRC Press, Boca Raton, FL, 1986, p. 101.
9. Karickhoff, S.W., Brown, D.S., and Scott, T.A. Sorption of hydrophobic pollutants in natural sediments. *Water Res., 13*:241, 1979.
10. Bear, J. *Hydraulics of Groundwater*, McGraw-Hill, New York, 1978, p. 569.
11. Sudicky, E.A. A natural gradient experiment on solute transport in a sandy aquifer: spatial variability of hydraulic conductivity and its role in the dispersion process. *Water Resour. Res., 22*:2069, 1986.

12. Curtis, G.P., Roberts, P.V., and Reinhard, M. A natural gradient experiment in solute transport in a sandy aquifer. 4. Sorption of organic solutes and its influence on mobility. *Water Resour. Res., 22*:2059, 1986.

13. Ahmed, M., and Focht, D.D. Degradation of polychlorinated biphenyls by two species of *Achromobacter. Can. J. Microbiol., 19*:47, 1973.

14. Furukawa, K., Tomizuka, N., and Kamibayashi, A. Effect of chlorine substitution on the bacterial metabolism of various polychlorinated biphenyls. *Appl. Environ. Microbiol., 38*:301, 1979.

15. Clark, R.R., Chian, E.S.K., and Griffin, R.A. Degradation of polychlorinated biphenyls in mixed microbial cultures. *Appl. Environ. Microbiol., 37*:680, 1979.

16. Brown, J.F., Jr., Wagner, R.E., Eng, H., Bedard, D.L., Brennen, M.J., Carnahan, J.C., and May, R.J. Environmental dechlorination of PCBs. *Environ. Toxicol. Chem., 6*:579, 1984.

17. Brown, J.F., Jr., Bedard, D.L., Brennan, M.J., Carnahan, J.C., Eng, H., and Wagner, R.E. Polychlorinated biphenyl dechlorination in aquatic sediments. *Science, 236*:709, 1987.

18. Quensen, J.F., Tiedje, J.M., and Boyd, S.A. Reductive dechlorination of polychlorinated biphenyls by anaerobic microorganisms from sediments. *Science, 242*:752, 1988.

19. Foster, S.S.D. The Chalk groundwater tritium anomaly—a possible explanation. *J. Hydrol., 25*:159, 1975.

20. Young, C.P., Oakes, D.B., and Wilkinson, W.B. Prediction of future nitrate concentrations in ground water. *Ground Water, 14*:426, 1976.

21. Day, M. Analysis of movement and hydrochemistry of groundwater in fractured clay and till deposits of the Winnipeg area. M.Sc. Thesis, Dept. of Earth Sciences, Univ. Waterloo, Waterloo, Ontario, 1977, unpublished.

22. Wilke, C.R., and Chang, P. *Am. Inst. Chem. Eng. J., 1*:264, 1955.

23. Bonoli, L., and Witherspoon, P.A. Diffusion of aromatic and cycloparaffin compounds in water for 20°C to 60°C. *J. Phys. Chem., 72*:2532, 1968.

24. Tang, D.H., Frind, E.O., and Sudicky, E.A. Contaminant transport in fractured porous media: analytical solution for a single fracture. *Water Resour. Res., 17*:555, 1981.

25. Hoek, E., and Bray, J.W. *Rock Slope Engineering*, Inst. Mining and Metallurgy, Imperial College, London, UK, 1977.

26. Feenstra, S., Cherry, J.A., Sudicky, E.A., and Haq, Z. Matrix diffusion effects on contaminant migration from an injection well in fractured sandstone. *Ground Water, 22*:307, 1984.

27. McIlewain, T.A., Jackman, D.W., and Beukema, P. Characterization and remedial assessment of DNAPL PCB oil in fractured bedrock: a case study of the Smithville, Ontario site. Proc. Eastern Regional Ground Water Conference, National Water Well Association, Kitchener, Ontario, 1989.

28. Miller, M.M., Wasik, S.P., Huang, G.L., Shiu, W.Y., and Mackay, D. Relationships between octanol–water partition coefficient and aqueous solubility. *Environ. Sci. Technol., 19*:522, 1985.

29. Schwille, F. *Dense Chlorinated Solvents in Porous and Fractured Media*: Model Experiments, Lewis, Chelsea, MI, 1988.

17

Estimating the Fate and Mobility of CFC-113 in Groundwater: Results from the Gloucester Landfill Project

Richard E. Jackson

Intera Inc., Austin, Texas

Suzanne Lesage

National Water Research Institute, Environment Canada, Burlington, Ontario, Canada

M. W. Priddle

University of New Brunswick, Fredericton, New Brunswick, Canada

CFC-113 (1,1,2-trichloro-1,2,2-trifluoroethane) is an organic chemical that is vital to the electronics industry as a solvent for cleaning circuit boards and semiconductors. It is also widely known as Freon-113, the DuPont product name. The leakage of CFC-113 and other solvents from underground storage tanks has caused significant groundwater contamination in Silicon Valley, California. Recent research at the Gloucester landfill in Canada has shown that it degrades to a much more toxic product, CFC-1113 (1-chloro-1,2,2-trifluoroethene), an analogue of vinyl chloride. The transport rate of CFC-113 in groundwater appears to be similar to that of 1,1,1-trichloroethane, both having retardation factors of approximately 10 in a sand and gravel aquifer with an organic carbon content of 0.06%. Sorption to mineral surfaces also appears to be important in retarding the transport of CFC-113 in anoxic groundwaters. CFC-1113 is likely to be more mobile on account of its less halogenated state. The first-order half-life transformation rate of CFC-113 in anoxic landfill leachate was 5.3 days; that of CFC-1113 was 42 days.

I. INTRODUCTION

Approximately 160–170 million lb/yr of CFC-113 is consumed in the United States, in particular by the electronics industry, where it is used in the process of dry plasma etching during the manufacture of semiconductor chips [1] and is widely employed in the vapor degreasing and cold immersion cleaning of microelectronic components and circuit boards. CFC-113 is also frequently used as an electronics product cleaner and as a "hand wipe solvent" in the aerospace industry for surface cleaning prior to and during adhesive bonding and riveting of parts [2]. Furthermore, it is used in recent model automobiles to clean the fuel emission control chip.

CFC-113, either alone or as an azeotropic mixture with dichloromethane, acetone, or other solvents, is also used as a refrigerant in commercial and industrial air conditioning; as a solvent for the removal of organic and inorganic materials from plastics, metals, and films; and as a reagent in analytical organic chemistry. However, because of the implication of chlorine from CFCs in the destruction of atmospheric ozone [3], CFC-113 production is expected to be phased out by the year 2000. So far, no substitute for it has been developed [4].

The widespread industrial use of CFC-113 is a result of its many advantagenous properties, including nonflammability, low toxicity, excellent thermal and chemical stability, and high density coupled with low boiling point, low viscosity, and low surface tension. These physical and chemical properties and others of environmental importance in groundwater contamination are listed in Table 1.

The leakage of CFC-113 from underground storage tanks and its improper disposal have resulted in several documented cases of its contamination of groundwater (e.g., [12–16]), although many others are known to have occurred, particularly at electronics plants throughout the United States, but have not been made public.

II. CFC-113 IN GROUNDWATER

It is generally the case that solvents such as CFC-113, 1,1,1-trichloroethane (TCA), and trichloroethene (TCE) enter groundwater as denser-than-water, nonaqueous phase liquids, or DNAPLs, that is, in the form in which they are sold and used, rather than as aqueous solutions. Because of its density in both the liquid (1.565) and vapor (4.46) phases, CFC-113 rapidly sinks through soils, penetrates the water table, and pools on low permeability units known as *aquitards* or *aquicludes*. The residual solvent left behind the advancing front of solvent is retained by capillary forces as droplets or "ganglia" within the pores of the aquifer. The process by which halogenated DNAPLs

Table 1 Physical, Chemical, and Environmental Properties of Three Common Solvents Used in Electronics and Aerospace Manufacturing

	CFC-113	111-TCA	TCE	Ref.
Formula	$CCl_2 - CClF_2$	$CCl_3 - CH_3$	$CCl_2 = CClH$	5,8
Mol. wt.	187.38	133.4	131.5	5,8
BP (°C)	47.57	74	87	5,8
Density, liq. (g/cm³)	1.565	1.35	1.46	5,8
VP (mmHg) @ 25°C	482	100	58	5,8
RVD	4.46	1.47	1.27	5,8
Henry's law C_f (atm·m³/mol)	0.76	0.013	0.0071	5,8
Viscosity liq. (cP)	0.68 @ 25°C	0.84 @ 20°C	0.57 @ 20°C	5,8
Solubility (mg/L) @ 10°C	136.1	1707.	1061	7
Solubility (mg/L) @ 20°C	157.5	1554.	1083	7
IFT, @ 20°C (dyn/cm)	47.2	40.4	41.7	6,8
Log K_{ow}	3.16	2.49	2.42	9
Log K_{oc}	2.57	2.19	2.09	10
LD_{50} (g/kg)	43	10.3	4.9	11

RVD, relative vapor density (see Schwille [5], p. 131); IFT, interfacial tension (after method of Donahue and Bartell [6, pp. 21-2-21-7]) with solubilities from Horvath [7]. 1,1,1-TCA = 1,1,1-trichloroethane; TCE = trichloroethene.

migrate in the subsurface and are subsequently detected as dissolved components in monitoring wells is shown in Fig. 1.

The concentrations of halogenated solvents such as CFC-113, TCA, and TCE measured in monitoring wells are almost invariably less than their aqueous solubility. This is partly a result of hydrogeological phenomena and partly that of the multicomponent solubility of DNAPLs. The concentration of an organic solute in groundwater in equilibrium with a DNAPL is given by the analogy with Raoult's law [18]:

$$C_{GW} = a_i X_i C_{SOL} \qquad (1)$$

This indicates that the equilibrium solubility of an ideal organic solute in groundwater is controlled by the ratio of the activity coefficients, a_i, of component i in the DNAPL phase to that in the groundwater, the mole fraction, X_i, and its aqueous solubility, C_{SOL}. However, because the monitoring well used to collect samples of groundwater is probably somewhat removed from the DNAPL source, the concentration measured at the monitoring well will be substantially less than even that predicted by Eq. (1) as a result of dilution,

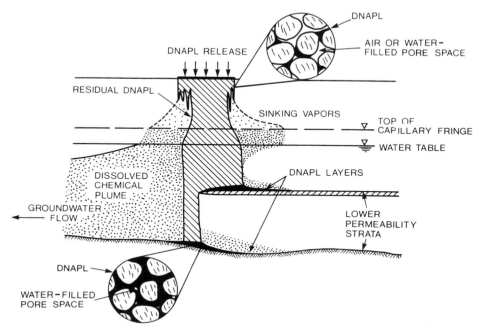

Figure 1 Behavior of dense, nonaqueous phase liquids or DNAPLs following a release. The DNAPL migrates not only as a dense, sinking liquid but also as a dense, sinking vapor. Dissolution of both vapor and liquid can cause groundwater contamination. (After Feenstra and Cherry [34].)

sorption, biotransformation, and dispersion between the DNAPL source and the well. This has been observed to be the case at several hazardous waste sites; thus it is likely that a component of a mixed DNAPL will be present in groundwater at a nearby monitoring well in concentrations much less than its aqueous solubility, for example, 0.1–10% of its solubility.

For example, concentrations of CFC-113 measured in groundwaters beneath the Gloucester landfill, near Ottawa, Ontario, Canada [15] varied from <5 μg/L to 2725 μg/L, or 2% of CFC-113's aqueous solubility, whereas a well in Silicon Valley, California had a mean concentration of 19 mg/L during 1986, or 12% of the compound's aqueous solubility. The farther the monitoring well is from the DNAPL source, the lower the measured concentration is likely to be.

Regulatory compliance levels for CFC-113 vary widely. The recommended drinking water action level in California is 1200 μg/L or ppb. In New York

State, the maximum contaminant level for all halogenated alkanes, for example, CFC-113, in drinking water supplies is 5 μg/L [19]. While New York is suspected to have CFC-113 DNAPLs contaminating groundwaters beneath only a few electronic manufacturing plants, hydrogeologists knowledgeable of the situation in Silicon Valley report that CFC-113 is "ubiquitous" or "pretty widespread." While not legally enforceable, a maximum permissible concentration of 0.19 μg/L (190 ppt) for all CFCs in drinking water has been proposed by Dr. W. George of Tulane University, based upon a toxicological assessment of organic solvents in drinking water (see Table 1 in Ref. 20).

The mobility of CFC-113 in groundwater is of particular interest. The plume of CFC-113 and TCA presumed to originate from an IBM facility in south San Jose, California reported by Esau and Chesterman [14] is thought to have migrated 4 miles at velocities of up to 30 ft/day (9 m/day) and to have perhaps contaminated the Tully well field in San Jose [21]. However, Roy and Griffin [10] concluded that CFC-113 had only "medium mobility" on the basis of a computed log K_{oc} = 2.57, implying an octanol–water partition coefficient, log K_{ow} = 2.73. Thus CFC-113 would fall in the same category as such other medium mobility solvents as tetrachloroethene (PCE), TCE, and TCA. Perhaps the San Jose example is an extreme case due to the coarseness of the gravels in the Edenville Gap area of Silicon Valley, and CFC-113 is indeed of similar mobility to the common solvents. However, this is a point requiring confirmation.

The question also arises as to whether CFC-113 will undergo biotransformation in anoxic groundwaters and produce more mobile and perhaps more toxic metabolites in the same way that PCE, TCE, and TCA can biotransform to produce chloroethene or vinyl chloride. An indication of the likelihood of this occurring is given by the behavior of another fully halogenated ethane, hexachloroethane (HCE), during the Stanford/Waterloo field experiment. In the course of this experiment, HCE was injected with other organic compounds into a shallow sand aquifer at Camp Borden, Ontario. HCE, having a solubility of 50 mg/L and log K_{ow} = 3.64, rapidly biodegraded to PCE with a half-life of 40 days in 10°C groundwater [22]. This occurrence strongly suggests that CFC-113 might also biotransform rapidly. However, CFC-113 has persisted in wells in Silicon Valley for many years, suggesting either slow degradation rates or DNAPL dissolution rates that exceed the degradation rate.

Thus, the mobility and fate in groundwater of one of the most vital and common solvents used by the electronics industry and an important groundwater contaminant are uncertain. Whereas field data from California suggest very high mobility and long-term persistence, physical chemical parameters suggest moderate mobility, and a field experiment suggests that CFC-113 might biotransform readily. Fortunately, a long-term study of the organically contaminated groundwaters beneath the Gloucester landfill near Ottawa, Ontario offers insight into these matters.

III. CFC-113 IN ANOXIC GROUNDWATERS AT THE GLOUCESTER LANDFILL

A. Site Description

Between 1969 and 1980, the Government of Canada disposed of hazardous wastes, principally organic chemicals from its laboratories in Ottawa, at the nearby Gloucester landfill. The disposal operations took place within a "special waste compound," hereafter referred to as the SWC. These operations involved the excavation of trenches (12 m × 3 m × 3 m) into which bottles of liquid chemical wastes were placed and subsequently combusted by detonation of explosive charges set within them. The wastes consisted of nonchlorinated solvents (56.4 m³), unspecified wood preservatives (30 m³), chlorinated solvents (8.6 m³), and smaller amounts of inorganic and other wastes.

Detailed soil coring and hydrogeological testing have resulted in the identification of the five hydrostratigraphic units shown in Fig. 2. Unit A, the

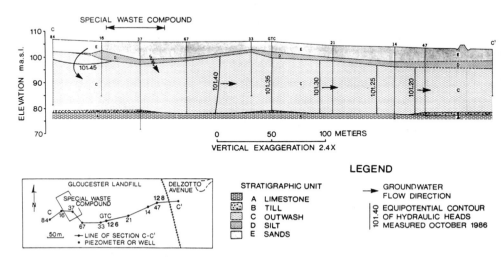

Figure 2 Hydraulic head contours (in meters above mean sea level) in the Gloucester outwash aquifer system. The cross section is defined in the inset. Squiggly arrow beneath the SWC indicates the potential for leakage through confining unit D. Note vertical exaggeration. Refer to text for description of hydrostratigraphy.

fractured limestone bedrock, is encountered at depths between 25 and 30 m and is partly covered by a till (unit B) of relatively low hydraulic conductivity. Unit C, overlying the till and/or bedrock, is a thick (up to 25 m) sequence of silts, sands, gravel, and boulders known as glacial outwash and forms a semi-confined aquifer beneath the landfill. This outwash is overlain by a discontinuous layer of silt, unit D. The surficial aquifer, unit E, may be up to 10 m thick and is composed of sands and gravels.

The outwash aquifer, unit C, is composed of feldspar (>50%), quartz (20%), and minor amounts of mica, calcite, dolomite, and hornblende. The organic carbon content of the aquifer sediments, f_{oc}, averages 0.0006. The aquifer has a mean hydraulic conductivity of 10^{-4} m/sec, a porosity of 0.30–0.35, and an average linear groundwater flow velocity of 0.05 m/day. The average groundwater temperature is 10°C [23,24].

The groundwater flow pattern beneath the site is shown in Fig. 2. Downward groundwater flow occurs from the surficial aquifer, unit E, to the confined outwash aquifer, unit C. Some seepage appears to occur through permeable windows in the silty confining layer, unit D. Once solutes enter the outwash aquifer, the vertical equipotential lines dictate that they are transported horizontally to the east. Monitoring of the observation well network established along the cross section shown in Fig. 2 has verified that contaminant transport is indeed in this direction.

B. Migration and Fate

Jackson and Patterson [5] reported that the confined aquifer appeared to be uncontaminated by "free product," that is, dense nonaqueous-phase liquids (DNAPLs) such as PCE, TCE, or CFC-113. However, samples from some wells that were analyzed by gas chromatography with a mass-selective detector (GC-MSD) that was equipped with a cryogenic valve to analyze low boiling point compounds such as the chlorofluorocarbons identified the presence of CFC-113 and its possible biotransformation products. These are shown as the early time peaks of the total ion chromatogram in Fig. 3. At multilevel sampling point 67M-11, some 20 m south of the SWC, the concentration of CFC-113 was approximately 2% of its aqueous solubility at 10°C, a percentage indicative of the probable presence of DNAPLs.

This analysis suggests that, following disposals of CFC-113, the solvent may have migrated as viscous fingers (see figures in Ref. 5) through the water table aquifer (unit E) and, by some pathway, through or around the aquitard (unit D) and into the outwash aquifer (unit C). A thin, clayey gravel layer was identified in drilling logs at an elevation immediately below sampling point 67M-11. This indicates that a small pool or extended ganglia of CFC-113 may well exist at this location and be sitting on a low permeability stratum quite common to this type of glaciofluvial sediment. Furthermore, this monitor

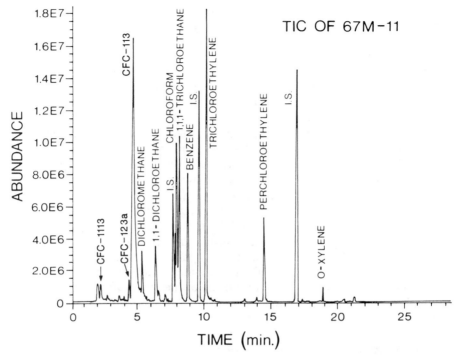

Figure 3   Total ion chromatogram of a groundwater sample from multilevel sampler 67M-11 showing early time chlorofluorocarbon peaks identified as CFC-1113, etc. The analytical column was a 30-m DB-624 directly interfaced into the source of a HP 5970 mass-selective detector. The GC was cooled to $-5°C$ with liquid CO_2 to allow analysis of the CFCs.

also contains the maximum concentrations of other volatile organic chemicals—1,2-dichloroethane, chlorobenzene, TCA, TCE, and PCE—observed at the site, as shown in Table 2. These other solvents were either part of the same disposal or were from some previous disposal and were retained as ganglia in the upper aquifer (unit E) and then mobilized by the later disposal. The most probable date for such a disposal or mobilization was May 1978, when 1 ton of chlorinated solvents was disposed of at the SWC. Thus, the data in Table 2, from samples collected in May 1988, show conditions 10 and 11 years after the probable disposal date.

Nowhere else in the outwash aquifer do CFC-113 concentrations exceed 150 μg/L; therefore, there is further reason to suspect that the area around

Table 2 Concentrations of Six Volatile Organic Chemicals (μg/L) in Multilevel Sampling Point 67M-11 and in Monitoring Well 125P, Sampled at the Gloucester Landfill, near Ottawa, Canada

Site	1,2-DCA	TCA	TCE	PCE	CFC-113	CLB
1988						
67M-11	58	520	583	105	2725	134
125P	14	193	505	60	200	68
1989						
67M-11	67	843	757	320	903	194
125P	9	231	242	56	234	69

1,2-DCA = 1,2-dichloroethane; CLB = chlorobenzene; see text for other abbreviations.
Note: Field measurements at 67M-11 in May 1988 indicated that pH = 7.1, Eh = 170 mV, and O_2 < 0.3 mg/L.

67M is a source of DNAPL ganglia that are dissolving to produce the groundwater pollution shown in Table 2. Concentrations of CFC-113 are much lower in multilevel monitors (see Fig. 4) that are downgradient of 67M, for example, 120 μg/L in 97M-15 and 115 μg/L in 97M-17, both only 20 m away. However, CFC-113 was not observed in 1988 in multilevel 54M, 70 m downgradient.

These findings prompted an assessment of the mobility of CFC-113 in a laboratory column made up of Gloucester aquifer sediments. The procedure (see Ref. 25) involved the use of uncontaminated Gloucester aquifer groundwater that was mixed with CFC-113 and TCA from a syringe pump and then pumped through the column as shown in Fig. 5. Eluant fractions were collected in Tedlar minibags to prevent volatilization of the CFC-113 and TCA. The measured retardation factors were 12 for CFC-113 and 7 for TCA. On the assumption that all sorption and therefore retardation of dissolved organics in groundwater is due to solid-phase organic carbon within the aquifer, such a value for CFC-113 implies a fractional organic carbon content (f_{oc}) of 0.005 rather than the mean measured value of 0.0006. This discrepancy suggests that mineral sorbents as well as organic substrates are effective in retarding CFC-113, a conclusion first reached for six other volatiles used during the Stanford/Waterloo field test at Borden [26].

If no column test had been conducted to estimate the retardation factor, it would have been necessary to estimate it on the basis of empirical relationships involving the f_{oc} value and the octanol–water partition coefficient (log K_{ow}). A frequently used expression is that of Schwartzenbach and Westall [27]:

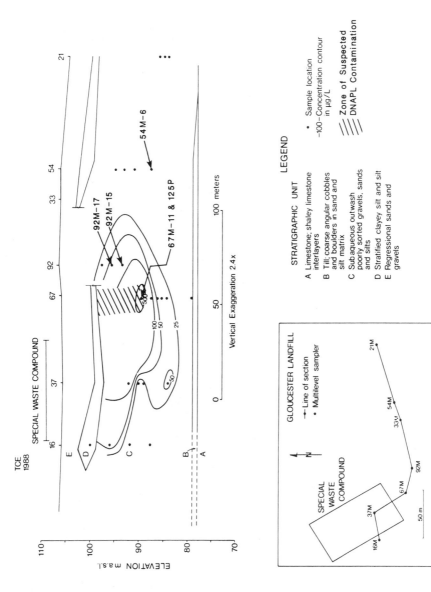

Figure 4 Solvent plume in the outwash aquifer, May 1988, as indicated by concentrations of TCE in $\mu g/L$. The CFC-113 center of mass appears to be in approximately the same location as the TCE center of mass. Multilevel monitors (e.g., 67M-11) and monitoring wells (e.g., 125P) are identified.

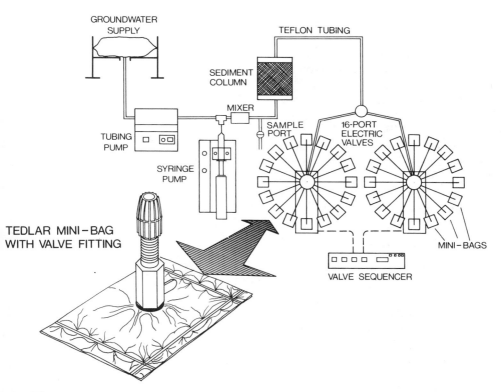

Figure 5 Schematic of laboratory column test apparatus for the measurement of the retardation factors of volatile organic contaminants (from Priddle and Jackson [25]). Eluants are collected in Tedlar minibags that were evacuated prior to the column test.

$$K_p = 0.72 \log K_{ow} + \log f_{oc} + 0.49 \qquad (2)$$

where K_p and K_{ow} are the contaminant partition and octanol–water partition coefficients of the organic compound in question. The retardation factor can then be obtained [28]:

$$R_f = 1 + (\rho_b/n) K_p \qquad (3)$$

where ρ_b is the bulk density of the aquifer material and n its porosity.

Using Eqs. (2) and (3), a retardation factor of 2.7 was estimated for CFC-113 given $f_{oc} = 0.0006$ and $\log K_{ow} = 3.16$. This fourfold difference in retardation factors (i.e., 2.7 vs. 12) yields a tremendous difference in migration distance over a 10-year period. Domenico and Robbins' [29] continuous source analytical solution to the solute transport equation [28] was used to estimate the extent of advance of the 5-μg CFC-113/L (5 ppb) front over 10 years from the source region around 67M-11. For an estimated retardation factor of 2.7, the 5-ppb front is found at approximately 190 m from the source. Such a front would have resulted in CFC-113 being present and readily measurable in multilevel samplers at 54M and 21M (see Fig. 4). If the measured retardation factor of 12 is used in the simulation, the 5-ppb front would have advanced less than 70 m from 67M. The absence of CFC-113 in multi-level monitor 54M, 70 m away from 67M, and the presence of CFC-113 in monitor 92M tend to confirm the higher value of the retardation factor, $R_f = 12$. Given the apparent importance of sorption of CFC-113 on mineral surfaces, the calculation of a K_{oc} value seems meaningless.

C. Biotransformation of CFC-113

The presence of CFC-113 and several other two-carbon chlorofluorocarbons in the chromatograms of the May 1988 samples (see Fig. 3) suggested the biotransformation of CFC-113 in these groundwaters. Lesage et al. [15] deduced that the biotransformation of CFC-113 occurs by a sequential process, shown in Fig. 6, in which CFC-113 is transformed first to 1,2-dichloro-1,2,2-trifluoroethane, or CFC-123a, by reductive dechlorination, and then to CFC-1113 by elimination. Subsequent experiments at Johns Hopkins University and at the Oklahoma Medical Research Foundation have confirmed the degradation of CFC-113 to these metabolites under sulfate-reducing and methanogenic conditions. Lesage et al. also reported that CFC-113 appeared to be the most persistent of all solvents disposed of at the Gloucester site, including PCE, TCA, and TCE, although it was not possible to estimate its half-life in anoxic groundwater.

However, this apparent persistence may merely reflect the fact that the rate of dissolution of the CFC-113 DNAPL may be significantly faster than the rate of transformation of CFC-113 to CFC-123a. A field study on the fate of organic contaminants in landfill leachate [32] has shown that CFC-113 is transformed to CFC-123a and CFC-1113 in 200-L storage tanks at the landfill. Recent microcosm experiments under anoxic conditions indicate a half-life of approximately 5.3 days at 21°C [33]. Furthermore, it was interesting to note that, under methanogenic conditions, no CFC-1113 was formed from CFC-123a, but further reductive dechlorination to CFC-133 and CFC-133b occurred instead (Fig. 7). It is thus possible that the elimination reaction producing CFC-1113 is affected by a different consortium of bacteria or occurs abiotically, as it was also formed in phosphate buffer.

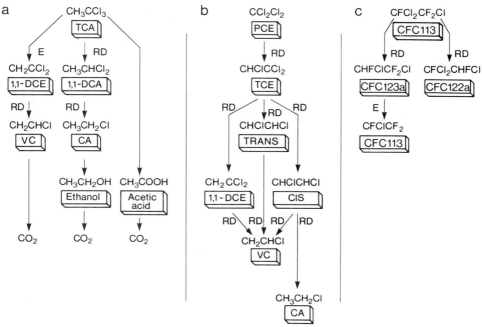

Figure 6 Proposed anoxic biotransformation pathways for three common solvents: (a) TCA, (b) PCE, and (c) CFC-113. The dehalogenation reactions shown are of two kinds: reductive dehalogenation (RD) in which a hydrogen atom substitutes for a halogen and elimination (E) in which the molecule has a net loss of H and Cl atoms, thereby forming an ethene. (Reproduced from Lesage et al. [15]. The pathways for TCA and PCE are from Refs. 30 and 31, respectively.)

Although CFC-113 may not be particularly toxic, there is real concern with the toxicity of CFC-1113, an analog of vinyl chloride, which Sax [17] classifies as highly toxic (268 mg/kg oral). Being less halogenated than CFC-113, the transformation products CFC-123a and CFC-1113 are more mobile. CFC-123a, for example, is found in 54M-6 (see Fig. 4). Thus the presence of CFC-113 in anoxic groundwaters used for public water supply purposes raises concerns based upon the toxicity of its degradation products rather than those due to its own toxicity.

CFC-123a and CFC-1113 are both more stable than CFC-113 in landfill leachate with measured half-lives of approximately 6.3 and 42 days, respectively. CFC-1113 is, however, hydrolyzed rapidly (in less than a day) in sulfide-containing buffer [33].

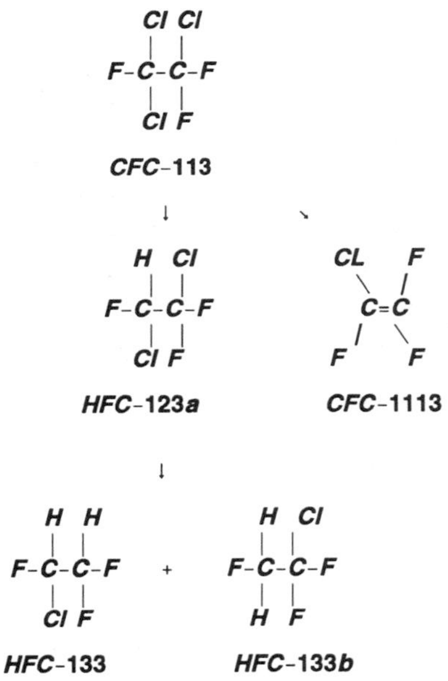

Figure 7 Observed degradation pathway under methanogenic conditions in microcosms containing landfill leachate or with reduced hematin in a cysteine/sodium sulfide buffer. (From Lesage et al. [33].)

IV. SUMMARY AND CONCLUSIONS

Recent field and laboratory studies of the transport and fate of CFC-113 in anoxic groundwaters beneath the Gloucester landfill near Ottawa, Canada are summarized. It has been shown that the relatively nontoxic CFC-113 appears to be persistent in groundwater but biotransforms to a vinyl chloride analogue of considerable toxicity, CFC-1113. Furthermore, CFC-113 is moderately mobile in the low organic carbon content sands and gravels (mean $f_{oc} = 0.0006$) having a retardation factor of 12, similar to that of TCA ($R_f = 7$). It appears that sorption of CFC-113 by mineral as well as organic sorbents is required to explain the high value of the retardation factor. The mobility of CFC-1113 is unknown but is likely to be greater than that for CFC-113 as CFC-1113 has fewer halogen atoms.

REFERENCES

1. LaDou, J. The not-so-clean business of making chips. *Technol. Rev., May/June*: 22–36, 1984.

2. Evanoff, S.P. Hazardous waste reduction in the aerospace industry. *Chem. Eng. Progr., April*:51–61, 1990.
3. McFarland, M. Chlorofluorocarbons and ozone. *Environ. Sci. Technol., 23*: 1203, 1989.
4. Derra, S. CFCs: no easy solutions. *R&D Mag., May*:56–66, 1990.
5. Schwille, F. *Dense Chlorinated Solvents in Porous and Fractured Media*: *Model Experiments* (translated from the German by J.F. Pankow), Lewis, Chelsea, MI, 1988.
6. Lyman, W.J., Reehl, W.F., and Rosenblatt, D.H. *Handbook of Chemical Property Estimation Methods*, American Chemical Society, Washington, DC, 1990.
7. Horvath, A.L. *Halogenated Hydrocarbons*, Marcel Dekker, New York, 1982.
8. DuPont. *Freon Fluorocarbons. Properties and Applications*, E.I. Du Pont de Nemours & Co. (Inc.), Freon Products Division, Wilmington, DE, 1985.
9. Howard, P.H. *Handbook of Environmental Fate and Exposure Data for Organic Chemicals*, Lewis, Chelsea, MI, 1990.
10. Roy, W.R., and Griffin, R.A. Mobility of organic solvents in water-saturated soil materials. *Environ. Geol. Water Sci., 7*:241, 1985.
11. Hughes, T.H., Brooks, K.E., Norris, B.W., Wilson, B.M., and Roche, B.N. A descriptive survey of selected organic solvents. Environmental Institute for Waste Management Studies, Univ. Alabama, Tuscaloosa, AL, Open File Rep. No. 1, 1985.
12. HLA (Harding Lawson Associates). Remedial investigation report, RI/FS Middlefield–Ellis–Whisman Area, Mountain View, CA, 1987.
13. Lewcock, T. Santa Clara Valley (Silicon Valley), California, case study. In *Planning for Groundwater Protection* (G.W. Page, ed.), Academic, Orlando, FL, 1987, pp. 299–324.
14. Esau, R.R., and Chesterman, D.J. Ground water contamination issues in Santa Clara County, California: a perspective. In *Hazardous Waste Site Management*: *Water Quality Issues*, National Academy Press, Washington, DC, 1988, p. 129.
15. Lesage, S., Jackson, R.E., Priddle, M.W., and Riemann, P.G. Occurrence and fate of organic solvent residues in anoxic groundwater at the Gloucester landfill, Canada. *Environ. Sci. Technol., 24*:559, 1990.
16. Semprini, L., Hopkins, G.D., Roberts, P.V., and McCarty, P.L. In-situ biotransformation of carbon tetrachloride, 1,1,1-trichloroethane, Freon-11 and Freon-113 under anaerobic conditions. *EOS, 71*(43):1324, 1990.
17. Sax, N.I. *Dangerous Properties of Industrial Materials*, 6th ed., Van Nostrand Reinhold, New York, 1984.
18. Stumm, W., and Morgan, J.J. *Aquatic Chemistry*: *An Introduction Emphasizing Chemical Equilibria in Natural Waters*, Wiley, New York, 1981.
19. New York State Department of Health. *Organic Chemical Standards, Summary of Code Revision*. Released by M.E. Burke, Director, Bureau of Public Water Supply Protection, Nov. 18, 1988; effective Jan. 9, 1989.
20. Roy, W.R., Griffin, R.A., Mitchell, J.K., and Mitchell, R.A. Limitations and feasibility of the land disposal of organic solvent-contaminated wastes. *Environ. Geol. Water Sci., 13*:225, 1989.

21. Iwamura, T.I. Santa Clara Valley Water District, San Jose, CA, personal communication, 1989.
22. Criddle, C.S., McCarty, P.L., Elliott, M.C., and Barker, J.F. Reduction of hexachloroethane to tetrachloroethylene in groundwater. *J. Contam. Hydrol., 1*:133, 1986.
23. Jackson, R.E., Patterson, R.J., Graham, B.W., Bahr, J., Belanger, D., Lockwood, J., and Priddle, M.W. *Contaminant Hydrogeology of Toxic Organic Chemicals at a Disposal Site, Gloucester, Ontario. 1. Chemical Concepts and Site Assessment.* IWD Sci. Ser. No. 141, Environment Canada, Ottawa, Canada, 1985.
24. Jackson, R.E., and Patterson, R.J. A remedial investigation of an organically polluted outwash aquifer. *Ground Water Monit. Rev., 9*:119, 1989.
25. Priddle, M.W., and Jackson, R.E. Laboratory column measurement of VOC retardation factors and comparison with field values. *Ground Water, March-April*, 1991, pp. 260–266.
26. Curtiss, G.P., Roberts, P.V., and Reinhard, M.V. A natural gradient experiment on solute transport in a sand aquifer. 4. Sorption of organic solutes and its influence on mobility. *Water Resour. Res., 22*:2059, 1986.
27. Schwarzenbach, R.P., and Westall, J. Transport of non-polar organic compounds from surface water to groundwater. Laboratory sorption studies. *Environ. Sci. Technol., 15*:1360, 1981.
28. Freeze, R.A., and Cherry, J.A. *Groundwater*, Prentice-Hall, Englewood Cliffs, NJ, 1979.
29. Domenico, P.A., and Robbins, G.A. A new method of contaminant plume analysis. *Ground Water, 23*:476, 1985.
30. Vogel, T.M., Criddle, C.S., and McCarty, P.L. Transformations of halogenated aliphatic compounds. *Environ. Sci. Technol., 21*:722, 1987.
31. Barrio-Lage, G., Parsons, F.Z., Nassar, R.S., and Lorrenzo, P.A. Sequential dehalogenation of chlorinated ethenes. *Environ. Sci. Technol., 20*:96, 1986.
32. Lesage, S., Riemann, P.G., and McBride, R.A. Degradation of organic solvents in landfill leachate. In *Proceedings of the Ontario Ministry of the Environment Technology Transfer Conference*, Toronto, Ontario, Canada, Vol. 2, 1989, pp. 88–97.
33. Lesage, S., Brown, S., and Hosler, K.R. Degradation of CFC-113 using reduced hematin and methanogenic landfill leachate. Presented at In-Situ and On-Site Bioreclamation Symposium, Mar. 19–21, 1991, San Diego, CA.
34. Feenstra, S., and Cherry, J.A. Subsurface contamination by dense, non-aqueous liquid (DNAPL) chemicals. Proc. Int. Groundwater Symposium, Int. Assoc. of Hydrogeologists, Halifax, Nova Scotia, Canada, 1988.

Contributor Biographies

Mary Jo Baedecker is a Research Chemist with the Water Resources Division of the U.S. Geological Survey in Reston, Virginia. She has a B.A. in chemistry from Vanderbilt University, an M.S. in organic chemistry from the University of Kentucky, and a Ph.D. in geochemistry from the George Washington University. She has been with the U.S. Geological Survey for 16 years where she has worked on the degradation of organic material in recent sediments and the fate of organic compounds in groundwater. Previously, she was at the Department of Geophysics and Planetary Geophysics, University of California, Los Angeles, Los Angeles, California where she worked on the organic geochemistry of marine sediments. Recently she worked on several field sites in the Toxic Substances Hydrology Program of the U.S. Geological Survey to investigate the effect of contaminants on the geochemistry of aquifers.

Chris Barber is a Senior Principal Research Scientist for the CSIRO, Division of Water Resources in Perth, Australia. He is responsible for coordinating an extensive research program in contamination and salinity of groundwater and surface waters. He completed his doctoral and post-doctoral studies in geochemistry in England. His particular skills are in the assessment of the impact and behavior of pollutants in groundwater and surface waters. Dr. Barber joined the U.K. Water Research Centre in 1973 to investigate the effects and behavior of hazardous wastes in landfills. This, and other research, sponsored by the Department of the Environment, formed the basis for U.K. legislation over the disposal of hazardous wastes on land. He was also involved in the treatment of landfill leachates and groundwater monitoring. Chris joined the CSIRO, Division of Water Resources in 1983 as a Principal Research Scientist. His responsibilities have included improving techniques for monitoring and assessing groundwater contamination and pollution in urban, industrial, mining, and agricultural areas. His current research includes a study of the impact of urbanization on groundwater with particular emphasis on sources of organic contaminants. His project team is also working on techniques for regional assessment of aquifer vulnerability. Dr. Barber is highly familiar with the problems of sampling of groundwater and contaminated soils. His project team has developed soil-gas sampling techniques to monitor volatile chlorinated organics as indicators of soil and aquifer contamination.

Larry B. Barber II is a research geochemist with the National Research Program of the U.S. Geological Survey, Water Resources Division. Dr. Barber received his B.S. in geology from the University of Arkansas, and his M.S. and Ph.D. in geology from the University of Colorado. His research interests include the application of analytical chemistry to evaluate the biogeochemical fate of organic contaminants in natural waters, the effect of sediment geochemical properties on the sorption of organic compounds, and the influence of geochemical heterogeneity on the transport of contaminants in aquifers.

David John Briegel is a Senior Technical Officer for the Division of Water Resources, CSIRO in Perth, Australia. He is responsible for coordinating field work for several projects in the Contaminant Hydrology Program in Perth. His work involves field studies into pollutant transport in groundwater, industrial production from landfills, hydrocarbons from underground storage tanks, and consultancies on the use of disused landfill sites. David has a broad scientific background of more than 30 years. His primary expertise is in the original design and innovative adaption of ideas and techniques.

Terri L. Bulman has over ten years experience specializing in remediation of contaminated soil, contaminated industrial site assessment, and land application of wastes. She is the senior soil scientist and site remediation specialist with Campbell Group Limited in West Perth, Australia. She is certified by the State of Victoria as an Environmental Auditor for contaminated sites and provides assistance to industry, state governments, and the Australian and New Zealand Environmental Council for the development of soil criteria and remediation plans. Prior to residence in Australia, Mrs. Bulman was a Soil Scientist/Project Manager with Environment Canada where she was responsible for a research program assessing the fate of toxic organic chemicals in soil and land application of municipal and industrial wastes. She served as chairman of the Technical Working Group—Industrial Site Decommissioning which was responsible for development of Canadian national cleanup guidelines and the AERIS risk assessment model.

Geneviève Couture received the B.Sc. in chemistry in 1983 and the M.Sc. in physical chemistry in 1987. Her work has included research in soil decontamination and fungicide extraction from biomass. She is now conducting research projects in waste and biomass pyrolysis and in soil and water decontamination.

Isabelle M. Cozzarelli is a Hydrologist with the Water Resources Division of the U.S. Geological Survey in Reston, Virginia where she has been since 1985. She received a B.S. in geomechanics from the University of Rochester and an M.S. in environmental sciences from the University of Virginia, where she is currently working toward a Ph.D. Her present research interests are focused on the geochemical fate and effects of organic contaminants in subsurface environments.

Stanley Feenstra is a Research Associate and Ph.D. candidate in hydrogeology at the Waterloo Centre for Groundwater Research, University of Waterloo, Waterloo, Ontario, Canada. Since 1987 he has been involved in a major research program on the behavior of chlorinated solvents in groundwater. He received B.Sc. and M.Sc. degrees in earth sciences from the University of Waterloo in 1978 and 1980, respectively, and was designated a Certified Ground Water Professional by the Association of Ground Water Scientists and Engineers in 1989. Since 1980, he has been a groundwater consultant with Golder Associates in Mississauga, Ontario and Zenon Environmental in Burlington, Ontario, and is currently President of Applied Groundwater Research Ltd. in Mississauga, Ontario. Mr. Feenstra specializes in the hydrogeochemical evaluation of groundwater contamination at waste disposal facilities and chemical spill sites.

P. Gélinas studied geological engineering at the Université Laval, Quebec, Canada and graduated in 1968. After graduate studies in hydrogeology and geotechnical engineering, he obtained a Ph.D. in civil engineering at The University of Western Ontario, London, Canada in 1974. Since 1972, he has been involved in teaching and research in hydrogeology and engineering geology at University of Ottawa and from 1976 at Université Laval in Quebec City. Current research interests include aquifer decontamination and restoration, acid rock drainage, and regional hydrogeology in developing countries.

He was chairman of the Department of Geological Engineering at Laval for six years.

Robert D. Gibbons received his Ph.D. in statistics and psychometric theory from the University of Chicago in 1981. His first academic appointment was Assistant Professor of Biostatistics at The University of Illinois at Chicago in 1981, where he is currently an Associate Professor. His work in the environmental area was stimulated by the original US EPA RCRA regulation which utilized statistical methods for groundwater detection monitoring at hazardous waste disposal facilities. He has since published extensively on the topic and much of this work is utilized in the revised RCRA regulation. Dr. Gibbons is currently working on a book on statistical methods for the evaluation of groundwater quality. In addition to work in the environmental sciences, Dr. Gibbons has published extensively on statistical models for the behavioral and biological sciences. His statistical research is currently supported by the Office of Naval Research, National Institutes of Health, and the MacArthur Foundation.

Robert W. Gillham is a professor in the Department of Earth Sciences of the University of Waterloo, and is the Director of the Waterloo Centre for Groundwater Research. His major research concerns transport and transformation of contaminants in the vadose and groundwater zones, and physical processes in the vadose zone. Dr. Gillham has a particular interest in remediation technologies including vapor extraction, plume extraction by pumping, abiotic degradation of halogenated organic compounds, and biodegredation of aromatic compounds in surface soils.

Donald F. Goerlitz is a Research Chemist with the U.S. Geological Survey at Menlo Park, California. He received his B.S. and M.S. degrees at the University of Missouri, Columbia. His major was analytical chemistry with emphasis on chromatographic analysis. Mr. Goerlitz joined the Survey in 1962 to study organic acids occurring in surface and groundwater. In 1968 he was made chief of the organics project. The main thrust of the project shifted to analytical methods development, and applications to field studies. The incumbent wrote the first analytical manual for the determination of organics in water and sediments published by the Survey. The pesticide analytical procedures in the manual were developed and tested by the organics project. Further, the phenoxy acid herbicide methods were adopted as standards by the American Society for Testing and Materials and *Standard Methods*. Since 1975, Mr. Goerlitz has been studying the occurrence, behavior, and fate of organic contaminants in aquifers throughout the United States.

Janis Kaye Hosking is an Experimental Scientist for the CSIRO, Division of Water Resources in Perth, Australia. She is responsible for the management of the Chemistry Laboratory. Her particular skills are in the analytical chemistry of inorganic constituents. She joined CSIRO in 1965 and for the last 15 years has been involved with chemical analyses for projects related to nutrient studies, salinity and contamination of groundwater. Her responsibilities have included development and improvement of methods for highly saline waters.

Bernadette M. Hughes (now Bernadette Hughes Conant) is currently a hydrogeologist with Environmental Project Control, Inc. in Grafton, Massachusetts and is also involved in a chlorinated solvents research project with the Waterloo Centre for Groundwater Research. She specializes in the hydrogeological analysis of sites contaminated by organic chemicals and the assessment of innovative technologies for monitoring and remediation. She received a B.Sc. in earth sciences and an M.Sc. in hydrogeology from the University of Waterloo, Waterloo, Ontario.

Richard E. Jackson, a co-editor of this volume, was one of the first undergraduates in the hydrology program at the University of Arizona in the 1960s and one of the first graduate students in the hydrogeology program at the University of Waterloo in the 1970s. His Ph.D. thesis was on the geochemical processes controlling radionuclide migration in groundwater at the Chalk River Nuclear Laboratories. He joined the Canadian Department of the Environment in 1975 and studied the migration and fate of radionuclides, organic solvents, and pesticides in various hydrogeological environments across Canada. In 1986, he established the Groundwater Contamination Section at the National Water Research Institute of Canada at Burlington, Ontario. He left NWRI in 1989 to become Manager of Chemical Hydrogeology at Intera Inc. in Austin, Texas, where he is involved in nuclear waste disposal and DNAPL remediation projects. He is certified by the American Institute of Hydrology as a Professional Hydrologist (Ground Water).

Richard L. Johnson is an associate professor in the Department of Environmental Science and Engineering at the Oregon Graduate Institute in Beaverton, Oregon. Dr. Johnson received his Bachelor of Science degree in chemistry from the University of Washington, a Master of Science degree in environmental science from the Oregon Graduate Institute, and a Doctor of Philosophy degree in environmental science with a specialty in organic contaminant transport in 1984 from the Oregon Graduate Institute. He was a post-doctoral fellow at the Institute for Groundwater Research at the University of Waterloo from 1984–1986. Since 1986 he has been a member of the faculty at the Oregon Graduate Institute. Dr. Johnson's research interests include the transport and fate of organic contaminants in aqueous and atmospheric systems. His current research activities include investigations of groundwater contamination due to dense nonaqueous phase liquids and the movement of solvents and petroleum hydrocarbons in the subsurface. Dr. Johnson's research involves both laboratory and field investigations, as well as the simulation of subsurface processes in large physical models. He is the director of the Oregon Graduate Institute's Large Experimental Aquifer Program (OGI/LEAP). This program is dedicated to the use of very large physical models to characterize chemical, physical, and biological processes in the subsurface.

Helmūt Kerndorff is a Research Scientist at the Institute for Water, Soil and Air Hygiene of the Federal Health Office in Berlin, Germany since 1981. He received his B.Sc. (1972), M.Sc. (1976), and Ph.D. (1980) in chemistry and geosciences from the Johannes Gutenberg University in Mainz, Germany. In 1978/1979 he received a scholarship from the German Academic Exchange Service (DAAD). He worked in the laboratory of Dr. M. Schnitzer at the Chemistry and Biology Research Institute in Ottawa, Canada on the relevance of humic substances for transport and fixation of pollutants in aquatic systems. His principal research interests for the past ten years have focused on the detection and characterization of groundwater contamination by hazardous waste sites and on the development of groundwater monitoring strategies and evaluation concepts. He published 54 papers on these subjects in various national and international journals.

Suzanne Lesage, a co-editor of this volume, obtained a B.Sc. in biochemistry from the University of Ottawa, Canada in 1973 and a Ph.D. in chemistry from McGill University in Montreal in 1977. She conducted research on fungicides residues at the Harrow Research Station of Agriculture Canada until 1980 when she joined the Wastewater Technology Center of Environment Canada in Burlington, Ontario. She organized their organic analytical laboratory, which she headed for six years. This laboratory conducted most of Environment Canada's surveys of municipal and industrial effluents. She then moved to the Groundwater Contamination Section of the National Water Research Institute, where she conducted research on the analysis and the fate of organic contaminants in groundwater. She is author or coauthor of 17 journal articles and 19 articles in proceedings and reports, in addition to four book chapters. Her research interests currently include such diverse topics as the degradation of chlorofluorocarbons under anaerobic conditions and the use of porphyrins as tracers in environmental contamination.

Jacques Locat studied geology at Université du Quebec, Montreal; quaternary geology (M.Sc.) at the University of Waterloo, Waterloo, Ontario; engineering geology at the University of Alberta; and geotechnical engineering (Ph.D.) at the University of Sherbrooke, Quebec. Since 1981, he has been involved in research and teaching engineering geology at the Université Laval where his main field of interest is marine geotechnical engineering and mass movements (submarine or subaerial). Other interests are in the fields of hydrogeology, rock mechanics, and quaternary geology. He is the current Chairman of the Engineering Geology Division of the Canadian Geotechnical Society (until 1992) and the President of the Canadian Geoscience Council.

Anne Masson received a Bachelor's degree in engineering geology in 1984 at the Université Laval in Quebec, Canada. She has worked for seven years as a research assistant in contaminant hydrogeology (landfill, organic industrial waste, and acid mine drainage contamination problems) at Université Laval. Her main responsibilities concern field and laboratory work including installation of equipment and all well monitoring activities.

R. David McClellan received a B.Sc. in water resources engineering from the University of Guelph and an M.Sc. in earth sciences specializing in contaminant hydrogeology from the University of Waterloo. He specializes in soils and groundwater remediation and is currently manager of the Toronto office of Intera Information Technologies (Canada) Ltd.

Gerald Milde received the Dr. rer. nat. degree in 1960 and the Dr. rer. nat. habil. degree in 1966 from the Mining Academy at Freiberg, Germany. He was Full Professor for Hydrogeology at the Mining Academy (1967–1975). Dr. Milde was Chief Geologist with Rheinbraun Consulting (1967–1979), at Cologne and from 1979–1990 was Department Head of the Institute for Water, Soil and Air Hygiene of the Federal Health Office. He is currently professor of hydrogeology at the University of Bonn. Dr. Milde has more than 200 publications in the field of applied hydrogeology.

Hooshang Pakdel was born in Iran and received his B.Sc. degree in chemistry there. After receiving his postgraduate diploma in petrochemical and hydrocarbon science from the University of Manchester, Institute of Science and Technology (UMIST) and his Ph.D. in fossil fuels from the University of Bradford, U.K., he immigrated to Canada in 1983 and worked as a postdoctoral fellow on biological markers in the Alberta Tar Sand at the University of Alberta for over a year. In February 1985, he left Alberta for Quebec and spent about seven months at the Université de Sherbrooke, Department of Chemical Engineering, working on the characterization of biomass pyrolysis oils, he then joined Petro-Sun Inc., as a National Science Engineering Council of Canada (NSERC) Industrial Research Fellow to work on synthetic fuels from solid wastes. In January 1988, he joined the Department of Chemical Engineering at the Université Laval as a research associate. He has remained there, working on pyrolysis of solid wastes, the characterization of pyrolysis oils, recovery of valuable chemicals from the pyrolysis oils and recycling of solid wastes. He has been awarded the NSERC operating grant. He has been involved with the Mercier, Quebec Aquifer decontamination for toxic organics project since 1989. Dr. Pakdel is an environmental analytical scientist and consultant. He has published over 40 scientific publications and has presented several articles at international conferences.

Jerry L. Parr is an organic chemist with 20 years of experience in the environmental field. He has conducted numerous research projects relative to groundwater contamination and was the primary author of an EPA report to the U.S. Congress on the adequacy of EPA's methods. He is currently working on efforts within EPA to develop performance-based consensus methods.

Russell H. Plumb, Jr. is a Project Manager in the Geosciences and Biotechnology Section at Lockheed Engineering & Sciences Company, Las Vegas, Nevada. He has more than 20 years experience evaluating the behavior and significance of chemical residues in various environmental media including surface water, groundwater, and sediments. His current research interests have focused on the development of groundwater monitoring strategies and more effective utilization of monitoring data in the site evaluation/assessment process. For the past eight years, he has been working on an EPA funded project to evaluate the technical aspects of the Resource Conservation and Recovery Act (RCRA) groundwater monitoring program and developing industry-specific groundwater monitoring strategies. Dr. Plumb has a B.S. degree in chemistry from Baldwin-Wallace College and an M.S. degree and a Ph.D. degree in water chemistry with a distributed minor in hydrogeology from the University of Wisconsin—Madison.

Terrence R. Power is a Senior Technical Officer for the CSIRO, Division of Water Resources, Perth, Australia. His particular skills are in identifying point sources of organic pollutants in groundwater, and mapping the resulting plume. Mr. Power joined CSIRO in 1965 and for the last 20 years has been involved with a wide range of chemical analyses on plant material, soils and water, including both inorganic and organic analyses. His responsibilities have included development of methods for analyses of trace levels of volatile organics in water.

M. W. Priddle obtained a B.Sc. from the University of Waterloo in 1985. From 1986 to 1990 he was a groundwater chemist with the National Water Research Institute of Environment Canada. He is presently completing an M.Sc.E. in civil engineering (groundwater studies) at the University of New Brunswick. His research interests include groundwater sampling and analysis methods and the migration and fate of organic chemicals in groundwater.

Christian Roy is a professor of chemical engineering at Université Laval He received a B.Sc.A. at Université de Sherbrooke in 1974 and a M.Sc. in engineering from McGill University in 1976. He obtained his Ph.D. from Sherbrooke in 1981. He has been the recipient of numerous grants and awards.

He is the author or coauthor of 42 journal articles, 62 proceedings articles, and several book chapters. He owns eight patents. He is the inventor, researcher, and promoter of a new process called vacuum pyrolysis which is applied in the transformation of solid wastes to products of higher value.

Ruprecht Schleyer was educated at the Geological Institute of the Free University of Berlin where he earned his first diploma and doctoral degree. He was also employed there from 1979 to 1983 as a research geologist. Since 1983, he has been conducting research on the detection and evaluation of groundwater pollution by contaminated sites at the Institute for Water, Soil and Air Hygiene of the Federal Health Office. Since 1989, he is Chief of the section on Human Health Aspects in Hydrogeology and Hydrogeochemistry.

Kathleen C. Swallow earned a B.S. in chemistry from Westhampton College, University of Richmond, in 1970. She was an industrial analytical chemist before beginning graduate work and was granted a Ph.D. in analytical chemistry from the Massachusetts Institute of Technology in 1978. Dr. Swallow was an assistant professor of chemistry at Wellesley College. She later founded and managed an analytical testing laboratory primarily involved in industrial effluent analysis and quality control. She established and managed the analytical laboratory for Modar, Inc., an engineering firm developing supercritical water oxidation as a hazardous waste treatment technology, and continues to serve actively as their chemical consultant. Dr. Swallow was a senior associate with Gradient Corp., an environmental consulting firm. She is currently an associate professor of chemistry at Merrimack College where she teaches analytical and environmental chemistry and instrument analysis.

Gary Walters is a technical director with Enseco Incorporated's Rocky Mountain Analytical Laboratory. He has 14 years of experience in the analysis of trace organics using GC/MS and GC. He has previously been employed at Jordan Laboratories and Shell Development Company. He received B.S. degrees in biology from Stephen F. Austin State University in 1977 and in chemistry from Corpus Christi State University in 1984.

Gary K. Ward is presently the director of Technology and Quality Assurance for Enseco Incorporated, one of the largest environmental analytical laboratories in the U.S. He has over 20 years experience in environmental analyses and served as lab director of NOAA's Marine Chemistry Laboratories in Washington, D.C., and as the national program manager for EPA's Contract Laboratory Program (CLP). While at EPA, Dr. Ward developed the Superfund inorganic CLP protocols and was a member of the EPA workgroup that established RCRA's SW-846 analytical methods.

Index